THE STRUCTURE
AND FUNCTION OF MUSCLE

SECOND EDITION

VOLUME I

Structure   Part 1

CONTRIBUTORS

H. L. ATWOOD

E. B. BECKETT

G. H. BOURNE

M. N. GOLARZ DE BOURNE

DONALD A. FISCHMAN

GEOFFREY GOLDSPINK

H. E. HUXLEY

R. D. LOCKHART

HUGUES MONOD

MARGARET R. MURRAY

J. W. S. PRINGLE

JACK ROSENBLUTH

THE STRUCTURE
AND FUNCTION OF MUSCLE

Second Edition

VOLUME I
Structure   Part 1

Edited by

Geoffrey H. Bourne

Yerkes Regional Primate Research Center
Emory University
Atlanta, Georgia

ACADEMIC PRESS   1972   *New York and London*

ACADEMIC PRESS, INC.
111 Fifth Avenue, New York, New York 10003

United Kingdom Edition published by
ACADEMIC PRESS, INC. (LONDON) LTD.
24/28 Oval Road, London NW1

LIBRARY OF CONGRESS CATALOG CARD NUMBER: 72-154373

PRINTED IN THE UNITED STATES OF AMERICA

CONTENTS

1. Anatomy of Muscles and Their Relation to Movement and Posture

R. D. Lockhart

2. How Muscles Are Used in the Body

Hugues Monod

3. Development of Striated Muscle

Donald A. Fischman

4. Histochemistry of Developing Skeletal and Cardiac Muscle

E. B. Beckett and G. H. Bourne

Revised by G. H. Bourne and M. N. Golarz de Bourne

5. Postembryonic Growth and Differentiation of Striated Muscle

Geoffrey Goldspink

6. Skeletal Muscle in Culture

Margaret R. Murray

7.   Molecular Basis of Contraction in Cross-Striated Muscles

H. E. Huxley

8.   Obliquely Striated Muscle

Jack Rosenbluth

9.   Crustacean Muscle

H. L. Atwood

10. Arthropod Muscle

J. W. S. Pringle

LIST OF CONTRIBUTORS

Numbers in parentheses indicate the pages on which the authors' contributions begin.

H. L. ATWOOD, *Department of Zoology, University of Toronto, Toronto, Canada* (421)

E. B. BECKETT, *Department of Histology, The University, Liverpool, England* (149)

G. H. BOURNE, *Yerkes Regional Primate Research Center, Emory University, Atlanta, Georgia* (149)

M. N. GOLARZ DE BOURNE, *Yerkes Regional Primate Research Center, Emory University, Atlanta, Georgia* (149)

DONALD A. FISCHMAN, *Department of Biology and Anatomy, The University of Chicago, Chicago, Illinois* (75)

GEOFFREY GOLDSPINK, *Department of Zoology, University of Hull, Hull, England* (179)

H. E. HUXLEY, *MRC Laboratory of Molecular Biology, University Postgraduate Medical School, Hills Road, Cambridge, England* (301)

R. D. LOCKHART, *Anthropological Museum, Marischal College, University of Aberdeen, Aberdeen, Scotland* (1)

HUGHES MONOD, *Laboratoire de Physiologie du Travail du C.N.R.S., Faculté de Médecine Pitié-Salpetrière, Paris, France* (23)

MARGARET R. MURRAY, *College of Physicians and Surgeons, Columbia University, Morningside Heights, New York* (237)

J. W. S. PRINGLE, *University of Oxford, Department of Zoology, Oxford, England* (491)

JACK ROSENBLUTH, *Departments of Physiology and Rehabilitation Medicine, New York University Medical Center, School of Medicine, New York, New York* (389)

PREFACE

Twelve years have elapsed since the first edition of this work was published in 1960, and studies on muscle have advanced to such a degree that a second edition has long been overdue. Although the original three volumes have grown to four, we have covered only a fraction of the new developments that have taken place since that time. It is not surprising that these advances have not been uniform, and in this new edition not only have earlier chapters been updated but also areas in which there was only limited knowledge before. Examples are the development of our knowledge of crustacean muscle (172 of 213 references in the reference list for this chapter are dated since the first edition appeared) and arthropod muscle (205 of 233 references are dated since the last edition). Obliquely striated muscle, described in 1869, had to wait until the electron microscope was focused on it in the 1960's before it began to yield the secrets of its structure, and 33 of 43 references dated after 1960 in this chapter show that the findings described are the result of recent research. There has also been a great increase in knowledge in some areas in which considerable advances had been made by the time the first edition appeared. As an example, in Dr. Hugh Huxley's chapter on "Molecular Basis of Contraction in Cross-Striated Muscles," 76 of his 126 references are dated after 1960.

The first volume of this new edition deals primarily with structure and considers muscles from the macroscopic, embryonic, histological, and molecular points of view. The other volumes will deal with further aspects of structure, with the physiology and biochemistry of muscle, and with some aspects of muscle disease.

We have been fortunate in that many of our original authors agreed to revise their chapters from the first edition, and it has also been our good fortune to find other distinguished authors to write the new chapters included in this second edition.

To all authors I must express my indebtedness for their hard work and patience, and to the staff of Academic Press I can only renew my confidence in their handling of this publication.

Geoffrey H. Bourne

PREFACE
TO THE FIRST EDITION

Muscle is unique among tissues in demonstrating to the eye even of the lay person the convertibility of chemical into kinetic energy.

The precise manner in which this is done is a problem, the solution of which has been pursued for many years by workers in many different disciplines; yet only in the last 15 or 20 years have the critical findings been obtained which have enabled us to build up some sort of general picture of the way in which this transformation of energy may take place. In some cases the studies which produced such rich results were carried out directly on muscle tissue. In others, collateral studies on other tissues were shown to have direct application to the study of muscular contraction.

Prior to 1930 our knowledge of muscle was largely restricted to the macroscopical appearance and distribution of various muscles in different animals, to their microscopical structure, to the classic studies of the electro- and other physiologists and to some basic chemical and biochemical properties. Some of the latter studies go back a number of years and might perhaps be considered to have started with the classic researches of Fletcher and Hopkins in 1907, who demonstrated the accumulation of lactic acid in contracting frog muscle. This led very shortly afterward to the demonstration by Meyerhof that the lactic acid so formed is derived from glycogen under anaerobic conditions. The lactic acid formed is quantitatively related to the glycogen hydrolyzed. However, it took until nearly 1930 before it was established that the energy required for the contraction of a muscle was derived from the transformation of glycogen to lactic acid.

This was followed by the isolation of creatine phosphate and its establishment as an energy source for contraction. The isolation of ADP and ATP and their relation with creatine phosphate as expressed in the Lohmann reaction were studies carried out in the thirties. What might be described as a spectacular claim was made by Engelhart and

Lubimova, who in the 1940's said that the myosin of the muscle fiber had ATPase activity. The identification of actin and relationship of actin and myosin to muscular contraction, the advent of the electron microscope and its application with other physical techniques to the study of the general morphology and ultrastructure of the muscle fibers were events in the 1940's which greatly developed our knowledge of this complex and most mobile of tissues.

In the 1950's the technique of differential centrifugation extended the knowledge obtained during previous years of observation by muscle cytologists and electron microscopists to show the differential localization of metabolic activity in the muscle fiber. The Krebs cycle and the rest of the complex of aerobic metabolism was shown to be present in the sarcosomes—the muscle mitochondria.

This is only a minute fraction of the story of muscle in the last 50 years. Many types of discipline have contributed to it. The secret of the muscle fiber has been probed by biochemists, physiologists, histologists and cytologists, electron microscopists and biophysicists, pathologists, and clinicians. Pharmacologists have insulted skeletal, heart, and smooth muscle with a variety of drugs, *in vitro*, *in vivo*, and *in extenso;* nutritionists have peered at the muscle fiber after vitamin and other nutritional deficiencies; endocrinologists have eyed the metabolic processes through hormonal glasses. Even the humble histochemist has had the temerity to apply his techniques to the muscle fiber and describe results which were interesting but not as yet very illuminating—but who knows where knowledge will lead. Such a ferment of interest (a statement probably felicitously applied to muscle) in this unique tissue has produced thousands of papers and many distinguished workers, many of whom we are honored to have as authors in this compendium.

Originally we thought, the publishers and I, to have *a book* on muscle which would contain a fairly comprehensive account of various aspects of modern research. As we began to consider the subjects to be treated it became obvious that two volumes would be required. This rapidly grew to three volumes, and even so we have dealt lightly or not at all with many important aspects of muscle research. Nevertheless, we feel that we have brought together a considerable wealth of material which was hitherto available only in widely scattered publications. As with all treatises of this type, there is some overlap, and it is perhaps unnecessary to mention that to a certain extent this is desirable. It is, however, necessary to point out that most of the overlap was planned, and that which was not planned was thought to be worthwhile and was thus not deleted.

We believe that a comprehensive work of this nature will find favor with all those who work with muscle, whatever their disciplines, and that although the division of subject matter is such that various categories of workers may need only to buy the volume which is especially apposite to their specialty, they will nevertheless feel a need to have the other volumes as well.

The Editor wishes to express his special appreciation of the willing collaboration of the international group of distinguished persons who made this treatise possible. To them and to the publishers his heartfelt thanks are due for their help, their patience, and their understanding.

Emory University, Atlanta, Georgia   Geoffrey H. Bourne
October 1, 1959

CONTENTS OF OTHER VOLUMES

1

ANATOMY OF MUSCLES AND THEIR RELATION TO MOVEMENT AND POSTURE

R. D. LOCKHART

I. Muscle Structure

Despite the variety of form and function throughout divergent phyla, it is an interesting reflection that the motive power throughout the animal kingdom with few exceptions is muscle (Pantin, 1956). The muscular system confers the grace of the human body in movement or at rest. The paramount virtue of muscle, its contractility, is the basis of physical culture, a fundamental principle in many aspects of treatment in orthopedic surgery and invaluable in the examination of the nervous system, when the clinician, by testing the power of various muscle groups, reveals the state of their nerve supply as intact or defective. Again, happiness or depression, vigor or weariness are not only emotionally registered by the facial muscles but are also depicted by the muscular system in the whole poise of the individual.

A. *Importance of Studying Muscles in the Living Body*

Since many of our muscles are superficial and easily seen and felt in action, no apology is made for stressing at the outset the elementary fact, often overlooked, that the beginner may learn much about the activity of muscles independent of all textbooks and literally at first hand merely by resisting the movements of a companion's limbs with one hand while the other hand identifies the muscles which alternately harden with the movement and soften with the reverse movement.

Neglect of this simple method caused Winslow (1732) to stress the need for experiment upon the muscles in the living subject, yet Duchenne (1867) had still to show the fallacies of study by scalpel upon the dead instead of by "electiization" upon the living, while Beevor (1904) repeated the insistence that the action of muscles must be studied upon the living body. But it was still necessary for the distinguished clinician F. M. R. Walshe (1951) to state in his foreword to the reprint of Beevor's "Croonian Lectures on Muscular Movements," issued by the publishers of the journal *Brain*, that Beevor's valuable studies are already to a large extent lost and that neither the clinical neurologist nor the experimental physiologist can afford this loss today. For the details of the action of individual muscles beyond the scope of this text, the reader will find a mine of interest in the brilliant works of Duchenne and Beevor.

B. Nomenclature

Muscles have been named from a variety of their features—e.g., from attachments (sternomastoid), action (supinator), direction (rectus), situation (gluteus), structure (triceps), size (magnus), and shape (trapezius). Many are designated by composite terms such as flexor digitorum profundus. It is interesting to note, in passing, that few muscles received special names until the early eighteenth century. Previously, numbers were given to muscles in regional groups by Galen and Vesalius, while Leonardo da Vinci applied letters in his illustrations.

C. Design

Apart from phylogenetic inheritance, the form of a muscle is determined by its function, which, in turn, requires compromise between power, speed, and range of movement. The more one regards the formation of the body, the more one appreciates the beauty of the design—the mass of the muscle bellies placed to avoid interference with movement and tapered to tendons or blended with aponeurotic sheets where joints must be free or where muscle is unnecessary.

The striated, voluntary, or skeletal muscles, comprising about 40% of the weight of the body, are made up of bundles of fibers, each fiber embedded in fine connective tissue (the endomysium), each bundle in a sheath of perimysium, and all the bundles of the complete muscle invested by the sheath of the muscle, the epimysium.

D. Length, Thickness, and Number of Muscle Fibers

After treatment of a muscle with a little weak alkali or acid or by boiling to soften the connective tissue, individual fibers may be patiently dissected out using hedgehog spines and traced from the tendon of origin to the tendon of insertion. They may be more than 30 cm long in the sartorius muscle, as shown by Lockhart and Brandt (1938). These authors never found several fibers interdigitating in the length of a muscle between the tendons of origin and insertion, an arrangement that had frequently been described. Muscle fibers never gain direct attachment to bone, as tendon always intervenes. It is the endomysium, perimysium, and epimysium which blend with the tendon; indeed, there is not enough area of bone to afford attachment for all the muscle fibers.

Fiber length increases by appositional growth at each end not by interstitial growth (Kitiyakara and Angevine, 1963). Thickness of fibers varies from 0.01 mm to 0.1 mm, even in the same muscle. Physical training makes an obvious and astonishing increase in the muscles exercised by thickening the fibers, not by adding to their number which remains constant soon after birth. Bridge and Allbrook (1970) compare muscles from a marsupial's fore and hind limbs, showing the remarkable increase in the number of fibers before and after birth, all the more interesting in view of the important task of the fore limb at birth compared with the hind limb, and the subsequent reversal of their roles.

E. Tendons and Fiber Arrangement

Where short muscle fibers are adequate for the movement required, tendon saves the more specialized muscle tissue. Again, the use of a long tendon cord allows a muscle belly to exert its full power upon a distal point, and further, the tendon may ply around a pulley whereby the muscle exerts a line of pull divergent from the axis of its own length. Tendons may run along or into the body of a muscle, the muscle fibers being attached to one side in the unipennate types of muscle, to both sides in the bipennate, and giving rise to multipennate forms where several tendons invade the muscle (Fig. 1).

So much attention has been accorded to muscle, that there is apt to be neglect of the importance of tendon in its role as buffer guarding the limbs against the danger of sudden strain and speed that muscle alone would incur. Again, a stretched tendon by its elastic recoil increases the muscular action (Hill, 1951).

Although the term "origin" is usually applied to the proximal or less mobile point of a muscle's attachment and the term "insertion" to the distal and more mobile attachment, the action of a muscle may be reversed. For example, in climbing up a rope the body is pulled toward the arms and the insertions of the muscles into the arms are the more fixed points. Then again, in the foot and leg, the terms are unfortunate since the muscles usually act when the foot is applied to the ground.

Usually, in the limbs, a muscle is inserted just beyond the joint it moves, thereby gaining speed but with loss of the power a longer lever would have obtained. At the elbow, for example, a wide and rapid movement of the hand results from a comparatively small muscular contraction.

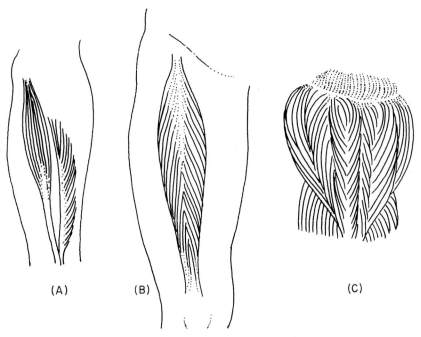

(A)   (B)   (C)

Fig. 1. Types of muscle fiber and tendon arrangement. (A) Fusiform and unipennate: flexor carpi radialis and flexor pollicus longus. (B) Bipennate: rectus femoris. (C) Multipennate: deltoid.

F. Motor or Neuromotor Unit; Tonus

The nerve supplying a muscle contains both motor and sensory fibers, almost half of them sensory. A single motor nerve fiber, an axon, is responsible for the innervation of a definite group of muscle fibers, sending a filament to each muscle fiber of the group. In a large muscle such as gluteus maximus, there might be 200 muscle fibers in each group, but only 5 muscle fibers in the groups comprising a small muscle moving the eyeball. An impulse in a single axon would activate the smallest number of muscle fibers that can be engaged. Accordingly, an eye muscle axon stimulating only 5 muscle fibers will produce a more delicately precise action than a gluteal muscle axon stimulating 200 muscle fibers at the least. Each muscle is a company of neuromotor units, with each unit comprising a nerve cell, its axon, and its group of muscle fibers.

In moderate activity of a muscle, the units work in relays providing continued action without fatigue. The more powerful the action required, the more units act at the same time. The smaller the part of the body to be moved, the smaller the muscle, the more rapid its action, and the smaller the unit. The activity of the motor unit system may be responsible to a certain extent for the tension or tone always present in a muscle even when it is at rest, although precise and delicate electromyographic tests do not support this theory. Possibly the tone is due to the inherent elasticity of the muscle. All efferent nerves to voluntary muscle stimulate the muscle to contract; there are no efferent nerves to produce relaxation. Relaxation is achieved by reducing or terminating the neuromotor stimuli. Joseph (1960) advises discarding the term "tonus," which if used at all, should refer to the response of skeletal muscle to stretch. See also Section II,D.

G. The Neurovascular Hilum

Vessels and nerves enter a muscle, as a rule, at a definite cleft, the neurovascular hilum. These hila have been determined for the limb muscles by Brash (1955).

II. MUSCULAR ACTION

A. Amount of Contraction of a Muscle Fiber

It is possible for a muscle fiber fully stretched to contract to 57% of its length (Haines, 1934). This amount generally corresponds to the degree of movement permitted at the joint between the bones involved. However, Ramsey and Street (1940) consider that contraction to less than 70% of the resting length may cause permanent shortening and structural change.

B. Muscles of Short or Insufficient Action

The fibers of certain muscles, however, are so short that they do not execute the full movement possible between their attachments, as exemplified in the biceps femoris, semimembranosus, and gastrocnemius. When these muscles are fully contracted, the knee is not fully bent, and in further flexion of the joint, these muscles, described as of short action or insufficient action, are no longer taut. Again, some muscles may be unable to relax sufficiently to allow full movement to occur at a joint in certain conditions; for example, the hamstring muscles, when the knee is extended, will not relax enough to give full flexion at the hip, a state of passive insufficiency which will be discussed subsequently under the ligamentous action of muscle.

C. Use of the Terms Contraction, Relaxation, "Concentric" and "Excentric" Action, Active Shortening and Lengthening, Adaptive Shortening and Lengthening

It must be emphasized that a muscle may be described as contracted in the sense of being firm, tense, or active without implying shortening; indeed, a muscle may be exerting its greatest power when fully stretched.

When the arm is raised to the horizontal, the deltoid muscle (Fig. 2) is felt firmly contracted, and when it is lowered in the absence of resistance, the muscle is still felt firmly contracted, while under tension, it controls the gravitational descent of the arm as a crane lowers a weight. In the ascending arm it is actively shortening and in the descend-

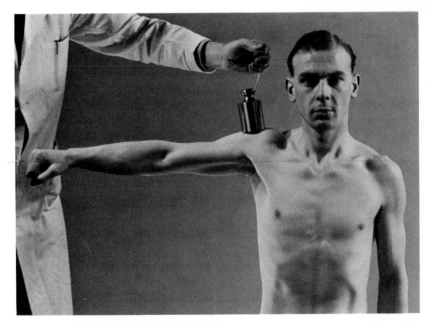

Fig. 2. Deltoid active, pectorals flaccid. Whether the arm is being raised or lowered, the weight rides upon the firm deltoid, which not only raises the arm but also controls its gravitational descent, respectively contracting or paying out under tension as a crane lowers a weight. Contrast with Fig. 3.

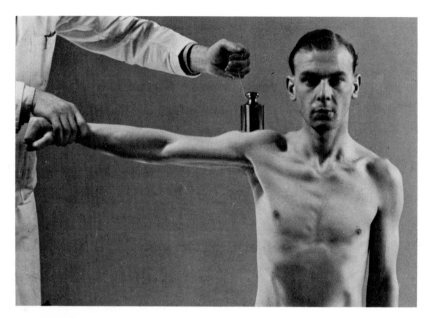

Fig. 3. Deltoid flaccid, pectorals active. When, however, the arm is pulled down against resistance, the weight sinks into the suddenly flaccid deltoid, and its antagonists, such as the pectoral and latissimus dorsi muscles, come into action. Reproduced from "Living Anatomy" by R. D. Lockhart (1970). (Courtesy of Faber & Faber, London; Oxford University Press, New York.)

ing arm it is actively lengthening or relaxing. These actions are sometimes, in physiotherapy and physical training colleges, described respectively as "concentric" and "excentric." During the ascent of the arm, the muscles that can oppose the movement must adaptively lengthen or relax, and during the descent they must adaptively shorten. When the descending arm encounters resistance, the deltoid (Fig. 3) becomes suddenly flaccid and adaptively lengthens, while the muscles pulling down the arm become firmly contracted in active shortening.

D. Muscles Maintaining Posture against Gravity

Muscles maintaining posture against gravity, sometimes termed antigravity or postural muscles, are under the control of the central nervous system through reflexes in the spinal cord activated by stretch stimuli from the muscles and stimuli from the skin, the eyes, and vestibular structures. This is surely too matter of fact a statement to appreciate the floating grace and exquisite poise of the ballerina, the skater, the gymnast, and the acrobatic stars, to say nothing of the precise muscular coordination of the golfer. Posture under reflex control has been described as the basis of movement, because movement starts from and terminates in a posture; indeed, the stand-at-ease position, according to Hellebrandt (1938), is in reality movement upon a stationary base. Sway is inseparable from the upright stance, and the altered balance causes new stretch stimuli. As long as the balance is steady, in the erect posture there is very little and in most persons no detectable activity in the trunk, thigh, and anterior leg muscles, but the calf muscles are active (Joseph *et al.*, 1955; Joseph and Nightingale, 1952, 1954, 1956; Joseph, 1960; Floyd and Silver, 1955). In man's upright stance, the center of gravity of each segment of his body is just above its supporting joint, the line through the center of gravity of the part above the hip joints passing behind these joints and a similar line through the segment above the knee joints passing in front of these joints, but the center of gravity of the body as a whole lies in front of the ankle joint. While the ligaments in front of the hip and behind the knee may momentarily take the strain before muscles are employed to retrieve the balance, the calf muscles are always active. The delicate precision with which the trunk muscles control balance is easily appreciated by feeling the spinal muscles contract when one merely raises an arm forward (cf. Fig. 4). Again, with one hand on the back and the other on the front of the trunk, alternate contraction and relaxation of the muscles is felt with the slightest to and fro movement of the trunk. The activity of a muscle maintaining the erect posture is no different from that in executing a movement. Although a muscle may frequently have such a postural role, it is no different from other muscles.

The importance of muscular action in relation to gravity concerns every muscle and movement. The sternomastoid muscle is usually described as flexing the head and neck but it never does so unless it is working against resistance or gravity as in a person rising from the supine position; otherwise flexion of the head and neck is due to active lengthening of the muscles on the back of the neck. To take an example from the upper limb, stretch the arm horizontally (palm up) and then slowly bend the elbow. The biceps will be felt firm and the triceps on the back of the upper arm soft until the forearm is vertical, but the moment flexion continues across the plumb line the biceps becomes soft and the triceps firm. If both flexion and extension movements are resisted, then obviously the biceps is active throughout the whole of flexion and the triceps throughout the whole of extension (Figs. 5 and 6).

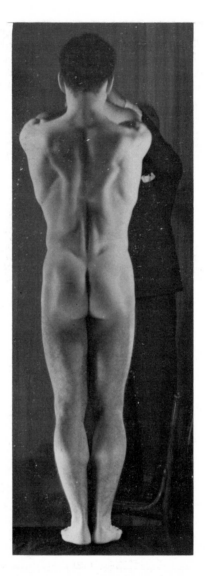

Fig. 4. Activity of trapezius and sarcospinalis. The trapezius muscles are raising the arms against resistance, while the two heavy ridges in the lower part of the back show the activity of the sacrospinalis muscles in maintaining the erect posture against the heavy leverage upon the arms. Reproduced from "Living Anatomy" by R. D. Lockhart (1970). (Courtesy of Faber & Faber, London; Oxford University Press, New York.)

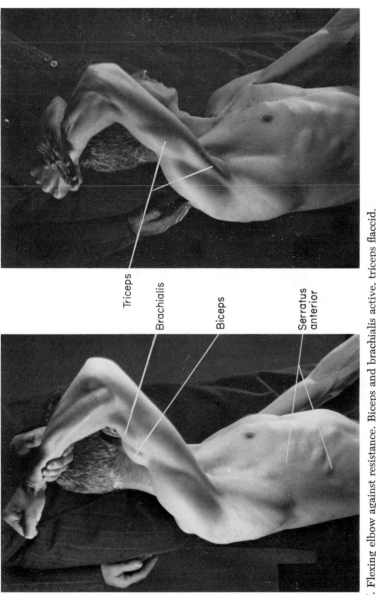

Fig. 5. Flexing elbow against resistance. Biceps and brachialis active, triceps flaccid.
Fig. 6. Extending elbow against resistance. Biceps and brachialis flaccid, triceps active. Serratus active in steadying the scapula in both figures. Reproduced from "Living Anatomy" by R. D. Lockhart (1970). (Courtesy of Faber & Faber, London; Oxford University Press, New York.)

Triceps

Brachialis

Biceps

Serratus
anterior

Alterations in posture precede the majority of movements to secure the best attitude for the best performance. Indeed, it is a cult assiduously practiced in games. To be caught on the wrong foot can be literally expressive. In rising from a chair, the first act is to bring the feet under the center of gravity. The starting attitude of the sprinter and the natural movement of a horse in first stepping backward in order to lean its weight forward in starting to pull a heavy load illustrate the same principle.

In the standing posture, electromyography, adequately delicate to detect activity even in a single neuromotor unit, finds only the calf muscles active. The anterior leg muscles are certainly felt firmer than in a leg hanging free and much firmer still if the erect person leans forward. The anterior tibial muscle must play its part in maintaining the arch of the foot. Because of the normal action of the calf muscles, plantar flexing the foot in the leg hanging free, dorsiflexion from this position, as in placing the foot upon the ground increases the tension in the calf muscles. With the foot on the ground, leg muscles pulling from the foot increase tension on the fascial stocking ensheathing the leg muscles.

In walking, the initial forward inclination of the trunk is made by relaxing the tension of the hamstring and calf muscles in both limbs. As the right foot leaves the ground, the trunk is balanced by a tilt to the left. The right limb swings forward flexing at knee and hip and extending as the heel touches the ground. The impetus in the push-off in walking, running, and jumping is given by the propelling limb extending at the knee and hip and plantar flexing at the ankle. The left limb now propels, the left heel is raised by the calf muscles, and the weight, at first carried by the whole foot, is shifted forward. The final take-off is from the metatarsal pad region, especially that of the great toe. The final impetus is derived from the flexors of the toes. Alternately during walking, at each push-off, the extending limb pushes downward and backward and the body is propelled upward and forward to be caught as its trajectory begins to fall by the opposite limb swinging forward in front of the body. In walking there is always a short supporting phase by both limbs, but in running there are moments when both feet are off the ground, the forward lean of the body is marked, its center of gravity already in front of the advancing foot before it touches the ground as the runner propels himself to keep up with his center of gravity, which keeps a path approximately parallel with the ground.

Using telemetering electromyography, Battye and Joseph (1966) recorded in men and women the active periods of leg muscles in walking. Tibialis anterior is active as the phase of the swing begins and ends, soleus during the latter half of the support phase, quadriceps just before and during the first half of the support phase, and again, in some cases, as the opposite heel meets the ground. Hamstrings are in action except while flexion of the knee occurs in the swinging phase. Hip flexors are active only as hip flexion ends. Gluteus maximus is active as the swinging phase ends and the supporting phase begins. The erector spinae has two short and marked active periods at the beginning and the end of the supporting phase.

E. Muscle Power, Range, and Speed of Movement

The muscle with the greater number of fibers is the more powerful, and the longer the fibers in a muscle the greater its range of movement. The long, narrow, straplike sartorius, with its parallel fibers, effects a wide movement, but it does not have the power of a short, multipennate type like the deltoid. As the strength of a muscle depends upon the sum of the areas of cross section of the separate fibers, weight for weight the muscle with the greater number of fibers is the stronger.

Speed is an intrinsic quality varying with individual muscles, inherent in the biochemical and physicochemical makeup of the muscle and not dependent alone upon the nerve supply. The smaller the structure to be moved, such as the eyeball, the more rapid its muscles, compared with the heavy gluteus maximus muscle moving the trunk and lower limb. According to Sotavalta (1947), the wing muscles of a gnat beat at the rate of $\frac{1}{1000}$ sec, while the sea anemone take 4 min to contract (Pantin, 1956).

Muscles are accurately designed for their usual work. A compromise exists between power, speed, and range, according to the differences in number of fibers, their length, intrinsic speed, position in relation to joints, and weights to be moved. This entails that a muscle's neuromotor units be endowed with different speeds of contraction and power and resistance to fatigue according to the type of fibers and their varying number in the separate units. In the cat's pale-fibered gastrocnemius muscle, there are three types of units, most of them strongly and swiftly contracting and quickly subject to fatigue, though a few are slow and some are very small. In contrast, the red-fibered soleus, designed with one type of fiber, has a wider range of speeds and is slower and resistant to fatigue. The references for this paragraph are the successive related papers by Henneman *et al.* (1965), Wuerker *et al.* (1965), and Henneman and Olson (1965).

The finest athletic achievements depend upon a remarkable efficiency in timing the various associated movements to a nicety. Hill (1951) cites the record throw of a cricket ball which leaves the fingers at a rate of about 82 miles per hour, a result achieved by the summation of the velocity of each successive part of the body's movement added to that of the previous parts—a principle probably more readily appreciated in the parallel of the cracking whip where the ultimate velocity of the tip breaks through the sound barrier.

F. Isometric and Isotonic Contraction

When a muscle fails to lift a weight beyond its strength, although its tension is maximal, its length is unaltered and its condition is described by the somewhat contradictory term "isometric contraction." In isotonic contraction, the length varies but the tone remains the same.

G. Voluntary Movement

Skeletal, or striated, muscle is frequently termed "voluntary muscle," but this is a figure of speech because only the movement is voluntary. We have no freedom in selecting the muscle that carries out the movement. Indeed, there are usually several muscles involved in a definite pattern of activity and timing, and we are incapable of altering the pattern of increasing or decreasing the numbers involved by even one member. In the simple action of jutting out the thumb laterally, the muscles on the front and back of the ulnar side of the forearm are seen and felt contracting to prevent the whole hand moving in the same direction. Even if the thumb muscles are paralyzed, the ulnar muscles contract the moment the movement is attempted. According to the parts they play—principal or supporting in a movement—muscles are described as prime movers, antagonists, fixation muscles, and synergists.

H. Prime Movers and Antagonists

The prime movers are the principal actors. They actively bring about the desired movement. The antagonists must relax to allow the prime movers to operate. See Figs. 5, 6. Despite its name, the antagonist assists the prime mover. It is as important in a good system of training to secure full relaxation as well as full contraction of a muscle. Indeed,

Fig. 7. Actions of latissimus dorsi and trapezius muscles contrasted. On the right the upper border of latissimus dorsi curving around the inferior angle of the scapula to reach the axilla shows in relief as the muscle pulls down the right arm against resistance, while on the left the trapezius is firmly contracted in raising the left arm against resistance.

in some people, the pectoral muscles are so tight that the arms cannot be raised erect, and calf muscles in most people prevent the full amount of dorsiflexion at the ankle. When the prime movers are contracting, the antagonists, although not stimulated, are immediately ready to steady the part as a guy rope does in securing precision of movement. As an act of volition, of course, we can set in firm tension both prime movers and antagonists at the same moment just as a professional strong man accentuates his muscles in a pose for exhibition. Prime mover and antagonist are seen in reversed roles on both sides of Fig. 7. See also Figs. 2, 3, and 10.

I. Fixation and Articular Muscles

Fixation and articular muscles fix the base upon which the movement carried out by the prime movers is made. During movements of the upper limb, the scapula is steadied by fixation muscles such as serratus

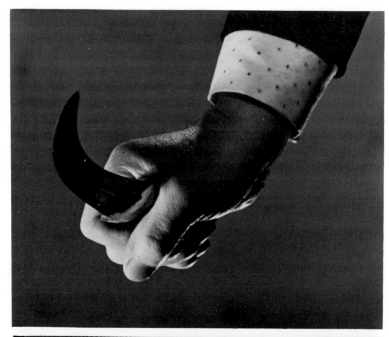

Figs. 8 and 9. See facing page for legend.

(Figs. 2 and 3), and the head of the humerus is kept in position on the scapula by small articular muscles which prevent the head of the humerus from sliding when the arm is raised—much in the way that a man places his foot upon one end of a long ladder while his companion raises the other end. Anyone who has had the exasperating experience of trying to raise a ladder too long for him single handed will have a keener appreciation of the value of a fixation muscle. The rectus abdominis muscles become painful in repeating the exercise of raising both legs with the body supine, not because they act directly on the limbs but because they must fix the pelvis in counteracting the heavy leverage of the limbs.

J. Synergists

Special instances of fixation muscles controlling movement at proximal joints and thereby enabling prime movers to act upon a distal joint illustrate muscular synergy, as explained in the legend to Figs. 8 and 9. However, both fixation muscles and synergists prevent loss of power and there is no rigorous distinction between them.

K. Group Action of Muscles

Prime movers, antagonists, fixation muscles, and synergist muscles perform together in most movements. Their cooperation is described as group action. But the term is also applied to a group of prime movers or antagonists where each group comprises a rival team. If the surgeon transplants the tendon of an antagonist to take the place of a paralyzed prime mover, then the transplanted muscle will continue to act with its old team in the accustomed pattern of reciprocal activity until a cooperative patient reeducates the muscle (Dunn, 1920).

Figs. 8 and 9. Muscular synergy or combined action. The firmly clenched fist is always bent backwards (extended) at the wrist, but the fingers lose their grip and open when the wrist is forcibly bent forwards (flexed), a method of compelling an assailant to drop his weapon. The extensor muscles cannot stretch far enough to allow the flexors full action both at wrist and fingers. Accordingly, these two rival muscle groups coordinate in a precise synergic or combined action whereby the extensors control, or fix, the proximal wrist joint, while the flexors clench the fingers at the distal joints. Reproduced from "Living Anatomy" by R. D. Lockhart 1970). (Courtesy of Faber & Faber, London; Oxford University Press, New York.)

L. Dual Action of Muscles

Dual action of muscles is exemplified by a muscle taking part in two distinct, but not opposite, movements. The biceps is both a flexor of the elbow and a supinator of the forearm, and, as the result of cortical injury, the power of one movement may be lost and the other movement retained. In conjugate movements of the eyeballs the medial rectus muscle of each eye may act with the lateral rectus of the opposite eye but fail to act in convergence of the eyes with the opposite medial rectus (Beevor, 1904). Apart from voluntary movements, as in the above examples, muscles may also be engaged in bilateral involuntary movements. Latissimus dorsi in its unilateral voluntary movement adducts the arm against resistance, but both latissimus dorsi muscles have also an involuntary bilateral action in coughing or sneezing, an expiratory function easily felt by placing the hands against the lower ribs and coughing. If the clinician finds the bilateral action retained and the unilateral lost, then he has valuable evidence that the lesion is cerebral and not spinal or peripheral (Beevor, 1904).

M. Ligamentous Action of Muscles

The ligamentous action of muscles is well seen in the inability of the hamstring muscles to relax sufficiently to allow full flexion at the hip joint when the knee is extended and accounts for inability to touch the toes and to maintain or even attain the high kick position. Pectoral muscles may be unable to stretch enough to allow the arms to be raised erect. To get the arms erect, the person bends the vertebral column backwards. Again, the calf muscles, always much heavier than the anterior leg muscles, a feature characteristic of man, tend to draw up the heel producing plantar flexion of the foot when the body is recumbent. The wearing of heels, even in men's shoes, has the same effect in ultimately allowing the calf muscles to become permanently shortened with the result that full dorsiflexion of the foot can no longer be obtained. In marked cases, the person may experience painful discomfort in walking with heelless shoes. This is one of the causes of flat foot and is easily remedied by resuming the customary heel. In the above examples the muscles are passively insufficient, a condition already discussed in Section II,B.

N. Associated or Cooperative Action

The terms associated or cooperative action may be quite properly applied to the group action of prime movers, antagonists, fixation muscles, and synergists, but usually it refers to more remote muscles that act in unison. For example, the head and often the trunk turn in the same direction as the eyes. In fact, it requires a little premeditation if not practice to move the eyes without automatically turning the head in unison.

If the pads of the thumb and forefinger of the right hand are pressed very lightly together the other three fingers can be easily moved to and fro by the left hand, but if the thumb and forefinger are tightly compressed the other fingers become immobile. Again, tight closure of the eyes by the orbicularis oculi muscles is accompanied by a faint drumming in the ear as the stapedius muscles contract in unison (Gowers, 1896).

O. Accessory Movements

This term is applied to movements that cannot be carried out actively at a joint but may be executed voluntarily against resistance; for example, the fingers cannot be rotated at the metacarpophalangeal joints, but the grasping of a firm object, a cricket ball for instance, allows the movement to occur.

P. Diametrically Opposed Movements

The fact that a muscle happens to be favorably placed for the execution of a movement and even the fact that it does so under electrical stimulation provide no guarantee whatever that in the normal state the muscle does carry out this movement. It has been stated that the sternomastoid, when the head is bent backwards, is able to induce further extension. But Beevor demonstrated a case of paralysis of the extensor muscles of the back of the neck in which, with the patient supine and the head well back in good position to be further extended, the patient actually relaxed the sternomastoid when asked to extend the head.

The pectoralis major, a most interesting muscle, has two parts a clavicular and a sternocostal (Fig. 10). Both act together in adduction of

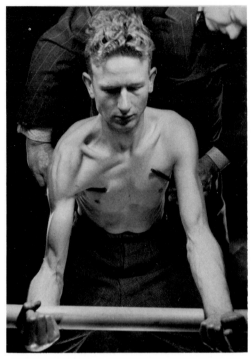

Fig. 10. Actions of pectoralis major muscle. The right arm is raised against resistance and the left arm lowered against resistance. The clavicular part of pectoralis major is active in the right arm and inactive in the left arm. The sternocostal part is inactive in the right arm, so that the pencil may be depressed into the muscle, and active in the left arm so that the pencil cannot be impressed into the muscle. The upper fibers of trapezius, the anterior fibers of deltoid, and the biceps are seen active on the subject's right side and flaccid on his left. Reproduced from "Living Anatomy" by R. D. Lockhart (1970). (Courtesy of Faber & Faber, London; Oxford University Press, New York.)

the arms. But in pulling down the arm against resistance, only the sterno-costal part is active; the clavicular part is inactive although well placed to depress the erect arm (Fig. 10). In raising the arm, however, the clavicular part is active and the sternocostal relaxed (Fig. 10). This is an instance where a knowledge of the two attachments, origin and insertion, of a muscle is no guide to the action.

Q. Muscular Fatigue

The fact that an air cuff when applied to stop the circulation in a limb exhausted by heavy exercise prevents the recovery of strength until the circulation recommences, seems to indicate that the muscle and not the nervous system is the site of the fatigue. The act of writing, an intricate and skilled but light exercise, becomes difficult within 2 min when the circulation is prevented, and within another minute impossible (Merton, 1956).

At the beginning of this chapter, the importance of determining the action of muscles in the living body was stressed. In conclusion, it must again be reiterated that despite the scalpel, electrical stimulation, and electromyography, the method of reinforcing the action of muscles by resisting the patient's movements, so admirably demonstrated by Charles Beevor, is simple, immediate, and provides both student and experienced physician with literally firsthand information.

REFERENCES

Battye, C. K., and Joseph, J. (1966). *Med. Biol. Eng.* 4, 125–135.

Beevor, C. E. (1904). "The Croonian Lectures on Muscular Movements and their Representation in the Central Nervous System." Adlard, London. (Edited and reprinted for publishers of *Brain*, Macmillan, New York, 1951.)

Brash, J. C. (1955). "Neurovascular Hila of Limb Muscles." Livingstone, Edinburgh.

Bridge, D. T., and Allbrook, D. (1970). *J. Anat.* 106, 285.

Duchenne, G. B. A. (1867). "Physiologie des mouvements démontrée a l'aide de l'expérimentation électrique et de l'observation clinique." Baillière et Fils, Paris. ["Physiology of Motion" (translated and edited by E. B. Kaplan). Lippincott, Philadelphia, Pennsylvania, 1949.]

Dunn, N. (1920). *J. Orthop. Surg.* 2, 554.

Floyd, W. F., and Silver, P. H. S. (1955). *J. Physiol. (London)* 129, 184.

Gowers, Sir W. (1896). *Lancet* 2, 1357.

Haines, R. W. (1934). *J. Anat.* 69, 20.

Hellebrandt, F. A. (1938). *Amer. J. Physiol.* 121, 471.

Henneman, E., and Olson, C. B. (1965). *J. Neurophysiol.* 28, 581–598.

Henneman, E., Somjen, G., and Carpenter, D. O. (1965). *J. Neurophysiol.* 28, 560–580.

Hill, A. V. (1951). *Lancet* 2, 947.

Joseph, J. (1960). "Man's Posture Electromyographic Studies." Thomas, Springfield, Illinois.

Joseph, J., and Nightingale, A. (1952). *J. Physiol. (London)* 117, 484.

Joseph, J., and Nightingale, A. (1954). *J. Physiol. (London)* 126, 81.

Joseph, J., and Nightingale, A. (1956). *J. Physiol. (London)* 132, 81.

Joseph, J., Nightingale, A., and Williams, P. L. (1955). *J. Physiol. (London)* 127, 617.

Kitiyakara, A., and Angevine, D. M. (1963). *Develop. Biol.* 8, 322.

Lockhart, R. D. (1970). "Living Anatomy, A Photographic Atlas of Muscles in Action and Surface Contours," 6th ed. Faber & Faber, London.

Lockhart, R. D., and Brandt, P. W. (1938). *J. Anat.* 72, 470.

Merton, P. A. (1956). *Brit. Med. Bull.* 12, 219.

Pantin, C. F. A. (1956). *Brit. Med. Bull.* **12**, 199.

Ramsey, R. W., and Street, G. (1940). *J. Cell. Comp. Physiol.* **15**, 11.

Sotavalta, O. (1947). *Acta Entomol. Fenn.* **4**, 177.

Walshe, F. M. R. (1951). In foreword to reprint of "Croonian Lectures on Muscular Movements," by C. E. Beevor (1904).

Winslow, J. B. (1732). "Exposition anatomique de la structure du corps humain." Desprez, Paris. (Translated by G. Douglas, Betterworth & Hitch, London, 1734; G. Douglas, R. Ware, and others, London, 1743.)

Wuerker, R. B., McPhedran, A. M., and Henneman, E. (1965). *J. Neurophysiol.* **28**, 85–99.

2

HOW MUSCLES ARE USED IN THE BODY

HUGUES MONOD

The muscular system is indispensable to all active life. It is by contraction of their muscles that animals swim, fly, crawl, jump, or walk, move from place to place, assume varied postures of rest or alertness, and obtain the nourishment necessary for their basic existence. Muscular work is a part of the daily life of man; it gives him, moreover, a certain independence with respect to the physical environment in which he

23

lives (freedom of movement), the ability to practise a trade with which he forms part of a society, and to accomplish or merely express his ideas. Finally, he may satisfy his inclination to indulge in sport or discovery of the world.

To describe how muscles are used or may be used by the body is to consider, what laws or general rules muscles follow in contracting to maintain postures or create movements. It is also to specify the limiting conditions of muscular function. While muscles may be activated in various ways—reflex, semiautomatic, or voluntary—essentially it is voluntary contraction in man which will be considered here, in view of the amount of investigations which it has occasioned. It should not be forgotten, however, that the molecular processes of contraction, the biochemical reactions from which the muscle draws its energy, the transmission of force engendered by contraction outside the muscle, differ little from one kind of muscular contraction to the other, and are identical whatever the type of nervous control.

I. The Muscle in the Body

A. Origin of Muscular Strength

1. STRENGTH OF THE ISOLATED MUSCLE

At the time of the activation of muscular fibers, the sliding of the filaments of actin on the filaments of myosin is accompanied by the production of a force that tends to bring together both extremities of a sarcomere or of a series of sarcomeres belonging to the same myofibril. The sum of the forces created in each myofibril determines the force produced by the muscle fiber that contains them. The force that appears at the extremities of a muscle does not, however, exactly represent the total mechanical energy produced by the different sarcomeres, myofibrils, or fibers of this muscle.

A very small part of the energy could be absorbed by the internal friction which opposes the sliding of the filaments of actin and of myosin, or to the shortening of the fibers.

An important part of the force of the fibers is lost in the pennate muscles, which are composed of fibers arranged obliquely with respect to the long axis of the muscle. This disposition gives to these muscles more strength but less ability to shorten, since the pennate muscles contain a larger number of fibers of smaller length than those of muscles with parallel fibers.

The force produced by the myofibrils is partially absorbed by the elastic components, in series or in parallel, constituting the sarcolemma of the fibers, the connective tissue surrounding the bundles of fiber, and the fascia and tendons of the muscle. A part, but not all, of the energy transmitted to the elastic elements is returned due to the hysteresis. The resulting muscular strength has therefore less magnitude than might be expected, and is dispersed in time.

In the case of muscle twitch, there is a physiological asynchronism of activation of diverse motor units in the muscle, and, to a lesser degree, a very slight delay in the activation of the different fibers of a single motor unit, which contribute also to the spreading in time of the resultant muscular force.

2. Action of Force in the Muscle When Operating in the Body

The force of a muscle is exerted by a bone lever on which it is inserted, one of the insertions being always considered as fixed with respect to the other. The contraction of the muscle produces a torque equal to the product of the force exerted and the perpendicular distance between the axis of articulation and the action line of the muscle (see Williams and Lissner, 1962). This torque T corresponds to the following equation:

$$T = F_m \, l_m \sin \alpha \tag{1}$$

in which F_m is the muscular force available for a given articular position, l_m is the anatomical distance between the articular axis and the point of insertion of the muscle, and α is the angle formed by the action line of the muscle and the direction of the lever arm on which this muscle is inserted.

The relation between the force of the muscle at its point of insertion and that which it can exert at a given point of application of the lever arm, at a distance l_x from the axis of articulation, is given by the formula

$$F_m \, l_m \sin \alpha = F_x \, l_x \sin \beta \tag{2}$$

in which β designates the angle formed by the direction in which the strength F_x is exerted and the lever arm that transmits the strength. It appears, therefore, that the available force at the muscle tendon resolves into a component that is directed toward the articular surfaces, and into a usable component which appears at the point of application of the strength. The available energy should therefore always be expressed in terms of torque and not of force. It is in fact possible to determine the length of the lever arms with anthropometric or radiological measurements (Ikai and Fukunaga, 1968).

Outside of biomechanical studies in which they are valid, such determinations become useless when one considers, instead of the absolute value of the exerted force, the relative force obtained by relating the exerted force to the maximal strength measured under the same experimental conditions. This is the reason for the large number of studies devoted to the measurement of maximal force.

3. Factors Determining the Force Exerted by the Muscle

The force at a point on the lever arm depends on numerous factors.

1. For given anatomical and biomechanical conditions, the strength is determined by the number of simultaneously activated motor units. This number varies according to circumstances (see Section B,II,5).

2. For a given frequency of motor unit firing, the force depends on the number of myofibrils contained in the muscle. When the activation is maximal, the strength is also maximal. It is admitted, as a first approximation, that the force at the tendon is then proportional to the surface of a muscle section, which is determined perpendicularly to the large axis of the muscle for the nonpennate muscles. For pennate muscles, the force depends, however, on the direction of the fibers in the muscle and on their length with respect to the length of the muscle. The old data of Weber concerning maximal strength of the quadriceps femoris were from 2.82 to 4 kg/cm², and Haxton (1944) indicates 3.9 kg/cm² for the calf muscles, while Ralston (1953) has 2.4–4 kg/cm² for the biceps brachii among amputees lacking an upper limb. For Hettinger (1961), the mean figure of 4 kg/cm² may be retained. However, some other authors, including Ikai and Fukunaga (1968), who used an ultrasonic method to determine the muscle section, obtain figures of up to 9 kg/cm² for the biceps brachii.

3. For a given muscle, the strength corresponding to the activation of a given number of motor units depends on the length of the muscle at the moment of contraction. The variations in maximal strength are generally represented on a length–force diagram (Fig. 1). The highest value of force is observed for a muscle length which is called resting length and is equal to 125% of the length of the disinserted muscle (the equilibrium length). The resting length of the inserted muscle corresponds quite often to an intermediate position of articulation.

4. Thus, for a maximal activation of the motor units, the maximal strength depends at the same time on the length of the muscle and

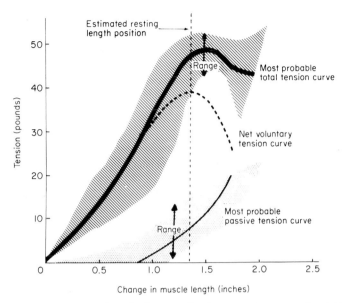

Fig. 1. Isometric tension–length diagram for human triceps muscle. (After University of California, Fundamental studies of human locomotion and other information relating to design of artificial limbs, Vol. 2, 1947.

on the length of the mechanical lever arm, as is indicated in Eq. (1); i.e., on the articular position. Concerning the length of the muscle, it is necessary to take into account the position of the neighboring articulations wherever muscles cross over several articulations. This is shown in the studies devoted to the biomechanical analysis of movements (see Govaerts, 1949; Vandervael, 1951; Donskoi, 1961; Dempster, 1965; Molbech, 1965; Dyson, 1967; Elftman, 1966).

B. The Active Muscle—Description, Definition, Terminology

There are varied types of muscular activity and little general agreement on the exact meanings of the terms that are used. These types must be carefully described with enough precision to avoid confusion (see Kroemer, 1970).

1. WAYS OF USING MUSCULAR FORCE

Muscular activity can be described in terms of at least four factors: (1) the force of contraction, maximal or submaximal, involving a greater or smaller number of motor units; (2) the muscle length over which the force is exerted, or the variation of length during the period of muscular activation; (3) the duration of the activity, which corresponds sometimes to a muscular twitch, but more often to a brief or prolonged tetanus; (4) the possible repetition of tetanic contractions which are separated by rests of various duration; when the repetition is regular, it is defined by its frequency and by its duration.

The effect of muscular activation depends on the relation between the muscular force F_m that is produced and the external force F_e, which may be exerted in the opposite direction by the bone lever where the muscle is inserted. If the muscular force and the external force are equal ($F_m = F_e$), the contraction operates at constant length and does not cause any movement. If the muscular force differs from the external force ($F_m \neq F_e$), the contraction involves movement of the bone lever, and there are then two possibilities to be considered: (1) should the muscular force prevail ($F_m > F_e$), contraction is achieved with shortening of the muscle which is then responsible for the movement; (2) should the external force be stronger ($F_m < F_e$), contraction is performed with a lengthening of the muscle which offers a resistance to the movement.

2. Isometric and Anisometric Contractions

The term "contraction" means literally the shortening of a muscle when it is activated and when the extremities are free. By extension of usage, "contraction" designates the active state of the muscle whether its length remains constant, diminishes, or increases. Although the mechanism of contraction is not fundamentally different within the muscle fibers, the terms isometric contraction and nonisometric (or anisometric) contraction are generally opposed.

A. Isometric Contraction. Isometric contraction is contraction that takes place with the muscle at constant length. The contraction includes, at least, a phase of increasing tension followed by a relaxation phase; the contraction is thus in this case anisotonic. These two phases may be separated by a more or less long period during which a constant or fluctuating tension is exerted. In such contractions, workers pay attention to either the rate of increase of the force, or to the constant maintenance of it.

Frequently used synonyms for isometric contraction are static contraction, sustained contraction, static exercise, isometric exercise, and continuous static work. Intermittent static work (or contractions) or repeated isometric contractions, involve a sequence of periods of contraction and relaxation, the absolute and relative durations of which may be different.

B. ANISOMETRIC CONTRACTION. Anisometric contraction is contraction that involves either lengthening or shortening in one direction or another of the muscle, with one of the extremities free with respect to the other. The expression "isotonic contraction" has long been common usage; but this would be incorrect, since the term presupposes that muscular force remains constant during the movement. "If a constant mass is to be moved, this does not even mean that a constant force is required at the mass to effect its motion; the resistance offered by the mass changes with the changing velocity" (Kroemer, 1970). Considering the tension–length curve of an active muscle, it is clear that, with only rare exceptions, the force is necessarily not constant during the movement, chiefly if the activation is constant. In order to explain a constant force, we would have to imagine an interaction of the different motor units so complicated that it could hardly be regulated. It should be recognized, therefore, that anisometric contraction is anisotonic as well.

The expression "dynamic contraction," although preferable to "isotonic" and very widely used, is not applicable either, because the adjective "dynamic" connotes the idea of force; and force is exerted in isometric contraction as well as in anisometric. Other terms used as synonyms are "phasic contraction," if the contraction is a single one, or in the case of repetition, "rhythmic contraction," "dynamic work," "kinetic contraction"; this latter is the preferable term to use because it implicates the notion of movement.

C. CONCENTRIC AND ECCENTRIC CONTRACTIONS. During repetitive anisometric contractions of a muscle, there are successive phases of shortening and lengthening. Thus there is an opposition between:

1. *Concentric contraction,* or contraction with a shortening that produces active or positive work when it is repeated

2. *Eccentric contraction,* or contraction with lengthening that produces, with repetition, resistant or negative work

The terms concentric and eccentric do not presuppose any direction of movement. Rather, they express the change of muscular length during the movement. Thus, if the forearm is raised toward the shoulder, the movement might involve either concentric contraction of the biceps brachii or eccentric contraction of triceps brachii resisting passive flexion of the forearm under the influence of an external force. The terms refer to the operation of individual muscles themselves.

3. WORK AND ENDURANCE

"Contraction" refers to a short event of which the temporal development may be studied; and it is thus somewhat opposed to the term "work," which generally indicates repetitive contractions within a certain period of time (this does not apply to continuous isometric contraction, which is both contraction and work). From the idea of a certain maximum possible duration comes "endurance," which designates both the quantity of work done and the time during which a force is repeated at a certain frequency, or maintained at a given level. The term "stamina" is used in physical education as a synonym for endurance.

The term "work," which has been criticized by some physicists, corresponds, with its physiological meaning, to several types of muscular activity.

A. MECHANICAL WORK AND PHYSIOLOGICAL WORK. The expression "static work" was first used in 1864 by Heindenhain in the sense of work that a muscle performs in the course of a continuous isometric contraction. This seems to have bothered an age in which physiology was, as a science, not so well developed as physics (see Monod, 1956; Kroemer, 1970). Among physicists, mechanical work is defined as the product of a force and displacement, or as the amount of change in potential or kinetic energy. The impossibility of quantifying the static maintenance holding of a force in physical units is one reason for the rejection of the so-called static work of the physiologists by the physicists (Solvay, 1904).

However, even if no external work is done, each fiber of the muscle produces during an isometric contraction a certain amount of internal work. This does not differ fundamentally from that which occurs in the course of dynamic contraction. In either instance, the muscle transfers energy, using the same biochemical reactions that give rise to electrical and chemical phenomena. In either instance, general adaptive reactions follow that indicate that muscular work depends up on the same logistical system of energy supply.

B. DIFFERENT TYPES OF MUSCULAR EXERCISE. The term "exercise" means any muscular activity, regardless of the circumstance in which it occurs (e.g., general basic activity of the body, laboratory tests, professional work, sports, etc.). The whole body is concerned in a battery of adaptive reactions (cardiac, circulatory, respiratory, metabolic, endocrine) characteristic of a given level of activity. In this sense, exercise is the opposite of "rest." Scherrer and Monod (1960) have suggested a classification of muscular work according to the total mass of muscles that go into action:

1. *Local work* is that which involves less than a third of all body muscles. It essentially includes continuous or intermittent static work and kinetic work done in the course of alternating movements of one or two joints. The muscles concerned are at most those of both the upper limbs or of only one lower limb. The limiting factors of the work are in this case mainly located in the muscle or in the central nervous system.

2. *General work* activates nearly all of the muscles in the body, and it is a matter of kinetic work. The intensity of the general adaptive reactions may be such that they themselves constitute a limiting factor of the work. "Physical fitness" concerns the ability to perform a general work.

3. *Regional work* activates more than a third and less than two-thirds of the muscles. Local and general factors may limit the muscular activity. The frontiers between regional work and local or general work are, in fact, rather artificial. Thus, a maximal amount of kinetic work obtained by cranking cannot be called local work, even though it involves only both the upper limbs.

C. The Three Principal Types of Local Work. The work that a muscle is able to do when it is activated includes the three following kinds of activity which will be discussed in detail later:

1. The repetition of a maximal or submaximal force—*brief repetitive isometric contractions*
2. The maintenance of a maximal force, or of a submaximal force during a time period that is maximal or voluntarily limited—*static work*
3. The repetitive displacement of a load—*dynamic or kinetic work*

A single or sporadically repeated exercise of maximal strength does not, properly speaking, constitute muscular work.

4. Fatigue and Exhaustion

The term "fatigue" is among the more difficult to define, as is demonstrated by the large number of studies devoted to the attempt (Ioteyko, 1920; Bartley and Chute, 1947; Ryan, 1947; Floyd and Welford, 1953; Bartley, 1957; Scherrer and Monod, 1960). A step toward clarification was made by J. Scherrer when he distinguished objective or physiological fatigue from subjective fatigue, proposing this definition: "Physiological fatigue is the decrease of activity in a living system, related only to the working of this system, occurring despite a constant flow of the adequate stimulus and disappearing after rest."

Muscular work does not always lead to fatigue, even if it is performed for a long time. In certain situations, however, a drop of performance takes place, progressively or suddenly, which allows one to define a relative or absolute threshold of exhaustion (the threshold would be the point at which it becomes impossible to exert even a minimal force, or a force greater than a given percentage of maximal). Thus, "exhaustion" has to do with the decrease in performance that is characteristic of objective fatigue, and it connotes, moreover, the main cause of the fatigue, which is to say the excessive utilization of energetic and functional reserves of the muscle or of the body. The definition of fatigue is linked with those of the various level of muscular activity:

1. *Cruising level* corresponds to a relatively low power (or force), allowing the muscle to contract for a long time without fatigue.

2. *Crest level* is the highest sudden power (or force) a muscle is able to produce during the course of a brief tetanic contraction.

3. *Critical level* is the highest cruising rate at which a muscle is able to work without fatigue. This level is measured by the *critical power* (dynamic contraction) or the *critical strength* (static contraction).

The term "critical level" has the same physiological meaning as "aerobic capacity," the maximal amount of oxygen the body can use in the time unit. "Critical level" is synonymous with "work capacity for a local work," whereas *"physical fitness"* refers to the work capacity for a general exercise.

5. Effort

This term is used as a synonym of exercise, work, or force in popular language. Used in this way, it is not entirely satisfactory. In fact, "effort" was initially defined by physiologists as the contraction of the abdominal muscles, with closed glottis, so that pressure is exerted on the thoracic and abdominal viscera. This pressure occurs in coughing, sneezing, defecation, vomiting, or parturition. Carrying out certain tasks with the upper limbs sometimes requires a good thoracic stance, which implies the closing of the glottis; in this case, muscular work would be accomplished, in the strict sense, with "effort," so that here, finally, "work" and "effort" are used interchangeably. Use of "static effort" has also been suggested by some authors who would like to use it to replace "static work." This would in reality add to the confusion.

By analogy with the view of some psychologists (see Gaultier, 1970) who consider "effort" to entail a consciousness of the work to be done and the extra demands it will make, a physiological sense may be given to "effort" insofar as it concerns an objective fact. The term may be defined as the supplementary activity that is necessary for the organism to begin working or to carry that work to its conclusion. In spite of subjective and objective signs of fatigue, an example of this effort would be the general cardiorespiratory adaptations at the beginning of a task of muscular work, as well as all the compensatory reactions that increase to maintain performance at a constant level. Physiological effort may then be understood in terms of cardiovascular, respiratory, electromyographic, and cortical changes that occur in the course of muscular work at a given rate of activity. In this sense, the physiological effort that retards the drop of performance constitutes a complex of objective signs of fatigue for an organism engaged in local or general work.

II. Exertion of Maximal Muscular Strength

The maximal strength is the greatest strength that a muscle can exert under given conditions in the course of a single voluntary or nonvoluntary contraction. The maximal muscular force is a result of the activation of the maximal number of motor units.

A. Measurement of Muscular Strength

1. APPARATUS

A. MEASUREMENT OF ISOMETRIC STRENGTH. The earliest measurements of muscular strength were by Graham, who opposed the action of the muscle to weights of increasing value until a maximum was reached. To Regnier, we owe the idea of the elliptical spring steel dynamometer designed in 1807 at the request of the naturalist Buffon, who wanted to study the variation of grip strength according to age. Collin later made a dynamometer based on the same principle. For more than 150 years, manual force has been regarded as a faithful index of the total body strength, although the muscles of the hand are more adapted to delicate and precise movements. The criticisms of Collin's dynamometer (e.g., its variable position in the hand, slippage aggravated by sweat, pain at the points of maximal pressure, etc.) led to devices better adapted to the form of the hand, such as the dynamometers of Smedley and Charles Henry. Whipple's accessories permitted the dynamometer to be used for other than grip strength measurement (see Fautrel, 1954; Hunsicker and Donnelly, 1955).

These methods of determining muscular strength evolved as it became necessary to obtain precise values for a large number of muscles, many subjects, and for segments of a member exactly located in space, corresponding to certain professional situations (e.g., the manual controls in an airplane cockpit). Different kinds of dynamometers could be placed between a fixed point and a body segment. For rapid measurements, the tensiometer is tending to replace the spring steel dynamometer (see Clarke, 1966).

When a record is desired, transducers are used; for example, the strain gauge is a metal piece with a wire of variable resistance coiled around it; the resistance is proportional to the forces of pressure or traction applied to the piece. For multiple measurements, the subject is seated within a framework that supports strain gauges at various points around him (Fig. 2) (Hunsicker, 1957; Asmussen *et al.*, 1959; Caldwell, 1959; Rohmert, 1960; Tornvall, 1963).

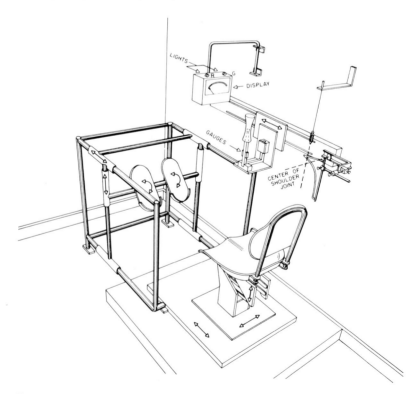

Fig. 2. Apparatus for isometric strength determination. An adjustable seat and foot rest permits placement of all subjects in the same anatomical position. (Caldwell, 1963.)

B. MEASUREMENT OF ANISOMETRIC STRENGTH. This measurement is made with a dynamometer placed in series between the body segment and a point of reference, which is moved at constant speed; the dynamometer operates by traction or pressure, depending on the type of contraction and movement under consideration. Asmussen *et al.* (1965) placed their dynamometer at the end of a piston which was inserted in an oil-filled cylinder. A valve allowed the oil to flow from one side of the piston to the other, controlling, by its adjustment, the speed of movement. Singh and Karpovich (1966) used a lever arm moved by a motor at constant angular speed of 17.4 degrees per second and at constant torque of 150 foot-pounds (20.7 m kg). Later on, Karpovich developed some controlled speed devices.

The two types of apparatus described above are used for the study of alternate movements. For the measurement of muscular force in the course of rotating movements, Kogi *et al.* (1965) placed strain gauges between the foot and the pedal of an ergocycle.

2. PROCEDURE FOR MEASUREMENT

The numerous causes of intra- or interindividual variations of maximal muscular strength (see Section II,B) make certain precautions necessary:

1. A general clinical examination should verify the anatomical integrity of the subject—the back strength measurement, for instance, may in some cases cause sciatic injuries.

2. The subject should be questioned about his state of health. He may be in physical training. On the other hand, he may be convalescent or suffering from minor troubles that would make him a poor subject. It is known, for instance, that pain, particularly in the splanchnic area diminishes the maximal strength.

3. Sufficient explanation should be given the subject concerning the apparatus and experimental procedure so that he knows what he is supposed to do; a demonstration and several trials should be performed. The nature of the information given to the subject (Kroemer, 1970) and the reasons told him for the measurements (Nelson, 1962) may modify the value of muscular force.

4. The positions of the subjects should be standardized with respect to the angles between the principal segments of the body at the time of measurement.

5. The calibration of the apparatus should be checked regularly.

6. The results ought to accord with a precise definition of maximal force—e.g., value attained during a single trial, mean value of several

trials, largest value of several trials. A checklist for the measurement of maximal muscular force has been proposed by Kroemer (1970).

3. Indexes of Strength

The results of maximal strength measurements are expressed either by its value in kilograms or in torque (meter-kilograms) or else across different indices. Knowledge of the maximal strength of a muscular group is useful to judge the physical fitness of a subject considering:

1. The changes in his condition caused by physical training and sports, recuperation after an accident, or a period of bedrest.
2. His ability with respect to other subjects of the same group.
3. His selection in terms of the physical work which is to be done.
4. His capacity and the organization of his professional work, so that he is able to carry it out without fatigue for an extended time.

A. Knowledge of Maximal Strength. The maximal strength is that which the subject exerts with one or more defined muscular groups when he is asked to contract the muscles as strongly as possible in a predetermined position. It is generally a matter of isometric strength. Knowledge of strength is especially important, since the muscular group tested should be fully involved in the activity for which the test was designed. If the muscular group tested does not figure importantly in the exercise under consideration, the measurement of its strength is next to useless. Tornvall (1963) has in fact shown that the correlations are small between the absolute values of the maximal strength of two muscles in the same subject (see Table I).

B. Calculation of Indexes. These calculations are intended to give a broader view of the strength of the individual. At least two or three muscles are compared in order to clarify the normal or abnormal proportion of their respective strength. Thus, the index of ambidexterity considers the relation between the strength of the same muscles on the right and left sides. The index of harmonious strength expresses the relation between the strength of different muscular groups; e.g., the strength of the back extensor muscles compared to the average grip strength of the two hands.

At the most, one can conceive an index that represents all of the muscles in the body, or at least those that play a dominant role in the performance of a given exercise. This is used by Tornvall (1963)

TABLE I

CORRELATION BETWEEN ISOMETRIC MUSCLE STRENGTH RECORDED IN DIFFERENT PARTS OF THE BODY (TORNVALL, 1963)

Groups: Neck — Upper extremities — Trunk — Lower extremities

	Neck flexion	Neck extension	Shoulder pull	Elbow flexion r.	Elbow flexion l.	Elbow extension r.	Elbow extension l.	Finger flexion r.	Finger flexion l.	Back forward flexion	Back backward flexion	Leg extension	Hip flexion r.	Hip flexion l.	Knee flexion r.	Knee flexion l.	Knee extension r.	Knee extension l.	Foot plant. flexion r.	Foot plant. flexion l.	Foot dors. flexion r.
Neck extension	0.25																				
Shoulder pull	0.08	0.23																			
Elbow flexion r.	0.21	0.41	0.48																		
Elbow flexion l.	0.25	0.35	0.42	0.92																	
Elbow extension r.	0.07	0.30	0.47	0.74	0.68																
Elbow extension l.	0.05	0.30	0.40	0.66	0.62	0.83															
Finger flexion r.	0.10	0.41	0.47	0.68	0.67	0.49	0.41														
Finger flexion l.	0.07	0.29	0.43	0.63	0.66	0.49	0.38	0.89													
Back forward flexion	0.21	0.19	0.29	0.37	0.38	0.33	0.23	0.37	0.41												
Back backward flexion	0.08	0.34	0.26	0.12	0.12	0.11	0.06	0.21	0.26	0.43											
Leg extension	0.23	0.43	0.43	0.46	0.46	0.29	0.28	0.55	0.45	0.43	0.32										
Hip flexion r.	0.25	0.46	0.46	0.38	0.35	0.37	0.33	0.33	0.29	0.43	0.49	0.47									
Hip flexion l.	0.14	0.43	0.43	0.40	0.37	0.36	0.33	0.29	0.26	0.58	0.48	0.43	0.93								
Knee flexion r.	0.14	0.40	0.29	0.30	0.25	0.32	0.28	0.26	0.26	0.60	0.51	0.24	0.60	0.52							
Knee flexion l.	0.14	0.53	0.35	0.42	0.39	0.36	0.27	0.35	0.36	0.37	0.52	0.29	0.63	0.55	0.87						
Knee extension r.	0.26	0.41	0.38	0.53	0.52	0.56	0.49	0.41	0.45	0.44	0.36	0.49	0.59	0.56	0.53	0.54					
Knee extension l.	0.29	0.44	0.27	0.53	0.57	0.43	0.42	0.42	0.45	0.46	0.34	0.51	0.55	0.56	0.46	0.55	0.87				
Foot plant. flexion r.	0.18	0.29	0.41	0.42	0.42	0.29	0.19	0.53	0.50	0.52	0.48	0.33	0.58	0.55	0.48	0.54	0.41	0.39			
Foot plant. flexion l.	0.09	0.30	0.40	0.34	0.35	0.28	0.16	0.50	0.48	0.38	0.44	0.43	0.58	0.56	0.53	0.53	0.37	0.35	0.92		
Foot dors. flexion r.	0.29	0.04	0.21	0.27	0.33	0.19	0.21	0.16	0.14	0.19	0.07	0.13	0.23	0.20	0.27	0.23	0.27	0.30	0.26	0.22	
Foot dors. flexion l.	0.25	0.11	0.24	0.23	0.28	0.14	0.13	0.14	0.10	0.19	0.08	0.14	0.25	0.23	0.34	0.25	0.34	0.24	0.29	0.24	0.93

according to Clarke and Lindegård in order to classify a subject among
a group:

$$I = \frac{\Sigma \left[(x - \bar{x})/s \right]}{n}$$

where n is the number of muscles tested; x the individual value of
the maximal strength for a given muscle; $\bar{x}$ the mean value for the
group; and s the standard deviation of individual values. Thus, this
index takes into account first the standard deviations of values with
respect to the mean, and second the mean of these standard deviations
for the entire group of muscles under consideration.

B. Value and Variations of Maximal Muscular Force

The absolute values of muscular force that are found in the literature
are useful as a reference only insofar as the point of application and
the direction of force transmitted by the measuring apparatus are given.
If not, values expressed in terms of torque may be used. In other cases,
interest is centered on the systematic variations between different subjects in identical conditions of measurement, or the same subject in
different conditions. There is a monograph on this topic by Hettinger
(1961). The variability of force was studied principally as a function
of age, sex, possible bilateral asymmetry, articular position, motivation
of the subject, and training.

1. CHANGES WITH AGE

Maximal strength increases linearly with age until puberty, when the
linear increase diminishes until the overall increase reaches its absolute
maximum. According to Hunsicker (1955), the maximal strength at 6
years is about 20% what it will be at maturity, and at 20 years it is
about 80%. Over 25 years, the strength has a tendency to decrease, at
65 not exceeding 75% of the maturity maximum. The data confirm those
measured between 12 and 79 by Burke *et al.* (1953). The overall force
diminution, however, may be retarded even above 40 years, if the muscles undergo sufficient activity. The decrease in the strength plotted
against the age varies with the muscular group under consideration
(Asmussen and Heebøll-Nielsen, 1961), perhaps with respect to the contribution of each of them to general activities. Some muscles certainly
retain their strength longer than others as a result of regular exercise,

since the aerobic capacity of subjects is always lowered by age above 40 years.

2. Sex Differences

Whipple (1924) has shown that of 3000 boys and 3000 girls, the manual force, measured with a Smedley dynamometer, increased by 4 kg per year until the age of 18 among the boys, and 3 kg per year to the age of 15 among the girls. According to Rich (1960) the sexual differences may be ignored between 8 and 13 years, although after puberty the rate of increase in strength is maintained among the boys, while it declines among the girls. This sexual difference is related to the secretion of androgen and estrogen (Danowski and Wratney, 1959).

In women, the maximal strength is established at about 55–80% (depending on the muscles) of the strength in men of the same age (Hettinger, 1961). Asmussen and Heebøll-Nielsen (1961) have shown, however, that of 360 men and 250 women, the sex differences were smaller than is generally believed, based on the comparison of 25 muscular groups. The relative strength in women, which was established at 65% of that in men, is actually 75–80% if the comparison is made not according to age, but according to the same body height. These last figures could be explained by the correlations between certain anthropometric dimensions and muscular force (see Roberts *et al.*, 1959).

3. Bilateral Comparison

Right and left muscles in a normal subject have a muscular strength with a slight predominance of one side over the other. Asmussen and Molbech (1958) undertook a study of the normal limits of these variations—a study that has practical application in the tracking down of unilateral muscular deficiencies such as those that occur in poliomyelitis or traumatic injuries of the limbs. According to these authors, the differences between the two sides are of the order of 5–6% for muscle groups in the upper limbs, which figure was recently confirmed by Toews (1964) in an inquiry based on industrial workers. In the lower limbs, the differences reach 8–9%. The dispersion of differences between the stronger and the weaker side at twice the standard deviation are of the order of 15–20% for the upper limb, 30% for the lower limb; and larger differences than these figures are indicative of a functional abnormality.

According to these same authors, the muscles of the upper limb on the right side are generally the strongest, but less often than is usually thought—55–66% of the subjects have stronger right arms; 21–34% are left-handed. Conversely, the muscles of the lower left limb predominate in 51–55% of the cases, and of the lower right limb in 38–40%. This difference is attributed to bending or leaning reactions which cause the lower left limb to be used when the upper right limb is used. The correlations between the maximal strength of similar muscles are relatively high—0.71–0.88 for Asmussen and Heebøll-Nielsen (1961) and Heebøll-Nielsen (1964), except for three muscle groups out of 25, and 0.83–0.93 for Tornvall (1963).

4. Variations with Position of Articulation

A considerable number of studies have been devoted to the dependence of muscular strength upon the position of articulation (Fig. 3).

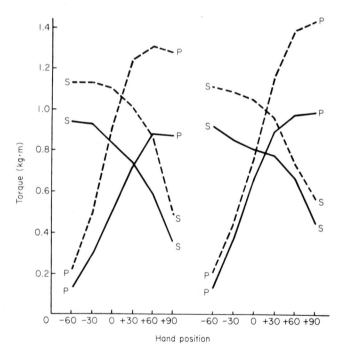

Fig. 3. Strength variations with articular position. Maximum isometric pronation (P) and supination (S) force in six different hand positions. Strength before training (—) and increases in strength after static (*right*) and dynamic (*left*) training in pronation. (Darcus and Salter, 1955.)

This depends on the properties of the muscle tissues themselves, which are shown in the tension–length diagram, and on the obliquity of the muscle with respect to the body segment by means of which the force is exerted.

It is well known that for many muscles the greatest maximal strength, following the activation of the greatest number of motor units, is measured when the muscle is at its resting length, corresponding to an intermediate position of articulation, which is between 90° and 110° for flexion of the elbow, extension of the knee, and abduction and extension of the shoulder (Hunsicker, 1955; Provins and Salter, 1955.

Other muscles, however, produce their maximal strength in an extreme position of the articulation, corresponding to a maximal stretching for the muscles inserted in the body. This is the case for extension of the elbow, flexion of the knee, abduction and flexion of the shoulder, and extension of the thigh (Houtz *et al.*, 1957; Clarke, 1966). The antagonistic muscles, then, do not produce their maximal strength at the same angle of articulation as the agonists.

Comparing curves obtained by different authors from the same muscle group, we find large differences, most often in muscles that cross several articulations, like the biceps brachii (for one head) and the flexor muscles of the knee. It is important to consider not only the angulation of the main articulation, but also the adjacent articulations; they contribute partly to fix the muscle length.

Precise analysis of articular angles loses its interest when muscular strength is exerted by the whole member in its totality, upper or lower, and when the authors are concerned only with establishing isodynamic curves of push or pull. For the practical use of such results, it is only necessary to indicate (a) the direction in which strength is exerted with respect to some anatomical point of reference, such as the acromial point for the upper members and (b) the distance between this and the point where the force is applied (see Caldwell, 1959).

In the evaluation of the force a muscle can exert in a given situation (sports or professional work), we should take into consideration (a) the nature of fulcrum, which gives stability to the whole body (Caldwell, 1962), and (b) the weight of the total body, or of the body segments, which are used when it lacks support.

5. Variations with Degree of Motivation

Everyone who has practiced measurements of maximal strength knows that they are satisfactory only if the subject concentrates his attention

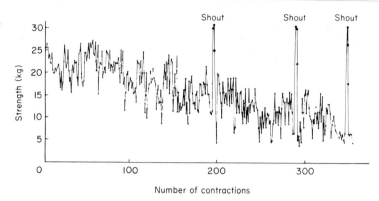

Fig. 4. Fatigue curve of the muscular strength of the arm flexor and effect of "shout" on the performance. (Ikai, 1962.)

on the task to be performed. This way, an effort of will may bring into play a very large number of motor units firing at a maximum rhythm. Rather often, this high value of maximal strength cannot be attained because of inhibiting phenomena in the central nervous system due, for instance, to sensory discharges or to a particular basic state. This explains the variability of maximal strength in a muscle when the measurements are repeated on the same individual on subsequent occasions.

Thus, it should be recognized that the greatest value of strength is not generally attained in the course of a maximal effort of will. A particularly strong internal or external incitement is needed to obtain an elevated response. Ikai *et al.* (1967) have shown that in the course of a series of voluntary maximal contractions of the biceps brachialis, leading after 300 repetitions to a fall of 25 to 5 kg in the strength of the biceps, the subject is still able by "shouting" to obtain peaks of maximal strength as high as 30 kg (Fig. 4). There are numerous examples of exceptional efforts, sometimes almost superhuman, during which a degree of muscular strength much higher than the usual values is produced. (Imagine a mother lifting an extremely heavy stone that has crushed the arm or leg of her child.) An increase in the frequency of the motor units firing may also be obtained by using various experimental procedures.

Elicitation of the myotatic reflex by a relatively light stretching in the course of a maximal voluntary contraction of the biceps brachii gives an augmentation in the force of the muscle. It arises from a volley of nerve impulses originating from the spindle and adding to the cortical outflow.

Habitual inhibitions may be removed by hypnosis. Ikai and Steinhaus (1961) have demonstrated that it is possible to increase or to decrease muscular strength under hypnosis.

6. ELECTRICAL AND VOLUNTARY CONTRACTIONS

Merton (1954) and Naess and Storm-Mathisen (1955) have shown that the maximal muscular strength obtained by an electrical stimulation is equal to that resulting from a voluntary contraction, however strong the motivation of the subject. Ikai *et al.* (1967) used, as did Merton, the adductor muscle of the thumb, demonstrating that experimental tetanus at sufficient voltage and a frequency of 50 per second produces a force 30% greater than does voluntary contraction. The difference is yet more evident in a muscle fatigued by 120 maximal voluntary contractions. The voluntary strength and the strength obtained by electrical stimulation are decreased, respectively, to 20% and 65% of the initial value; but in terms of the absolute values of strength remaining after relative exhaustion, it seems that electrical stimulation can obtain a strength 4 times as large the strength obtained by voluntary contraction (see Fig. 5).

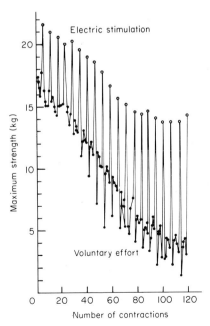

Fig. 5. Maximal strength during electric stimulation and voluntary contraction. (Ikai *et al.*, 1967.)

7. Effect of Training

Maximal strength is improved with training. The oldest and most illustrative example of such improvement is found in the story of Milo of Crotona who lifted the same calf each day until finally he could lift an ox. The augmentation in volume of muscles is to be attributed to the increase in muscle fiber diameter, the enlargement of the capillary area, and the development of interstitial connective tissue; and each kind of training modifies a particular one of these three factors (see Karpovich, 1959). Training usually has less effect on women than on men, since women tend to hypertrophy the interstitial tissue (Hettinger, 1961). The same fact has been noticed among men during excessively intense training, which has, as an end, the development of muscle form, but not exercise ability.

The possibilities of training vary with the muscles. It is in practice difficult to compare the increase in strength of muscles that have different functions but not the same aptitude for training. The research of Hettinger (1961) set out to clarify the best training conditions. The use of a strength lower than 20–30% of the maximal strength is insufficient, for training only takes place between 30–80% of maximal strength. Generally, training should make use of a greater strength than is exerted in the activities of everyday life, or of the specific muscular activity for which the training is undertaken. Hettinger (1961, 1964) believes that a single maximal contraction each day for several weeks can obtain a regular increase of strength before a maximal level is reached. A number of authors deny this point of view and have demonstrated that the repetition of twenty contractions a day gives better results than a single exercise of strength.

Truly, there is no single method for the training of muscular strength. Numerous procedures have been proposed, for example, the use of heavy weights, or of progressive resistance, or of working the muscle in elongation. In each case, there ought to be optimum combinations of force, duration, and frequency of repetition of the contractions. Training of force may be obtained as well by isometric contractions (which develop mainly strength) as by means of anisometric contractions (which develop primarily endurance). This has been described, for instance, by Darcus and Salter (1955) for an increase in the strength of pronation and supination (see Fig. 3).

Finally, we ought to mention that training of strength results not only in morphological changes in the muscle, but also in a change of its central control. A better synchronization of the motor units in the same muscle and a better coordination of the different muscles involved in the same activity, increase the available muscular strength.

C. Effect of Work on Isometric Maximal Strength

Three types of muscular activity have been studied for their effects on muscular strength: (1) brief repeated maximal isometric contraction, (2) maintained isometric contraction, and (3) dynamic work. The fall of muscular strength appears as one of the signs of muscular fatigue. Different kinds of quantification have been proposed for fatigue, and the simplest is the strength decrement index (SDI) of Clarke *et al.* (1954), which is the percentage loss in strength:

$$\text{SDI} = \frac{S_i - S_f}{S_i} \times 100$$

in which S_i is the initial maximal strength and S_f the final strength.

1. EFFECT OF A SERIES OF MAXIMAL ISOMETRIC CONTRACTIONS

The activity under consideration includes only increase followed by decrease of tension repeated at various frequencies to the exclusion of any isotonic contraction. Physiological work carried out in this way is as much more intense as the frequency of the contractions is increased. Similarly, a decrease in muscular performance is seen with low frequency of contraction. Thus, in the flexor muscles of the forearm, a decrease was noticed at frequencies of 0.1/min (Clarke *et al.*, 1954), 0.2–0.5/min (Pastor, 1959), and 0.1/min (Ikai and Steinhaus, 1961). The same conclusion is found in the work of Bourguignon *et al.* (1959) on the biceps brachii and Kroll (1968) on the extensors of the hand.

In addition, it seems that the loss in strength depends more on the number of contractions taking place than on their frequency (Bourguignon *et al.*, 1959). Since the blood flow in the muscle is satisfactory for the frequencies used by these authors (1–6/min), we may think that if the processes of energy mobilization are rapid, the complete restoration of muscular strength is a slow process, possibly beyond 10 min even after a short exercise (20–30 contractions).

Bourguignon *et al.* (1959) for low frequencies of contraction and Molbeck (1963) for larger frequencies (6–29/min) and for different muscle groups have both shown that after an initial fall, a maximal strength plateau occurs before the 10th min. Kroll (1965) noticed a stabilization of the maximal strength of hand extensor muscles at 77% of the initial strength after 20 trials carried out at a frequency of 1.7/min.

The rate of decrease in strength becomes faster when maximal contractions take place with a frequency equal to or greater than 30/min. Molbeck (1963) considers that this frequency should not be exceeded because of the time becoming too short to allow the muscle to manifest its available maximal strength (the strength increases as an exponential curve) and because of insufficient circulation in the muscle. The results of Rich (1960) and Clarke (1962) on grip strength and of Williams (1969) on the strength of the biceps brachii performed at the frequency of 30/min show a fall in the strength of 50–60% of its initial value at the sixth or seventh minute of activity. The exact maximal frequency of contraction allowing a muscle to be correctly perfused depends on its size; the larger the muscle, the lower maximal frequency.

Some contradictions appear at times in the results of different authors. The differences are due to the experimental procedure, the type of muscle studied, and its degree of training. It is well known, in fact, that all muscles are not composed of the same types of muscle fiber. The fast phasic fibers, including a large number of myofibrils, are less adapted to prolonged work than the slow phasic fibers. It is therefore normal that muscles including a larger proportion of fast fibers exhibit a more rapid decrease in their maximal strength or in their general work capacity. This explains the hypothesis of Tuttle *et al.* (1950) confirmed by Kroll (1968) and by the curare experiments of Molbeck and Johansen (1969), which assert that strong muscles become fatigued more rapidly than weak muscles. One is tempted to accept the idea that training that increases fibers diameter could concern specially the fast fibers into the muscle.

2. Effect of a Sustained Static Contraction

The decrease in strength following a continuous static contraction is evident when a subject is asked to produce a maximal contraction at once and to maintain it as long as possible. Fessard *et al.* (1933) using the ergometer of Charles Henry have proposed an "index of tenacity," taking account of the time in which the maximal strength diminishes to 50%. Tuttle *et al.* (1950) have considered the mean strength of a voluntary maximal contraction maintained for 1 min, which is established at about 60% of the initial strength.

Clarke (1962) showed that for contractions maintained for 2 min, the decrease in strength measured at intervals of 5 sec effectively follows an exponential curve of the time. This confirms the observations of Royce (1958) made on contractions of 90 sec, which showed that the fall of strength is greatly accelerated when the muscle works without blood circulation.

The fall of strength during an isometric contraction maintaining a constant submaximal force is not evident, since it is necessary to wait until a threshold of exhaustion in order to determine that the maximal strength after work has become inferior to the initial strength. Fessard *et al.* (1933) have shown with measurements of maximal strength at regular intervals that this decrease in strength also follows an exponential curve of the time.

3. EFFECT OF A SERIES OF DYNAMIC CONTRACTIONS

The decrease of maximal strength during or following dynamic work is apparent in the first ergographic studies of Mosso (1890), which were focused upon a fall in the amplitude of successive movements. Clarke *et al.* (1954), using the Kelso–Hellebrandt ergograph for dynamic work following Mosso's technique, showed that for lifting a charge equal to 37.5% of the maximal isometric strength, at a frequency of movement of 30/min, the strength after work is only 70% of the initial strength. The decrease in strength, of course, depends upon the power put into action; but for a given relative power, the decrease is the same whatever the age of the subject. (Power is here defined by the relative force exerted and the frequency of contraction.) The fall of force measured at regular intervals during dynamic work follows an exponential curve that Clarke (1962) proposes to define by the time in which a fall in strength occurs equal to 50% of the fraction that will be lost in 6 min (which latter has been called the "fatigable strength").

In the course of dynamic work carried out at constant power until exhaustion, the decrease in force is progressive but not regular, having a rapid first phase and a terminal drop proceeding arrest of the contractions (see Fig. 6). The fall of functional muscular capacity is manifested again by the failure of muscle which has just worked to maintain its actual maximal strength for more than 1 sec, while a nonfatigued muscle can maintain it for 5–10 sec. Bourguignon *et al.* (1959) specify, moreover, that the fall of force occurs even if the work is not exhaustive; the strength falls to 90–80% of its initial value and then is maintained at an almost constant level.

The diminution of maximal strength was used by Pastor (1959) to define a "threshold of fatigue" as being the smallest quantity of work in which a significant fall of strength becomes apparent. Using the Kelso–Hellebrandt ergograph with work of the biceps brachii, Pastor demonstrated that in the most favorable conditions of power (25% of the maximal strength, 38 times per min), the decrease in force is related to the number of successive contractions, and the threshold of fatigue is only attained after a minimum of 9 contractions.

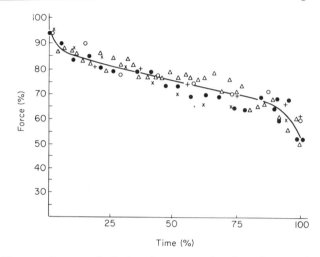

Fig. 6. Decrease in strength during dynamic work. The subject performed three tests of dynamic work until exhaustion with the triceps brachii; these tests are carried out at different powers. Decrease in the strength is plotted versus the time as a percentage of the limit time for each test. (Bourguignon *et al.*, 1959.)

4. Recovery of Muscular Strength after Work

The recovery of strength after work is accomplished at a rapid rate immediately after the work, and then more slowly, whatever the type of muscular activity. The exponential curves of restoration put forth by Müller in 1934 have not been confirmed by all authors. Bourguignon *et al.* (1959) give complex curves following dynamic work of the triceps brachii, which represent a transitory diminution of the force. Lind (1959), Clarke (1962), and Clarke and Smith (1966), using more systematic measurements on a larger number of subjects, showed that recovery takes place in an exponential manner and defined the two components—rapid and slow—of the curve.

Lind suggests that the initial rapid phase corresponds to the elimination of waste products, the second to the restoration of the energy reserves in the muscle. Recalling that the curves of blood flow in the muscle after work are also exponential, one would emphasize the role of the circulation in muscular recovery. A limitation of circulation considerably retards the restoration of muscular strength (Clarke and Smith, 1966).

Clarke (1962), studying the strength recovery after dynamic work and after a series of brief isometric contractions, recognized a similar exponential evolution in the two cases, although the restoration would be somewhat more rapid in static work. One may invoke the smaller expenditure of energy in isometric contraction (Monod, 1967) to explain this difference. In practice, any comparison between static and dynamic work is difficult, for lack of a common unit of measurement and lack of certainty that a similar level of local exhaustion has been attained.

A recovery of strength following an exponential pattern does not appear clearly in the results of Pastor (1959); but it is discussed by Kroll (1967), who recognizes it following work carried out at a frequency of 6/min. At lower frequency, restoration is accomplished linearly. The divergence between authors probably depends on different experimental conditions, on the quantity of work done, on whether or not exhaustion has been attained, and on the duration of the period of observation following work. Pastor (1959) has shown, for instance, that the recovery is as long as the work performed at constant power has been large. Bourguignon *et al.* (1959) have equally shown that the recovery of strength comes later when the subject has performed several periods of dynamic work separated by brief periods of rest. The frequency at which maximal strength is measured during the phase of restoration is certainly involved here, since the exercise of maximal strength, which is the test of recovery of muscle capacity, is also a cause for decreasing capacity, as we have seen above.

An interesting result of Kroll (1967) deserves mention. It concerns the phenomenon of overrecovery, noticed first by Lombard in 1890 and mentioned by Bourguignon *et al.* (1959) in relation to brief isometric contractions. Kroll groups subjects according to their maximal muscular force. After a series of 30 maximal isometric contractions of 5 sec separated by rests of various duration (5, 10, and 20 sec), the maximal strength is exerted each minute for 2 sec for an overall period of 10 min. At the end of the measurement the subjects with the lowest strength achieved a maximal strength which clearly surpassed its initial value. This phenomenon might correspond to facilitation which could be peripheral, reminiscent of post-tetanic potentiation, and/or central, due to better recruitment and a better synchronization of the motor units.

III. Exertion of Isometric Strength—Static Work

A number of everyday tasks include, in various forms, muscular activity with static components. At least, isometric contraction is responsible for the simple maintenance of the standing position and constitutes a factor in fatigue when the posture is prolonged. Besides this, postural constraints are imposed, either by the inconvenience of a working position insufficiently studied or unchangeable, or else by the need to stabilize the trunk during the activity of the upper limbs or to fix the head when a constant and precise visual control of the task is required. At the extreme, continuous or intermittent isometric contraction constitutes the principal muscular activity when the task involves holding for a long time a lever or a large load in a given position. In certain particular cases, such as that of fighter pilots, postural activity is exaggerated due to the considerable accelerations to which the body is subject; the reinforced contraction of the neck and back muscles leads to lumbar or dorsal stiffness, pains in the area of the cervical muscles and eventual deformations of the cervical column.

Under laboratory conditions, static work is obtained either by holding a constant load or by the prolonged exertion of a force measured with a dynamometer, eventually known by the subject and recorded as a function of time. For lack of expression in physical units, static work (kilogram-seconds) is considered as the product of the force (kilograms) and the time (seconds) it is maintained: $W = Ft$. The maximal capacity for static work is thus defined by both force and time.

A. Types of Static Work

Static work, continuous or intermittent, is performed by exerting a strength that is either maximal (using Mosso's technique) or submaximal, as is usually done now (see Fig. 11). All static muscular activities of everyday life occur as one of the three following types.

1. EXERTION OF MAXIMAL STRENGTH

When the muscle exerts a maximal strength by pulling on a dynamometer, this strength drops in a very short time (Section II,C,2). In holding a load, the decrease in contractile power of the muscle occurs after a certain time by a change in the angle of the articulation; the muscle is quite progressively lengthened. This eccentric contraction is considered as static work, for the lengthening of the muscle occurs very slowly. This kind of test is of little interest owing to the change in direction of the external force with respect to the lever arm and the change in length of the muscle.

The duration of the test may be unlimited or limited, whether by circumstances, or by the experimenter if it is a fitness test. Fessard *et al.* (1933) consider as an index the time in which a drop of 50% in the initial maximal strength occurs. Tuttle *et al.* (1950) study the drop of the maximal strength following a maintained force limited to

1 min; the work accomplished is then evaluated by the mean force exerted during the fixed time. Using the Henry mercury dynamometer, Petz (1964) expresses the work accomplished by the sum of the values of force read every 5 sec on the manometric tube.

2. Continuous Exertion of Submaximal Strength

Static work is sometimes performed with a given load that is identical for all subjects. This load thus represents a fraction of the maximal strength, which is, of course, different for each subject. In order to compare the performances of the subjects, it is preferable to impose a force that represents a given fraction of the maximal strength that has been previously measured. Holding a load with control by the subject of the articular position seems better than the adjustment of a constant strength in a position which is imposed on him.

The maximal capacity for static work at a given strength is defined by the maximal time (limit time) during which the strength is maintained. This time corresponds to the occurrence of a relative threshold of local exhaustion for the force under consideration (see Monod and Scherrer, 1965).

3. Intermittent Exertion of Submaximal Strength

This kind of work is like a continuous isometric contraction, since the muscle remains at constant length and is also close to dynamic contraction since the phases of intertrial rest permit sufficient irrigation of the muscle. Such work is accomplished with the aid of a support manipulated by an operator. It is placed under the load during the phases of rest and removed when the muscle has to contract. Automatic devices (mechanical, pneumatic, electromagnetic move the support (Fig. 7) or exert directly a constant or eventually sinusoidal force (Berthoz and Metral, 1970).

Intermittent static work is adequately defined by the durations of the phases of contraction and of rest, after which one may calculate the frequency of the contractions and the relative duration of the contractions during the work.

B. Relation between Force and Holding Time

It is known from individual experience that the greater the load, the shorter the time during which one can maintain it at a given level. The first studies of Dolgin and Lehmann, of Wachholder, and of Müller (see Scherrer and Monod, 1960) have shown that the maximal duration

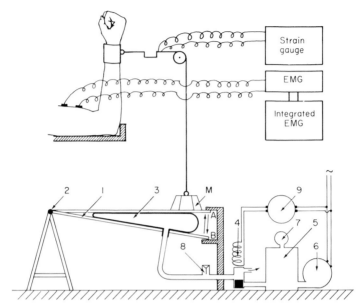

Fig. 7. Equipment for intermittent static work study. A small plate (1) turning around an axis (2) changes its position in relation to the content of a rubber bag (3). The plate can occupy two positions: (A) the weight load on the plate, the subject does no work; (B) the weight load free, muscles contract to maintain the load at the initial level. An electric valve (4) establishes communication between the bag and the atmosphere or a pneumatic machine (5, 6). Other elements are manometers (7), safety valve (8), and periodic command of the electromechanical valve (9).

of maintained contraction (or limit time) varies with the reciprocal of the force exerted and that this time may be very long if the strength is no more than 20% of the maximal strength. It is well known besides that postural muscles, which do not exhibit a high level of strength, are able to maintain posture for a very long time.

1. FORM OF RELATIONSHIP IN CONTINUOUS STATIC WORK

The relationship between strength and limit time has been determined by Monod (1956) in four different muscular groups, with the logarithms of limit time and of strength in a linear relation:

$$\log t = -n \log (F - f) + \log k \tag{3}$$

which may be written

$$t = k/(F/F_{\max} - f/F_{\max})^n \tag{4}$$

In this formula, t is the limit time, k a constant, F the strength exerted, F_{max} the maximal strength of the muscle, and f the strength for which the contraction time tends toward the infinite, and n is an index slightly higher than 2. The above formula is valid for all muscles if the strength is expressed as a percentage of the maximal strength. The value of f, called critical strength, is between 15 and 20% of the maximal strength. The concept of critical strength which appears in these studies is especially important, since it allows the delimitation of isometric contractions which include exhaustion, and those that may be performed without fatigue for a very long time.

The existence of an index n larger than 2 in the above formula allows the assumption that the exerted strength intervenes in at least two ways in the limitation of the time of contraction: (a) by exhausting the energy reserves of the muscle, and (b) by decreasing blood flow, which limits the supply of energy to the muscle. The separation between these two factors, although not expressly formulated by Dolgin and Lehmann (1929), appears in the results of these authors on static work performed under ischemia.

The results obtained by different authors working on this problem in the course of the last twenty years are in good agreement. Ikai (1962), referring to the experiments of Ishito, ends up with the same mathematical formulation that is also accepted by Kogi and Hakamada (1962). These two authors specify in addition that an analogous formula may be established between the strength of static contraction and the time at which the pain threshold occurs, which constantly precedes the threshold of local exhaustion. Rohmert (1960), Caldwell (1963), and Molbech and Johansen (1969) confirm the validity of the above relation for different muscle groups and for subjects of different sex (Fig. 8). For Rohmert, the relationship between limit time t and strength F may likewise be written

$$t = -1.5 + 2.1/F - 0.6/F^2 + 0.11/F^3 \tag{5}$$

Equations (4) and (5) do not noticeably differ from each other when the curves that correspond to them are compared. This is why the relation they describe is often cited as being the "law of Monod and Rohmert."

2. Variability of Limit Time

The variability of the maximal time of contraction is relatively low for high values of strength, although it is not the same when the strength exerted is less than 30% of the maximal strength. In fact, some authors

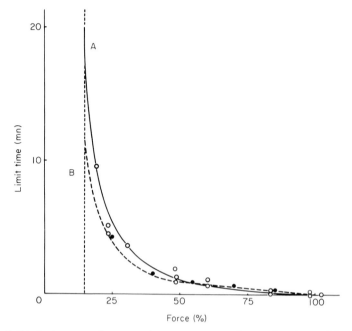

Fig. 8. Limit time as a function of isometric strength. Results obtained by various authors for continuous static work. Monod and Scherrer (1957) (curve A); Rohmert (1960) (curve B); Caldwell (1963) (average values—○); Phuon (1963) (individual values—●). (Monod and Scherrer, 1965.)

(Start and Holmes, 1963; Morioka, 1964) have drawn attention to a value of the critical strength greater than the 15–20% that is generally recognized, specifically for the biceps brachii. A possible explanation of this disagreement may be found in the procedure for measuring the maximal strength which varies with the authors; consequently the relative strengths exerted are not comparable (Kroemer, 1970). But there are some other factors of interindividual variations that should be discussed.

A. MOTIVATION. It may be supposed that the degree of motivation is not the same for all subjects. Carlson (1969) admits that it is easier for male populations to obtain a high degree of motivation than for female. The effect of motivation on the exertion of muscular force has already been discussed (Section II,B,4), and the factors examined in this regard should be recalled here. Scherrer *et al.* (1960) have shown that the value of the critical strength is clearly increased in the case of static work accomplished by subjects who know the performances of the rest of the group to which they belong when competition between the subjects exist.

B. PRESSURE OF OCCLUSION. It has been recognized since the results of Barcroft (see Barcroft *et al.*, 1963) that isometric contraction involves a limitation of the muscular blood flow when the relative strength surpasses 20%, which nearly corresponds to the value of the critical strength of the muscle. Humphreys and Lind (1963) have shown that the effect of occlusion increases with the force of contraction and that it becomes total if the relative force reaches 70%. Carlson (1969) suggests that the variability of the critical strength followed by that of the limit time may be linked to the value of the strength beyond which the blood supply begins to be limited in the muscle. This pressure of incipient occlusion would occur at the same value of the absolute force and consequently at a relative force that is lower for the muscles of strong subjects than for weak ones. The limit time may also depend on the intensity of the circulatory reactions which are opposed to the occlusion of the blood vessels (Section III,D,2).

C. STRUCTURE OF THE MUSCLE. Carlson (1969) emphasizes the modifications within the muscle that follow in the course of training (Section II,B,6). The diameter of the fibers is, proportionally, more increased than the capillary bed, so that the coefficient of perfusion (blood supply per unit of muscular volume) diminishes in the muscles of the most highly trained subjects. The threshold of exhaustion linked to muscle blood flow would thus be arrived at more rapidly in these subjects.

3. STRENGTH–LIMIT TIME RELATION FOR INTERMITTENT STATIC WORK

A. FORM OF THE RELATION. The strength–limit time relation, pointed out for the continuous static work, remains valid when isometric contractions are intermittent (Fig. 9). Evaluation of the capacity of intermittent static work is calculated with the following quantities:

F	strength of the contraction
t_w	duration of each isometric contraction, which is necessarily less than the limit time for a single contraction
t_r	duration of the rest following each contraction
$t_w + t_r$	duration of a work–rest period
$1/t_w + t_r$	frequency of contraction
p	time ratio of the duration of a contraction to the duration of a work rest period: $t_w/t_w + t_r$ or, if the periods are irregular $\Sigma\, t_w/\, \Sigma\, t_w + \Sigma\, t_r$

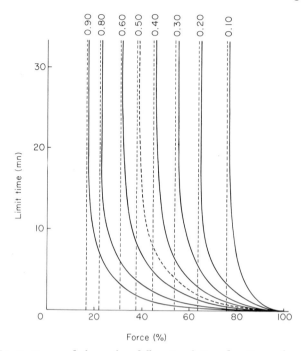

Fig. 9. Limit time and force for different relative durations of contraction. A better local blood flow enables an extra amount of work to be performed when the relative duration of the contraction decreases from 0.90 to 0.10 of the limit time.

The term $\Sigma t_w + \Sigma t_r$ defines the duration of intermittent static work and the limit time if the work is continued until exhaustion.

Pottier *et al.* (1969) have established strength–limit time curves for intermittent static work for various values of p between 10 and 80%. In each case, a curve of the same kind as that described for continuous static work was obtained. These curves differ from each other according to the increasing value of the critical strength when p diminishes. The introduction of pauses into the isometric contraction thus improves the circulatory conditions in the muscle, allowing it to produce a greater amount of work. The general formula relating strength and limit time, valid for both continuous and intermittent static work, is written:

$$t = \frac{k}{(F - f/F_{max})^{2.4p}} \tag{6}$$

There is an inverse relation between the relative duration of work p and the value of the critical force f, two parameters of the above equation (Fig. 10).

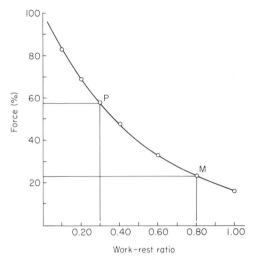

Fig. 10. Critical force versus the relative duration of contractions. Force is expressed as a percentage of the maximum strength. (M) Amount of work performed during intermittent static contractions, the force developed being about 0.22 of the maximum force. This work can be continued without exhaustion when the work's duration is not more than 0.80 of the total time. (P) When the contraction duration is 0.30 of the total time, the maximal strength that can be exerted without local exhaustion is not more than 60%.

B. Maximal Critical Level of Intermittent Static Work. The amount of work accomplished in the course of intermittent contractions with a given strength is expressed by the product of the strength of contraction and the cumulative duration of the isometric contractions:

$$W = F \, \Sigma \, t_w \tag{7}$$

If the exerted strength is greater than the critical strength, the duration of the work is limited; this is not so if it is equal or less. The definition of critical strength implies that the greatest quantity of work will be obtained by asking the subject to exert a force equal to the critical force:

$$W_{max} = f \, \Sigma \, t_w \tag{8}$$

This formula may be accepted only insofar as the absolute duration of each contraction t_w is not too large. Pottier *et al.* (1969) used periods of contraction not surpassing 24 sec. Petz (1964) had, in fact, shown a relative diminution of the work done when the duration of the contractions exceeded 45 sec. When the respective proportions of work and rest are fixed, it is preferable to have brief and frequent rests rather than long and infrequent rests.

The maximal critical level of intermittent static work is achieved when each contraction does not surpass 30 sec for a work and rest proportion of 50%, which corresponds to the use of a critical strength that is 40% of the maximal strength. The critical level (Eq. 8) is not very different (−5%), however, from this maximum for different combinations of p and f: the relative duration of work could be varied from 80% to 35% of the critical strength and the critical strength could be varied conversely from 25 to 50%. Intermittent contractions with relative forces greater than 50% are not consistent with a large amount of work.

C. Restoration of Capacity for Static Work

The definition of muscular fatigue implies that after a time of rest the muscle recovers the whole of its capacity to work. It has been seen (Section II,C,4) that the recovery of contractile capacity may be monitored by some measurements of force after work. This is best done by trials of continuous static work at various intervals after work.

Müller (1935) compares by the method of two successive trials of static work the speed of recovery for isometric contractions of maximal or submaximal duration, thus demonstrating that the accumulation of waste products during the first task could be responsible for a reduction of the limit time of the second. The results were confirmed by Monod (1956), who showed that for three consecutive tests the reduction of the limit time is much greater between the first and the second than between the second and the third; besides, the reduction of the limit time is relatively greater when the muscle exerts less force than when it exerts considerable force.

Lind (1959) showed that the restoration of the capacity for static work is accomplished first at a rapid rate immediately after stopping work, and more slowly beyond that, if the rest period is prolonged. Bujanovic *et al.* (1964), Caldwell (1965), and Krafft (1966) show also, by variations of the resting time between two trials of static work, that the work capacity is restored exponentially whether or not the task concludes in exhaustion. The time for a 90% recovery varies as the logarithm of the time during which a contraction of given strength has been maintained (Bujanovic *et al.* 1964).

D. Local and General Reactions during Static Work

The changes in the electromyogram, heart rate, arterial pressure, and pulmonary ventilation constitute some signs of fatigue with which the body tries to retard the approach of muscular exhaustion. These signs precede the arrest of the contractions and are particularly characteristic of continuous static work, their intensity being less in the course of intermittent static work, which does not, from this point of view, differ from dynamic work.

1. The Electromyogram

In the course of isometric contractions maintained until exhaustion, an increase occurs in the amplitude of the potentials recorded by surface electrodes, while the muscular force remains constant. This increase denotes the progressive recruitment of new motor units. Monod (1956) showed that the reinforcement of the muscular potentials does not appear immediately, but after a variable time corresponding to about two thirds of the limit time. Morioka (1964), using integrated electromyography, indicated that the increase in electrical activity may occur earlier. However, in the last third of the limit time, when the pain threshold is reached (Kogi and Hakamada, 1962), tremor follows, with relative difficulty in maintaining constant force and a definite articular position, imposing upon the subject continual readjustments; the electromyogram then records peaks of electrical activity and large potentials, which testify to firm synchronization of the motor units.

A study of Kogi and Hakamada (1962) covers the analysis of the frequencies of the electromyogram. Considering during a period of 15 sec the ratio of potentials of low and high frequency ($<$40/sec and $>$40/sec) called slow wave ratio, these authors showed that the latter increases when the muscle approaches the threshold of exhaustion.

It should be noted, finally, that very slight variations of posture may occur during prolonged isometric contraction, allowing the subject to retard the appearance of the pain threshold. These rotations between the fasciculi of the same muscle are not always perceptible in a surface electromyogram and should appear more clearly in records obtained with wire or needle electrodes. This may explain the diminution of amplitude sometimes observed by some authors in the electromyogram.

2. Circulatory Reactions

The mechanical pressure in the contracted muscle limits blood circulation when the relative force surpasses 20%. The organism responds with an increase of the heart rate, which raises the arterial pressure and so favors the circulation of blood in the muscle. This cardiac acceleration occurs from the beginning of static work and continues to the threshold of local exhaustion. The increase of heart rate is relatively greater for high value of strength than for medium or low strength, and this increase is of the same magnitude for the smallest muscles, such as the flexor of the medius, and for the large muscles, such as the quadriceps femoris. In the course of rapidly exhausting static contractions (less than 3 min), it is possible to observe heart rate surpassing 120/min, and systolic arterial pressure greater than 200 Torr (Monod, 1956; Humphreys and Lind, 1963). Contrary to what happens during dynamic work, diastolic pressure follows the changes in systolic pressure.

The mechanism of these circulatory reactions, quite different from those observed in local dynamic work, has some explanation. Alam and Smirk had concluded in 1933 that the elevation of arterial pressure should be attributed to the appearance in the blood of metabolites coming from the active muscles. Then Lind *et al.* (1966) showed that circulatory reactions are initiated by a reflex, which originates in the active muscles. In fact, these reactions persist after the circulation has been arrested in the muscle by inflating a cuff. The observation of a patient suffering from syringomyelia, who does not produce any cardiovascular reaction during static work, leads to suppose that this reflex uses an afferent pathway from the area of the muscles. Lind suggests that hypoxia is insufficient to elicit this reflex and explains it less by the appearance of lactic acid into the blood than by the liberation of potassium around the muscular fibers.

3. Maximal Permissible Time for Continuous Static Work

The intensity of the circulatory reactions observed in the course of static contraction represents a great physiological strain in the case of frequent repetition or if the subject suffers from minor vascular trouble. On the basis of changes in the electromyogram or the heart rate it was proposed that the time of static contraction be limited and that pauses of sufficient duration should be introduced.

Considering heart rate increases, Rohmert (1960) proposed a method of calculating the time of rest to be allowed after a single isometric contraction which is or is not performed until exhaustion. He takes into account the strength exerted F, the maximal strength of the active muscle F_{max}, the duration of the contraction t, and the maximal holding time t_{max} for the percentage of strength F/F_{max}. The minimal duration of rest t_r, related to the duration of the contraction t, is given by the formula

$$t_r/t = 18(t/t_{max})^{1.4} \times (F/F_{max} - 0.15)^{0.5} \qquad (9)$$

This formula makes it possible to calculate, for instance, that a relative force of 75% exerted for 24 sec requires a resting time of about 5 min and 36 sec. If such a contraction has to be repeated during an 8 hour day, one sees that it may only be every 6 min, or 10 times per hour. One notices that, in such a cycle, the time of work represents only 8% but the time of rest 92% of the total time.

To appreciate the minimal time of intertrial rest that must be allowed in the course of intermittent static work performed without exhaustion for at least 1 hour, the relation established between the critical strength and the total time during which the muscle contracts may also be taken into account. Figure 10 shows that the critical strength is higher as the relative time of contraction is shorter, which is to say that the resting time is longer. One sees (point *P* in the Fig. 10) that, if the force exerted is about 75%, the maximal duration of the work cannot surpass 16% of the total time. The resting time is then at least 84%. These proportions differ slightly from those given by the formula of Rohmert, but it should be noticed that this second kind of calculation of the respective times of work and rest do not include the recovery time to be allowed after stopping intermittent static work.

In the case of continuous static work, it was proposed to substitute for the maximal holding time, defined by the drop of mechanical performance, a maximal permissible time that takes into account the reactions which mark the arrival at the threshold of local exhaustion. Thus, for the strengths between 15 and 40% of the maximal strength, the time of work should not surpass two-thirds of the limit time in order to take into account electromyographic reactions. Above 40% of the maximal strength, the duration of isometric contractions should not surpass a third of the limit time, in order to avoid excessive cardiovascular reactions (Monod and Scherrer, 1964).

IV. Dynamic Work Output by Muscle

The production of movement plays a greater role in everyday life than does the holding of posture. Even if a static component is mingled with it, dynamic work is found in all everyday activities of the body, in the practice of an occupation (especially manual), and in indulging in sport. The evaluation of work is generally impossible in physical units due to the great complexity of the movements involved. At the most, it can be approached by considering the energy expenditure during the work and an average value for the efficiency (see Durnin and Pasmore, 1967; Scherrer, 1967). This is quite different for the dynamic work performed with an ergometer under laboratory conditions; thus, the general rules for the muscle considered as a living motor have been established.

A. Ways of Performing Dynamic Work

Different kinds of ergometers are used, whether for general work or for local work. In the second case, the muscular work may be done at decreasing power or at constant power.

1. GENERAL WORK

Work produced during a walk on a treadmill or pedaling an ergocycle is easily measurable. In the first case, the work takes into account the displacement of body weight. The analysis of the mechanical work done in walking was initially given by Fenn (see Karpovich, 1959) and more recently by Ralston and Lukin (1968) and Cavagna (1969). In the second case, the subject pedals against a resistance. But exact calibration of the ergocycles is not so easy, and thus, it is often difficult to compare the results of investigations that have used different ergocycles.

In the tests of physical fitness performed with treadmill or ergocycle, the subject should work either (a) from the beginning at a maximal or submaximal high level of power, or (b) at a power increased by successive steps until a maximal level is reached.

Measurement of the work done by swimmers has also been accomplished by attaching to them a weight, a dynamometer cable, or placing them in a current of a given speed.

2. LOCAL WORK

It is to Mosso and his students (see Scherrer and Monod, 1960; Clarke, 1966) that we owe the first studies of the dynamic work capacity of a muscle group. The first ergograph of Mosso, designed for the contraction of the flexor of the medius, was followed by those of Bidou, Johansson, and Kelso-Hellebrant and is useful for a greater variety of muscles. Mechanical work may be done in different ways:

A. MOVING A LOAD. This is the principle of Mosso's ergometer, which he used to establish his famous curves of fatigue. The subject should move a given load F over a certain distance of l and repose it before beginning again. The mechanical work done W in a time t is equal to the product of the load and the sum of the displacements.

$$W = F \Sigma l$$

Thus, if the frequency n of the movements is regular

$$W = Flnt$$

The units generally used are the kilogram-meter per minute (kgm/min) or the watt (W). Friction ergometers may be better than those with weights, since the former oppose the muscle with a constant resistance, and do not involve the acceleration of load.

B. ACCELERATING A WHEEL. This is the method used by Hill for his biomechanical studies of a single contraction. It allows the establishment of a force–speed relation in concentric contraction. Here, the measurements of performed work include the displacement speeds of the center of gravity of the body (jumping test, using the extensor muscles of the lower limbs) or only of a body segment. In the latter instance, the displacements may be accomplished either unloaded or with an additional load that augments the inertia of the body segment. The work is then equal to the product of the force and the displacement speed of its point of application; that is, to the product of the torque and the angular speed of the articulation which is moved.

C. STRENGTH EXERTED AGAINST A SPRING. The work is effected against a progressively increasing or decreasing external force.

D. EVALUATION OF NEGATIVE WORK. In dynamic work accomplished with a loaded ergometer or a spring ergometer, only mechanical work done by concentric contraction is taken into account. Physiologically, however, the negative work done by eccentric contraction should also be considered. For alternating movement, where the two phases are strictly symmetrical and of equal duration, the negative work represents about a third of the positive work to which it is linked. This is clear from investigations in which the criteria of energy, electromyography, and maximal work until exhaustion have been retained.

The suppression of the positive or the negative work is easily accomplished with pneumatic, hydraulic, or electromagnetic devices, that substitute their action for that of the muscle in one of the two phases of movement.

3. LOCAL WORK AT DECREASING POWER

According to Mosso (1890), ergometric work includes a more or less early drop in the amplitude of movement, the test ending when no more contraction is possible (Fig. 11). In this manner, different kinds of diminution in the work capacity of the muscle have been described under the term "curve of fatigue." Apart from the drop in available strength, these curves indicate possible variations in the motivation of the subjects from one moment to another, and they express the strategy adopted by subjects to accomplish the set task. They may have, then, some interest from the point of view of experimental psychology.

In physiology, the curves of Mosso are less interesting, for a muscle does not work in the same area of the tension–length diagram at the beginning and the end of a test. The biomechanical conditions of the movement are not constant in a cranking test where the maximal speed of rotation progressively diminishes nor in tests with the mercury ergometer of Henry, where the strength of the contractions decreases.

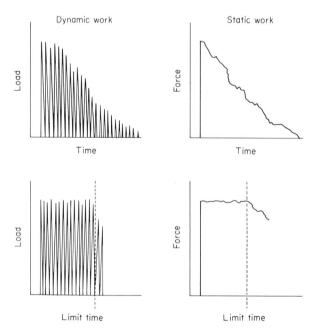

Fig. 11. Different kinds of dynamic and static work. The displacement of the load for dynamic work, or the force for static work are plotted against time. *Upper part:* Muscular work (according to Mosso 1890) performed at a decreasing power or force. Local exhaustion occurs at the moment (limit time) when the initial level of activity can no longer be maintained.

The tests proposed by Clarke (1957) partially avoid this last criticism. The index of isotonic endurance described by this author is the ability to raise and set down a submaximal weight at the frequency of 30/ min during a time limited to 2 min. The drop in the amplitude of movement is then only incomplete.

4. LOCAL WORK AT CONSTANT POWER

We owe to Scherrer *et al.* (1954) the proposition of dynamic work tests at constant power. The subject should keep constant, for successive liftings of a given load, both the initial frequency and amplitude of the movement. The test is stopped when it has led to a fall of either one or the other, which determines a threshold of local exhaustion for a given power (Fig. 11). The arrival at this level of local exhaustion is marked, in addition, by a drop in the muscular strength (Section II,C,3) and by a slowing of the movement, which indicates a diminution of the number of motor units that can be activated when the load to be displaced should be accelerated.

In order to compare tests accomplished by subjects with different morphologies, the experimenter should choose a load to be moved that corresponds to a given fraction of the maximal isometric strength. It matters little, however, whether the point of application of the muscular strength on the body segment is the same for all. In practice, only the torque is counted, since what the subjects with longer body segments gain in length they lose in strength. Finally, a calculation of mechanical work done should take into account the movement against gravity necessitated by the body segments themselves, for their mass is not negligible.

B. *Relation between Limit Work and Limit Time*

In the course of ergometric tests at constant power, a relation may be established between limit work and limit time. From this relation is drawn the value of the critical power of the muscle under consideration.

1. EXPERIMENTAL EVIDENCE OF A RELATION

When a subject is asked to perform a task of local dynamic work at a sufficiently high power, exhaustion occurs within the limit time

t_{lim} after a limit work W_{lim} has been done. The quotient of these two quantities defines the power P of the work.

$$P = W_{\mathrm{lim}}/t_{\mathrm{lim}} \tag{10}$$

It is easily seen that the greater the power, the sooner exhaustion will occur and the smaller will be the amount of work performed. There is a linear relation between the limit work and the limit time such that

$$W_{\mathrm{lim}} = a + bt_{\mathrm{lim}} \tag{11}$$

This relation, in which there are two constants, is verified for times included between 1 and 60 minutes (Fig. 12).

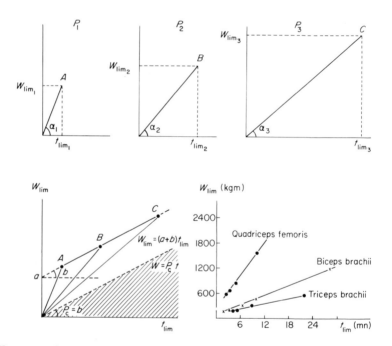

Fig. 12. The critical power of dynamic work. *Upper part:* Three dynamic work tests until exhaustion have been performed. For each, a limit work (W_{lim}) has been performed in a limit time (t_{lim}). *Lower part left:* points *A*, *B*, and *C* are situated on a straight line defined by the relation between limit work and limit time ($W_{\mathrm{lim}} = a + bt_{\mathrm{lim}}$). The slope of the straight line going through the origin of coordinates and parallel to the previous straight corresponds to the critical power (P_c) of the muscle. In the shaded zone, dynamic work will always be performed without fatigue. *Lower part right:* examples of work capacity measurement for three muscular groups.

2. Meaning of the Relation

The above equation means that the work that a muscle can produce before being exhausted is the sum of two terms:

1. An energy reserve available in the muscle and able to be mobilized swiftly corresponding both to the energy contained in the phosphorous components and that originating from the use of intramuscular glycogen and oxygen in aerobic reactions. This energy corresponds to factor a in Eq. (11); factor a is equal, for instance, to about 100 kg for the triceps brachii.

2. An energy supply established at a constant rate during work. This supply depends on the power at which the muscle works but cannot surpass a maximal value. This rate is represented in Eq. (11) by the factor b, associated with the blood flow in the muscle, and is about 30 kg/min for the triceps brachii.

Two arguments may be advanced in support of this point of view, one analogical and the other experimental.

1. An identical interpretation has been given for general muscular work, following the results of numerous authors and especially those of A. V. Hill (1927). The maximal work performed includes the utilization of the maximal aerobic capacity, the production of a maximal oxygen debt, and a maximal rate of energy use (Wilkie, 1960).

2. When a subject is asked to perform dynamic work during ischemia, the amount of work done is always represented by the factor a, whatever the power, and consequently, the duration of the work. The arterial cuff stops all energetic supply to the muscle and slows the elimination of wastes; the factor b becomes equal to zero. The influence of circulatory restriction on the time of exhaustion has already been mentioned for the static contraction.

3. Critical Power of the Muscle Group

The maximal value of the factor b corresponds to the critical power of the muscle. When the imposed power surpasses this critical power, the muscle must use its energy reserves, and exhaustion occurs when these have been depleted, in a time which may be calculated by a combination of Eqs. (10) and (11):

$$t = a/(P - b)$$

The critical power thus delimits work which is performed with or without fatigue. It may be determined from three or four tests to exhaustion, leading to Eq. (11), which gives the value of the critical power, and allows the calculation of the limit time, and consequently, the limit work for a test of a power P which is greater than the critical power. The critical power varies with the size of the muscle and the fitness of the subject. It may serve as an index to follow the course of physical training or to compare subjects with each other.

C. Factors Linked to Muscle Work Capacity

The expression "capacity for dynamic work" is used to mean (a) the maximal amount of work by unit of time that the muscle can perform without fatigue (i.e., its critical power) or (b) the maximal quantity of work done in a given time, generally short, arbitrarily chosen in the form of a physical fitness test. The work capacity is thus considered either as power or as amount of work. Whatever the sense in which the term is used, work capacity has been studied in relation to the strength of the muscle, its speed of contraction, and the energy supply that it receives.

1. Work Capacity and Strength

There is a close relation between the maximal amount of work done in a given time and the isometric strength. Ikai (1966) indicates a correlation coefficient of 0.71–0.83 between the two quantities, for the femoral quadriceps. Differences, meanwhile, are evident between subjects; some have a low work capacity despite a high strength, and vice versa. This explains why the measurement of muscular strength alone is not sufficient to assess work capacity (Pierson and Rasch, 1963).

It is not evident *a priori* that strength measured isometrically is entirely usable for an anisometric contraction. This assumption may be accepted, however, when high correlations have been shown by investigations into the comparison of isometric strength and anisometric strength (Asmussen *et al.*, 1965; Singh and Karpovich, 1966; Carlson, 1970). For the same articular position, the concentric strength is always less than the isometric strength, while the eccentric strength is greater than it. To some degree these differences are as much more marked with an increase in movement speed. These results are in good agreement with the well known Hill's force–velocity relation.

The factors of variation in strength must of necessity affect work capacity (they have been examined in Section II,B). Among these factors it is important to mention again the role of the subject's motivation for the imposed task. Nelson (see Clarke, 1966) showed that in a group of 25 subjects carrying out dynamic work with the biceps brachii, the ensuing decrease in strength depends on the instructions given by the observer. The encouragement he gives during the tests or an indication of interest in the subject's score does not significantly modify the results with respect to normal instructions. On the other hand, knowledge of the score obtained by other members of the same group, or the idea that the score may be used to select the subject, increases the work done and diminishes the remaining strength at the end of the test. The greater amount of work performed is also evidenced by a slower recovery of muscular strength.

2. Work Capacity and Speed of Contraction

A. Continuous (Mean) Power. The ergometric test takes into account the strength, the length of the movement, and the frequency of the contractions, quantities from which the average power at which the muscle works may be calculated. Except in cases where the frequency of contraction is so high that there is no adequate resting time, little importance is attached to the ways in which each contraction is accomplished. It is true, however, that in a measurement where the subject has a choice, he spontaneously adopts a speed that represents about two-thirds of the maximal speed of the movement (see Scherrer and Monod, 1960). This speed suddenly drops when the muscle is about to reach the level of local exhaustion. In work involving exhaustion, there seems to be a specific combination of load and frequency for which the work done is maximal (see Clarke, 1966).

B. Explosive Power. Numerous studies led to discussion of the maximal speed of contraction from which the power developed in a single movement may be calculated. The speed of a movement is easily measured by photographic or cinematographic techniques, or by attaching the member to a potentiometer. Wilkie (1960) considers that for movements of no more than a second in duration, the explosive power can reach 6 HP, while the average power (including rests between contractions) does not exceed 0.5–2 HP in activities with a maximal duration of 5 min, and 0.2 HP in activities of several hours. The figures of continuous or explosive power are only of interest if one knows the mass of the active muscles. The use of his maximal aerobic capacity allows a subject to produce during a steady state a mechanical power of 0.5 HP (Glencross, 1965).

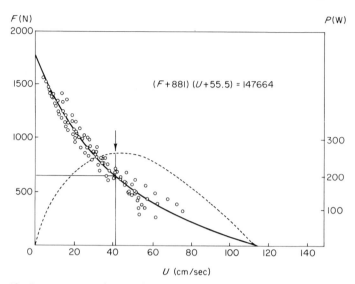

Fig. 13. Instantaneous force–velocity relation. Maximal concentric contractions of the biceps brachii have been performed by the same subject against six different inertia obtained by loading the forearm. Shortening velocity (U) is plotted versus the exerted force (F). Equation of the hyperbola is given. The dotted line indicates the instantaneous power (P) as the product FU. The speed corresponding to the maximal power is shown by the arrow. (Pertuzon and Bouisset, 1971.)

C. INSTANTANEOUS POWER. During a monoarticular movement toward one of the two possible positions, the strength and the speed vary with the displacement. It may be considered that at each time the power is equal to the product of the strength and the instantaneous speed. Pertuzon and Bouisset (1971) recorded these two parameters in the course of the maximal elbow flexion; they showed that the instantaneous power has a maximal value for a given angular position (Fig. 13). The flexion speed corresponding to the peak of instantaneous power is equal to about 40% of the maximal.

3. WORK CAPACITY AND EFFICIENCY

Hill showed that each muscle has two optimum speeds, one corresponding to the maximal explosive or instantaneous power, the other to its maximal energetic efficiency; the efficiency is established by relating the mechanical work done to the total energy expended for performing this work. Studies of oxygen consumption (see Monod, 1967) have shown that:

1. The net efficiency is nearly constant when the work is done at a power inferior to the critical power of the muscles, while it becomes

lower when this is exceeded. The same statement is true of isometric contraction when the exerted strength is above or below the critical strength.

2. For a given muscle group, and for the same mechanical work—i.e., for a constant mean power—the efficiency diminishes when the speed is either too high or too low. The optimal speed at which the energy expenditure is least is greater if the muscle is small.

3. For a given muscle group working at constant power, the efficiency diminishes when the three factors that define the power assume extreme values. Thus, for movements of small amplitude, or those accomplished with light load, the frequency of the contractions is necessarily high (with this goes an increase in the frequency of active states), and of displacement of the involved body segment. A high frequency of movement involves increases in the speed of contraction and, conversely, the moving of heavy weights (representing more than 40–50% of the isometric strength of the muscle) involves a reduction of the contraction speed.

Any diminution of efficiency, whatever the reason, implies a decrease of the muscle's work capacity if one admits that there is a maximum rate to its energy supply.

ACKNOWLEDGMENTS

I would like to express my thanks to Professor Jean Scherrer for the useful suggestions he gave me when I was writing this chapter. Many ideas expressed here are common to both of us. I wish also to thank Dr. Brian Greenwood for his kindly help in preparing this English text.

REFERENCES

Alam, M., and Smirk, F. H. (1937). *J. Physiol. (London)* 89, 372.
Asmussen, E., Fredsted, A., and Ryge, E. (1965). *Comm. Dan. Nat. Ass. Infant Paral.* No. 20.
Asmussen, E., and Heebøll-Nielsen, K. (1961). *Comm. Dan. Nat. Ass. Infant. Paral.* No. 11.
Asmussen, E., Heebøll-Nielsen, K., and Molbech, S. (1959). *Comm. Dan. Nat. Ass. Infant Paral.* No. 5.
Asmussen, E., and Molbech, S. (1958). *Comm. Dan. Nat. Ass. Infant Paral.* No. 2.
Barcroft, H., Greenwood, B., and Whelan, R. F. (1963). *J. Physiol. (London)* 168, 848.
Bartley, S. H. (1957). *Physiol. Rev.* 37, 301.
Bartley, S. H., and Chute, E. (1947). "Fatigue and Impairment in Man." McGraw-Hill, New York.
Berthoz, A., and Metral, S. (1970). *J. Appl. Physiol.* 29, 378.
Bourguignon, A., Marty, R., and Scherrer, J. (1959). *J. Physiol. (Paris)* 51, 93.

Bujanović, R., Bujas, Z., Petz, B., and Vidaček, S. (1964). Acta Inst. Psychol. Univ. Zagreb. **40**, 43.
Burke, W. E., Tuttle, W. W., Thompson, C. W., Janney, C. D., and Weber, R. J. (1953). J. Appl. Physiol. **5**, 628.
Caldwell, L. S. (1959). U.S. Army Med. Res. Lab., Rep. **411**.
Caldwell, L. S. (1962). Hum. Factors **4**, 125.
Caldwell, L. S. (1963). J. Eng. Psychol. **2**, 155.
Caldwell, L. S. (1965). J. Eng. Psychol. **4**, 22.
Carlson, B. R. (1969). Ergonomics. **12**, 429.
Carlson, B. R. (1970). Arch. Phys. Med. Rehabil. **51**, 176.
Cavagna, A. G. (1969). J. Physiol. (Paris) **61**, Suppl. 1, 3.
Clarke, D. H. (1957). Arch. Phys. Med. Rehab. **38** (9) 584.
Clarke, D. H. (1962). Res. Quart. **33**, 349.
Clarke, D. H. (1966). "Muscular Strength and Endurance in Man." Prentice-Hall, Englewood Cliffs, New Jersey.
Clarke, H. H., Shay, C. T., and Mathews, D. K. (1954). Arch. Phys. Med. Rehabil. **35**, 560.
Clarke, D. H., and Smith (1966). J. Assoc. Phys. Ment. Rehab. **20**, 123.
Danowski, T. S., and Wratney, M. J. (1959). Arch. Phys. Med. Rehabil. **40**, 516.
Darcus, H. D., and Salter, N. (1955). J. Physiol. (London) **129**, 325.
Dempster, W. T. (1965). Arch. Phys. Med. Rehabil. **46**, 49.
Dolgin, P., and Lehmann, G. (1929). Arbeitsphysiologie **2**, 248.
Donskoi, D. (1961). "Biomechanik der Körperübungen." Sportverlag, Berlin.
Durnin, J. V. G. A., and Passmore, R. (1967). "Energy, Work and Leisure." Heinemann, London.
Dyson, G. (1967). "The Mechanisms of Athletics." Univ. of London Press, London.
Elftman, H. (1966). J. Bone Joint Surg., Amer. Vol. **48**, 363.
Fautrel, M. (1954). Bull. Cent. Etud. Rech. Psychotech. **3**, 3.
Fessard, A., Laugier, H., and Nouel, S. (1933). Trav. Hum. **1**, 32.
Floyd, W. F., and Welford, A. T. (1953). "Symposium on Fatigue." Lewis, London.
Gaultier, C. (1970). 3rd cycle, Thesis, Tours.
Glencross, D. J. (1965). Phys. Educ. **57**, 18.
Govaerts, A. (1949). "Notion sur l'analyse des mouvements musculaires." Off. Int. Libr., Brussels.
Haxton, H. A. (1944). J. Physiol. (London) **103**, 267.
Heebøll-Nielsen, K. R. (1964). Comm. Dan. Nat. Ass. Infant Paral. No. 18.
Heidenhain, R. (1864). "Mechanische Leistung. Wärmeentwicklung" (Breitkopf and Härtel, eds.), vol. 1 Leipzig.
Hettinger, T. (1961). "Physiology of Strength." Thomas, Springfield, Illinois.
Hettinger, T. (1964). "Isometrisches Muskeltraining." Thieme, Stuttgart.
Hill, A. V. (1927). "Muscular Movement in Man." McGraw-Hill, New York.
Houtz, S. J., Lebow, M. J., and Beyer, F. R. (1957). J. Appl. Physiol. **11**, 475.
Humphreys, P. W., and Lind, A. R. (1963). J. Physiol. (London) **166**, 120.
Hunsicker, P. A. (1955). Wright Air Development Center, Tech. Rep. No. 54-548.
Hunsicker, P. A. (1957). Wright Air Development Center, Tech. Rep. No. 57-586.
Hunsicker, P. A., and Donnelly, R. L. (1955). Res. Quart. **26**, 408.
Ikai, M. (1962). Res. J. Phys. Educ. **6**, 1.
Ikai, M. (1966). J. Sports Med. Phys. Fitness **6**, 100.
Ikai, M., and Fukunaga, T. (1968). Int. Z. angew. Physiol. einschl. Arbeitsphysiol. **26**, 26.
Ikai, M., and Steinhaus, A. H. (1961). J. Appl. Physiol. **16**, 158.

Ikai, M., Yabe, K., and Ischii, K. (1967). *Sportarzt Sportmed.* **5**, 197.

Ioteyko, J. (1920). "La fatigue." Flammarion, Paris.

Karpovich, P. V. (1959). "Physiology of Muscular Activity." Saunders, Philadelphia, Pennsylvania.

Kogi, K., and Hakamada, T. (1962). *Inst. Sci. Labour* **60**, 27.

Kogi, K., Müller, E. A., and Rohmert, W. (1965). *Int. Z. angew. Physiol. einschl. Arbeitsphysiol.* **20**, 465.

Krafft, R. (1966). *Actes Congr. Ergon. Langue Fr. 2nd.* 1964 p. 171.

Kroemer, K. H. (1970). *Hum. Factors* **12**, 297.

Kroll, W. (1965). *Int. Z. angew. Physiol. einschl. Arbeitsphysiol.* **21**, 224.

Kroll, W. (1967). *Int. Z. angew. Physiol. einschl. Arbeitsphysiol.* **23**, 331.

Kroll, W. (1968). *Res. Quart.* **39**, 106.

Lind, A. R. (1959). *J. Physiol. (London)* **147**, 162.

Lind, A. R., McNicol, G. W., and Donald, K. W. (1966). *In* "Physical Activity in Health and Disease" (K. Evang and K. Lange-Andersen, eds.), pp. 38–63. Universitetsforlaget, Oslo.

Lombard, W. P. (1890). *Arch. Ital. Biol.* **13**, 371.

Merton, P. A. (1954). *J. Physiol. (London)* **123**, 553.

Molbech, S. (1963). *Comm. Dan. Nat. Ass. Infant Paral.* No. 16.

Molbech, S. (1965). *Acta Morphol. Neer-Scand.* **6**, 171.

Molbech, S., and Johansen, S. H. (1969). *J. Appl. Physiol.* **27**, 44.

Monod, H. (1956). M.D. Thesis, Foulon, Paris.

Monod, H. (1967). *In* "Physiologie du Travail. Ergonomie" (J. Scherrer, ed.), Vol. I, pp. 154–208. Masson, Paris.

Monod, H., and Scherrer, J. (1957). *C. R. Soc. Biol.* **151**, 1358.

Monod, H., and Scherrer, J. (1964). *Ergonomics, Proc. IEA Congr., 2nd, 1964* p. 45.

Monod, H., and Scherrer, J. (1965). *Ergonomics* **8**, 329.

Morioka, M. (1964). *Ergonomics, Proc. IEA Congr., 2nd, 1964* p. 35.

Mosso, A. (1890). *Arch. Ital. Biol.* **13**, 123.

Müller, E. A. (1935). *Arbeitsphysiologie* **8**, 62, 72.

Naess, K., and Storm-Mathisen, A. (1955). *Acta Physiol. Scand.* **34**, 351.

Nelson, J. K. (1962). Doctoral Dissertation, University of Oregon.

Pastor, P. J. (1959). *Arch. Phys. Med. Rehabil.* **40**, 247.

Pertuzon, E., and Bouisset, S. (1971). *In* "Biomechanics II" (J. Vredenbregt and J. Wartenweiler, eds.), pp. 170–173.

Petz, B. (1964). *Arch. Hig. Rada Toksikol. Jugoslav.* **15**, 183.

Phuon, M. (1963). Thèse Doctorale Méd., Paris.

Pierson, W. R., and Rasch, P. J. (1963). *Amer. J. Phys. Med.* **42**, 205.

Pottier, M., Lille, F., Phyon, M., and Monod, H. (1969). *Trav. Hum.* **32**, 271.

Provins, K. A., and Salter, N. (1955). *J. Appl. Physiol.* **7**, 393.

Ralston, H. J. (1953). *Abst. Commun., Int. Physiol. Congr., 19th, 1953* p. 692.

Ralston, H. J., and Lukin, L. (1968). Nasa CR 1042, April, 1968.

Rich, G. Q. (1960). *Res. Quart.* **31**, 485.

Roberts, D. F., Provins, K. A., and Morton, R. J. (1959). *Hum. Biol.* **31**, 334.

Rohmert, W. (1960). *Int. Z. angew. Physiol. einschl. Arbeitsphysiol.* **18**, 175.

Royce, J. (1958). *Res. Quart.* **29**, 204.

Ryan, T. A. (1947). "Work and Effort." Ronald Press, New York.

Scherrer, J. (1967). "Physiologie du Travail. Ergonomie." Masson, Paris.

Scherrer, J., Bourguignon, A., and Monod, H. (1960). *Rev. Pathol. Gen. Physiol. Clin.* **60**, 357.

Scherrer, J., and Monod, H. (1960). *J. Physiol. (Paris)* **52**, 419.

Scherrer, J., Samson, M., and Paleologue, A. (1954). *J. Physiol. (Paris)* **46**, 887.

Singh, M., and Karpovich, P. V. (1966). *J. Appl. Physiol.* **21**, 1435.

Solvay, E. (1904). *C. R. Acad. Sci., Ser. D* **138**, 121.

Start, K. B., and Holmes, R. (1963). *J. Appl. Physiol.* **18**, 804.

Toews, J. V. (1964). *Arch. Phys. Med. Rehabil.* **45**, 413.

Tornvall, G. (1963). *Acta Physiol. Scand.* **58**, Suppl. 201.

Tuttle, W. W., Janney, C. D., and Thompson, C. W. (1950). *J. Appl. Physiol.* **2**, 663.

Vandervael, F. (1951). "Analyse des mouvements du corps humain." Maloine, Paris.

Whipple, O. M. (1924). "Manual of Mental and Physical Tests. I-Simpler Processes; II-Complex Processes." Warwick and York, Baltimore, Maryland.

Wilkie, D. R. (1960). *Ergonomics* **3**, 1.

Williams, M. H. (1969). *Res. Quart.* **40**, 831.

Williams, P. M., and Lissner, H. R. (1962). "Biomechanics of Human Motion." Saunders, Philadelphia, Pennsylvania.

DEVELOPMENT OF STRIATED MUSCLE

DONALD A. FISCHMAN

I. Introduction

As prepared for the first edition (1960) of this book, the chapter by Professor Boyd on the "Development of Striated Muscle" was largely concerned with the embryologic origin and migration of cells involved in myogenesis. The first three sections of the former chapter were entitled: I. Source of Mesoderm Concerned in the Development of Striated Muscle; II. Regional Differences in Origin of the Striated Musculature; and III. Developmental History of the Myotome. Boyd admirably reviewed a large body of information principally collected during the first half of the twentieth century. In the ten years that have elapsed since publication of the first edition, there has been relatively little research on the topics listed above. More recent studies have focused on the metabolic, cytogenetic, and ultrastructural events pertinent to the cytodifferentiation of muscle rather than on questions of cell lineage or migration. In this review, major emphasis will be placed on the more current work in an attempt to place in some perspective the field as it exists today. For the sake of completeness, some of the studies covered by Professor Boyd will be reviewed briefly here; readers interested in a more thorough coverage of the early literature are referred to the first edition of this treatise.

II. Origin of Cells Forming Striated Muscle

With one or two exceptions,[1] skeletal muscle is derived from lateral plate and paraxial (somitic) mesoderm. In the head, striated muscle has two distinct embryonic sources. One set, of somitic (myotomic) origin, includes the extraocular and tongue musculature and is innervated by cranial nerves III, IV, V, VI, and XII. The second set, of visceral or branchial arch mesodermal origin, comprises the muscles of mastication, the laryngeal and pharyngeal musculature and the superficial muscles of the face, auricle, and scalp, and certain muscles of the shoulder (Huber, 1931). This second group of muscles is innervated by cranial nerves V, VII, IX, X and XI. The striated muscle in the wall of the esophagus (postpharyngeal striated gut musculature) appears to be derived from the splanchnopleuric subdivision of lateral plate mesoderm, but as emphasized by Boyd (1960), this has not been proven unequivocably. Musculature of at least the dorsal third of the body wall, including the paravertebral muscles, are of somitic origin (Straus and Rawles, 1953); the precise derivation of the remaining trunk musculature remains a controversial matter (Straus and Rawles, 1953; Detwiler, 1955; Liedke, 1958; Theiler, 1957), possibly reflecting real differences between the different orders of the vertebrates. The principal

---

[1] The sphincter pupillae of birds and reptiles appear to be of ectodermal rather than mesodermal origin (Boyd, 1960).

point of disagreement is whether the ventral two-thirds of the trunk musculature is of lateral plate or myotomic origin. In their careful carbon marking and intracoelomic graft experiments performed with chick embryos, Straus and Rawles (1953) concluded that lateral plate and not myotome is the source for at least the ventral half of the trunk musculature. In contrast, Detwiler (1955), based on his extensive studies with salamanders, decided that the ventrolateral musculature is of somitic origin. Detwiler's conclusions have been supported in frogs by Liedke (1958) and in human embryos by Theiler (1957). At the present time, the discrepancies remain unresolved.

As for the origin of the limb musculature, there is now little doubt that it develops *in situ* by the differentiation of mesenchyme derived from lateral plate (somatopleuric) mesoderm (Detwiler, 1934, 1955; Hamburger, 1938, 1939; Rudnick, 1945; Saunders, 1948; Tschumi, 1957). Except for the fin musculature of nontetrapods, which is probably of myotomic origin, it is fairly certain that limb musculature in most, if not all, tetrapods is of lateral plate and not myotomic derivation. Dalcq and Pasteels (1954) have presented evidence for an exception to this rule based on their studies of reptilian limb musculature.

III. Epithelial–Mesenchymal Interactions Relevant to Muscle Development

It has been appreciated for some time that the main mass of muscle of somitic origin arises from the inner aspects of each myotome or from cells proliferated from such regions. Yet cells immediately adjacent to these muscle-forming regions of each somite differentiate into cartilage and bone (sclerotome) or dermis (dermatome). No satisfactory explanation of this important phenomenon has yet been put forward.

The prospective somitic material in a number of vertebrates has been mapped in the ectoblast prior to gastrulation. In experiments conducted with amphibia, T. Yamada (1937, 1939) showed that if such prospective muscle-forming areas of the ectoblast are transplanted into the midventral belly region or grown in epidermal jackets, the cells do not form muscle but differentiate into pronephric-like tubules. Based on experiments of this type, it was suggested that some environmental factor(s) acting upon the prospective somite region was necessary for muscle differentiation. The subsequent search for such a factor(s) pinpointed two possible sources for its elaboration: the notochord (T. Yamada, 1939; Muratori; 1939; Smithberg, 1954; Muchmore, 1958) and the spinal cord (Strudel, 1955; Holtzer and Detwiler, 1953; Holtzer *et al.*, 1956; Avery *et al.*, 1956). Neither, however, has proven to have a true inductive relationship with the myogenic tissue, at least not in the usual embryological meaning of this term. In a recent review of this subject, Muchmore (1968) concluded that "presumptive somite mesoderm . . . can undergo differentiation in the absence of neural tissue." However, such muscle begins to degenerate 10–13 days after explantation, and by 2 months postsurgery, little or no trace of muscle can be found in the graft. In contrast, somites transplanted with spinal cord initially exhibited the same degree of muscle formation but did not undergo subsequent involution or degeneration. It appeared that development of somitic muscle can occur in the total absence of notochord or nerve, but that continued maintenance of the differentiated myofibers is in some way dependent on neural tissue. Muchmore (1968) suggested that this influence is related to the innervation of the muscle, but definitive studies on this point have not been reported.

These observations with the developing somite are significant in that they lend further support to the important concept, first clearly stated by Harrison (1904), that "the constructive processes involved in the production of the specific structure and arrangement of the muscle fibers take place independently of stimuli from the nervous system" (p. 218). Harrison's conclusions have been amply supported by subsequent studies with almost all embryonic muscle systems that have been examined. Most impressive have been the *in vitro* studies of myogenesis initiated by Lake (1916) and Lewis and Lewis (1917) and confirmed by scores of later investigators. See reviews by Konigsberg (1965), Yaffe (1969), Holtzer (1970), and Murray (1972). Clonal studies of myogenesis, first pursued by Konigsberg (1963) with chick muscle and later extended to rat (Yaffe, 1968) and reptilian muscle (Simpson and Cox, 1967; Cox, 1968), have shown conclusively that single myogenic cells, when plated out in monolayer culture, are capable of giving rise to colonies of myoblasts that subsequently fuse and differentiate into characteristic cross-striated myofibers. Most dramatic have been the investigations of Yaffe (1968) in which several lines of myoblasts have been isolated and serially propagated as dividing, mononucleated cells for over 2 years *in vitro*. Such cells have retained the capacity to differentiate into muscle even after these extended periods *in vitro*, thus providing an excellent example of the stability of cell lineage within the muscle precursor population. One might claim that such mononucleated cells were already "determined" or "programmed" to form muscle prior to removal from the animal, i.e., any inductive interaction with another tissue (e.g., nerve) might have occurred prior to dissociation of the myogenic tissue for cell culture. The same objection might be raised about Muchmore's (1968) study discussed above. Exactly when an undetermined mesenchymal cell becomes committed to form muscle or cartilage remains one of the central, unanswered problems in this field. This question has been discussed in detail by Holtzer (1970) and Holtzer and Bischoff (1970) and will be returned to later in this chapter when myoblast terminology is reviewed.

Using clonal studies of myogenesis, Hauschka and Konigsberg (1966) have demonstrated that collagen substrates promote the development and differentiation of myogenic cells. The effects of embryo extract, used in the culture medium, can be duplicated in large part by purified collagen. In recent studies Hauschka and White (1972) has shown that denaturation of collagen does not destroy this myogenic-promoting effect; gelatin will support muscle differentiation in clonal cultures. Furthermore, when examining cyanogen bromide fragments of collagen, he has found that certain sections of the molecule retain this property of supporting myogenesis, other peptide fragments lack this property. Further studies of collagen interactions at the myogenic cell surface should prove extremely interesting.

Although muscle differentiation is independent of neural control, there are many studies that indicate that myofibers, once formed, are profoundly altered by subsequent innervation. Not only are nerves required for the normal growth and maintenance of muscle (Zalena, 1962), but its physiological properties, (Buller *et al.*, 1960; Eccles, 1967), ultrastructural characteristics) Page, 1965; Padykula and Gauthier, 1970), myofibrillar ATPase activity (Bárányi *et al.*, 1965; Guth, 1968; Robbins and Engel, 1969; Prewitt and Salafsky, 1970), and regenerative capacity (Singer, 1952, 1965) are all markedly influenced, if not directly controlled, by the specific nerves that contact the myofibers. This poorly understood trophic influence of nerve on muscle is discussed in more detail by Hines in Volume III, Chapter 5.

IV. Résumé of Myogenesis

The general scheme of muscle development is now well understood (also see reviews by Konigsberg, 1965; Betz *et al.*, 1966; Holtzer, 1970; Herrmann *et al.*, 1970; Fischman, 1970). A mononucleated cell population exhibiting prominent mitotic activity accumulates at certain sites destined to form muscle (somites, limb rudiments., etc.) (Kityakara, 1959). One of the initial signs of myogenesis is an elongation of certain cells which assume a bipolar, spindle shape in contrast to the more

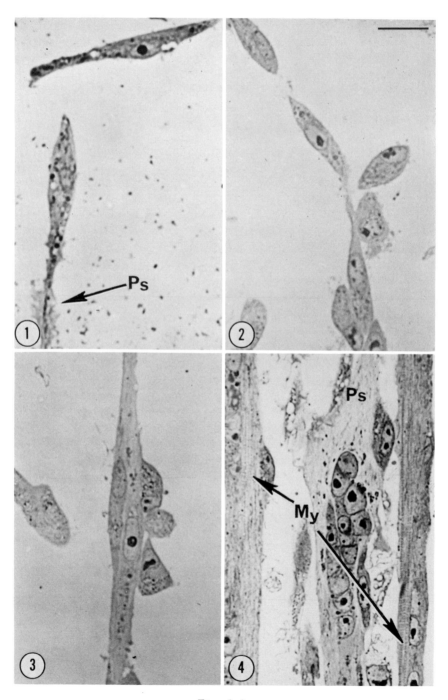

Figs. 1–4.

flattened, polygonal appearance of the adjacent mesenchymal population. (Figs. 1–4). Within the somite, some of these spindle-shaped cells become exceedingly long (almost 400 μm), and myofibrils can be detected within their cytoplasm by fluorescent antibody (Holtzer *et al.*, 1957) and electron microscopic techniques (Przybylski and Blumberg, 1966). These cross-striated mononucleated cells do not incorporate thymidine-^{3}H into nuclear DNA, nor do they exhibit mitotic activity (Stockdale and Holtzer, 1961). Such cells have been termed myoblasts by Holtzer and co-workers (Okazaki and Holtzer, 1965, 1966; Holtzer, 1970) to distinguish them from the mitotically active myogenic cell population, which contain no detectable myofibrils and have been termed presumptive myoblasts. Although cross-striated mononucleated cells are observed infrequently in cultures of leg muscle from 10–12 day old chick embryos (Coleman *et al.*, 1966) and can be induced chemically by treating myogenic cultures of breast muscle with 3.2×10^{-6} M bromodeoxyuridine (BUdR) (Bischoff and Holtzer, 1970), such cells are uncommon in later stage embryos in which myofibrillar protein synthesis is usually limited to the multinucleated cell population (Fischman, 1967). This difference between limb and somitic myogenesis remains unexplained, although Konigsberg (1965) has suggested that cytoplasmic fusion may simply occur earlier in the life cycle of a myoblast within the limb than in the somite.

The next discrete morphological step in myogenesis is the formation of the multinucleated syncitia (myotubes) by the cytoplasmic fusion of myoblasts. As will be discussed below, there is now strong evidence that cells undergoing fusion have ceased to synthesize DNA and can be considered a postmitotic cell population (Okazaki and Holtzer, 1965;

Figs. 1–4. Light micrographs of myogenic cells in monolayer culture. Cells fixed in glutaraldehyde and osmium tetroxide, embedded in Araldite, sectioned at 1 μm thickness and stained with alkaline toluidine blue. Calibration bar = 20 μm. See facing page for descriptions.

Fig. 1. Spindle-shaped cells in a muscle cell culture after 24 hours *in vitro*. It is likely these cells are presumptive or true myoblasts (see text). Note the prominent pseudopodial process (Ps) of the cell at the left.

Fig. 2. Cells lined up in a multicellular chain after 48 hours in culture. This alignment and aggregation of myogenic cells is a characteristic phenomenon prior to cellular fusion and myotube formation.

Fig. 3. A myotube with apposed mononucleated cells after 72 hours *in vitro*. Electron microscopy is required to establish whether or not these cells are undergoing cytoplasmic fusion.

Fig. 4. Myotubes within a 96 hour culture. The cells on the right and left contain myofibrils (My). Nuclei within the myotube, at the center of the field, are located near the pseudopodial (Ps) extremity of that cell.

Marchok and Herrmann, 1967; Yaffe, 1969; Bischoff and Holtzer, 1969).
Bulk synthesis of actin and myosin begins after cell fusion (Fischman,
1967; Coleman and Coleman, 1968), although the fact that some mononucleated cells contain myofibrils may indicate that fusion of myoblasts
is not an absolute requirement for such synthesis to be initiated. Subsequent development of the muscle cell entails the coordinated assembly
of myofibrils and the sarcotubular system, mitochondrial proliferation,
the deposition of glycogen, and lastly, innervation. Each of these stages
of myogenesis will now be examined in some detail.

V. The Myoblast

Although widely used for almost 70 years, there has never been a
definition for the term "myoblast" fully satisfactory to all investigators
of muscle development. Godlewski (1901, 1902), who introduced the
term, considered a myoblast to be a cell morphologically similar, if
not identical, to other mesenchymal cells, but destined or committed
to form a muscle fiber. The obvious ambiguities of such a definition
led Tello (1917, 1922) to propose a definition which was modified bv
Boyd (1960) to read as follows:

1. The promyoblast (termed myoblast by Tello) is the primordial
muscle cell not distinguishable from neighboring mesenchyme.
2. The myoblast (termed myocyte by Tello) is the spindle-shaped
mononucleated or multinucleated cell without demonstrable myofibrils.
3. The myotube is the elongated, multinucleated cell containing
peripherally disposed myofibrils.
4. The myofiber, or mature muscle cell, contains peripherally placed
nuclei with the mass of cytoplasm largely filled by myofibrils.

This definition has proven more acceptable, but the use of the term
"myoblast" to include both mono- and multinucleated cells has not
proven satisfactory. Most workers have restricted the term myoblast
to mononucleated cells. Furthermore, the demonstration by Holtzer *et
al.* (1957) that mononucleated cells in the chick somite can contain
myofibrils has made the distinction between Tello and Boyd's classification of myoblast and myotube untenable.

Based on their combined analyses of mitosis and contractile protein
synthesis in the myogenic cell population (see review by Holtzer, 1970),
Holtzer and colleagues have proposed the following scheme, which,

in my opinion, is the most reasonable thus far suggested. Myoblasts are defined as "postmitotic, mononucleated cells capable of fusion and of synthesizing contractile proteins. Myoblasts are the immediate descendants of presumptive myoblasts. Presumptive myoblasts do not fuse, and do not synthesize myosin." Holtzer would consider both myoblasts and presumptive myoblasts to be "differentiated cells in the myogenic lineage," but makes no claim as to the number of possible intermediary cell types or stages that might separate the presumptive myoblast from the progenitor, undetermined mesenchymal cell, if indeed such an undetermined cell ever exists. These precursor cells of the presumptive myoblast are termed myogenic α and myogenic β cells for want of better terms (see Holtzer and Bischoff, 1970). The existence of a myogenic cell (myogenic β) distinct from the presumptive myoblast is based on the following experiment (Holtzer, 1970). Thymidine-^{3}H-labeled mononucleated cells from 8-day or 10-day-old chick embryo breast muscle will fuse with unlabeled myoblasts from 10-day-old embryos, but similarly labeled cells from 5-day-old embryos are not incorporated into myotubes when interspersed with myogenic cells from 10-day-old embryos. Okazaki and Holtzer (1965) had demonstrated previously that nonmyogenic cells (e.g., chondrocytes, liver cells, kidney cells) will not fuse with myoblasts and are never incorporated into myotubes.[2] Presumably, the cell surfaces of these heterotypic cells are sufficiently distinct from myoblasts to prevent their fusion with the myogenic cells, although this remains to be proven. Similarly, Holtzer (1970) has suggested that myoblasts from the 10-day-old breast muscle possess different surface properties than the 5-day-old cell, leading to a sorting out of these respective cell populations and the absence of mutual cell fusion. The elucidation of these biochemical differences at the cell periphery of different myogenic cells should prove extremely important in the years to come.

In the definition of the myoblast proposed by Holtzer and his colleagues, great emphasis is placed on the relative exclusivity between the replication of DNA and the synthesis of the myofibrillar proteins. It has been stated (Holtzer, 1970, p. 480) that "no cell going through S, G_2, or M is making, or has made, these contractile proteins," and that "only cells in some phase of G_1 *may* initiate the synthesis of the contractile proteins." This concept, which has received wide support

---

[2] An exception to this statement has been claimed by Maslow (1969) in which the specificity of this fusion process was altered by prior treatment of cells with actinomycin D. This important finding, if confirmed, might provide an experimental approach to the problem of cell recognition, particularly as this relates to cell fusion and myotube formation.

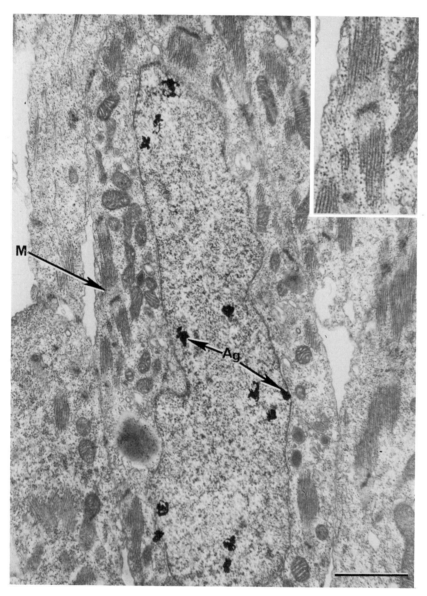

Fig. 5. Electron microscopic autoradiogram provided by Weinstein and Hay (1970) illustrating nuclear DNA synthesis within cross-striated embryonic chick heart cells. Embryonic hearts were incubated in the presence of thymidine-^{3}H and, after fixation, were prepared for electron microscopy and autoradiography. The dark silver grains (Ag) over the nucleus indicate the presence of radioactive thymidine, almost certainly within DNA, at this site. Since this same cell contains myofibrils (M), it is unlikely that myofibrillar protein synthesis and DNA synthesis are mutually exclusive phenomena in an embryonic myocardial cell. Calibration bar = 2 μm.

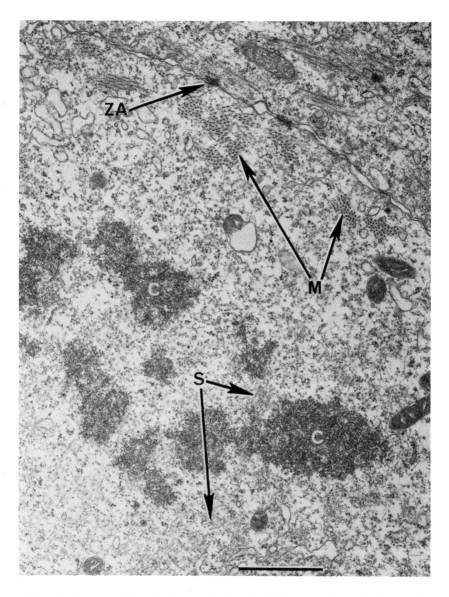

Fig. 6. A myocardial cell in metaphase of cell division from the heart of a 7-day-old chick embryo. This micrograph illustrates that mitosis can take place in embryonic heart cells that contain myofibrils (M). Note the absence of a nuclear envelope, the compact chromosomes (C), and adjacent spindle tubules (S) in this cell which has retained attachment to a neighboring heart cell via zonulae adherentia (ZA). Calibration bar = 1 μm.

from studies of skeletal muscle development, cannot be extended without some modification to cardiac muscle (Figs. 5 and 6). Mitotic activity (Mark and Strasser, 1966; Rumyantsev, 1963; Manasek, 1968) and DNA synthesis (Rumyantsev, 1966, 1968; Weinstein and Hay, 1970) have been demonstrated conclusively in myocardial cells that contain well organized myofibrils. Thus, although myocardial cells in adult birds and mammals do not divide, a substantial percentage of these contractile cells can replicate during embryogenesis. Are such cells to be classified as myoblasts? To avoid confusion, it seems advisable at this time to restrict the usage of the term myoblast to skeletal muscle and not apply it without qualification to the embryonic myocardial cell. Of obvious importance in these studies of the developing heart is the clear demonstration that DNA synthesis and myofibrillar protein synthesis are not necessarily exclusive phenomena within a given cell. In the adult mammalian heart, however, DNA synthesis is not observed in myocardial cells even under conditions of extreme cardiac hypertrophy (Morkin and Ashford, 1968; Grove *et al.*, 1969a,b). After injury, the adult heart muscle does not regenerate (Rona and Kahn, 1967) but is replaced by a connective tissue scar. Elucidation of the factor or factors involved in this repression of DNA synthesis in the differentiated myocardial cell is of enormous clinical significance and deserves more intensive analysis than has heretofore been the case.

A problem with most definitions of the myoblast, as Konigsberg (1965, p. 340) has emphasized, is that they "beg the question of the absence of a truly covert phase of differentiation." For instance, one might consider the hypothetical situation in which a myogenic cell has initiated the transcription of messenger ribonucleic acid (mRNA) for myosin but not begun its translation. In recent years, much emphasis has been placed upon developing sensitive cytochemical (Holtzer *et al.*, 1957) and biochemical (Coleman and Coleman, 1968) techniques for the detection of characteristic gene products in the differentiating cells. With such methods, attempts have been made to correlate the synthesis of the respective myofibrillar proteins with embryonic age, cellular proliferation, cytological structure, and myotube formation. Unfortunately, most *in vivo* myogenic systems do not lend themselves readily to biochemical analysis. None of these systems contains a homogeneous population of cells, nor are the stages of myogenesis ever synchronous. It is not emphasized enough that muscle development *in vivo* is an asynchronous phenomenon; in a given region of embryonic muscle, all stages of development ranging from presumptive myoblasts through differentiated myofibers are observed. Synthesis of muscle proteins must be considered on a cellular and not a tissue level. As such, cytochemical rather than bulk biochemical methods are to be preferred. The situation is better, although not perfect, in muscle cell cultures (Coleman and Coleman, 1968; Yaffe, 1969), for under appropriate *in vitro* conditions, myogenesis is more synchronous than within the animal. Interpretation of any chemical assay obviously depends upon the sensitivity of the method and the degree of precision desired. In the case of the myoblast, any definition based upon a cytochemical assay will of necessity be relevant only in terms of that assay or another which is comparably sensitive. As more precise methods are developed, the labels applied to a given cell probably will have to change. Furthermore, there is a strong degree of arbitrariness in choosing which proteins are differentiative determinants. For instance, if actin was synthesized at an appreciably earlier period than myosin, how could one decide at which stage the cell was differentiated. Although this is an arbitrary decision, it is not usually stated as such. Bulk synthesis of the contractile proteins actin, myosin, and tropomyosin may be coupled, but this has not been proven definitively. In fact, contrary evidence exists (Heywood *et al.*, 1967). This problem has been further complicated by the recent demonstration of actin-like filaments in nonmuscle cells within chick embryos (Ishikawa *et al.*, 1969). Thus, our assumption that the synthesis of contractile proteins denotes muscle differentiation may have to be reevaluated in light of the fact that other cells, not usually considered contractile, may synthesize proteins similar, if not identical, to those previously considered specific to muscle.

Another approach to the classification of myogenic cells has been to catalog and compare their ultrastructural features. Unfortunately, this has not proven of great value in elucidating the covert phases of myogenesis discussed above, but it has provided some insight into later stages of muscle development within the myotube. Electron micrographs of mononucleated cells, presumed to be myoblasts, have been published for amphibian (Hay, 1963; Lentz, 1969a), avian (Allen and Pepe, 1965; Przybylski and Blumberg, 1966; Fischman, 1967), and mammalian muscle (A. M. Kelly and Zacks, 1969a). It is apparent from all of these studies, that in the absence of myofibrils or characteristic myofilaments, the ultrastructural features of the myoblast are not unique or in any way specific to this cell type. As Hay (1963) has emphasized, it is likely that a spindle-shaped cell closely apposed to a myotube is a myoblast, but this cannot be considered certain. In adult muscle (Fig. 7), such a mononucleated cell lying beneath the basal lamina of an adjacent myofiber is termed a satellite cell (Mauro, 1961; Ishikawa,

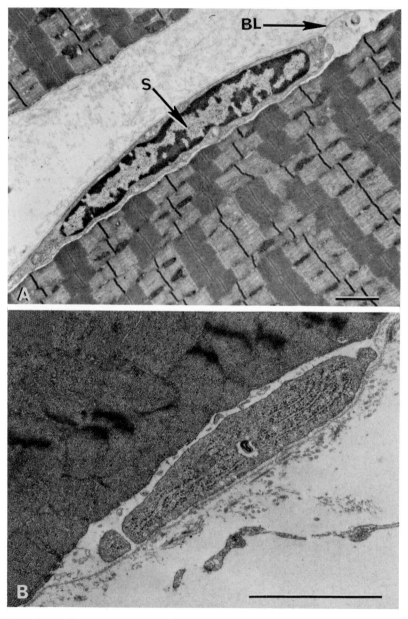

Fig. 7. Satellite cells (S) within adult rat skeletal muscle, provided by Dr. M. Vye (University of Illinois Medical School). Note the presence of each satellite cell beneath the basal lamina (BL) of the adjacent myofiber. Typically, the satellite cell nuclei are heterochromatic (Fig. 7a) in contrast with the euchromatic appearance of most embryonic myoblasts (see Fig. 8). Calibration bars = 2 μm.

1966; Mauro *et al.*, 1970). Such cells have been considered the adult form of the myoblast arrested in G_1 and a possible source of myogenic cells during muscle regeneration following trauma (Holtzer, 1970). In early postnatal life of birds and mammals, when the number of muscle nuclei has been shown to increase (MacConnachie *et al.*, 1964; Enesco and Puddy, 1964; Moss, 1968), the population of cells which continue to divide are the satellite cells (Moss and Leblond, 1970).

As illustrated in Fig. 8, the myoblast contains a large, fairly euchromatic nucleus, with the prominent nucleolus(i) typical of a cell active in RNA synthesis. Most cytoplasmic ribosomes are nonmembrane-bound, although a varying quantity of rough-surfaced endoplasmic reticulum is apparent in all myoblasts. The relative sparcity of rough-surfaced endoplasmic reticulum is usually considered important when distinguishing myoblasts from fibroblasts (Hay, 1963). Generally, this is a useful criterion; however, cases arise (Fig. 9) in which typical spindle-shaped cells exhibit a myoblastic morphology at one pole, while the other end of the cell contains an extensive region of rough-surfaced endoplasmic reticulum (i.e., a fibroblastic morphology) (Fischman, 1970).

Beneath the plasma membrane and often extending up to the fingerlike extensions at each end of the myoblast are sheets or bundles of thin filaments approximately 5 nm in diameter (Fig. 10). Such filaments are not unique to myoblasts, being found in many cells that are undergoing cytoplasmic elongation, ameboid movement, or cytokinesis (Carter, 1967; Baker and Schroeder, 1968; Schroeder, 1969; Szollosi, 1970; Spooner and Wessels, 1970; Wrenn and Wessells, 1970; K. M. Yamada *et al.*, 1970; Wessells *et al.*, 1971). Filaments of this type have been termed cortical filaments, and they appear to bind heavy meromyosin (HMM) rather specifically (Ishikawa *et al.*, 1969; Pollard *et al.*, 1970), suggesting certain similarities to actin (Fig. 11).

However, the studies of Wessells and co-workers (1971) have demonstrated a separate class of cytofilaments (~5 nm diameter) at the growth cone and microspike region within the pseudopodia of migrating cells both *in vivo* and *in vitro*. This latter set of filaments is termed the "microfilament network" by Wessells and co-workers to distinguish them from the cortical filaments. Within the microfilament network, the length and precise arrangement of the individual filaments is difficult to establish, for they exhibit a complex interlocking network without any substantial parallel alignment of the adjacent filaments. Wessells and collaborators have shown that in the presence of the drug cytochalasin B, the growth cones and microspikes of cultured neurons, fibroblasts, etc. are rapidly retracted with a concomitant disappearance of the microfilament network. Removal of the drug permits a reformation of the

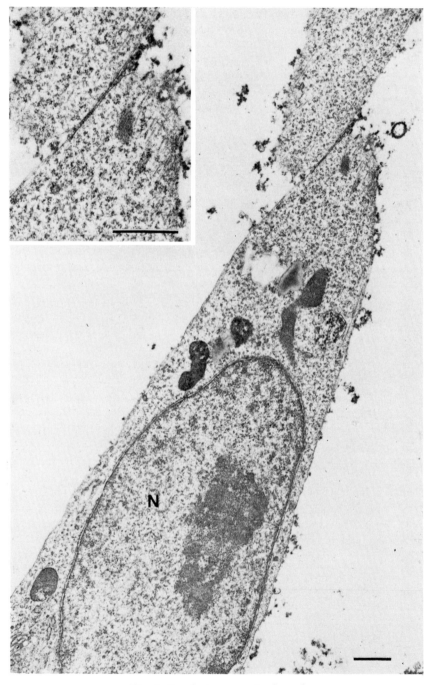

Fig. 8. See facing page for legend.

complete pseudopod with its apical microfilament network. Actin filaments within myofibrils appear to be structurally unaffected by cytochalasin B (Holtzer *et al.,* 1971; Sanger *et al.,* 1971). At the present time, it is not known if the cortical filaments are directly related to myofibrillogenesis or to the microfilament network. Until the individual classes of filaments have been isolated and the respective proteins compared, this subject is likely to remain ambiguous and largely incomplete.

In addition to the 5 nm diameter cortical filaments and the microfilament network, Ishikawa *et al.* (1968, 1969) and Rash *et al.* (1968) have described a class of filaments, 8–11 nm in diameter, which they have termed "intermediate filaments." This latter group of filaments do not bind HMM (Ishikawa *et al.,* 1969) and are found in many other cell types including fibroblasts, chondrocytes, and macrophages. The composition of these intermediate filaments is unknown, but a possible relationship to microtubules is suggested by the fact that cells exposed to colchicine or Colcemid, under conditions in which microtubules disassemble, exhibit a marked increase in the number of cytoplasmic intermediate filaments (Ishikawa *et al.,* 1968).

Identification of myoblasts has been achieved most successfully by Konigsberg (1963) in monolayer cultures. By following the progeny of cloned cells, isolated from embryonic chick leg muscle, Konigsberg was able to establish that bipolar, spindle-shaped cells gave rise to muscle colonies, whereas, cells with a more flattened, triangular or polygonal appearance formed fibroblastic colonies. These important observations have been confirmed with chick material in our own laboratory (Fischman and Yaffe, 1971) and by Yaffe (1969) with rat muscle. Of particular interest is the fact that such myogenic cells retain a spindle shape after many cell divisions *in vitro* (Yaffe, 1968, 1969). This cellular shape presumably represents a fundamental differentiative characteristic of the presumptive or true myoblast, although this has not been established unequivocally. It would be of interest to know when in the determination of the myogenic cell line such a cell shape is assumed by the mesenchymal cells destined to form muscle. Furthermore, the relationship,

---

Fig. 8. Micrograph of a characteristic spindle-shaped cell, presumably a myoblast or presumptive myoblast, within a chick muscle culture, 24 hours *in vitro.* The nucleus (N) is largely euchromatic, with a prominent granular nucleolus. Microtubules and microfilaments course in the long axis of the cell, extending to its distal extremities. In contrast to Fig. 9, this cell has very few membrane-bounded ribosomes; most are free in the cytoplasm. Calibration bar = 1 μm. The insert is an enlargement of the upper portion of the micrograph and illustrates the contact region with an adjacent myoblast. This particular contact appears to be a gap junction. Calibration bar for the insert = 1 μm.

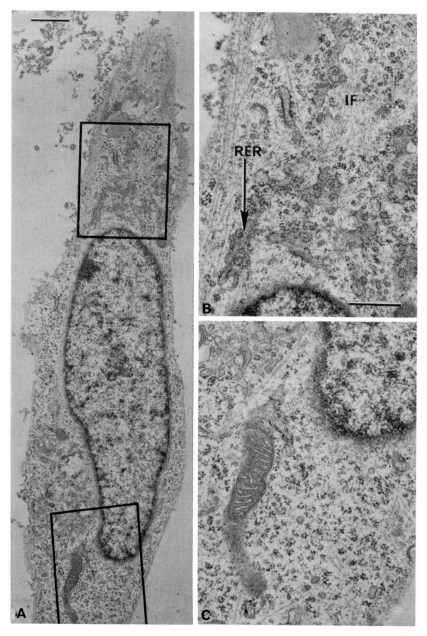

Fig. 9. (a) Micrograph of a bipolar, spindle-shaped cell from an embryonic chick skeletal muscle culture (Konigsberg, 1963) after 1 day *in vitro*. Calibration bar = 1 μm.   (b) and (c) Enlargements of the upper and lower boxed-off

if any, of myoblast shape to subsequent cellular differentiation is unknown. Recently, Strohman and Paterson (1971) have shown that myoblasts grown *in vitro* in the presence of EGTA (ethyleneglycolbis(aminoethyl ether)-*N*, *N*-tetracetic acid) will not fuse until calcium is replaced in the culture medium. These EGTA-inhibited myoblasts assume an extraordinarily long, pencil-like shape and in the electron microscope are seen to contain large numbers of cortical filaments and microtubules coursing in the long axis of the cell (Fig. 12). Such cells retract their microspikes within minutes after exposure to 5 μg/ml of cytochalasin B (Maker and Fischman, 1972).

In summary, the multinucleated skeletal muscle fiber is formed by the cytoplasmic fusion of postmitotic, spindle-shaped, mononucleated cells termed myoblasts. Such myoblasts are derived from an actively dividing, spindle-shaped population of cells called presumptive myoblasts, which must be considered determined in that presumptive myoblasts continue to form muscle colonies even after many cell divisions *in vitro*. Except for the occasional presence of myofibrils, the ultrastructural characteristics of the myoblast are not unique to this cell type and cannot be relied upon to classify or identify it with certainty. Although myoblasts from the chick somite (Holtzer *et al.*, 1957), limb bud (Coleman *et al.*, 1966), and rat (Fambrough and Rash, 1972) are capable of synthesizing myofibrillar proteins, in most other myogenic systems of higher vertebrates, bulk synthesis and accumulation of actin and myosin only begins after myotube formation. In skeletal muscle development, mitotically active cells do not synthesize myofibrillar proteins. This is in contrast to embryonic myocardial cells in which DNA synthesis, mitosis, and myofibrillar assembly have all been observed in the same cell. It is presumed that the myogenic cell population is ultimately derived from undetermined mesenchymal cells, but our understanding of this early phase of myogenesis is totally inadequate.

---

areas, respectively. One portion of the cell (b) exhibits the cytological characteristics of a cell synthesizing proteins for export (extensive rough-surfaced endoplasmic reticulum, RER) while another region of cytoplasm (c) mainly contains free ribosomes (R) more typical of a myoblast. Note the large number of 10 nm diameter filaments (IF) in the region of the cell containing granular ER (b), but relatively few in the zone of nonmembrane-bound ribosomes (c). Both microtubules and microfilaments run in parallel toward the extended tip of the cell (a), where they merge imperceptibly with the plasma membrane. Calibration bar for (b) and (c) = 0.5 μm.

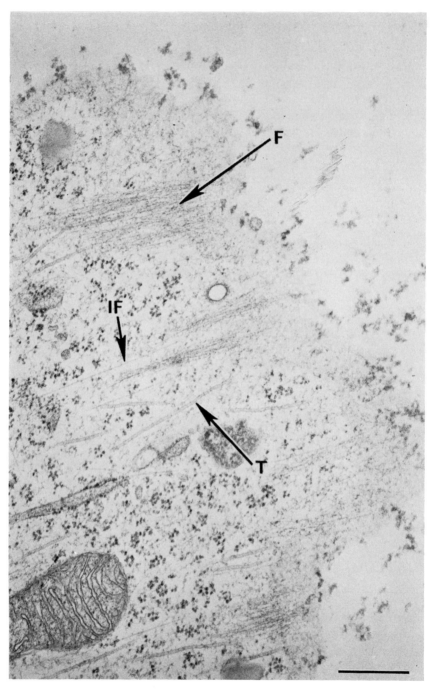

Fig. 10.

VI. Mitosis, Cell Fusion, and Myogenesis

Nuclei within the multinucleated syncitia of the myotube or myofiber are inactive in DNA synthesis and can be considered postmitotic. Multinucleation results from the cytoplasmic fusion of myogenic cells and not from amitosis or some other process involving the multiplication of myotube nuclei without cytokinesis. Overwhelming evidence for this statement has come from five principal types of experiments: (1) direct morphological analysis by light and electron microscopy; (2) quantitative DNA measurements by microspectrophotometry of Feulgen-stained nuclei; (3) autoradiographic analysis of nuclear DNA synthesis using thymidine-^{3}H as a labeled precursor of DNA; (4) inhibitor studies in which DNA synthesis is blocked by X rays, nitrogen mustards, folic acid antagonists, colchicine, etc; (5) fusion studies using genetic markers.

A. Morphological Analysis of Cell Fusion

In the older histological literature, cleft or dumbbell-shaped nuclei within myotubes were often presented as evidence for amitosis. However, when such nuclei were observed in cell cultures of muscle and followed by time-lapse cinematography, the amitotic theory received no support (Cooper and Konigsberg, 1961). Deformations of nuclear shape, which might have simulated division figures in fixed material, were found to be transient in nature and unrelated to nuclear cleavage. On the contrary, careful time-lapse studies have demonstrated that muscle syncitia form by the cytoplasmic fusion of myogenic cells (Capers, 1960). Electron microscopic analysis of myotube formation, either *in vivo* (Hay, 1963) or *in vitro* (Shimada *et al.*, 1967; Fischman, 1970; Shimada, 1971; Lipton and Konigsberg, 1971) are in complete agreement with the time-lapse studies (Figs. 13 and 14). Fusion can occur between two myoblasts, between a myoblast and a myotube, or at the apposed surfaces of two myotubes. Although micrographs illustrat-

---

Fig. 10. The distal extremity of a spindle-shaped cell within 1-day-old chick muscle culture. Cortical filaments (F) approximately 5 nm in diameter merge gradually with the plasma membrane. In addition, microtubules (T) and intermediate filaments (IF) are present in this field. In contrast to the cortical filaments, which form arrowheadlike complexes with HMM, the intermediate filaments and microtubules appear not to bind HMM. Calibration bar = 0.5 μm.

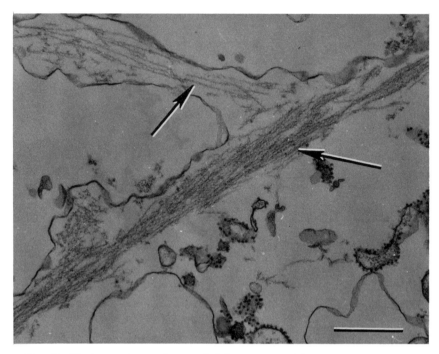

Fig. 11. Electron micrograph provided by Dr. H. Ishikawa (Tokyo University) of cortical filaments (arrows) within a cultured chick fibroblast. This particular preparation has been treated with glycerol and then reacted with heavy meromyosin (HMM) as described by Ishikawa et al. (1969). Note the arrowheadlike configuration of these filaments after HMM-binding (compare with Fig. 10). Calibration bar = 0.5 μm.

ing cellular fusion have appeared in the literature, electron microscopy has contributed relatively little to our fundamental understanding of the dynamic events occurring in this process. When muscle cultures are examined by electron microscopy during the period of active cell fusion (second through fourth days *in vitro*), small adherent and focal tight junctions (Trelstad *et al.*, 1967) are often observed between adjacent myoblasts and myotubes (Fischman, 1970). Similar intercellular junctions have been observed in embryonic mammalian muscle developing *in vivo* (Kelly and Zachs, 1969a). In neither case, however, is it understood how these junctions directly relate to the actual process of cell fusion. At clearly defined areas of cell fusion (Figs. 13 and 14), the apposed plasma membranes of the two adjacent cells can be shown to be continuous with one another, and cytoplasmic confluence is demonstrable. Often, there are multiple sites of membrane interruption along the adjacent cell surfaces, suggesting that cytoplasmic fusion does not proceed in a zipperlike fashion, but occurs simultaneously at many spots along the membrane. Membrane vesicles that are presumably remnants of the interrupted cell boundaries are observed at the zone of fusion, but the mechanism of membrane vesiculation remains obscure. Nothing is known about the underlying chemical events involved in this process, nor have any enzymic reactions relevant to fusion been established. Our understanding of this basic process in muscle formation is totally inadequate.

B. *Microspectrophotometric Experiments*

Quantitative measurements of the Feulgen DNA content within regenerating mouse muscle (Lash *et al.*, 1957), cultured chick muscle (Firket, 1958; Basleer, 1962), and reptilian muscle (Cox and Simpson, 1970) reveal that all nuclei within myotubes have DNA values equivalent to the $2n$ amount. Adjacent mononuclear cells, however, have bimodal distributions of DNA content; both $4n$ and $2n$ values are observed in the nonsyncitial cell population. These studies demonstrate the absence of DNA doubling within myotube nuclei; the obvious conclusion is that nuclear replication must occur exclusively in the mononucleated cells prior to their fusion.

C. *Autoradiographic Studies*

All autoradiographic studies of DNA synthesis have shown that thymidine-^{3}H is not incorporated into the nuclear DNA of myotubes (Firket, 1958; Hay and Fischman, 1961; Bintliff and Walker, 1960; Stockdale and Holtzer, 1961; Basleer, 1962; Zhinkin and Andreeva, 1963; Okazaki and Holtzer, 1966; Marchok and Herrmann, 1967; Coleman and Coleman, 1968; Yaffe, 1969; Bischoff and Holtzer, 1969; Reznik, 1969). When a myogenic system is pulse-labeled with thymidine-^{3}H and fixed soon afterward, an autoradiogram contains silver grains exclusively over nuclei of mononucleated cells. But if the fixation is delayed for progressively longer periods after the isotopic pulse, radioactive nuclei begin to appear within myotubes. The precise timing of this phenomenon is species and age dependent, for it reflects the component phases of the mitotic cycles (S, G_2, M, and G_1) through which the cells must pass. An elegant analysis of this problem has been presented by Holtzer and his colleagues (Okazaki and Holtzer, 1966; Bischoff and Holtzer, 1969; Holtzer

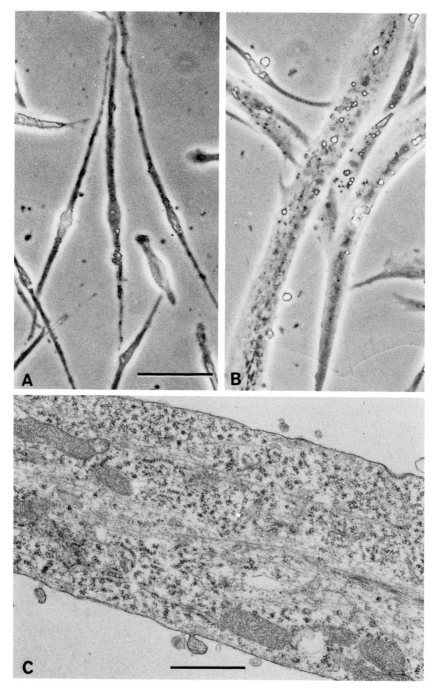

Fig. 12. See facing page for legend.

and Bischoff, 1970). When cultures of embryonic chick breast muscle were pulse-labeled with thymidine-^{3}H and chased for varying lengths of time in nonradioactive media, labeled nuclei were first observed 8–10 hr after exposure to the isotope. Assuming thymidine incorporation into the DNA was limited to the S period, it could be concluded that fusion of the myoblasts was delayed at least 8–10 hr after the preceding DNA synthetic period (S). Knowing that the mean durations of S, G_2, and M periods were 5, 1–2, and 1 hr, respectively, it was apparent that cell fusion must occur in the G_1 phase of the mitotic cycle and that a minimum of 5–8 hr must elapse between the end of mitosis and the completion of fusion. Holtzer and colleagues suggest that the final mitosis prior to cell fusion is a critical step in which the presumptive myoblast becomes committed to cell fusion (Holtzer and Bischoff, 1970). After completion of this division, termed a "quantal mitosis" by these authors, the myoblast enters a prolonged G_1 phase, and presumably initiates the metabolic events required for cell fusion. Although somewhat speculative, this hypothesis has had the important effect of focusing attention on the mitotic and postmitotic period immediately preceding fusion, for it now appears certain that critical regulatory steps occur during this brief 3–4 hr period. Space limitations do not permit a more complete analysis of Holtzer's quantal mitosis theory; interested readers are referred to the recent publications of Holtzer (1970) and Holtzer and Bischoff (1970) where this important and provocative proposal is discussed in detail.

These autoradiographic studies of DNA synthesis correlate nicely with biochemical assays of DNA polymerase activity in cultured muscle (O'Neill and Strohman, 1969). At a stage *in vitro* when cytoplasmic fusion is most prominent, there is a sharp decline in DNA polymerase activity within the cultures. It remains to be demonstrated, however,

---

Fig. 12. Micrographs of myogenic cells within an embryonic chick muscle culture after EGTA treatment (Strohman and Paterson, 1971). (A) Very long spindle-shaped cells are typically observed in these cultures after a 24 hr exposure to 1.9 mM EGTA (see text) which reversibly blocks cellular fusion. (B) When 2mM calcium chloride is added to the culture shown in (A), cytoplasmic fusion occurs and myotubes form. The myotubes shown in (B) have been photographed 24 hr after Ca^{2+} addition to an EGTA-containing culture. (C) Electron micrograph of a spindle-shaped cell from the preparation shown in (A). These cells contain many cytofilaments (both cortical and intermediate varieties) and microtubules. Long polyribosome chains are apparent, most of which are nonmembrane-bound. When exposed to cytochalasin (10 μg/ml), such cells exhibit a marked retraction of the long, cell extensions which is reversed upon removal of this drug (Maker and Fischman, 1972). Calibration bars = 50 μm (A and B), 1 μm (C).

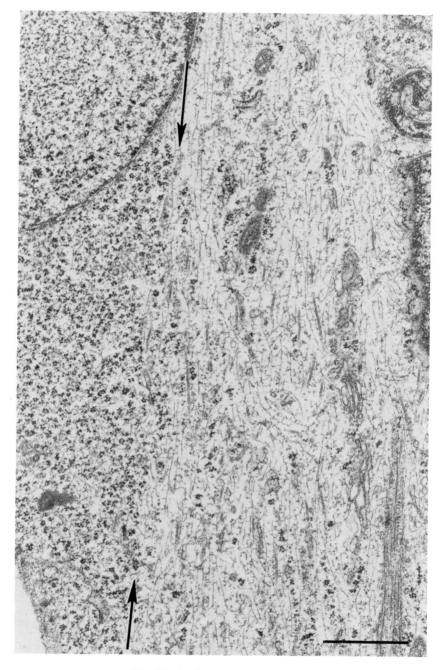

Fig. 13. See facing page for legend.

that DNA polymerase is the rate limiting enzyme that controls DNA synthesis in the myoblast prior to fusion.

D. Inhibitor Studies

Many attempts have been made to block specific metabolic pathways before or after myotube formation. For instance, if multinucleation were hypothesized to arise by the replication of myotube nuclei in the absence of cytokinesis, then blockade of DNA synthesis should prevent the formation of such syncitia. The experimental literature is quite clear regarding this point. Inhibitors of DNA synthesis when added to a myogenic system, either *in vivo* or *in vitro*, will prevent replication of the mononucleated cells but not prevent myotube formation. Such studies have been performed with the nitrogen mustard methylbis(β-chlorethyl)amine (Konigsberg *et al.*, 1960), the alkylating agent Myleran (Basleer, 1962), and 5-fluorodeoxyuridine (FUdR), the inhibitor of thymidylate synthetase (Coleman and Coleman, 1968; Bischoff and Holtzer, 1970). Of these drugs, FUdR seems the inhibitor of choice, for its blockage of DNA synthesis can be made nearly 100% effective, yet its effects can be reversed with exogenous thymidine. Furthermore, at doses that block DNA synthesis, it exhibits little effect on RNA or protein synthesis for at least 5 days *in vitro* (Coleman and Coleman, 1968). If fact, through the use of this base analog, the Colemans have been able to selectively destroy mononucleated cells in muscle cultures that contain a mixed population of mono- and multinucleated cells. Thus, these authors were able to examine the biosynthetic activities of muscle fibers without the added background complexities introduced by mononucleated cell metabolism. This selective use of FUdR should prove extremely useful in future studies.

Another base analog, bromodeoxyuridine, has been shown to inhibit the differentiation of myogenic cultures even at concentrations that do

---

Fig. 13. The region of a myotube exhibiting unequal distribution of cytoplasmic organelles, including ribosomes, cytofilaments, and myofibrils in a chick muscle culture, 6 days *in vitro*. It is suggested that this partition of cytoplasmic material reflects a recent fusion event between a myoblast (on the left) and a myotube (on the right). All remnants of intervening plasma membranes have been resorbed, but admixture of cytoplasm has not occurred. The nucleus (N) at the upper left presumably will migrate to a centrotubular position, with the nuclei at the right of the micrograph. Most of the free cytofilaments (F) are considered to be intermediate filaments (Ishikawa *et al.*, 1969). Calibration bars = 1 μm. Electron micrograph supplied by Dr. Y. Shimada, Chiba University, Japan, published in Shimada (1971).

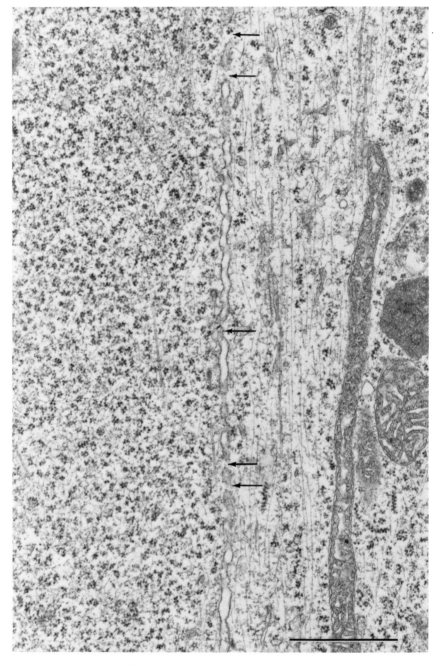

Fig. 14. See facing page for legend.

not block cell replication (Stockdale *et al.*, 1966; Coleman *et al.*, 1970; Bischoff and Holtzer, 1970). In the presence of BUdR at approximately 10^{-5}–10^{-6} M, muscle cultures develop few or no myotubes, but mononucleated cells continue to divide with the resulting formation of confluent cultures that contain epithelioid-like mononucleated cells. Such cells differ markedly from the characteristic bipolar myoblasts seen in control cultures (Bischoff and Holtzer, 1970). When added to muscle cultures that have already differentiated, BUdR apparently has little or no effect on further development of myotubes; only the mononucleated cells appear to be sensitive to this compound. Under appropriate conditions (Bischoff and Holtzer, 1970), removal of BUdR will permit the differentiation of a blocked culture, but this probably requires two or three cell divisions in the absence of the base analog. The mechanism of action of BUdR is, at the present time, unknown.

In other systems, it has been demonstrated that exposure of cells to BUdR results in the incorporation of bromodeoxyuracil (BU) into DNA as a replacement for the base thymine (Eidinoff *et al.*, 1959; Djordjevic and Szybalski, 1960; Simon, 1963). Holtzer and colleagues have suggested that the effect of BUdR on myogenesis is a reflection of this insertion of BU into the DNA of the presumptive myoblasts, presumably altering transcriptive processes at the nuclear level. Evidence for this hypothesis rests on the following experiments. Growth of myogenic cells in BUdR for at least one or two rounds of cell division results in an increase in the density of the DNA when examined by isopycnic centrifugation in cesium chloride gradients (Stockdale *et al.*, 1966). Based on the fact that three DNA peaks were observed in the cesium chloride gradients, Stockdale *et al.* (1964) proposed that both single and double strand replacement of BU for thymine could occur, depending upon the number of cell divisions which take place in the presence of BUdR. Later studies by Bischoff and Holtzer (1970), however, have shown that myogenesis can be suppressed by BUdR after a single round of DNA synthesis, and they have concluded that insertion of BU into one strand of DNA is sufficient to alter this differentiative

Fig. 14. Another example of presumed cytoplasmic fusion between a myoblast (on the left) and a myotube (on the right) in a chick muscle culture, 6 days *in vitro*. In this case, the plasma membranes have not been resorbed, and multiple interruptions in the fused membranes have been indicated by arrows. It is suggested that this micrograph represents an earlier stage of fusion than Fig. 13. Note the many intermediate filaments in the myotube cytoplasm. One obtains the impression that long polyribosome chains are more frequently encountered in the myotube than the adjacent myoblast. Calibration bar = 1 μm. Electron micrograph supplied by Dr. Y. Shimada, Chiba University, Japan, published in Shimada (1971).

process. Holtzer and co-workers have estimated that 50% replacement of thymine by BU in DNA is optimal for a nonlethal suppression of myogenesis by BUdR.

Two other lines of evidence suggest that BUdR exerts its effect through a substitution for thymine in DNA. First, the action of BUdR is prevented if thymidine is added simultaneously to the culture medium. Second, BUdR will initiate mitosis in cells that have been blocked by FUdR (Bischoff and Holtzer, 1970). Additional support for an action of BUdR at the chromosome level has been derived from autoradiographic studies (Coleman *et al.*, 1970; Bischoff and Holtzer, 1970) in which it was shown that myogenic cells grown in the presence of 5-bromodeoxyuridine-^{3}H exhibit label almost exclusively over the cell nuclei. Thus, in the myogenic system, most of the published work supports the hypothesis that BU substitution for thymine in nuclear DNA, in some unknown way, prevents the expression of genes required for cell fusion and the synthesis of the myofibrillar proteins. However, Schubert and Jacob (1970), while investigating a neuroblastoma cell line, found that axonal outgrowth by the neurons in culture could be altered by BUdR without nuclear DNA synthesis or cell division. These authors concluded that differentiative traits might be modified by BUdR at a post-transcriptional level, perhaps by a direct modification of the surface properties of these and other cells. Further work is clearly necessary before this study can be directly related to the muscle system, but it does illustrate the important possibility that BUdR may have more than one biochemical effect when applied to a muscle culture.

Another group of metabolic inhibitors, the antimitotic agents colchicine and Colcemid, have been widely employed by Holtzer and colleagues (Okazaki and Holtzer, 1965; Bischoff and Holtzer, 1968, 1970) to clarify the relationship between mitosis and myogenesis. These studies have shown that only mononucleated cells are arrested in metaphase, and furthermore, these mitotically arrested cells do not participate in cell fusion. However, those cells that have completed their terminal cell division prior to fusion continue to fuse even in the presence of these alkaloids. In other words, colchicine or Colemid do not interfere with the fusion process per se, but prevent cells from entering the G_1 phase in their cell cycle in which they are competent to fuse (Holtzer, 1970). Colchicine and Colcemid have been demonstrated also to have a dramatic effect on myotube shape, which should be distinguished from the antimitotic events in the mononucleated cell population (Ishikawa *et al.*, 1969). This effect on myotube shape will be discussed later when myofibrillar assembly is considered.

Presumptive myoblasts also can be inhibited from dividing and differentiating without the use of metabolic inhibitors. When suspensions of myogenic cells are cultured in the presence of heterotypic cells (nerve, liver, heart, cartilage, etc.), or grown in contact with already differentiated myotubes, the mononucleated cells will not fuse, and their cell division is suppressed (Shimada, 1968; Nameroff and Holtzer, 1969). This inhibition of myogenesis requires cell contact between the presumptive myoblasts and a second population of cells; the effect is not diffusion mediated, for separation of the two cell populations by a Millipore filter permits muscle differentiation to proceed normally. It has been suggested (Nameroff and Holtzer, 1969) that this phenomenon represents an embryological example of cell contact-mediated inhibition of DNA synthesis well-documented in various cell lines (see review by Stoker, 1967). As with colchicine, this inhibition of DNA synthesis in the presumptive myoblast presumably prevents that population of cells from entering the G_1 period in which fusion takes place.

Finally, a number of other experimental conditions have been shown to block the fusion of myogenic cells. The respiratory poisons antimycin A (Konigsberg, 1964), 2,4-dinitrophenol, cyanide, and azide all produce a predominantly irreversible inhibition of fusion. Variations of pH (Strehler et al., 1963), inhibition of protein synthesis by cycloheximide (Holtzer, 1970), and reduction of extracellular Ca^{2+} (Shainberg et al., 1969; Paterson and Strohman, 1970; Holtzer, 1970) all produce a reversible inhibition of cell fusion. It has been suggested by Holtzer (1970) that these inhibitors block myotube formation by interrupting postmitotic events occurring in the G_1 period immediately preceding fusion.

These inhibitor studies strongly suggest that fusion must be preceded by at least two discrete steps. The first step requires DNA synthesis and cell division, following which the myogenic cell enters a unique G_1 period. This first step is interrupted by any inhibitor of DNA synthesis or cell division. The second step takes place in this G_1 period and presumably results in a modification of the myoblast cell surface which makes it competent to fuse. This latter phase is interrupted by calcium depletion and inhibitors of protein synthesis. With judicious use of these reversible inhibitors, one may now dissect pre- and postfusion events and explore in more detail than has heretofore been possible the biochemical processes controlling muscle differentiation.

E. Cytogenetic Studies of Muscle Cell Fusion

As discussed above, cell fusion is quite specific, i.e., myoblasts will not fuse with cells from other tissues (Moscona, 1957; Okazaki and Holtzer, 1965; Yaffe and Feldman, 1965) but apparently will fuse with myoblasts from other species (Yaffe and Feldman, 1965; Yaffe, 1969).

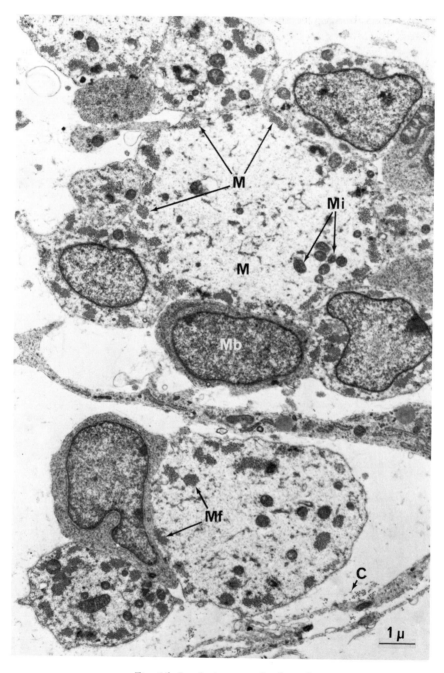

Fig. 15. See facing page for legend.

This observation is yet another example of the important principle first stated by Moscona (1955, 1965) that mixed populations of cells will sort out and mutually aggregate according to their tissue, rather than species of origin.

Perhaps the most elegant demonstration of myoblast fusion was the set of experiments performed by Mintz and Baker (1967) with allophenic mice. These investigators aggregated blastomeres *in vitro* from two genetically distinct strains of mice, and after uterine implantation of the chimeric embryos, raised mice to maturity that contained two sets of nuclei, one of each parental genotype. In this particular experiment, each strain of mouse was homozygous for one of the two subunits of the enzyme isocitric dehydrogenase. Each of the isozymes could be easily identified and separated by gel electrophoresis. After formation of the chimeric mice, a third, intermediate isozymic band was identified in homogenates of muscle but not in other tissues. The authors reasonably concluded that this new isocitric dehydrogenase band represented a hybrid enzyme formed in muscle fibers that contained both nuclei in a common cytoplasm.

VII. The Myotube

The myotube is here defined as the multinucleated syncitium which results from the cytoplasmic fusion of myoblasts. The term is purely descriptive, for it only implies an immature muscle fiber in which the myofibrils are, in general, circumferentially distributed within the cell, and the nuclei occupy the core or central zone of the syncitium (Fig. 15). As nuclei migrate to the periphery and occupy a subsarcolemmal position, myofibrils become more evenly distributed throughout the cytoplasm, and the cell is then termed a myofiber. This definition obviously has its ambiguities, for there is no sharp cut-off between these two stages of muscle development. In fact, the same cell can exhibit characteristics of a myofiber at one region along its length, while at a more distal zone appear less differentiated and could properly be labeled a myotube. This ambiguous situation results from the fact that develop-

Fig. 15. Low magnification cross section through 12-day-old, embryonic chick leg muscle (Fischman, 1967). This electron micrograph illustrates the characteristic aggregation of myogenic cells surrounded by a circumferentially oriented fibroblastic sheath. These myogenic cells exhibit different stages of muscle differentiation; some myoblasts (Mb) have no visible myofibrils, whereas, the myotubes (M) generally contain a peripheral ring of myofibrils (MF); Collagen = C; mitochondria = Mi. Calibration bar = 1 μm.

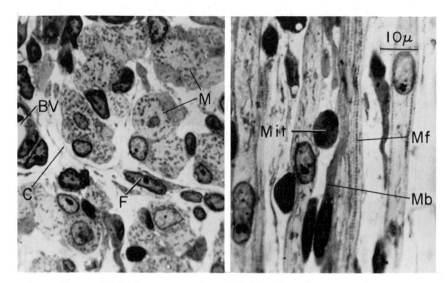

Figs. 16 and 17. Light micrographs of 12-day-old embryonic chick leg muscle. Fig. 16 is a cross section, Fig. 17 a longitudinal section of the tissue. A mitotic figure (Mit) is seen in Fig. 17. Such cells are observed exclusively in the mononucleated cell population. Abbreviations: BV = blood vessel; C = collagen; F = fibroblast; M = myotubes; Mb = myoblast; Mf = myofibril. Calibration bar = 10 μm.

ment of a muscle fiber is not synchronous along the whole length of the cell; fusion is more prominent at the ends of the syncitium (Kitiyakara and Angevine, 1963) leading to a gradient of differentiation along the cell with the most mature regions being the middle part of the cell, the least mature being its two distal ends. The term myotube arose from the light microscopic appearance of paraffin-embedded embryonic muscle when examined in cross section. At low resolution, the peripherally disposed myofibrils confer a tubelike configuration on these young muscle fibers. In 0.5 μm thick plastic-embedded material (Figs. 16 and 17), the individual myofibrils can be resolved, but this is difficult in the usual paraffin sections cut at 5 or 10 μm thickness.

Following fusion of the myoblasts, there are some obvious changes in cytoplasmic organelles and metabolic activities of the myogenic cells. These cellular modifications are reviewed below.

A. Mitochondria

There is a marked increase in the number of mitochondria in myotubes as compared with myoblasts, and this difference can be detected both in electron micrographs (Allen and Pepe, 1965) and by histochemical reactions for enzymes known to be localized in this organelle (Cooper and Konigsberg, 1961; Emmart *et al.*, 1963). Probably related to this increase in mitochondrial number is an increased reliance on oxidative phosphorylation as an energy source; myotubes are destroyed by antimycin A at concentrations that do not interfere with the survival and proliferation of myoblasts (Konigsberg, 1964). Not only do mitochondria increase in number within myotubes, but there occurs a progressive structural differentiation of these organelles in parallel with the development of the muscle cell. Mitochondria in myoblasts and young myotubes are, in general, shorter and less compact than those mitochondria within more mature muscle cells (cf. Figs. 18 and 19). There is an increased number and tightness of packing of the cristae as development procedes within the myotube. These changes presumably reflect a differentiative change within the mitochondrion consonant with the increased oxidative metabolism of the cell (Gregory *et al.*, 1967; Lennie *et al.*, 1967; A. C. Walker and Bert, 1969). In analogy with other systems (Luck, 1963, 1965; Roodyn and Wilkie, 1968; Swift and Wolstenholme, 1969; Rabinowitz and Swift, 1970), the increased numbers of mitochondria probably arise from mitochondrial division within the myotubes, but this has never been studied in detail. In young myotubes, mitochondria are distributed preferentially near the core of the syncitium. As myofibrils are assembled, the mitochondria shift to the interfibrillar location seen in the mature muscle fiber (Larson *et al.*, 1970).

B. Ribosomes

As stated above, ribonucleoprotein granules are quite densely packed within the cytoplasm of the myoblast, conferring upon this cell its characteristic basophilic appearance. Although myotubes are also basophilic, the greater dispersal of ribosomes in the larger cell volume of the syncitia results in the reduced basophilia of the myotubes when compared to the neighboring mononucleated cells (see Fig. 17). After cell fusion, the cytoplasmic contents of the myoblast are diffused within the sarcoplasm of the myotube. We have almost no information bearing on the kinetics of this dispersal process nor of the regional distribution over which the myoblast cytoplasm is distributed within the myotube. In electron micrographs of cells in fusion (Figs. 13 and 14), a gradient of ribosomal concentration can be appreciated. It has been suggested by Fischman *et al.* (1967) and Shimada (1971) on the basis of such micrographs that transfer of myoblast ribosomes into the myotube may be a slow enough process to be feasibly analyzed.

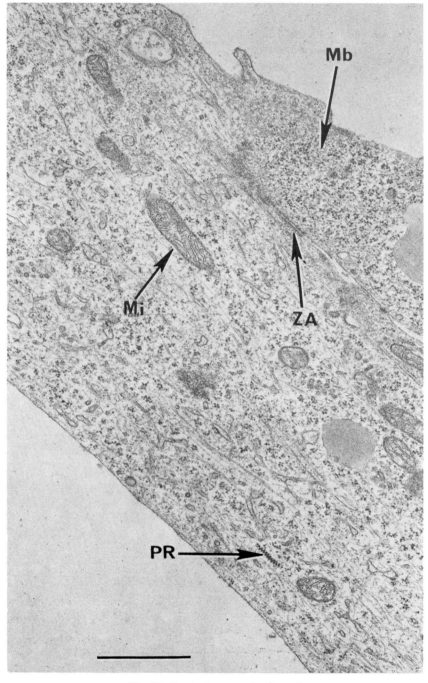

Fig. 18. See facing page for legend.

Largely through the efforts of Heywood and co-workers, there has recently been a considerable improvement in our understanding of the translational events surrounding the biosynthesis of myosin, actin, and tropomyosin in embryonic muscle. In the first paper of that series (Heywood *et al.*, 1967) a reproducible method was presented for the isolation and sucrose gradient analysis of polysomes from embryonic chick leg muscle. Using acrylamide gel electrophoresis to identify nascent peptides, it was possible to demonstrate that myosin is synthesized on a class of polyribosomes which contain 50–60 ribosomes. From the size of the polysomal cluster it was inferred that the polypeptide chain being synthesized has a molecular weight of 170,000–200,000 daltons. This peptide size is consistent with the established molecular weight for the heavy subunit of myosin (Frederiksen and Holtzer, 1968; Lowey *et al.*, 1969; Paterson and Strohman, 1970). Next, a cell-free system was established in which the syntheses of myosin, actin, and tropomyosin could be demonstrated *in vitro* (Heywood and Rich, 1968). It was shown that each of these proteins is made on a separate class of polysome, each of a size consistent with the known subunit sizes of the proteins. It was concluded, therefore, that the synthesis of each of these myofibrillar proteins is directed by a monocistronic mRNA. In the subsequent papers of this series (Heywood and Nwagwu, 1968, 1969), attention was directed to the isolation and partial characterization of myosin mRNA. By first separating the large polysomes (previously shown to synthesize myosin) from the remaining cellular ribosomes and then isolating RNA from this fraction, it was possible to test this RNA in a cell-free, protein synthesizing system. The results indicated that a species of RNA of approximately 26 S coded for the heavy subunits of myosin. Independent studies by Sarkar and Cooke (1970) have shown that the low molecular weight subunits of myosin (Dreizen *et al.*, 1966, 1967) are synthesized on a separate class of polysome consistent with the size of these protein subunits (15,000–30,000). Thus, the synthesis of the subunits of myosin is occurring on at least two clearly separate classes of polyribosome, and

---

Fig. 18. Longitudinal section of a myotube with an adjacent myoblast (Mb) 48 hr after initiating an embryonic chick monolayer culture. A zonula adherens (ZA) between apposed myoblasts and myotubes is observed frequently in early stages of myogenesis, an initial stage of cell attachment prior to cytoplasmic fusion. At a phase when thick (~16 nm in diameter) cytoplasmic filaments appear in the myotube, long helical chains of ribosomes (PR) are evident. Note the absence of a distinct basal lamina around the myoblast and the myotube at this stage of differentiation. Typically, the mitochondria (Mi) within these young myotubes are short and contain few cristae, which are loosely packed and irregularly positioned (compare with Fig. 19). Calibration bar = 1 μm.

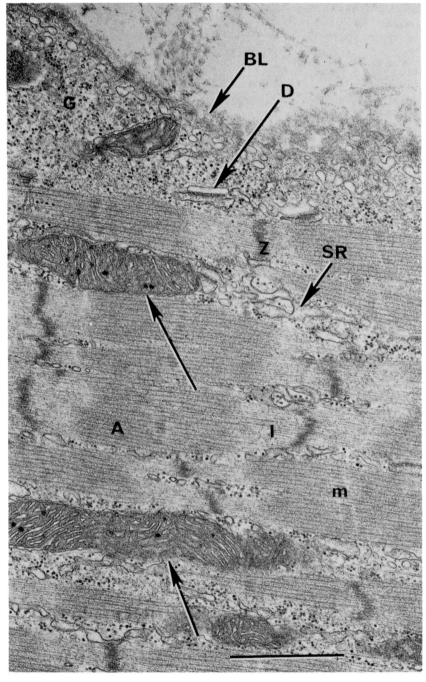

Fig. 19. See facing page for legend.

completion of the whole molecule probably does not occur until release of the nascent subunits into the sarcoplasm. It would be of interest to establish whether or not these independent classes of myosin polysomes are in close physical proximity within the embryonic muscle cell.

It has been demonstrated by Johnson *et al.* (1967) that myosin and actin isolated from adult rabbit muscle contains the unusual amino acid 3-methylhistidine. Although actin in embryonic muscle contains this amino acid, 3-methylhistidine is not present in myosin isolated from the rabbit fetus. After birth, the amount of this amino acid in myosin increases until it reaches the adult value (Trayer *et al.*, 1968). Very little is known of the mechanism underlying this methylation of the histidine residues of myosin. Recently, Hardy and Perry (1969) have demonstrated that homogenates of back muscle from 5-day-old rabbits are capable of catalyzing the methyl transfer of $^{14}CH_3$-labeled S-adenosylmethionine to the histidine residues of myofibrillar proteins. These authors suggest that the methylation of myosin occurs after peptide bound formation.

Considering the fact that the large subunits of myosin are approximately 200,000 daltons, it is conceivable that nascent peptide subunits on adjacent ribosomes interact and perhaps form coiled coils before release from the polysome chain. However, attractive as this suggestion may be, it still awaits experimental confirmation. Recent studies by Obinata (1969; Obinata and Hayashi, 1972) have demonstrated that when myosin is extracted from embryonic chicken muscle, two forms of the protein can be isolated. The major myosin fraction consists of a protein with the ATPase and sedimentation properties of adult myosin and presumably contains the two identical heavy subunits characteristic of the mature protein. A second fraction of lower ATPase activity can be isolated from embryonic muscle which also binds to and is slightly activated by actin and sediments at a rate consistent with a molecular weight one-half that of adult myosin. Acrylamide gel electrophoresis of this ATPase in the presence of SDS has shown that it contains a large subunit of identical molecular weight to that in adult myosin

Fig. 19. A myotube in a 12-day-old embryonic chick muscle culture. In comparison with a newly formed myotube (see Fig. 18), this relatively mature myofiber contains well formed myofibrils with A, I, M, and Z bands. The sarcotubular system is composed of both a sarcoplasmic reticulum (SR) and transverse tubular network (T system). As with embryonic muscle differentiated *in vivo*, the T-system is often oriented longitudinally rather than transversely and makes dyadic (D) contacts with the SR. Glycogen (G) granules are seen more frequently in the mature fibers, and mitochondria (arrows) are large, long and tightly packed with well aligned, densely staining cristae. The basal lamina (BL) is a well-defined layer outside the plasma membrane. Calibration bar = 1 μm.

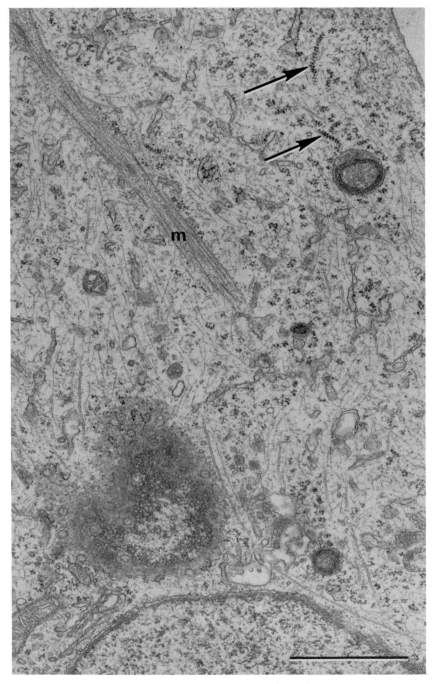

Fig. 20. See facing page for legend.

3. Development of Striated Muscle

(Obinata and Hayashi, 1972). Presumably, some myosin exists within the cytoplasm of embryonic chick muscle containing only one heavy subunit. However, it has not been ruled out that a proportion of embryonic myosin is dissociating into component subunits during extraction of the protein from muscle. Another possibility is that nascent peptide is being released during myosin isolation; i.e., some of the polysomes are being stripped of nascent peptide, and this is the source of Obinata's unusual myosin fraction. The problem might be clarified with pulse-chase experiments using radioactively labeled amino acids.

The studies of Huxley (1963), Kaminer and Bell (1966), Kaminer (1969), and Harrington and Josephs (1968) have shown conclusively that thick filament formation can occur *in vitro* by dialyzing monomeric myosin against potassium chloride solutions at physiological pH and ionic strength. Structured cellular organelles are not required for filamentogenesis to occur *in vitro*, and at the present time, there is no compelling reason to propose a role for the polyribosome in the formation or length regulation of myosin filaments (see, however, Cedergren and Harary, 1964; Larson *et al.*, 1969).

In our own laboratory, we have found no consistent relationship between the cellular location of very large polyribosomes and the position of thick filaments (Fischman, 1967, 1970). This is true for both embryonic chick skeletal and cardiac muscle. Within the myotube, the ribosomes are not distributed preferentially in any one cross-sectional zone of the cell, yet fibrils are deposited predominantly beneath the sarcolemma (Figs. 20 and 21). Furthermore, if polysome sizes are examined in these electron micrographs, there does not appear to be any correlation between polysome size and their spatial distribution within the myotube; i.e., large polyribosomes are not observed any more frequently in the cortex than in the core of the myotube. Occasionally, linear or spiral polyribosome chains appear contiguous with thick filaments, but this is not a consistent observation. In most cases, thick filaments are unattached to polyribosomes. Whatever factors regulate the length of the myosin filament, it is the conclusion of this author that the physical configuration or spatial disposition of the polyribosome chain is not one of them.

Fig. 20. Myotube from a 3-day-old chick muscle culture. An incompletely assembled myofibril (m) is present at a considerable distance from the long polyribosome chains (arrows). There does not appear to be any consistent relationship between the location of these long polysomes believed to synthesize the heavy subunits of myosin. Note the absence of any sarcoplasmic reticulum in the immediate vicinity of the myofibril, suggesting that myofibrillar assembly is not dependent upon prior elaboration of the SR. Calibration bar = 1 μm.

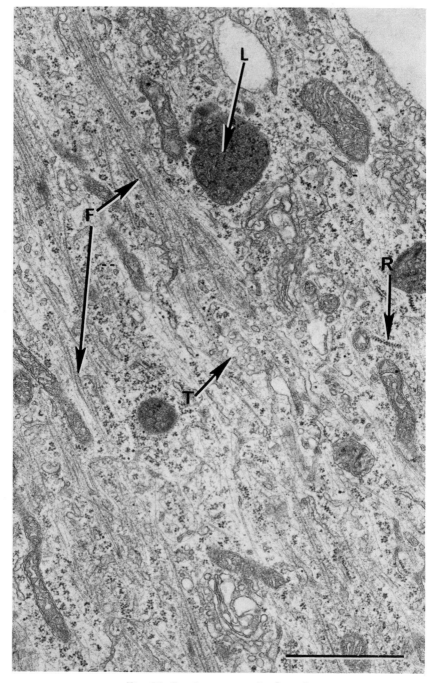

Fig. 21. See facing page for legend.

A question often raised is whether transcription of the myosin message (mRNA) and formation of the large polysomes for myosin peptide synthesis occurs within myoblasts prior to cytoplasmic fusion. Studies by Heywood (Herrmann *et al.*, 1970, see p. 227) indicate that the appearance of myosin-synthesizing polysomes parallels myotube formation in embryonic chick skeletal muscle. A more direct analysis of this problem will require the isolation of a pure population of myoblasts and both sucrose density gradient and electron microscopic study of the polysomes within these cells. As discussed above (Section V), Strohman and Paterson (1971) have shown that EGTA will inhibit the fusion of chick myoblasts in cell culture. Electron microscope analysis of such cultures (Maker and Fischman, 1972) has revealed the presence of large polyribosome chains in mononucleated cells that do not contain thick filaments (Fig. 12). If these large ribosome chains are the myosin-synthesizing polysomes (see below), the results might be interpreted as indicating that transcription of myosin mRNA can occur prior to cellular fusion. As stated above, Holtzer *et al.* (1957), Przybylski and Blumberg (1966), and Fambrough and Rash (1972) have all presented evidence that myofibrils can be detected in mononucleated cells. Thus, although the bulk of myofibrillar protein synthesis occurs in multinucleated cells, there is now fairly good evidence that both transcription and translation of the mRNAs involved in the synthesis of these proteins can take place before fusion.

C. Myofilaments and Myofibrils

Probably no aspect of muscle differentiation has elicited as much interest with biologists as the problem of myofibrillar biosynthesis and assembly. Unfortunately, most of the histological interpretations of embryonic material prior to 1955 were severely limited by our inadequate understanding of myofibrillar chemistry and structure both in adult and embryonic muscle. Within the past 15 years, there has been extraordinary

Fig. 21. Myotube after 3 days in monolayer culture exhibiting many early features of myofibrillogenesis. Some aggregates of thick and thin myofilaments (F) show no definite Z band density. Characteristically, myofilaments aggregate and align in parallel before Z bands are evident. Elaborate membranous networks (T) are apparent which Ishikawa (1968) has shown to be part of the T system by ferritin diffusion studies (see text). Note the long polyribosome chains (R) at a considerable distance from the thick filaments. Lysosomes (L) are encountered at times in these cells. Compare the mitochondria with those in Figs. 18 and 19. Calibration bar = 1 μm.

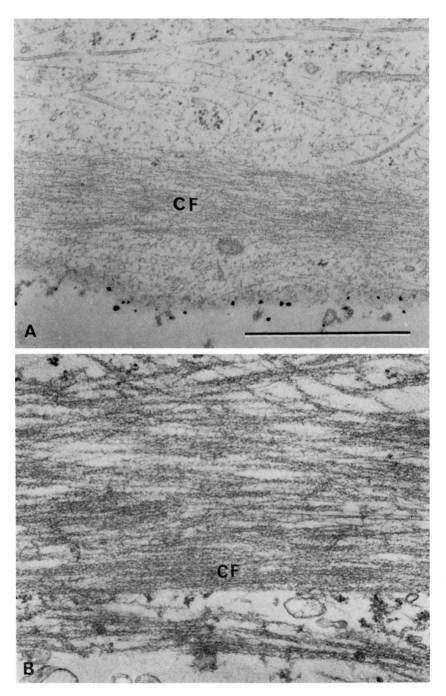

Fig. 22. See facing page for legend.

progress in the field of myofibrillar chemistry, largely based upon the integrated application of microscopic, crystallographic, and biochemical approaches (see Chapter 7 by H. E. Huxley in this volume, and recent reviews by Perry, 1967; Hanson, 1968; Ebashi and Endo, 1968; Young, 1969; Pellegrino and Franzini-Armstrong, 1969). Of all the organelles in eucaryotic cells, there is probably none better understood than the myofibril in terms of its physiology, composition, and ultrastructure. As will be apparent below, this information gained with adult material has had a profound influence on the interpretations of embryonic muscle. Nevertheless, our understanding of myofibrillogenesis remains fragmentary and largely incomplete.

1. CYTOFILAMENTS WITHIN MYOTUBES

Bundles or sheets of thin filaments approximately 5–6 nm in diameter of undetermined length are located beneath the plasma membrane, particularly near the ends of the myotube (Fig. 22). This class of filaments appears identical in structure to the cortical filaments discussed above (see Section V) and are found in nonmyogenic as well as myogenic cells. Cortical filaments appear to bind HMM (Fig. 11) (Ishikawa et al., 1969). As discussed above in Section V, it is uncertain how or if the cortical filaments relate to myofibrillogenesis. If the evidence gathered with unicellular systems (Pollard and Ito, 1970; Pollard et al., 1970; Nachmias et al., 1970; Pollard and Korn, 1971) can be safely extended to embryonic muscle, one must suspect the involvement of such thin filaments in both the ameboid movement and cytoplasmic streaming of myoblasts and myotubes. As suggested by Fischman (1967), such cytoplasmic streaming may be important in the axial alignment of myofilaments prior to myofibrillar assembly.

A second class of cytofilaments approximately 8–11 nm in diameter has been analyzed in some detail by Ishikawa (1968) and termed intermediate filaments. Rash et al. (1968) have described a similar class of filaments within developing cardiac muscle. There are several characteristics of these filaments that distinguish them from actin filaments. Many filaments are greater than 10 nm in diameter and longer than 2 μm

Fig. 22. Cortical filaments (CF) in myotubes cultured for 2 days in vitro. In (a) the cells have been fixed and sectioned in a routine manner. The cell in (b) has been processed by the method of Ishikawa et al. (1969) and then reacted with heavy meromyosin (HMM). The cortical filaments can be seen to decorate with HMM and form arrowheadlike complexes (Huxley, 1963). The relationship of these cortical filaments to myofibrillogenesis is unknown. Calibration bar = 1 μm.

in length. In contrast to the rather precise 5–6 nm diameter of actin filaments, the intermediate filaments exhibit a fairly wide size dispersion. These filaments do not bind HMM (Ishikawa *et al.*, 1969), they increase markedly in number after exposure of cells to Colcemid or colchicine (Fig. 23) (Ishikawa *et al.*, 1969), and they are found in many non-myogenic cells. It is doubtful whether this class of filaments participates directly in myofibrillar assembly (Ishikawa, 1968). Present evidence supports the interpretation that these filaments are more specifically related to the assembly or depolymerization of microtubules, but such an interpretation must await experimental verification, presumably by direct isolation and characterization of intermediate filaments. Evidence has been presented by Holtzer *et al.* (1971) that intermediate filaments

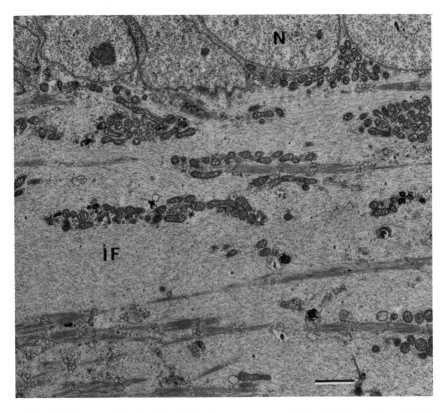

Fig. 23. A myotube *in vitro* following treatment with colchicine. Microtubules disappear, and many cytofilaments (~10 nm diameter) are observed (IF). The filaments appear identical to the intermediate filaments and do not bind HMM. Micrograph courtesy of H. Ishikawa. Calibration bar = 1 μm.

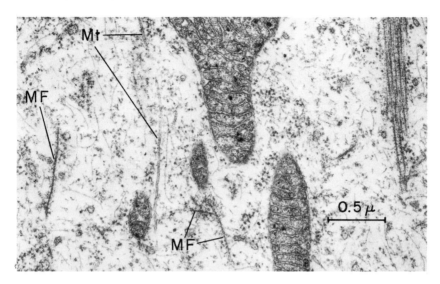

Fig. 24. Longitudinal section of a myotube within the leg muscle of a 12-day-old chick embryo. Filaments (MF) of approximately 16 nm diameter are observed frequently apart from any myofibrils. It is assumed these filaments are myosin filaments soon to be incorporated within a fibril. Microtubules (Mt), mitochondria, myosin filaments and myofibrils all are oriented in the long axis of the myotube. This orientation can be disrupted by sufficient concentrations of colchicine which induce rounding of the myotube (Ishikawa et al., 1968). Calibration bar = 0.5 μm.

are not depolymerized by cytochalasin B at concentrations that cause the disappearance of the microfilament network.

Thick (∼ 16 nm diameter) filaments, which appear identical in structure to the primary or myosin filaments of the myofibril, usually appear within embryonic vertebrate muscle after cytoplasmic fusion of the myogenic cells (Figs. 24 and 25). Such filaments have not been observed by electron microscopy in nonmyogenic embryonic cells. It is fairly certain that these thick filaments can be equated with myosin filaments (Fischman, 1967), although there has never been a detailed chemical or structural comparison made between such filaments isolated both from adult and embryonic material. The appearance of these filaments in cultured muscle occurs at the same stage of development when myotubes first react positively with fluorescein-conjugated antimyosin (Okazaki and Holtzer, 1966) and when myosin first can be detected biochemically (Coleman and Coleman, 1968; Strohman and Paterson, 1971). It is at this same period of myogenesis that long polyribosomes

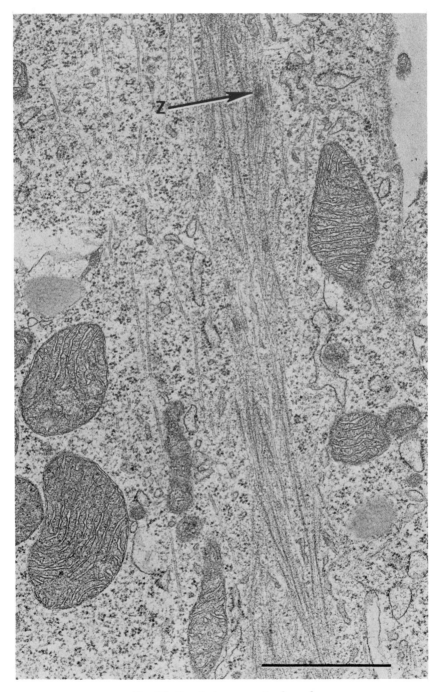

Fig. 25. See facing page for legend.

containing 50–60 monosomes are observed by electron microscopy (Allen and Pepe, 1965; Allen and Terrence, 1968). Since the studies of Heywood and co-workers (1967; Heywood and Nwagwu, 1968) have demonstrated that the heavy subunits of myosin (200,000 daltons) are synthesized on polysomes of this size, it seems reasonable to conclude that the detection of this class of large polyribosomes and thick filaments by electron microscopy reflects the biosynthesis of myosin and the posttranslational assembly of myosin filaments. The myosin filaments, once assembled, align in the long axis of the myotubes, principally beneath the sarcolemma. Their length does not exceed 1.5–1.6 μm, and they contain the same crossbridge and noncrossbridge-bearing zones characteristic of myofibrillar, A band filaments (Huxley, 1963; Fischman, 1967). It is not known if nascent thick filaments contain the M band cross bridges (Franzini-Armstrong and Porter, 1964a; Knappeis and Carlsen, 1968) or M band protein (Masaki et al., 1968; Kundrat and Pepe, 1971), nor has it been shown whether or not additional proteins participate in thick filament assembly in vivo. In 1963, H. E. Huxley demonstrated that the aggregates of myosin that formed when the protein was placed in 0.05–0.1 M potassium chloride at neutral pH possessed a bipolar, tapered, filamentous shape remarkably similar to native thick filaments. The length of these synthetic filaments can be adjusted predictably by varying the ionic strength and pH during the polymerization reaction (Kaminer and Bell, 1966; Harrington and Josephs, 1968). When formed in salt solutions that approximate the presumed intracellular conditions, synthetic myosin filaments have a mean length of about 1.5 μm, very close to the natural A band dimension. In contrast to the relative constancy of A band and thick filament length in vertebrate striated muscle, there is a considerable variability of these same structures in invertebrates (Franzini-Armstrong, 1970). It is fairly certain that an additional protein, paramyosin, is located within some invertebrate thick filaments and may be involved in their length determination (Ikemoto and Kawaguti, 1967). In developing insect flight muscle, the A band filaments increase in length, even after incorporation into a sarcomere (Shafiq, 1963; Auber, 1969a,b). Models of thick filament assembly and fine struc-

---

Fig. 25. A myotube within a 3-day-old muscle culture which exhibits an early stage of myofibrillar assembly. Thick and thin filaments have aggregated in parallel, but sarcomere structure has not yet been formed. Darkly staining dense bodies (Z) reminiscent of smooth muscle are seen. These dense bodies are the precursors of the Z bands. Note the absence of any polyribosome chains in contiguity with the thick filaments. Compare the mitochondria with those in Figs. 18 and 19. Calibration bar = 1 μm.

ture have been reported (Huxley, 1963; Pepe, 1967; Huxley and Brown, 1967) but await verification with embryonic material.

A fourth class of filaments 5–6 nm in diameter can usually be observed in close association with thick filaments in embryonic myotubes. These filaments bind HMM (Fig. 26) (Ishikawa *et al.*, 1969) and are not more than 1 μm in length (Fischman, 1967). It is likely these filaments are identical to the secondary or actin filaments of the myofibril (Obinata *et al.*, 1966). Both cortical filaments (see above) and actin filaments bind HMM (Fig. 26), but thick filaments are not observed in association with the cortical filaments in embryonic muscle (Fischman, 1972). Presumably, an actinlike protein is present in the cortical filaments; it now must be determined how these filaments differ chemically from true actin filaments, and their relationship to myofibrillogenesis must be explored.

Length determination of the actin filaments remains a puzzling phenomenon. Although myosin polymer length appears to be regulated by

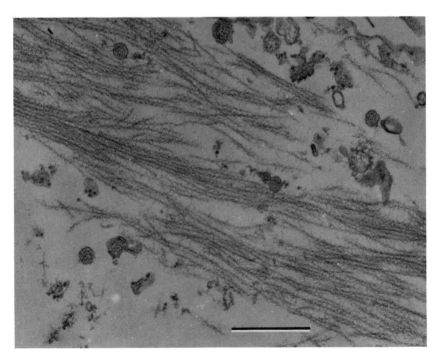

Fig. 26. Arrowhead formation along the thin filaments within an immature myofibril after reaction of the glycerinated myotube with HMM. Micrograph courtesy of H. Ishikawa. Calibration bar = 0.5 μm.

the ionic strength and pH of the bathing solution, polymerization of actin (G → F transformation) under identical conditions results in a polymer of indeterminate length (Hanson and Lowy, 1963). In other words, the myosin molecule appears to contain most, if not all, of the information necessary to establish the length of its polymer. Such information, if contained in the G-actin molecule, is not expressed under the usual *in vitro* polymerization conditions. There is now good evidence that proteins additional to actin are present in the I band filaments. These include, at a minimum, tropomyosin and troponin (Ebashi and Endo, 1968). It has been suggested that I band filaments form by the copolymerization of actin with these additional proteins (Hanson and Lowy, 1963; Huxley, 1963), and a 1 μm length is determined by a minimization of total free energy in this copolymer. So far, no one has successfully repolymerized thin filaments to an exact 1 μm length. It is interesting that in myofibrils of varying A band length, there is a proportional adjustment of thin filament length (Franzini-Armstrong, 1970). Apparently, within a sarcomere, there is a direct correspondence between the length of the cross-bridge bearing zone of a thick filament and the length of the thin filaments which interact with it. This might suggest that the eventual size of a thin filament is regulated, or at least influenced, by the thick filaments over which it slides. If actin polymerization occurred *in vivo* along an already formed thick filament, then thin filament length might be determined by the size of the myosin polymer. Myosin filaments have been demonstrated to influence F-actin length (Kikuchi *et al.*, 1969; Kawamura and Maruyama, 1969). When homogeneous populations of synthetic myosin filaments are prepared, the F-actin polymers formed in their presence are shorter and more uniform in length than F-actin prepared without myosin. It remains to be demonstrated how actin filament length is regulated in developing muscle.

In an interesting recent study, Hitchcock (1970) has shown that embryonic actomyosin from day 14 and earlier has a reduced Ca^{2+} sensitivity and calcium binding ability when compared with native actomyosin isolated from later stage embryos or adult muscle. Since the calcium sensitivity of this Mg^{2+}-activated actomyosin ATPase can be restored by addition of tropomyosin–troponin to the actomyosin, the author has concluded that the relaxing system is deficient in early embryonic muscle and becomes an integral component of the actomyosin system after 14 days of incubation. It would be of interest now to isolate native thin filaments from 10–12 day muscle and compare these with similar filaments isolated from 18–20 day embryos. Can a 40 nm tropomyosin–troponin repeat be demonstrated along both sets of filaments (Ebashi and

Endo, 1968)? Are tropomyosin and troponin actually synthesized late in embryonic development, or are the two regulatory proteins present but in an inactive form?

2. MYOFILAMENT ALIGNMENT

Within myotubes, free myofilaments exhibit an axial alignment parallel to the long axis of the myotube (Fischman, 1967). This observation is true for all filaments and asymmetric structures within the sarcoplasm. A similar orientation is evident for microtubules, polyribosomes, mitochondria, and nuclei. Although axial asymmetry is one of the most characteristic features of muscle and is evident early in myogenesis, we have little direct information bearing upon it. Myoblasts orient in parallel with neighboring myoblasts (Figs. 1–4) both *in vitro* and *in vivo*. Organellar alignment similar to that in myotubes is found in myoblasts even before fusion.

The orientation and shape of myogenic cells has been shown to be affected by both intra and extracellular factors. First, the orientation of the fibrous substrate (collagen) upon which myoblasts and myotubes develop has a direct bearing on the directionality of muscle growth. Myogenic cells move and elongate in parallel with the underlying collagen fibers (Fischman, 1972) in a manner similar to the examples of substrate-mediated cellular orientation shown so beautifully by Garber (1953) and Weiss (1958, 1961). The orientation of myoblasts along a preformed extracellular matrix must have an important bearing upon the subsequent orientation of the myotubes which they form. Unfortunately, there has been very little study of the extracellular matrix in embryonic muscle (Low, 1967), and its importance to muscle development requires elucidation.

Myoblast shape is profoundly yet reversibly altered by cytochalasin B (Sanger *et al.,* 1971; Maker and Fischman, 1972). Bipolar cells in muscle cultures retract considerably when exposed to cytochalasin B at a concentration of 2×10^{-6} μg/ml but retain slender residual cell processes which contain many intermediate (∼10 nm) filaments and microtubules. After removal of the drug, the cells reassume a bipolar, spindle-shaped configuration. Based upon other studies with this drug (see review by Wessells *et al.,* 1971), it is likely that modification of cell shape in the myoblasts is a reflection of the breakdown of thin (∼5–6 nm diameter) cytofilaments within the spindle-shaped extremities of the cell. It is not known if this drug affects the orientation of free myofilaments or other organelles within myotubes. Cross-striated myofibrils can develop within myotubes grown in the presence of cytochalasin B (Sanger and Holtzer, 1970; Sanger *et al.*, 1971) at concentrations that inhibit the ameboid movement of both myotubes and myoblasts.

Addition of colchicine or Colcemid to myogenic cultures results in the rounding up and pinching off of myotube segments, forming structures termed "myosacs" by Bischoff and Holtzer (1968). Within these myosacs, myofibrils are displaced and often misaligned (Ishikawa *et al.*, 1968). Axial asymmetry of the myogenic cells is profoundly altered, and it is likely this effect is a reflection of the disruption of microtubules by these alkaloids (Borisy and Taylor, 1967). Similar retraction of long cellular processes after exposure to colchicine have been noted with the protozoan *Actinosphaerium* (Tilney, 1965), cultured neurons (Bunge and Bunge, 1968), and microvilli (Tilney and Cardell, 1970), to name just three examples. Presumably, a depolymerization of microtubules is common to each of these experimental systems. Although colchicine has a profound effect on muscle cultures, Warren and Porter (1969) have shown that myotubes within the ventral abdominal musculature of the larval insect *Rhodnius prolixis* are not altered in shape after exposure to this drug at concentrations that depolymerize microtubules. These authors have suggested that the insertions of the myotubes into the cuticle by strong myoepithelial attachments maintain cellular shape after microtubular depolymerization. Of particular importance in this study was the demonstration that colchicine did not impair filament alignment or myofibrillogenesis. Apparently, microtubules are not essential for these processes, at least in developing insect musculature. There have been no studies reported on the effects of colchicine on myogenesis after administration to vertebrate embryos *in vivo*.

In summary, there is insufficient evidence to explain how myofilaments or other cellular organelles are aligned in the long axis of myotubes. At least three structures affect cellular asymmetry: the extracellular collagen network, cytoplasmic filaments sensitive to cytochalasin B, and microtubules, but the interrelationship or relative importance of each is uncertain at this time. The alignment of myofilaments appears secondary to the alignment of myotubes and not the reverse.

3. Myofibrillogenesis

Almost as soon as thick ($\sim$15 nm diameter) filaments can be identified within myotubes, clusters of both thick and thin ($\sim$6 nm diameter) filaments are observed within the cortical regions of the cell (Fig. 27). These aggregates of filaments increase both in girth and length and soon exhibit the sarcomeric band pattern of the myofibril (Holtzer *et al.*,

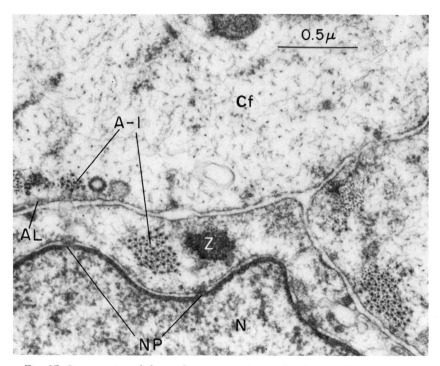

Fig. 27. Cross section of three adjacent myotubes within leg muscle of the 12-day-old chick embryo (Fischman, 1967). Myofibrils, even at early stages of assembly, exhibit an hexagonal lattice of myofilaments. Some fibrils have been cut through the region of A–I overlap, one is transected in the Z band region. A narrow amorphous layer (AL) is often seen between the myofibrils and the adjacent plasma membrane. Many cytofilaments (CF) are present in the core of the myotube, predominantly oriented in the long axis of the cell. NP = nuclear pore; N = nucleus. Calibration bar = 0.5 μm.

1957). Neither the transverse nor longitudinal forces underlying this construction of the myofibril are well understood; there have been no successful reports of myofibrillar assembly in cell-free systems. Although electron microscopic studies of myofibrillogenesis have been performed on insect (Shafiq, 1963; Beinbrech, 1968; Toselli and Pepe, 1968; Auber, 1969a), fish (Waterman, 1969), amphibian (Hay, 1963; D. E. Kelly, 1969; Lentz, 1969a), chick (Allen and Pepe, 1965; Przybylski and Blumberg, 1966; Fischman, 1967, 1970; Spiro and Hagopian, 1967), and mammalian (Bergman, 1962; Heuson-Stiennon, 1964, 1965; A. M. Kelly and Zacks, 1969a) muscle, there is still considerable disagreement about the sequence and mechanics of this process.

Within a mature sarcomere, the myofilaments are positioned in a precise double hexagonal lattice (see Chapter 7 by H. E. Huxley). Three distinct sets of transverse interfilament linkages exist within a sarcomere which might underly construction and/or stabilization of this hexagonal lattice. These are: (1) the Z band material linking adjacent thin filaments; (2) the M band cross bridges linking thick filaments in the middle of the sarcomere (Knappeis and Carlsen, 1968; Kundrat and Pepe, 1971); and (3) the myosin cross bridges linking thick and thin filaments in the zone of A-I overlap. Most of the evidence at hand supports the hypothesis that the third set of cross bridges, namely, the myosin to actin connections, determines the hexagonal lattice.

Myofibrils in some of the slow muscles of amphibia lack M bands, yet exhibit an hexagonal dual filament lattice (Page, 1965). Similar observations have been made by Fischman (1967) with embryonic chick muscle. When both M and Z bands are chemically extracted from myofibrils, the hexagonal myofilament lattice is unaffected (Stromer *et al.,* 1969). In embryonic chick (Fischman, 1967) and regenerating amphibian (Lentz, 1969a) muscle, newly formed myofibrils have been described which exhibit an hexagonal packing of myofilaments but lack, or contain at best, a diffuse Z band. At a stage in myofibrillogenesis when the myofilaments are positioned in a well formed hexagonal lattice, the Z band is a poorly developed structure which lacks the square or rhombic configuration characteristic of adult myofibrils (Knappeis and Carlsen, 1962; Reedy, 1964). Additional evidence that Z bands are not necessary for construction of the hexagonal lattice of myofilaments has recently been obtained by Fischman (1970), and Fischman and Zak (1971) with developing cardiac muscle. When embryonic chick hearts (6–8 days of incubation) are dissociated into a single-cell suspension by incubation in trypsin (Moscona, 1961), most of the preexisting myofibrils are disrupted intracellularly into a pool of randomly aligned thick and thin myofilaments (Fig. 28). Z bands are almost totally disrupted during this process. If such cells are incubated for 3 hours (see legend of Fig. 29) in the presence of 250 μg/ml of cycloheximide (a concentration which completely inhibits protein synthesis in this tissue) and then fixed for electron microscopy, the myofilaments are seen to have aligned in parallel bundles which exhibit hexagonal symmetry in cross section. Most of these semicrystalline arrays of myofilaments lack Z bands and are one A band in length. This evidence, obtained with heart muscle, strongly suggests that packing of the single filaments into the hexagonal lattice is not dependent upon Z band formation and can occur in the absence of protein synthesis. Growth of the myofibril in length, however, does require

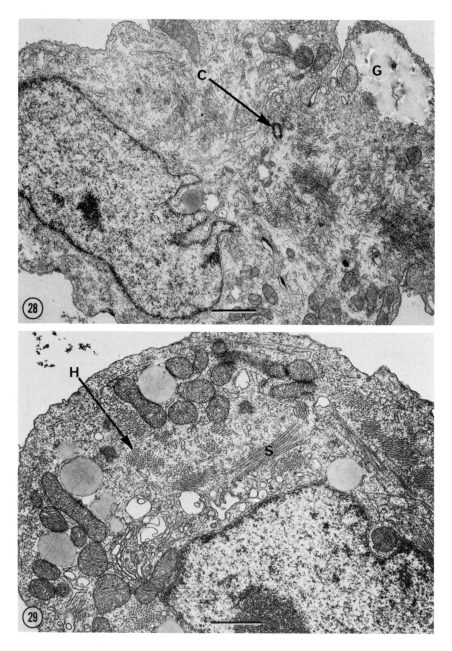

See facing page for legends.

Z band material. It seems reasonable to conclude that the hexagonal symmetry of the dual filament lattice depends upon the transverse fibrillar structures linking thick and thin filaments, i.e., the heavy meromyosin cross bridges. The elegant X-ray diffraction studies of Huxley and Brown (1967) and electron microscopy of Reedy (1968) have demonstrated that cross bridges project from the thick filaments with sixfold symmetry. It appears likely that when myosin molecules aggregate to form a thick filament, the packing of each molecule within the core of the filament automatically establishes the 6/2 axial surface lattice of cross bridges projecting from that filament. In other words, the primary amino acid sequences of the myosin subunits that determine the tertiary structure of the whole molecule, and which in turn determines the polymeric structure of myosin aggregates, must contain the requisite information to establish the sixfold axial helix of the thick filament. Since these cross bridges contain the actin-combining sites of the myosin molecules, thin (actin) filaments with which they interact will tend to be positioned in a hexagonal array determined by the sixfold symmetry of the thick filaments. In this manner, radial growth of the double filament lattice would continue until the sum total of forces tending toward fibrillar splitting counterbalance the attractive forces between myofilaments. Evidence that myofibrillar enlargement occurs by the apposition of newly synthesized proteins at the periphery of fibrils has recently been obtained by Morkin (1970). Using light and electron microscopic autoradiography, he was able to demonstrate that myofibrillar proteins are synthesized during the *in vitro* incubation of diaphragmatic muscle, and such proteins are added at the peripheral, rather than the central zones, of preexisting myofibrils. If myofilaments are added appositionally to myofibrils, there remains the obvious problem of how a relative constancy of myofibrillar diameter is maintained.

Fig. 28. A myocardial cell in suspension after trypsin-induced dissociation of the ventricles of 7-day-old chick embryo hearts (Fischman and Zak, 1971). More than 90% of the single cells exhibit almost total disarray of the myofibrils. The double hexagonal array is lost, and most myofilaments are randomly positioned. Z band material is not seen in most cells. G = glycogen pool partially extracted by *en bloc* aqueous uranyl acetate staining; C = centriole. Calibration bar = 1 μm.

Fig. 29. A myocardial cell similar to that in Fig. 28 but incubated in rotation culture (Moscona, 1961) for 3 hours in the presence of 250 μg/ml of cycloheximide. In the absence of protein synthesis, the previously random myofilaments have been repositioned into the characteristic hexagonal lattice of the myofibril (H). In most cells, Z bands are not evident, and fibrillar reassembly is limited to one sarcomere segments (S). These results provide strong support for the hypothesis that hexagonal lattice formation is not dependent upon prior Z band assembly. Calibration bar = 1 μm.

The studies of Goldspink (1970a,b) have provided considerable insight into this problem. Goldspink demonstrated that some myofibrils exhibit a longitudinal split down the mid section of the cylinder. Such splitting is observed most frequently in myofibrils, which, on the average, are twice as large as nonsplit myofibrils. The measurements indicate that myofibrils divide approximately down the middle once they attain a critical size. Thus, Goldspink suggests that proliferation of myofibrils during development and muscle hypertrophy occurs by the longitudinal cleavage of myofilament lattices that have exceeded a stable diameter. These considerations of myofibrillar splitting and enlargement are of some importance, for they highlight an aspect of muscle structure and metabolism of which we know very little, i.e., the molding, turnover, and replacement of myofibrils in mature muscle. It now appears likely that myofibrils, similar to the osteons of bone, are in a continual state of growth and remodeling in response to metabolic and physical demands placed upon the muscle fibers.

In addition to their radial growth, myofibrils have been demonstrated to increase in length. This is particularly striking in embryonic and postnatal development. In vertebrate muscle, this increase in myofibrillar length is not accompanied by any parallel increase in myofilament length (Fischman, 1970; Goldspink, 1970b), but must be accounted for by the addition of sarcomeres to the myofibrils. Since half or partially completed sarcomeres are not oberved along the shaft of a myofibril, except in pathological conditions, it is reasonable to conclude that sarcomeres are added at the ends of myofibrils. Although Z bands are unlikely to be formative structures in the appositional growth of the myofilament lattice, they are probably essential for longitudinal myofibrillar growth by unit sarcomere addition. In the cardiac myofibrillar reconstruction studies referred to above (Fischman and Zak, 1971) (Fig. 29), radial myofibrillar growth has been separated from longitudinal growth by preventing Z band biosynthesis during myofilament aggregation. Under these conditions, double hexagonal lattices of thick and thin filaments are formed in the absence of Z band material, which is apparently degraded during trypsin-induced dissociation of the heart tissue, but there is little or no longitudinal myofibrillar growth (Fig. 29). For the back-to-back attachment of thin filaments in adjacent sarcomeres, proteins in addition to actin and myosin (perhaps α-actinin and other unidentified proteins—Stromer *et al.*, 1969) are required. There is relatively little understanding of this phase of fibrillar growth. It is puzzling that extraction of Z bands from adult myofibrils (Stromer *et al.*, 1969) does not cause myofibrillar breakdown. Exactly how myofibrils are held together after Z band extraction requires explanation.

D. The Sarcotubular System

Striated muscle contains an elaborate smooth-surfaced membrane system (Porter and Palade, 1957; Andersson-Cedergren, 1959) which can be subdivided into two components: (1) the transverse tubular system (T system) and (2) the sarcoplasmic reticulum (SR). The structure and function of this system is reviewed by Franzini-Armstrong in Chapter 9 of Volume II (also see recent reviews by Pellegrino and Franzini-Armstrong, 1969; Peachey, 1970). There is now little doubt that the lumen of the T system is in direct continuity with the extracellular space (Franzini-Armstrong and Porter, 1964b; Huxley, 1964; Endo, 1964; Eisenberg and Eisenberg, 1968; Peachey and Schild, 1968). The studies of Ezerman and Ishikawa (1967) and Ishikawa (1968) have demonstrated that the T system in cultured chick muscle develops by an invagination of the sarcolemma of the myotube. By including ferritin in the culture medium as a marker of the extracellular space, these authors were able to show that the T system is derived from caveolar inpocketings of the plasma membrane (Fig. 30). In contrast, the SR is derived from the granular endoplasmic reticulum, which, in turn, probably develops from the outer membrane of the nuclear envelope (Ezerman and Ishikawa, 1967). The cavities of the SR are not confluent with the extracellular space as shown by the failure to fill the SR with markers of this space. These studies by Ishikawa and colleagues suggest that the SR and T systems develop independently and are subsequently apposed to form the characteristic triads around the myofibrils.

The relationship of myofibrillogenesis to the development of the SR and T systems is of some interest. Although cytoplasmic membranes, both rough and smooth-surfaced, are present in early myotubes, these membranes are not extensive, nor is there any indication of the membranous collar which subsequently surrounds the myofibrils. There is no evidence that arrangement of the myofilaments into the hexagonal lattice of the myofibril is determined by a membranous scaffold. On the contrary, newly formed myofibrils are devoid of a well organized SR. It appears likely that the SR is elaborated around the myofibril and not the reverse (Fischman, 1967, 1970).

The morphogenesis of the SR and triads in mammalian muscle has been examined in detail by Walker and Schrodt (1968), Schiaffino and Margreth (1969), and Edge (1970). At early stages, the SR forms a lacelike network of tubules which encircles the myofibril and does not exhibit the sarcomeric repetition so characteristic of mature myofibers. Initially, there are no signs of terminal cisternae, fenestrated collars,

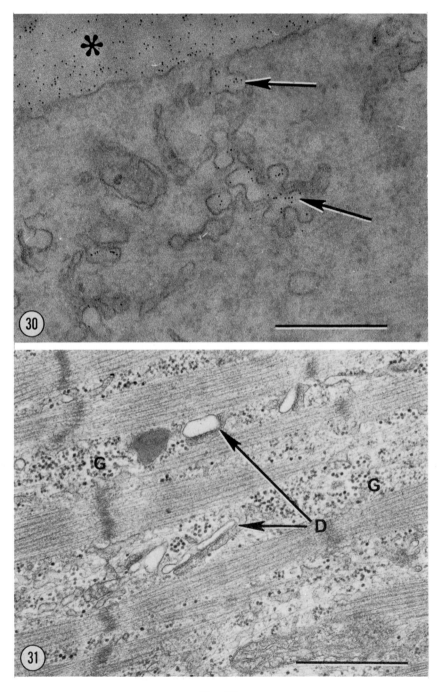

Fig. 30.

or other differentiated subdivisions of the SR. Upon ingrowth of the T system, there occur appositional contacts between SR and T tubules, with the subsequent formation of dyad and triad structures at irregular intervals along the myofibrils. The T system in the embryonic and early postnatal period usually runs in a longitudinal or oblique direction within the muscle cells. As a result, triads in muscle of this stage exhibit a longitudinal rather than transverse orientation (Fig. 31). In mice, transverse alignment of the T system may not occur until 10 days to 2 weeks after birth. Longitudinal orientation of the triads has been observed in adult muscle but is usually associated with muscle pathology (see Pellegrino and Franzini-Armstrong, 1969). The physiological manifestations of this aberrant alignment of the triads in embryonic and early postnatal muscle have not been studied in detail. Recent investigations by Chaplin *et al.* (1972) have shown with newborn rat extensor digitorum longus and soleus muscles that the time between the beginning of the action potential and the recorded development of tension, the excitation–contraction latency, declines rapidly from 6 msec to 2.8 msec during the first 10–15 days after birth. In view of the fact that triads change in orientation during this same period, and these structures play an important role in excitation–contraction coupling (Sandow, 1965), the shortened excitation–contraction latency may reflect this modification in alignment of the sarcotubular system.

Expansion or ballooning of the SR at the triads to form the terminal cisternal occurs after contacts between SR and T tubules have formed (Edge, 1970). Swollen or enlarged SR tubules are not observed in developing muscle except at contact points between the SR and the T tubules or between SR and the sarcolemma. Edge (1970) has suggested that apposition of the T and SR tubules sets in train the series of morphogenetic changes which result in development of the mature SR.

Fig. 30. The transverse tubular system forms by inward extensions of caveolar inpocketings of the plasma membrane (Ezerman and Ishikawa, 1967). By incubating cultured muscle in the presence of ferritin, the electron-dense protein granules serve as a marker of the extracellular space (asterisk) or of channels confluent with this space. Arrows indicate ferritin granules within caveolae. Micrograph provided by H. Ishikawa. Calibration bar = 0.5 μm.

Fig. 31. A myotube within a 12-day-old muscle culture. The T system has formed dyads (D) with the sarcoplasmic reticulum. Note the longitudinal or oblique orientation of the T system. Transverse orientation of the T system with a conversion of almost all dyads to triads is a later differentiative transition of the sarcotubular system and generally occurs after birth. G = glycogen granules. Calibration bar = 1 μm.

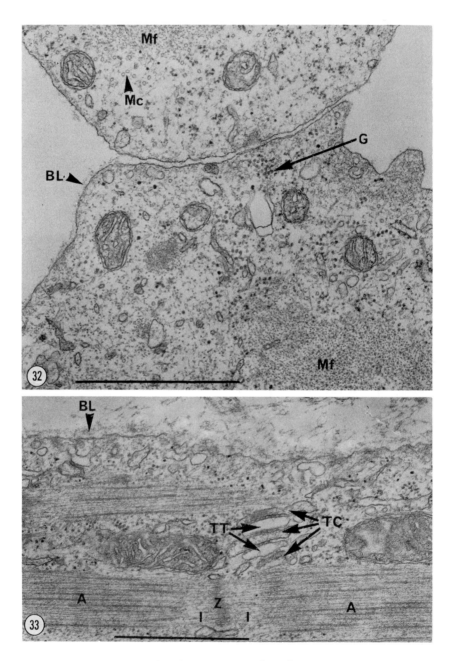

See facing page for legends.

E. The Basal Lamina

Surrounding the plasma membrane of the mature myofiber is a well developed glycocalyx (Bennett, 1963), termed the basal lamina or basement membrane. Collagen fibrils of the endomysium merge imperceptibly with this cell coat of mucopolysaccharide, and it must play an important role in the transmission of tension between muscle and tendon during contraction and relaxation of the myofibers. If the structure, composition, and function of this layer are poorly understood in adult muscle, its morphogenesis in embryonic tissue is even less clear. Low (1967) has examined the surface coat of myogenic cells in chick embryos with the aid of the electron microscope. A fine fibrillar layer was demonstrated on the surface of myotubes, but it was impossible to relate the synthesis of this surface fuzz to the fusion of myoblasts, since myogenesis is so asynchronous *in vivo*. Fischman (1970), in his fine structural studies of myogenesis *in vitro*, noted that signs of the basal lamina are first observed at the fifth and sixth days of cell cultures (Figs. 32 and 33). Myoblasts are devoid of a fibrillar surface coat when examined by electron microscopy after routine glutaraldehyde–osmium tetroxide fixation and lead citrate staining. A tenuous fibrillar network is seen outside the plasma membrane of myotubes after approximately 120 hr *in vitro*. This surface coat is found along the shaft of the myotube but is absent at either end of the cell, where ameboid movement and cell fusion are still observed. Since the decline in cell fusion occurs during this same period in culture when the basal lamina first appears, Fischman (1970) has suggested the two events might be causally related. With the sensitive and reliable methods now available for the cytochemical demonstration of acid and neutral mucopolysaccharides at the cell surface (Luft, 1966; Revel and Karnofsky, 1967; Rambourg and Leblond, 1967; Bernhard and Avrameas, 1971), it would be enlightening to reexamine the development of the basal lamina during myogenesis *in vitro*.

Fig. 32. Transverse section of two adjacent myotubes in a 6-day-old culture of embryonic muscle. A thin fuzzy coat (BL) appears on myotube surfaces approximately 5–6 days *in vitro*. This surface coat represents the first indication of basal lamina formation. Numerous microtubules (Mc) are evident, but show no consistent relationship to the myofilaments (Mf). G = glycogen granules. Calibration bar = 1 μm.

Fig. 33. A myotube in a 10-day-old culture. Note the increased thickness of the basal lamina (BL) over that observed in Fig. 32. A pentad composed of three sacks of terminal cisternae (TC) and two T tubules (TT) is often seen in these cultured myotubes. Z = Z band; I = I band; A = A band. Calibration bar = 1 μm.

VIII. Innervation of Developing Muscle

As discussed above (see Section III), there is conclusive evidence
that myofibers can differentiate in the total absence of innervation.
Nevertheless, the maintenance of the differentiated state and the continuing functional growth of muscle in the postnatal period requires such
innervation. Although initial myofibrillar and sarcotubular differentiation
is independent of nerves, denervation of muscle immediately after birth
can cause marked muscle atrophy. The degree of atrophy is related
to the predominant fiber type within the muscle (Engel and Karpati,
1968). Recently, Schiaffino and Settembrini (1970) have shown that
postnatal denervation of rat muscle results in a differential atrophy of
various organelles within the myofibers. At a stage when myofibrillar
development is severely retarded or blocked by denervation at birth,
there occurs a normal if not hypertrophic development of the sarcotubular system. These morphological observations support the biochemical
studies of Shapira *et al.* (1950) in which it was shown that rabbit muscles denervated during development exhibit an impaired growth of contractile proteins, but an increased content of sarcoplasmic proteins.
Schiaffino and Settembrini (1970) suggest that various muscle cell components are differentially sensitive to neuronal influence which becomes
apparent upon denervation. They found no evidence to support the
concept that denervation of muscle induces a dedifferentiation of the
myofibers (Eccles, 1963; Muscatello *et al.*, 1965).

Motor end-plate development *in vivo* has now been examined by electron microscopy in amphibians (Lentz, 1969b, 1970), birds (Hirano,
1967), and mammals (Teravainen, 1968; A. M. Kelley and Zacks, 1969b).
In addition, a number of *in vitro* models have been established by which
muscle innervation can be followed in tissue culture. Using mouse and
rat material, Bornstein *et al.* (1968) and Petterson and Crain (1970)
have shown that slices of spinal cord combined with isolated explants
of muscle in Maximow preparations will form neuromuscular junctions
after 3–4 weeks *in vitro*. Verification of these synaptic contacts has been
established by electrophysiology (Crain, 1970) and electron microscopy
(Pappas *et al.*, 1971). With chick material, also grown in organ culture,
James and Tresman (1969a,b) have demonstrated morphological contacts between nerve and muscle which suggest synaptogenesis. Finally,
cell cultures produced by the recombination of trypsin-dissociated chick
spinal cord and leg muscle will develop neuromuscular junctions under
appropriate conditions (Shimada *et al.*, 1969a,b). These preparations
exhibit electrical activity characteristic of this synapse, including excita-

tory junctional potentials (Fischbach, 1970), sensitivity to *d*-tubo-curarine (Robbins and Yonezawa, 1971), and the quantal, focal release of transmitter (ibid). Robbins and Yonezawa have concluded that "non-contractile myotubes with as few as three nuclei can participate in chemical synaptic transmission."

The morphological analysis of myoneural synaptogenesis has revealed the following facts. At the initial contact points between nerve and muscle, the apposing membranes lack junctional specializations; i.e., there is no evidence for pre- and postsynaptic membrane thickenings or postsynaptic infoldings antedating intercellular contact. After contact has been established, the staining density of pre- and postsynaptic membranes increases, and synaptic vesicles are observed in the nerve termi-

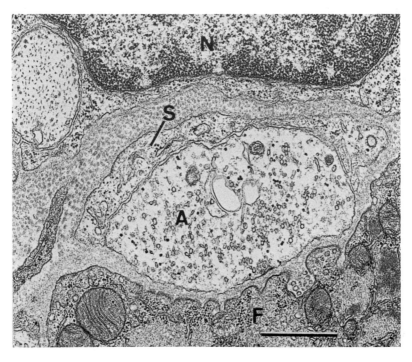

Fig. 34. A neuromuscular junction in a 33-day-old rat organ culture containing spinal cord slices and teased muscle fibers, grown in a Maximow assembly (see Pappas *et al.*, 1971). The axon (A) contains many presynaptic vesicales. Small indentations are seen in sarcolemma of the muscle fiber (F) at the postsynaptic site, and presumably represent an early stage in the development of synaptic clefts. A Schwann cell process (S) surrounds the axon except at the junctional contact area. N = nucleus of an adjacent Schwann cell ensheathing a small neurite. Micrograph provided by Dr. G. Pappas, Albert Einstein College of Medicine, New York. Calibration bar = 1 μm.

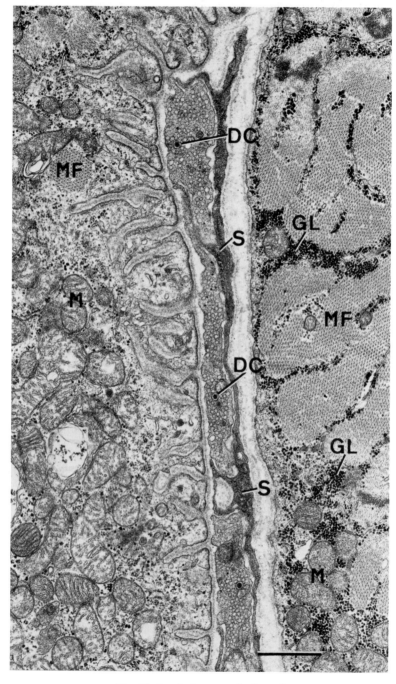

Fig. 35. See facing page for legend.

nal. At the synaptic region, the nerve is unmyelinated, is only partially ensheathed by Schwann cell, and is separated from the muscle fiber by a definable basement membrane (Fig. 34). The muscle fiber at this site is not surrounded by a satellite cell. Thus, the initial synaptic complex contains a bulbous nerve ending applied to a smooth surfaced muscle fiber. It has a similarity to the *en grappe* type ending when viewed in the light microscope.

Postsynaptic infoldings of the muscle fiber with a flattening out of the nerve terminal occur during the subsequent differentiation of the synapse (Fig. 35). This latter change results in the *en plaque* configuration seen by light microscopy. There have been no correlated studies by electron microscopy and electrophysiology in which these morphological transitions at the synapse have been given a functional interpretation. Furthermore, we have relatively little understanding of how or if the development of the synapse relates to the terminal differentiation of the sarcotubular system. Finally, there remains no anatomical explanation for the trophic interaction between nerve and muscle.

IX. Addendum

Several articles pertinent to this review have appeared after submission of the manuscript to the publishers. Konigsberg (1971) has demonstrated that myogenic cells secrete a factor of greater than 300,000 daltons, which promotes the fusion of myoblasts in cell cultures. The identification and characterization of this factor should be of great importance. Using an improved culture system in which the cell fusion of myoblasts can be synchronized, O'Neill and Stockdale (1972) have shown that after a thymidine-^{3}H pulse labeled nuclei appear within myotubes as early as 3 hours after the pulse. The results strongly suggest that the G_1 period immediately preceding fusion is not prolonged and the 5–8 period observed for the interval by Bischoff and Holtzer (1969) may reflect asynchrony of development rather than an alteration in the mitotic cycle. Finally, Hosick and Strohman (1971) have examined the ribosome and

---

Fig. 35. A portion of two muscle fibers in cross section from an 87-day-old muscle-cord organ culture. A section of a motor end-plate transversing the entire micrograph can be seen. The axonal terminals are filled with presynaptic vesicles, some of which are of the dense core variety (DC). The postsynaptic or subjunctional infoldings of the sarcolemma characteristic of neuromuscular junctions are found along the entire region of synaptic contact. The axon terminals are ensheathed with Schwann cell processes (S), except at the synaptic contact surface. M = mitochondria; GL = glycogen particles; MF = myofilaments. Micrograph provided by Dr. G. Pappas. Calibration bar = 1 μm.

142 Donald A. Fischman

polyribosome populations in embryonic chick muscle after trypsin dissociation of the tissue and after periods of monolayer culture. There is an
extensive degradation of polyribosomes during trypsinization suggesting
possible entry of proteolytic enzyme into the dissociated myoblasts. A
normal polysome to monosome ratio is restored within 5 hours *in vitro*.

ACKNOWLEDGMENTS

The author expresses his sincere appreciation to Drs. A. Moscona, H. Swift, and
H. Holtzer for reading this manuscript and suggesting helpful additions and correc-
tions. Additional appreciation is extended to Miss Rita Yambot and Mrs. Dorothy
Brown for technical assistance with this work. Dr. Y. Shimada, Dr. H. Ishikawa, Dr.
E. Hay, Dr. M. Vye, and Dr. G. Pappas have kindly provided plates for inclusion in
this chapter which have been cited in the respective figure legends. Parts of this study
have been generously supported by research grants from the National Science
Foundation (No. GB-7591), National Institutes of Health (IROI HE13505-CI),
Chicago and Illinois Heart Association (No. N69-32), and the Harry Levine Memorial
Foundation.

REFERENCES

Allen, E. R., and Pepe, F. (1965). *Amer. J. Anat.* 116, 115.
Allen, E. R., and Terrence, C. F. (1968). *Proc. Nat. Acad. Sci. U.S.* 60, 1209–1215.
Andersson-Cedergren, C. (1959). *J. Ultrastruct. Res.* 1, Suppl. 1.
Auber, J. (1969a). *J. Microsc. (Paris)* 8, 197–232.
Auber, J. (1969b). *J. Microsc. (Paris)* 8, 377–390.
Avery, G., Chow, M., and Holtzer, H. (1956). *J. Exp. Zool.* 132, 409.
Baker, P. C., and Schroeder, T. E. (1968). *Develop. Biol.* 15, 432.
Bárányi, M., Bárányi, K., Reckard, T., and Volpe, A. (1965). *Arch. Biochem. Biophys.*
    109, 185.
Basleer, R. (1962). *Z. Anat. Entwicklungsgesch.* 123, 184–205.
Beinbrech, G. (1968). *Z. Zellforsch. Mikrosk. Anat.* 90, 463–494.
Bennett, H. S. (1963). *J. Histochem. Cytochem.* 11, 14.
Bergman, R. A. (1962). *Bull. Johns Hopkins Hosp.* 110, 187.
Bernhard, W., and Avrameas, F. (1971). *Exp. Cell Res.* 64, 232–236.
Betz, E. H., Firket, H., and Reznik, M. (1966). *Int. Rev. Cytol.* 19, 203–227.
Bintliff, S., and Walker, B. E. (1960). *Amer. J. Anat.* 106, 233–245.
Bischoff, R., and Holtzer, H. (1968). *J. Cell Biol.* 36, 111.
Bischoff, R., and Holtzer, H. (1969). *J. Cell Biol.* 41, 188–203.
Bischoff, R., and Holtzer, H. (1970). *J. Cell Biol.* 44, 134.
Borisy, G. G., and Taylor, E. W. (1967). *J. Cell Biol.* 34, 525.
Bornstein, M. B., Iwanami, H., Lehrer, G. M., and Breitbart, L. (1968). *Z. Zellforsch.*
    Mikrosk. Anat. 92, 197–206.
Boyd, J. D. (1960). In "The Structure and Function of Muscle" (G. H. Bourne,
    ed.), 1st ed., Vol. 1, pp. 63–85. Academic Press, New York.
Buller, A. J., Eccles, J. C., and Eccles, R. M. (1960). *J. Physiol. (London)* 150, 417.
Bunge, R., and Bunge, M. (1968). *Anat. Rec.* 160, 323.
Capers, C. R. (1960). *J. Biophys. Biochem. Cytol.* 7, 559–566.
Carter, S. B. (1967). *Nature (London)* 213, 261.
Cedergren, B., and Harary, I. (1964). *J. Ultrastruct. Res.* 11, 428–442.
Chaplin, E. R., Nell, G. W., and Walker, S. M. (1972). *Exp. Neurol.* (in press).

Coleman, J. R., and Coleman, A. W. (1968). *J. Cell. Comp. Physiol.* **72**, Suppl. 1, 19.
Coleman, J. R., Coleman, A. W., and Roy, H. (1966). *Amer. Zool.* **6**, 234.
Coleman, J. R., Coleman, A. W., Kankel, D., and Werner, I. (1970). *Exp. Cell Res.* **59**, 319.
Cooper, W. G., and Konigsberg, I. R. (1961). *Anat. Rec.* **140**, 195.
Cox, P. G. (1968). *J. Morphol.* **126**, 1–18.
Cox, P. G., and Simpson, S. B., Jr. (1970). *Develop. Biol.* **23**, 433–443.
Crain, S. M. (1970). *J. Exp. Zool.* **173**, 353–369.
Dalcq, A., and Pasteels, J. (1954). *In* "Traité de Zoologie" (P.-P. Grassé, ed.), Vol. 12, p. 35. Masson, Paris.
Detwiler, S. R. (1934). *J. Exp. Zool.* **67**, 395.
Detwiler, S. R. (1955). *J. Exp. Zool.* **129**, 45–76.
Djordjevic, B., and Szybalski, W. (1960). *J. Exp. Med.* **112**, 509.
Dreizen, P., Hartshorne, D. J., and Stracher, A. (1966). *J. Biol. Chem.* **241**, 443.
Dreizen, P., Gershman, L. C., Tratta, P. P., and Stracher, A. (1967). *J. Gen. Physiol.* **50**, 85.
Ebashi, S., and Endo, M. (1968). *Progr. Biophys. Mol. Biol.* **18**, 125–183.
Eccles, J. C. (1963). *In* "The Effect of Use and Disuse on Neuromuscular Functions" (E. Gutmann and P. Hník, eds.), p. 549. Elsevier, Amsterdam.
Eccles, J. C. (1967). *In* "Exploratory Concepts in Muscular Dystrophy and Related Disorders" (A. T. Milhorat, ed.), pp. 151–160. Excerpta Med. Found., Amsterdam.
Edge, M. B. (1970). *Develop. Biol.* **23**, 634–650.
Eidinoff, M. L., Cheong, L., and Rich, M. A. (1959). *Science* **129**, 1551.
Eisenberg, B., and Eisenberg, R. S. (1968). *J. Cell Biol.* **39**, 451.
Emmart, E. W., Komenz, D. R., and Miguel, J. (1963). *J. Histochem. Cytochem.* **11**, 207–217.
Endo, M. (1964). *Nature (London)* **202**, 1115.
Enesco, M., and Puddy, D. (1964). *Amer. J. Anat.* **114**, 235.
Engel, W. K., and Karpati, G. (1968). *Develop. Biol.* **17**, 713–723.
Ezerman, E. B., and Ishikawa, H. (1967). *J. Cell Biol.* **35**, 405–417.
Fambrough, D., and Rash, J. E. (1972). *Develop. Biol.* (in press).
Firket, H. (1958). *Arch. Biol.* **69**, 1–166.
Fischbach, G. D. (1970) Science *169*:1331.
Fischman, D. A. (1972). Unpublished observations.
Fischman, D. A. (1967). *J. Cell Biol.* **32**, 557.
Fischman, D. A. (1970). *Curr. Top. Develop. Biol.* **5**, 235–280.
Fischman, D. A., and Yaffe, D. (1971). Unpublished observations.
Fischman, D. A., and Zak, R. (1971). *J. Gen. Physiol.* **57**, 245.
Fischman, D. A., Shimada, Y., and Moscona, A. A. (1967). *J. Cell Biol.* **35**, 39A.
Franzini-Armstrong, C. (1970). *J. Cell Sci.* **6**, 559–592.
Franzini-Armstrong, C., and Porter, K. R. (1964a). *J. Cell Biol.* **22**, 675.
Franzini-Armstrong, C., and Porter, K. R. (1964b). *Z. Zellforsch. Mikrosk. Anat.* **61**, 661.
Frederiksen, D. W., and Holtzer, A. (1968). *Biochemistry* **7**, 3935–3950.
Garber, B. (1953). *Exp. Cell Res.* **5**, 132.
Godlewski, E., (1901). *Krakauer Anz.* **10**, 15.
Godlewski, E. (1902). *Arch. Microsk. Anat. Entwicklungsmech.* **60**, 1.
Goldspink, G. (1970a). *J. Cell Sci.* **6**, 593–603.
Goldspink, G. (1970b). *In* "The Physiology and Biochemistry of Muscle as a Food"

144   *Donald A. Fischman*

(E. J. Briskey, R. G. Cassens, and B. B. Marsh, eds.), pp. 521–536. Univ. of Wisconsin Press, Madison.
Gregory, D. W., Lennie, R. W., and Birt, L. M. (1967). *J. Roy. Microsc. Soc.* [3] **88**, 151.
Grove, D., Nair, K. G., and Zak, R. (1969a). *Circ. Res.* **25**, 463.
Grove, D., Zak, R., Nair, K. G., and Aschenbrenner, V. (1969b). *Circ. Res.* **25**, 473.
Guth, L. (1968). *Physiol. Rev.* **48**, 645.
Guth, L., Watson, P. K., and Brown, W. C. (1968). *Exp. Neurol.* **20**, 52.
Hamburger, V. (1938). *J. Exp. Zool.* **77**, 379.
Hamburger, V. (1939). *J. Exp. Zool.* **80**, 347.
Hanson, J. (1968). *Quart. Rev. Biophys.* **1**, 177.
Hanson, J., and Lowy, J. (1963). *J. Mol. Biol.* **6**, 46.
Hardy, M. F., and Perry, S. V. (1969). *Nature (London)* **223**, 300–302.
Harrington, W. F., and Josephs, R. (1968). *Develop. Biol., Suppl.* **2**, 21–62.
Harrison, R. G. (1904). *Amer. J. Anat.* **3**, 197–220.
Hauschka, S. D., and Konigsberg, I. R. (1966). *Proc. Nat. Acad. Sci. U.S.* **55**, 119–126.
Hauschka, S. D., and White, H. K. (1972). In "Research in Muscle Development and the Muscle Spindle" (B. Q. Banker, R. J. Przybylski, J. Van der Muelen, and M. Victor, eds.), p. 53. Excerpta Medica, Amsterdam.
Hay, E. D. (1963). *Z. Zellforsch. Mikrosk. Anat.* **59**, 6–34.
Hay, E. D. (1968). In "Epithelial-Mesenchymal Interactions" (R. Fleischmajer and R. Billingham, eds.), pp. 31–55. Williams & Wilkins, Baltimore, Maryland.
Hay, E. D., and Fischman, D. A. (1961). *Develop. Biol.* **3**, 26.
Hay, E. D., and Revel, J. P. (1963). *J. Cell Biol.* **16**, 29–51.
Herrmann, H., Heywood, S. M., and Marchok, A. (1970). *Curr. Top. Develop. Biol.* **5**, 181–234.
Heuson-Stiennon, J. A. (1964). *J. Microsc. (Paris)* **3**, 229–239.
Heuson-Stiennon, J. A. (1965). *J. Microsc. (Paris)* **4**, 657–678.
Heywood, S. M., and Nwagwu, M. (1968). *Proc. Nat. Acad. Sci. U.S.* **60**, 229–234.
Heywood, S. M., and Nwagwu, M. (1969). *Biochemistry* **8**, 3839.
Heywood, S. M., and Rich, A. (1968). *Proc. Nat. Acad. Sci. U.S.* **59**, 590.
Heywood, S. M., Dowben, R. M., and Rich, A. (1967). *Proc. Nat. Acad. Sci. U.S.* **57**, 1002.
Hines, M. (1972). In "The Structure and Function of Muscle" (G. H. Bourne, ed.), Volume III, Chapter 5, 2nd ed. Academic Press, New York.
Hirano, H. (1967). *Z. Zellforsch. Mikrosk. Anat.* **79**, 198.
Hitchcock, S. E. (1970). *Develop. Biol.* **23**, 399–423.
Holtzer, H. (1970). In "Cell Differentiation" (O. Schjeide and J. de Vellis, eds.), pp. 476–503. Van Nostrand-Reinhold, Princeton, New Jersey.
Holtzer, H., and Bischoff, R. (1970). In "The Physiology and Biochemistry of Muscle as a Food" (E. Briskey, R. Cassens, and B. Marsh, eds.), Vol. 2, pp. 29–51. Univ. of Wisconsin Press, Madison.
Holtzer, H., and Detwiler, S. R. (1953). *J. Exp. Zool.* **123**, 335–370.
Holtzer, H., Lash, J., and Holtzer, S. (1956). *Biol. Bull.* **111**, 303.
Holtzer, H., Marshall, J. M., and Finck, H. (1957). *J. Biophys. Biochem. Cytol.* **3**, 705.
Holtzer, H., Sanger, J. W., and Ishikawa, H. (1971). *J. Gen. Physiol.* **57**, 245.
Hosick, H. L. and Strohman, R. C. (1971). *J. Cell. Physiol.* **77**, 145–156.
Huber, E. (1931). "Evolution of Facial Musculature and Facial Expression." Johns Hopkins Press, Baltimore, Maryland.

Huxley, H. E. (1963). *J. Mol. Biol.* **7**, 281.
Huxley, H. E. (1964). *Nature* (*London*) **202**, 1067.
Huxley, H. E., and Brown, W. (1967). *J. Mol. Biol.* **30**, 383.
Ikemoto, N., and Kawaguti, S. (1967). *Proc. Jap. Acad.* **43**, 974–979.
Ishikawa, H. (1966). *Z. Anat. Enwicklungsgesch.* **125**, 43–63.
Ishikawa, H. (1968). *J. Cell Biol.* **38**, 51.
Ishikawa, H., Bischoff, R., and Holtzer, H. (1968). *J. Cell Biol.* **38**, 538–558.
Ishikawa, H., Bischoff, R., and Holtzer, H. (1969). *J. Cell Biol.* **43**, 312–328.
James, D. W., and Tresman, R. L. (1969a). *Z. Zellforsch. Mikrosk. Anat.* **100**, 126–140.
James, D. W., and Tresman, R. L. (1969b). *Z. Zellforsch. Mikrosk. Anat.* **101**, 598–606.
Johnson, P., Harris, C. I., and Perry, S. V. (1967). *Biochem. J.* **105**, 361.
Kaminer, B. (1969). *J. Mol. Biol.* **39**, 257–264.
Kaminer, B., and Bell, A. T. (1966). *J. Mol. Biol.* **20**, 391–401.
Kawamura, M., and Maruyama, K. (1969). *J. Biochem.* (*Tokyo*) **66**, 619.
Kelly, A. M., and Zacks, S. I. (1969a). *J. Cell Biol.* **42**, 135–153.
Kelly, A. M., and Zacks, S. I. (1969b). *J. Cell Biol.* **42**, 154–169.
Kelly, D. E. (1969). *Anat. Rec.* **163**, 403.
Kikuchi, M., Noda, H., and Maruyama, K. (1969). *J. Biochem.* (*Tokyo*) **65**, 945.
Kitiyakara, A. (1959). *Anat. Rec.* **133**, 35.
Kitiyakara, A., and Angevine, D. M. (1963). *Develop. Biol.* **8**, 322–340.
Knappeis, G. G., and Carlson, F. (1962). *J. Cell Biol.* **13**, 323.
Knappeis, G. C., and Carlson, F. (1968). *J. Cell Biol.* **38**, 202.
Konigsberg, I. R. (1963). *Science* **140**, 1273–1284.
Konigsberg, I. R. (1964). *Carnegie Inst. Wash., Yearb.* **63**, 517.
Konigsberg, I. R. (1965). *In* "Organogenesis" (R. L. DeHaan and H. Ursprung, eds.), pp. 337–358. Holt, New York.
Konigsberg, I. R. (1971). *Develop. Biol.* **26**, 133.
Konigsberg, I. R., McElvain, N., Tootle, M., and Herrmann, H. (1960). *J. Biophys. Biochem. Cytol.* **8**, 333–343.
Kundrat, E., and Pepe, F. A. (1971). *J. Cell Biol.* **48**, 340–347.
Lake, N. C. (1916). *J. Physiol.* (*London*) **50**, 364.
Larson, P. F., Hudgson, P., and Walton, J. N. (1969). *Nature* (*London*) **222**, 1168–1169.
Larson, P. F., Jenkison, M., and Hudgson, P. (1970). *J. Neurol. Sci.* **10**, 385–405.
Lash, J. W., Holtzer, H., and Swift, H. (1957). *Anat. Rec.* **128**, 679–693.
Lennie, R. W., Gregory, D. W., and Bert, L. M. (1967). *J. Insect. Physiol.* **13**, 1745.
Lentz, T. L. (1969a). *Amer. J. Anat.* **124**, 447.
Lentz, T. L. (1969b). *J. Cell Biol.* **42**, 431–443.
Lentz, T. L. (1970). *J. Cell Biol.* **47**, 423–436.
Lewis, W. H., and Lewis, D. M. (1917). *Amer. J. Anat.* **22**, 169–194.
Liedke, K. B. (1958). *Anat. Rec.* **131**, 97–117.
Lipton, B. H., and Konigsberg, I. R. (1971). *Anat. Rec.* **169**, 368a.
Low, F. N. (1967). *Anat. Rec.* **159**, 231–238.
Lowey, S., Slayter, H. S., Weeds, A. G., and Baker, H. (1969). *J. Mol. Biol.* **42**, 1–29.
Luck, D. J. L. (1963). *J. Cell Biol.* **16**, 483.
Luck, D. J. L. (1965). *J. Cell Biol.* **24**, 461.
Luft, J. H. (1966). *Fed. Proc., Fed. Amer. Soc. Exp. Biol.* **25**, 1773.

MacConnachie, H. F., Enesco, M., and Leblond, C. P. (1964). *Amer. J. Anat.* **114**, 245.

Maker, B. A., and Fischman, D. A. (1972). In preparation.

Manasek, F. J. (1968). *J. Cell Biol.* **37**, 191.

Marchok, A. C., and Herrmann, H. (1967). *Develop. Biol.* **15**, 129.

Mark, G. E., and Strasser, F. F. (1966). *Exp. Cell Res.* **44**, 217.

Masaki, T., Takaiti, S., and Ebashi, S. (1968). *J. Biochem.* (*Tokyo*) **64**, 909–910.

Maslow, D. E. (1969). *Exp. Cell Res.* **54**, 381.

Mauro, A. (1961). *J. Biophys. Biochem. Cytol.* **9**, 493.

Mauro, A., Shafiq, S. A., and Milhorat, A. T. (1970). "Regeneration of Striated Muscle and Myogenesis." Excerpta Med. Found., Amsterdam.

Mintz, B., and Baker, W. W. (1967). *Proc. Nat. Acad. Sci. U.S.* **58**, 592.

Morkin, E. (1970). *Science* **167**, 1499–1501.

Morkin, E., and Ashford, T. P. (1968). *Amer. J. Physiol.* **215**, 1409.

Moscona, A. A. (1955). *Exp. Cell Res.* **9**, 377.

Moscona, A. A. (1957). *Proc. Nat. Acad. Sci. U.S.* **43**, 184.

Moscona, A. A. (1961). *Exp. Cell Res.* **22**, 455.

Moscona, A. A. (1965). *In* "Cells and Tissues in Culture" (E. N. Willmer, ed.), Vol. 1, pp. 489–529. Academic Press, New York.

Moss, F. P. (1968). *Amer. J. Anat.* **122**, 555.

Moss, F. P., and Leblond, C. P. (1970). *J. Cell Biol.* **44**, 459–461.

Muchmore, W. B. (1958). *J. Exp. Zool.* **139**, 181–188.

Muchmore, W. B. (1968). *J. Exp. Zool.* **169**, 251–258.

Muratori, G. (1939). *Anat. Anz.* **87**, 430.

Muscatello, V., Margreth, A., and Aloisi, M. (1965). *J. Cell Biol.* **27**, 1–24.

Nachmias, V. T., Kessler, D., and Huxley, H. E. (1970). *J. Mol. Biol.* **50**, 83.

Nameroff, M., and Holtzer, H. (1969). *Develop. Biol.* **19**, 380.

Obinata, T. (1969). *Arch. Biochem. Biophys.* **132**, 184–197.

Obinata, T., and Hayashi, T. (1972). In preparation.

Obinata, T., Yamamoto, M., and Maruyama, K. (1966). *Develop. Biol.* **14**, 192–213.

Okazaki, K., and Holtzer, H. (1965). *J. Histochem. Cytochem.* **13**, 726.

Okazaki, K., and Holtzer, H. (1966). *Proc. Nat. Acad. Sci. U.S.* **56**, 1484.

O'Neill, M., and Strohman, R. C. (1969). *J. Cell Physiol.* **73**, 61–68.

O'Neill, M. and Stockdale, F. E. (1972). *J. Cell. Biol.* **52**, 52.

Padykula, H. A., and Gauthier, G. F. (1970). *J. Cell Biol.* **46**, 27.

Page, S. G. (1965). *J. Cell Biol.* **26**, 477.

Pappas, G. D., Peterson, M. A., Masurovsky, E. B., and Crain, S. M. (1971). *Ann. N.Y. Acad. Sci.* **83**, (W. S. Fields, ed.), pp. 33–45.

Paterson, B., and Strohman, R. C. (1970), *Biochemistry* **9**, 4094–4105.

Peachey, L. D. (1970). *In* "The Physiology and Biochemistry of Muscle as a Food" (E. J. Briskey, R. D. Cassens, and B. B. Marsh, eds.), Vol. 2, pp. 273–310. Univ. of Wisconsin Press, Madison.

Peachey, L. D., and Schild, R. F. (1968). *J. Physiol.* (*London*) **194**, 249.

Pellegrino, C., and Franzini-Armstrong, C. (1969). *Int. Rev. Exp. Pathol.* **7**, 139–226.

Pepe, F. A. (1967). *J. Mol. Biol.* **27**, 203–225.

Perry, S. V. (1967). *Progr. Biophys. Mol. Biol.* **1**, 327–381.

Petterson, E. R., and Crain, S. M. (1970). *Z. Zellforsch. Mikrosk. Anat.* **106**, 1–21.

Pollard, T. D., and Ito, S. (1970). *J. Cell Biol.* **46**, 267–289.

Pollard, T. D., and Korn, E. D. (1971). *J. Cell Biol.* **48**, 216–218.
Pollard, T. D., Shelton, E., Weihing, R. R., and Korn, E. D. (1970). *J. Mol. Biol.* **50**, 91.
Porter, K. R., and Palade, G. E. (1957). *J. Biophys. Biochem. Cytol.* **3**, 269.
Prewitt, M. A., and Salafsky, B. (1970). *Amer. J. Physiol.* **218**, 69–74.
Przybylski, R., and Blumberg, J. M. (1966). *Lab Invest.* **15**, 836–863.
Rabinowitz, M., and Swift, H. (1970). *Physiol. Rev.* **50**, 376.
Rambourg, A., and Leblond, C. P. (1967). *J. Cell Biol.* **32**, 27.
Rash, J. E., Shay, J. W., and Biesele, J. M. (1968). *J. Ultrastruct. Res.* **24**, 181.
Reedy, M. K. (1964). *Proc. Roy. Soc., Ser. B* **160**, 458.
Reedy, M. D. (1968). *J. Mol. Biol.* **31**, 155.
Reporter, M. C., Konigsberg, I. R., and Strehler, B. L. (1963). *Exp. Cell Res.* **30**, 410–417.
Revel, J.-P., and Karnofsky, M. (1967). *J. Cell Biol.* **33**, C7.
Reznik, M. (1969). *Lab. Invest.* **20**, 353–363.
Robbins, N., and Engel, W. K. (1969). *Arch. Neurol.* **20**, 318–329.
Robbins, N., and Yonezawa, T. (1971). *Science* **172**, 395–398.
Rona, G., and Kahn, D. S. (1967). *Meth. Achievm. Exp. Pathol.* **3**, 200–249.
Roodyn, D. B., and Wilkie, D. (1968). "The Biogenesis of Mitochondria." Methuen, London.
Rudnick, D. (1945). *J. Exp. Zool.* **100**, 1.
Rumyantsev, P. P. (1963). *Folia Histochem. Cytochem.* **1**, 463.
Rumyantsev, P. P. (1966). *Folia Histochem. Cytochem.* **4**, 397.
Rumyantsev, P. P. (1968). *Experientia* **24**, 1234.
Sandow, A. (1965). *Pharmacol. Rev.* **17**, 265–320.
Sanger, J. W., and Holtzer, H. (1970). *J. Cell Biol.* **47**, 178a.
Sanger, J. W., Holtzer, S., and Holtzer, H. (1971). *Nature (London)* **229**, 121.
Sarkar, S., and Cooke, P. H. (1970). *Biochem. Biophys. Res. Commun.* **41**, 918.
Saunders, J. W. (1948). *J. Exp. Zool.* **108**, 363.
Schiaffino, S., and Margreth, A. (1969.) *J. Cell Biol.* **41**, 855–875.
Schiaffino, S., and Settembrini, P. (1970). *Virchous Arch., B* **4**, 345–356.
Schroeder, T. E. (1969). *Biol. Bull.* **137**, 413.
Schubert, D., and Jacob, F. (1970). *Proc. Nat. Acad. Sci. U.S.* **67**, 247.
Shafiq, S. A. (1963). *J. Cell Biol.* **17**, 363–373.
Shainberg, A., Yagil, G., and Yaffe, D. (1969). *Exp. Cell Res.* **58**, 163–167.
Shapira, G., Dreyfus, J., and Shapira, F. (1950). *C. R. Soc. Biol.* **144**, 829–832.
Shimada, Y. (1968). *Exp. Cell Res.* **51**, 564.
Shimada, Y. (1971). *J. Cell Biol.* **48**, 128–142.
Shimada, Y., Fischman, D. A., and Moscona, A. A. (1967). *J. Cell Biol.* **35**, 445.
Shimada, Y., Fischman, D. A., and Moscona, A. A. (1969a). *Proc. Nat. Acad. Sci. U.S.* **62**, 715–721.
Shimada, Y., Fischman, D. A., and Moscona, A. A. (1969b). *J. Cell Biol.* **43**, 382.
Simon, E. H. (1963). *Exp. Cell Res.* **9**, 263.
Simpson, S. B., Jr., and Cox, P. G. (1967). *Science* **157**, 1330–1332.
Singer, M. (1952). *Quart. Rev. Biol.* **27**, 169.
Singer, M. (1965). *In* "Regeneration in Animals and Related Problems" (V. Kiortsis and H. A. Trampusch, eds.), pp. 20–32. North-Holland Publ., Amsterdam.
Smithberg, M. (1954). *J. Exp. Zool.* **127**, 397.

Spiro, D., and Hagopian, M. (1967). In "Formation and Fate of Cell Organelles" (K. B. Warren, ed.), p. 71. Academic Press, New York.
Spooner, B. S., and Wessells, N. K. (1970). Proc. Nat. Acad. Sci. U.S. 66, 360.
Stockdale, F. E., and Holtzer, H. (1961). Exp. Cell Res. 24, 508–520.
Stockdale, F. E., Okazaki, K., Nameroff, M., and Holtzer, H. (1966). Science 146, 533.
Stoker, M. (1967). Curr. Top. Develop. Biol. 2, 107–128.
Straus, W. L., Jr., and Rawles, M. E. (1953). Amer. J. Anat. 92, 471–510.
Strehler, B. L., Konigsberg, I. R., and Kelley, J. E. T. (1963). Exp. Cell Res. 32, 232–241.
Strohman, R. C., and Paterson, B. (1971). J. Gen. Physiol. 57, 244.
Stromer, M., Hartshorne, D. J., Meuller, H., and Rice, R. V. (1969). J. Cell Biol. 40, 167–178.
Strudel, G. (1955). Arch. Anat. Microsc. Morphol. Exp. 44, 209.
Swift, H., and Wolstenholme, D. R. (1969). In "Handbook of Molecular Cytology" (A. Lima-de-Faria, ed.), p. 972. North-Holland Publ., Amsterdam.
Szollosi, D. (1970). J. Cell Biol. 44, 192–210.
Tello, J. F. (1917). Trab. Lab. Invest. Biol. Univ. Madrid 15, 101.
Tello, J. F. (1922). Z. Anat. Entwicklungsgesch. 64, 348.
Teravainen, H. (1968). Z. Zellforsch. Mikrosk. Anat. 87, 249.
Theiler, K. (1957). Acta Anat. 30, 842.
Tilney, L. G. (1965). Anat. Rec. 151, 426.
Tilney, L. G., and Cardell, R. R., Jr. (1970). J. Cell Biol. 47, 408–422.
Toselli, P. A., and Pepe, F. A. (1968). J. Cell Biol. 37, 445–461.
Trayer, I. P., Harris, C. I., and Perry, S. V. (1968). Nature (London) 217, 452.
Trelstad, R. L., Hay, E. D., and Revel, J. P. (1967). Develop. Biol. 16, 78–106.
Tschumi, P. A. (1957). J. Anat. 91, 149.
Walker, A. C., and Bert, L. M. (1969). J. Insect Physiol. 15, 305.
Walker, S. M., and Schrodt, A. G. (1968). J. Cell Biol. 37, 564–569.
Warren, R. H., and Porter, K. R. (1969). Amer. J. Anat. 124, 1.
Waterman, R. E. (1969). Amer. J. Anat. 125, 457–494.
Weinstein, R. B., and Hay, E. D. (1970). J. Cell Biol. 47, 310–316.
Weiss, P. (1958). Int. Rev. Cytol. 7, 1–30.
Weiss, P. (1961). Exp. Cell Res., Suppl. 8, 260–281.
Wessells, N. K., Spooner, B. S., Ash, J. F., Bradley, M. O., Luduena, M. A., Taylor, E. W., Wrenn, J. T., and Yamada, K. M. (1971). Science 171, 135–143.
Wrenn, J. T., and Wessells, N. K. (1970). Proc. Nat. Acad. Sci. U.S. 66, 904–908.
Yaffe, D. (1968). Proc. Nat. Acad. Sci. U.S. 61, 477.
Yaffe, D. (1969). Curr. Top. Develop. Biol. 4, 37–75.
Yaffe, D., and Feldman, M. (1965). Develop. Biol. 11, 300–317.
Yamada, K. M., Spooner, B. S., and Wessells, N. K. (1970). Proc. Nat. Acad. Sci. U.S. 66, 1206–1212.
Yamada, T. (1937). Wilhelm Roux' Arch. Entwicklungsmech. Organismen 137, 151.
Yamada, T. (1939). Okajimas Folia Anat. Jap. 18, 565.
Young, M. (1969). Annu. Rev. Biochem. 38, 913–950.
Zalena, J. (1962). In "The Denervated Muscle" (E. Gutman, ed.), pp. 103–126. Publ. House Czech. Acad. Sci., Prague.
Zhinkin, L. N., and Andreeva, L. F. (1963). J. Embryol. Exp. Morphol. 11, 353–367.

HISTOCHEMISTRY OF DEVELOPING SKELETAL AND CARDIAC MUSCLE

E. B. BECKETT and G. H. BOURNE

REVISED BY G. H. BOURNE and M. N. GOLARZ DE BOURNE

Although the morphological and histological study of embryonic development has been undertaken for many years now, little work has been carried out on the associated chemical aspects of differentiation, either

149

from the biochemical or from the histochemical point of view. Histochemical work, in particular, is of very recent date and has mainly
been concerned either with the development of the very early embryo
(Brachet, 1940; Moog, 1944; Buño, 1951, 1954; Buño and Mariño, 1952),
or with the processes involved in ossification (Horowitz, 1942; Lorch,
1947; Bevelander and Johnson, 1950; Pritchard, 1952).

A review of histochemical work on all fetal tissues was made by Rossi
and associates (1954), and from this it was clear that very few data
were available at that time concerning muscle at any but the very earliest
stages of embryonic development, and moreover, that information about
the histochemistry of developing muscle was almost nil. A moderate
number of papers concerned with histochemical and biochemical
changes in developing muscle have appeared since the first edition of
this treatise, and reference to these will be made in the appropriate
places in this chapter.

I. Succinic Dehydrogenase

An outline of the basic processes involved in the histochemical procedure for the demonstration of succinic dehydrogenase is given in Volume
III, Chapter 9 on adult skeletal muscle. Exactly the same technique
can be used for fetal tissues, but sometimes it is advantageous to use
a higher pH (pH 8.2 instead of the more usual pH 7.5), since this
enables regions of lower enzyme activity to be demonstrated (Rosa
and Velardo, 1954). In fetal tissues there is less succinic dehydrogenase
than in adult ones.

A. Cardiac Muscle

Succinic dehydrogenase activity in the hearts of developing chick and
rat was studied biochemically by Sippel (1954). In the rat, it was found
that there was a steep increase in enzyme activity between the tenth
and thirteenth days, followed by a leveling off until the sixteenth day
and then a further steep rise until the nineteenth day, i.e., 3 days before
term.

Although histochemical observations do not readily lend themselves
to quantitation, some interesting results have been obtained. Rossi and
associates (1954) report that in human fetuses ranging in size from
60 mm CR (crown–rump length) to 26 cm CR, there is a strong reaction

for succinic dehydrogenase. This has been confirmed in hearts of 4-month-old human fetuses (Beckett and Bourne, 1958). In a series of goat fetuses, of 5¼ inches CR to 15 inches CR, the reaction intensity was moderate in the youngest fetus and progressively increased throughout prenatal life. although never attaining adult levels. In fetal cardiac musculature, there was no evidence either of intercalated disks, which

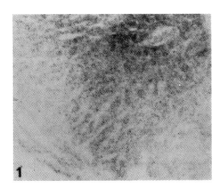 

Fig. 1. Succinic dehydrogenase in cardiac (ventricular) muscle from a 5¼ inch (64 day) goat fetus.

Fig. 2. Succinic dehydrogenase in ventricular muscle from a 15 inch (133 day) goat fetus.

 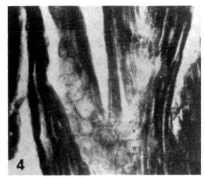

Fig. 3. Succinic dehydrogenase in adult goat ventricular muscle. Note the negative intercalated discs.

Fig. 4. Succinic dehydrogenase in Purkinje fibers in adult cardiac muscle. Note the positive reaction in the nucleus and at the periphery of each fiber.

in the adult are negative (Bourne, 1953; Beckett and Bourne, 1958), or of Purkinje fibers, which in the adult goat have a succinic dehydrogenase-positive nucleus and a positive reaction also in the peripheral myofibrils. Purkinje fibers are, however, morphologically differentiated by the 10½ inch CR stage in the goat, so that it seems they must acquire their typical distribution of succinic dehydrogenase activity some time after differentiation (See Figs. 1–4 for location of enzyme activity).

B. Skeletal Muscle

Shen (1949), studying homogenates of fetal rat skeletal muscle, found a steadily increasing amount of succinic dehydrogenase activity present during the later part of fetal life. Rossi and associates (1954), on the other hand, claim that there is a strong reaction for succinic dehydrogenase in skeletal muscle at all stages of development of the human fetus. The observations of Beckett and Bourne are not in accord with those of Rossi and his colleagues, since in about a dozen specimens of skeletal muscle taken from three human fetuses of 16 weeks gestation little succinic dehydrogenase activity could be found. In goat fetuses the reaction intensity progresses from weak to moderate during prenatal life. In goat fetuses, little positive reaction was obtained until about two-thirds term (i.e., about 100 days). During the next 30 days or so there was a considerable increase in enzyme activity, so that by about 130 days the diformazan granules were arranged in a typical linear fashion, and in tibialis anterior, biceps, and gastrocnemius, it was possible to distinguish between the fibers containing little succinic dehydrogenase activity and those containing more of the enzyme. In these muscles there was no difference in size between the fibers showing different degrees of enzyme activity. In the rectus femoris, however, all of the muscle fibers showed equal reaction intensity (Figs. 5–7).

Despite the increase in succinic dehydrogenase activity during the last third of prenatal life, the degree of activity shown by fetal muscle was always considerably less than that exhibited by adult skeletal muscle. There was no evidence of any tendency for the diformazan granules to be more concentrated at the edges of the muscle fibers, although this distribution is often seen in adult human and goat skeletal muscle.

In 1964, Diculesco and his colleagues studied not only succinic dehydrogenase, but also di- and triphosphopyridine nucleotide-linked dehydrogenase, and α- and β-glycerophosphate dehydrogenases in myoblasts in vitro. They found that the myoblasts gave a very intense reaction for all Krebs cycle enzymes, even though there were anaerobic glycolytic

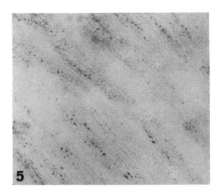
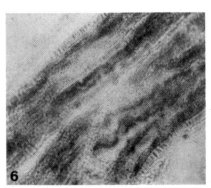
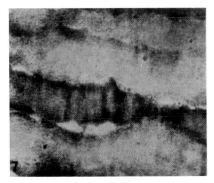

Fig. 5. Succinic dehydrogenase in skeletal muscle from a 10½ inch (100 day) fetus. The reaction is considerably weaker than in cardiac muscle.

Fig. 6. Succinic dehydrogenase in skeletal muscle of a 15 inch (133 day) fetus. Note the existence at this stage of fibers containing much and little enzyme activity.

Fig. 7. Succinic dehydrogenase in adult goat skeletal muscle.

processes going on in the myoblasts at that time, together with lipid synthesis and protein synthesis. They found, especially in the myoblasts that were forming white fibers, enzymes related to or concerned with anaerobic glycolysis, but no hexose shunt activity. In the same year, Wirsen and Larsson (1964) studied the distribution of lipids, glycogen phosphorylase, succinic dehydrogenase, and nonspecific esterases in embryonic mice from the fourteenth to the twenty-first day. Before birth, they found it was not possible to differentiate the muscle fibers on the basis of the succinic dehydrogenase and lipid reactions. They found esterase activity in the motor end plates and in nerve fibers in the fourteenth day of gestation. They also studied peroxidase activity in the fibers, and this activity presumably could be interpreted as indicating the degree of development in myoglobin and formation, and they could not demonstrate the presence of myoglobin by this method in the muscle fibers until after birth. Their periodic acid–Schiff (PAS)-positive substances, in this case glycogen, showed a uniform distribution in the myotubes in the fifteenth to the sixteenth day of gestation. There was a difference in the distribution of glycogen in the different muscle fibers which was seen from the seventeenth to the nineteenth day. They found that phosphorylase could be found in the myotubes in the sixteenth day; on the nineteenth day, sensory nerve fibers could be seen associated with the muscle fibers. Also on the nineteenth day, in addition to the two types of fibers that could be discriminated between on the basis of the enzymic reaction, a third type of fiber now appeared. This was a completely phosphorylase-negative fiber. The two other types of fiber, which had already developed, were small fibers that showed a weak phosphorylase reaction, and another type of bigger fiber that was associated with a high phosphorylase reaction. Just after the birth of the mice, the authors were not able to distinguish between the three fibers from the point of view of size, but they were all well developed from an enzymological point of view. They found that muscle spindles appeared in these muscles about the seventeenth to the eighteenth day of gestation.

In 1965, Dubowitz showed that there was a difference in the maturation of skeletal muscle in different species of animals (1965a,b). In the guinea pig, for example, he found that at birth, the muscle was already differentiated into two types which could be demonstrated by histochemical means—i.e., by oxidative enzyme reactions and also by phosphorylase activity—and that this differentiation between the two fibers commenced about half to three-quarters of the way through the pregnancy. In the rat and mouse, there was no differentiation between the fibers at birth, and in the rat, the full differentiation between the two fibers was not complete until as much as 2 weeks of age. He also found that the hamster and the rabbit showed some differentiation of the fibers at birth, but it was not as striking as that which was found in the guinea pig. Dubowitz (1965b) drew attention to the fact that there appeared to be a correlation between the degree of differentiation of the muscle at birth and the degree of maturity and the degree of mobility of the animals, which also appeared to be related to the length of gestation.

In 1966, Germino and his colleagues studied muscle development in chick embryos from the fourth to the twentieth day of development. They found that where the muscle buds were developing, there was a marked succinic dehydrogenase activity in the cytoplasm, especially around the nucleus in the developing myoblasts and also along the myofibrils and in the growing ends of the myoblasts. The succinic dehydrogenase in the first two areas, they thought, probably plays a part in the provision of energy for the synthesis of proteins, whereas they considered that the succinic dehydrogenase that was located between the myofibrils was more likely to be concerned with the mechanisms that supply the energy necessary for movement. From the seventeenth day, they found a differentiation between light- and dark-stained fibers. In the area of growth, they found that they could identify four types of muscle buds; (1) those that reacted intensely for succinic dehydrogenase, (2) those that reacted slightly, (3) those that reacted moderately, and (4) those that gave a strong positive action in some parts and a weak reaction in others. They also studied muscle-associated fibroblasts and found that the succinic dehydrogenase activity of fibroblasts in these muscle areas increased with the age of the embryo, but that cells such as macrophages had a succinic dehydrogenase activity that was unaffected by embryonic age. Brotchi and Mildadenov (1967) also studied dehydrogenases in the developing muscle of the chick embryo and found that the developing fibers in which a high activity of succinic dehydrogenase occurred were those that became fast fibers later in development; this is a rather surprising finding.

In 1969, Dorn studied a variety of enzymes in the muscle of developing guinea pigs. He studied not only succinic dehydrogenase, but also lactic dehydrogenase, glucose-6-phosphate dehydrogenase, glucose-1-phosphate dehydrogenase, cytochrome oxidase, and a number of other oxidative enzymes. He found that all the enzymes continuously increased their activity as development proceeded and that the intensity of their reactions compared with that of adult muscle at different stages of development for different enzymes. They noted the striking fact, which we have referred to earlier, that shortly after birth in the guinea pig, the distribution pattern and the activity of the various oxidative enzymes was identical with that of muscle in adults. As they put it, the muscle was topochemically adult in relation to the studied enzymes.

Kamieniecka (1968) found oxidative enzyme activity located in the central part of muscle fibers for the first time in human embryos 6–14 weeks old. The activity was seen to be uniform irrespective of the size of the fibers. On the other hand, ATPase activity was located around the periphery of the fibers and increased progressively with the growth of the fetus. The eleventh week of embryonic life was the period when phosphorylase activity first appeared in the fibers, and cholinesterase activity appeared with the onset of the ninth week of development. At the tenth week, cholinesterase activity was found in musculotendinous junctions, and muscle spindles were observed at about the same time. In the embryonic period from the sixth through fourteenth week, there still had not been any histochemical differentiation into the two metabolic fibers (fast and slow fibers) characteristic of adult muscle. According to the authors, this metabolic differentiation parallels the innervation of the individual fibers in the quadriceps femoris muscle; these events did not occur until 24 weeks of embryonic life. The authors claim that embryonic muscle sometimes bears similarities with the changes seen in some muscle diseases.

Nyström (1968) studied the development of cat muscles and found that he could distinguish between red and white fibers in kittens ranging from 10 to 20 days and also in adult cats. In newborn kittens, he found that the different types of muscle fiber could only be distinguished by ATPase staining. Later they could be distinguished by the phosphorylase and PAS reactions. After 14 days of postnatal life, they could also be distinguished by lipid staining. In 6–7-week-old kittens, the adult pattern of muscle fibers had been reached in all muscles.

II. Esterases

The available information concerning the distribution of simple esterases and lipases in the fetal heart and in the skeletal muscle of fetuses can be very rapidly summarized. Buño and Mariño (1952), using Gomori's Tween technique, obtained no reaction in the cells of cardiac and skeletal muscle of chick embryos of up to 18 days. Rossi and associates (1954) reported no esterase in the heart and blood vessels of the human fetus, and McKay et al. (1955) did not mention having obtained evidence of enzyme activity in cardiac or skeletal muscle fibers using the α-naphthyl acetate technique for esterases on an acetone-fixed 5 mm human embryo, so one must assume that there, too, these tissues were negative.

A. Cholinesterase in Motor End Plate Areas

Although the distribution of cholinesterase has occasionally been studied in the adult heart (see, for instance, Gomori, 1948), as far as we are aware distribution of this enzyme in the embryonic heart has not yet been investigated, and so we must confine our remarks in this chapter to the distribution of cholinesterase in skeletal muscle of the fetus (Figs. 8–13).

Kupfer and Koelle (1951), using the rat fetus as their subject, were the first to describe the development of cholinesterase in end plate areas. They were unable to find cholinesterase in significant amounts, either histochemically, or biochemically, until the sixteenth day of gestation (term 21–22 days). At this time, the amount of enzyme present in the end plate areas was low, since it required 50 min incubation to demonstrate it compared with 10 min needed for adult muscle. The description of development of cholinesterase in the subneural apparatus given by these authors is confusing. Their preparations must have shown the artifact of staining of muscle nuclei, and the authors seemed to be under the impression that the subneural apparatus was produced by the fusion of bodies resembling nuclei, but which showed more cholinesterase activity than the rest of the muscle nuclei. However, the picture that emerges from their observations is of cholinesterase being present at poorly defined points on the muscle fibers at 16 days, followed by the development of the subneural apparatus during the next few days, until it is recognizable, at 20 days gestation, as being similar to that of the adult. After this time, structural development continues both in prenatal life and post partum.

The investigations of Kupfer and Koelle also demonstrated that there was no innervation of the end plate areas until 21–22 days gestation, and it was suggested that cholinesterase might have a chemotactic effect on the nerve fibers. It is perhaps significant that East (1931) and Straus (1939) reported that the first movements of rat embryos in response to faradic stimuli occurred at 16 days. Opinions differ, however, about whether or not neuromuscular junction has occurred at this stage. Straus concurred with the view of Kupfer and Koelle that at this time no connection existed between nerve and muscle, whereas East thought that neuromuscular junction had just occurred. Both East and Straus considered that some change other than a morphological one was responsible for the appearance of contractility at the 16-day stage. Maybe the production of cholinesterase has something to do with this.

Goodwin and Sizer (1965) found in chicken embryos an increase in the cholinesterase curve of muscle after the fifteenth day and that this was due to the progressive development of motor end plates. Prior to this period in development, the cholinesterase activity of the developing muscle was contained in the myoblasts. In developing duck embryos Khera and Lohin (1965) saw end plates first on the nineteenth day of development and they increased in size from the twentieth to the twenty-first day when they reached a size of approximately (33×25 inches). The subneural apparatus in these end plates measured 5–12 inches wide. From the twenty-first day of incubation to hatching, the number of end plates increased. It is of interest that these results were obtained using a myristoylcholine as the substrate; when acetylthiocholine was used, motor end plates could only be seen in the muscles

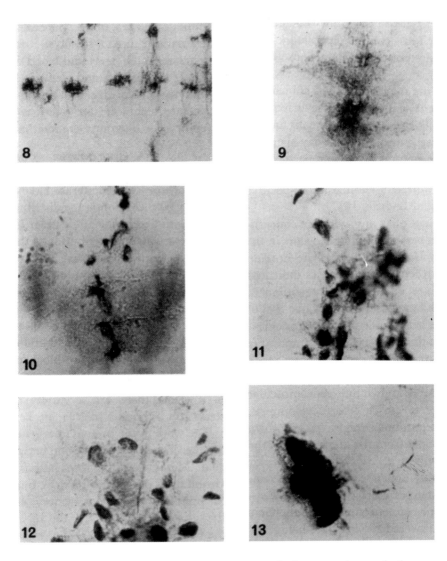

Fig. 8. Cholinesterase. Acetylcholinesterase in end plate areas in muscle from a 2½ inch (48 day) goat fetus.

Fig. 9. High power view of acetylcholinesterase in end plate areas of muscle taken from a 2½ inch (48 day) goat fetus.

Fig. 10. Acetylcholinesterase in motor end plates in muscle from a 5¼ inch (64 day) goat fetus.

Fig. 11. Acetylcholinesterase in end plate areas in muscle from a 10 inch (100 day) goat fetus.

after hatching. Filogamo and Gabella (1967) showed that in a wide variety of animals the development of motor nerve fibers showed that the development of the enzyme was characterized by three stages. In the first stage, the enzyme was diffused through the cytoplasm of the myoblasts. In the next stage, which is the period when the myoblasts are undergoing differentiation and growing in length, there was a great decrease in cholinesterase activity almost to the point of disappearance, and when enzyme activity is present, it appears to be located near the tip of each, differentiating fibers. In the third stage the enzyme reappears in increased intensity when the motor end plates begin to develop.

The pattern of development of the subneural apparatus in the fetal goat is apparently totally different from that in the rat (Beckett and Bourne, 1958), unless the difference can be explained on technical grounds. In a $2\frac{1}{2}$ inch CR goat fetus (i.e., about 48 days or one-third term), when the muscle fibers are at a very early stage of development with few myofibrils and with centrally placed nuclei, there is a moderate amount of cholinesterase which is demonstrable after $\frac{1}{4}$ hr incubation. This cholinesterase is situated at points in rows running at right angles to the long axis of the muscle fibers; i.e., the distribution of these areas of activity is the same as that for end plates in adult muscle. By the 64-day stage in the goat, cholinesterase activity in these areas has apparently risen to adult levels and this level is maintained for the rest of fetal life. This is in complete contrast to the slow development of enzyme activity in the rat.

When cholinesterase first appears in goat muscle fibers, there is little in the way of structure to be seen in the areas of enzyme activity, but occasionally there is evidence of some sort of rudimentary subneural apparatus. This, instead of being in the complex pattern observed in the adult of many animals, is in the form of a short straight tube or a crescent. At 64 days, the morphology of the subneural apparatus is reasonably clear. Some of these structures are still like a short tube or crescent, but others bear projections or outpicketings from the main tube. As prenatal life continues, the end plate areas become larger and more complex. The original short tube or crescent seems to bend round to form a circle, and other tubular structures stretch from side to side of the circle in an irregular pattern, although the complexity and size seen in the end plate of the adult goat are never attained before birth.

Fig. 12. Acetylcholinesterase in motor end plate areas of muscle from a 15 inch (133 day) goat fetus.

Fig. 13. Acetylcholinesterase in a single motor end plate from adult goat muscle. Note the enzyme activity in the nerve axon.

In the adult goat, the end plates are larger and even more complex than those usually seen in the rat.

The subneural apparatus of human end plates is present at 16 weeks and possibly earlier (Beckett and Bourne, 1958), since at 16 weeks the structure is like that of goat at 64 days. The rate of development, however, must be slower than in the goat, since Cöers (1955) found end plates with a simple subneural apparatus in children of up to 1 year old, and he also observed that the adult character of end plate structure emerged between 12 and 14 months of age. Similar observations have been made in other young animals by Gerebtzoff and co-workers (1954).

In a 6-month human fetus, the enzyme in the end plate area was found to be capable of splitting butyrylthiocholine as well as the acetyl compound (Beckett and Bourne, 1958) and it was eserine sensitive. Experiments to see if this was also true of other fetal human and goat muscle specimens were not carried out, but is it perhaps interesting that the only end plates in the adult human that contain an enzyme that can split butyrylthiocholine are the "classic" end plates, i.e., those similar in form to the ones described in other vertebrates. In human adult muscle, a wide variety of different morphological forms of end plate exists, whereas in the fetus there is apparently only one form, i.e., the "classic" type. Whether all of the classic end plates present in the fetus remain in that form until adult life and other structures present in the adult arise *de novo* after birth, or whether some of the classic end plates in fetal muscle are converted to other forms later on, perhaps during childhood, cannot be said at the present moment, but it is a fascinating problem. Certainly by the age of 9 years, there are a variety of different cholinesterase-positive structures in human muscle (Beckett and Bourne, 1957).

The work of (Beckett and Bourne) cannot offer any evidence concerning the time at which junction between nerve and muscle occurs in the goat. Methylene blue preparations by Dr. Rubinstein of the London Hospital on one of the 16-week human fetuses of their series indicated the presence of rudimentary nerve endings, and as has been mentioned, the subneural apparatus of the 16-week human fetus is very like that of the 64-day goat fetus. No studies have been made of the neurological aspect of formation of motor end plates in the goat, so that it is possible to say whether or not cholinesterase is present before the nerves reach the end plates and whether the enzyme might have a chemotactic effect on the growing nerve fibers. If it were known that neuromuscular junctions occurred at a definite stage of muscle fiber development, examination of hematoxylin–eosin preparations might have thrown some light on the problem, but is it obvious from the literature that the stage of development of the muscle fibers at which neuromuscular junction occurs is variable and probably depends on the species concerned. No work has been done on goat fetuses that correlates development of muscle fibers with the appearance of neuromuscular junctions, and so the question must remain unanswered for the time being.

B. Cholinesterase in Sites Other Than Motor End Plates

1. NERVE FIBERS

In the course of the work of Beckett and Bourne, (1958) cholinesterase-positive nerve axons were occasionally observed in fetal goat muscle and frequently in that of the adult of this species. In contrast to this, they did not detect cholinesterase in axons of human fetal muscle, and saw it very rarely in nerves of adult human muscle. This apparent difference in enzyme activity of nerve fibers might be caused by differences in permeability of the myelin sheath to the reagent used to detect cholinesterase, or it might be a true difference in levels of enzyme activity. It seems unlikely that the positive reaction seen in these structurs is due to a backward diffusion of cholinesterase from the motor end plates, since the enzyme activtiy is sometimes observed in nerve fibers situated some distance away from the motor end plates, and moreover, cholinesterase is not situated in the neural portion of end plates.

2. MUSCULOTENDINOUS JUNCTIONS

Cholinesterase at musculotendinous junctions was first observed by Couteaux in 1953 in frogs, mice, and fish (see Volume II, Chapter 8). Gerebtzoff and his colleagues have since demonstrated the same thing in a variety of other animals including man (Gerebtzoff, 1956), but the development of these cholinesterase-positive structures does not yet seem to have been described by authors other than ourselves.

In goat fetuses (Beckett and Bourne, 1958), structures at musculotendinous junctions make their appearance at about two-thirds term (i.e., at about 100 days), which is considerably later than the time at which motor end plates are first visible. At the 100-day stage, the cholinesterase-positive material at musculotendinous junctions is in the form of small dots of structure similar to that of "gutters" of the subneural apparatus of end plates, and it contains little cholinesterase activity.

During the time that follows, the musculotendinous junctions rapidly gain their adult form; i.e., the dots of "gutter" structure fuse together to form fingerlike projections into the muscle substance, and then equally rapidly acquire their full complement of cholinesterase activity. For the remaining period of prenatal life, they increase in size but do not become more complex or apparently gain any more enzyme activity.

In the human fetus, the structures at musculotendinous junctions are present in the early "dotted" form at 4 months, but at 6 months have changed very little in appearance, so that it seems that their development proceeds more slowly than in the goat.

III. Phosphatases

A. Alkaline Phosphatase

The various histochemical methods for the demonstration of alkaline phosphatase have been described elsewhere (Volume III, Chapter 9). It seems that in most cases the Gomori-Takamatsu technique has been used for the study of fetal muscle, but just occasionally, workers have used the α-naphthyl phosphate method (see, e.g., McKay et al., 1955).

1. Alkaline Phosphatase in Cardiac Muscle

The development of alkaline phosphatase of the heart appears to have been followed most closely in that of the human. McKay and his co-workers (1955) studied the heart of a 5 mm human embryo using both types of histochemical technique for the demonstration of alkaline phosphatase. They found that the auricular muscle gave apositive reaction with the Gomori technique and that the walls of capillaries contained alkaline phosphatase that could be demonstrated by both methods. In addition, alkaline phosphatase could be demonstrated in the endothelium of the auricle by both techniques, while that of the ventricle gave a positive reaction only with the α-naphthyl phosphate method. The endothelium of aorta also contained alkaline phosphatase.

Rossi and associates (1951) studied older human embryos (23 mm CR to term), and during the next few years they extended their observations to the 9 mm CR stage (Rossi et al., 1954). They found that apart from a transitory alkaline phosphatase activity in the intima of the aorta and other large vessels in the 23 mm CR embryo, this enzyme was restricted to the endothelium of the small vessels of the myocardium. Other work, using the Gomori technique (Beckett and Bourne, 1958), showed that in the hearts of 3- and 4-month fetuses alkaline phosphatase activity was frequently to be found in the endothelial linings of the larger blood vessels of the myocardium and occasionally also in the endothelia of myocardial capillaries. There was, in addition, an occasional patchy reaction in the endocardium and in scattered areas of undifferentiated tissue. No activity was observed in any of the vessels entering or leaving the heart.

The development of alkaline phosphatase in the heart of the chick, which was studied histochemically with Gomori's technique by Moog (1944), appears to proceed in rather a different fashion from that in the human. At 72 hr, Moog obtained a positive reaction (after $1\frac{1}{2}$–4 hrs incubation) in the endothelial lining of ventricle and atrium, and in the walls of the bulbus arteriosus, the ductus Cuvierii, and the ventral aorta. The cardinal veins were negative, and so also were the muscular walls of the atrium and ventricle. The nuclei of the myocardium showed transient activity only on the fourth day.

From the fourth to the eighth day of incubation, the nucleoli only of the myocardium were positive. The endothelial lining of the heart remained positive throughout this period, and so also did the linings of the bulbus arteriosus and all vessels entering or leaving the heart. A high level of alkaline phosphatase activity was present in the septa, including the cushion septa.

The picture seen in fetal goat hearts appears to be very similar to that in the fetal human (Beckett and Bourne, 1958). At the $2\frac{1}{2}$ inch CR stage (the earliest goat fetus studied), there was a reaction in the pericardium near the atrioventricular junction, but by the $5\frac{1}{4}$ inch CR stage, this had disappeared. In the hearts of these early fetuses, there was alkaline phosphatase activity also in the endothelial linings of the myocardial blood vessels, but there were usually fewer positive vessels in the atria than in the ventricles. In older goat fetuses (up to term), only ventricular muscle was studied, and here, as in the human hearts, most of the activity appeared to be centered in the endothelial linings of the vessels larger than capillary size. The number of positive capillaries was very variable. There was a slight tendency for more of the myocardial blood vessels to acquire alkaline phosphatase activity as prenatal life progressed, but this tendency was ill defined, and one could not be certain of it, because the number of vessels giving a positive reaction was so very variable at a given age.

The cardiac muscle of the adult goat gave no reaction at all under the incubation conditions used for the goat fetuses in the course of the work described above. This is perhaps interesting, since earlier work had indicated that in cardiac muscle of mouse, rat, and guinea pig (Zorzoli and Stowell, 1947) and of rat (Bourne, 1953), the intercalated disks were alkaline phosphatase-positive. However, the different results may be the result of differing incubation periods, since for the work on goats 15–30 min was used, in contrast to the much longer periods used by Zorzoli and Stowell and by Bourne.

2. Alkaline Phosphatase in Skeletal Muscle

Although little work has been done on the subject of the distribution of alkaline phosphatse in embryonic skeletal muscle, it is fairly certain that its localization is essentially similar to that in both fetal cardiac muscle and adult skeletal muscle.

McKay et al. (1955) noted the presence of alkaline phosphatase in the fibrils of the myotomes of a 5 mm human embryo, but in later fetuses it appeared that it was only the endothelial cells of blood vessels that contained this enzyme. Kabat and Furth (1941) observed a positive reaction in the capillary walls of embryonic mouse muscle, while the other authors (Beckett and Bourne, 1958) found alkaline phosphatase in the endothelium of capillaries and of larger vessels in skeletal muscle of both goat and human fetuses. In the human fetus, however, the reaction in the intima of larger vessels was more common than that in the capillaries and there was slight alkaline phosphatase activity in the connective tissue between the striated muscle fibers. As in adult man, only a certain percentage of the total number of blood vessels of fetal muscle gave a positive alkaline phosphatase reaction. In fetal goats, this percentage and the total number of vessels which showed enzyme activity was variable in the extreme. There appears to be no correlation between the number of positive vessels and the age of the fetus from which the muscle was taken. In this, skeletal muscle closely resembled cardiac muscle.

In addition to the enzyme activity displayed by blood vessel walks, alkaline phosphatase could also be detected in nerve axons of fetal goat muscle (Beckett and Bourne, 1958). No positive nerve axons were, however, found in muscle of 3- and 4-month human fetuses. Alkaline phosphatase activity had been previously noted in developing nerve fibers of Amblystoma embryos by Elftman and Copenhaver (1947) and in the peripheral nerves of a 5 mm human embryo by McKay et al. (1955).

B. Acid Phosphatase

Acid phosphatase, as demonstrated by the Gomori lead phosphate technique (Volume III, Chapter 9), has a much more widespread distribution than alkaline phosphatase, although, like this latter enzyme, it has been little studied in fetal striated musculature.

1. ACID PHOSPHATASE IN CARDIAC MUSCLE

A. CARDIAC BLOOD VESSELS. Rossi *et al.* (1952) reporting their findings on human fetuses of from 1 to 7 months gestation, stated that no acid phosphatase activity could be detected in blood vessels of any type even after 20–48 hr incubation with substrate. Later work by these authors showed that there was no acid phosphatase in the major blood vessels of the heart of a 9 mm CR human embryo, and that at later stages of development (up to 36 cm CH), some of the large blood vessels were negative, while others showed a positive reaction in their mesenchymal nuclei.

The results described by Rossi and his co-workers are in complete contrast to those obtained by other authors, e.g., Beckett and Bourne, (1958) using intact hearts of $2\frac{1}{2}$–$5\frac{1}{4}$ inch CR goat fetuses and of 12–16-week human fetuses, in addition to specimens of ventricular muscle taken from goat fetuses of 10–18 inch CR. Our incubation periods for the demonstration of acid phosphatase were 4–6 hr only.

In fetal goat hearts, the strongest reaction was in the walls of all blood vessels except capillaries. The reaction was located both in the fibroblasts and in the nuclei of all types of cell present in the blood vessel walls. The picture seen in the human fetal hearts was essentially the same, but in the human the most intense reaction was situated in cone-shaped groups of granules lying at the poles of the nuclei of the great vessels entering or leaving the heart. In addition, the nucleoli of blood vessels (which are always more strongly positive than the nuclei of these structures) were more conspicuous in human than in goat hearts.

There was a second striking difference between goat and human to be observed in these intact hearts. There were present beneath the endocardium of the human hearts, particularly in the atrial region, some large cells, squarish in shape and packed with large granules, which gave a most intense acid phosphatase reaction. These cells looked very much like mast cells, but could not be stained either metachromatically or with Gomori's (1950) aldehyde fuchsin, so that this morphological identification is likely to be incorrect.

B. PURKINJE FIBERS. The only observations on acid phosphatase in Purkinje fibers seem to be those of the present authors. This is perhaps not surprising, since most other work has made use of human fetal material, and we could find no trace of these fibers in the human fetal hearts of

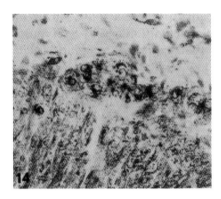 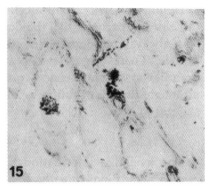

Fig. 14. Acid phosphatase in the Golgi apparatus of Purkinje fibers of fetal goat cardiac muscle.

Fig. 15. Acid phosphatase in the Golgi apparatus of Purkinje fibers of adult goat cardiac muscle.

our series. In fetal goats, a strikingly intense acid phosphatase reaction was present in the perinuclear zone of Purkinje fibers, i.e., in what probably represents the Golgi apparatus of these fibers. This very intense reaction was found in all specimens, except one of 2½ inch CR and one of 18 inch CR, and was also very clearly demonstrable in adult goat heart. The nucleus and peripheral myofibrils of the Purkinje fibers gave a light to moderate reaction intensity (Figs 14 and 15).

C. Myocardium. Rossi *et al.* (1952, 1953) reported that in their series of human fetuses, which ranged from 9 mm CR to 36 cm CH, the reaction in the myocardium was feeble and was limited to the nuclei of the muscle fibers. Other results, e,g., Beckett and Bourne (1958) showed rather a different picture. The reaction intensity was very variable, and there was no observable correlation with the age of the fetus concerned or with any known technical variation. One must therefore reserve judgment about whether the different reaction intensities reflected genuine variations in the level of enzyme activity or whether they were due to lack of reliability of the technique. In spite of the difficulties of interpretation arising from the variability of reaction intensity, something can be said about the distribution of acid phosphatase activity.

In general, in the fetal goat, the greatest reaction intensity was observed in the nucleoli of the myocardium. There was rather less activity in the nuclei, and a light to moderate reaction for acid phosphatase in the cytoplasm. Where the cytoplasmic activity rose to moderate levels, cross striations were visible, and in the adult, the intercalated disks were enzyme positive. There appeared to be no significant difference in enzyme activity between auricular and ventricular muscle in either the young fetal goat hearts or the human hearts.

In addition to the sites previously mentioned, acid phosphatase was sometimes also present in groups of granules at the poles of the nuclei. These granules have in the past been designated either as the Golgi apparatus (Beams, 1929; MacDougald, 1936; Eastlick, 1937; Moog, 1944) or as perinuclear sarcosomes (Kisch, 1951; Cleland and Slater, 1953). They were strongly positive in all four human hearts examined, but they were not a constant feature of the fetal goat cardiac muscle. In fetal goat tissue there were no enzyme-positive perinuclear granules at the $2\frac{1}{2}$ inch CR stage, but they made their first appearance at $5\frac{1}{4}$ inch CR (64 days). After this, acid phosphatase activity in the granules increased, reaching a maximum at about 100–110 days, and then decreased again, reaching zero at 120–130 days, and staying at this level until term (150 days). As this sequence has been observed only in one series of 21 specimens, however, one cannot be certain whether or not it represents a true picture of events. It will be necessary to make many further observations before this apparent rise and fall in acid phosphatase activity can be accepted as being a normal part of embryonic development. If, indeed, the acid phosphatase activity in these perinuclear granules does fall away to zero during the last part of fetal life, it must be regained at some later stage, since the cardiac muscle of the adult goat again shows these enzyme-positive granules.

2. Acid Phosphatase in Skeletal Muscle

As in the case of fetal cardiac muscle, very few observations have been made on the distribution of acid phosphatase in fetal skeletal muscle, but the pattern of enzyme activity appears to be very similar in the two tissues, and very similar also to that observed in adult skeletal muscle.

In the chick, there appears to be little acid phosphatase in muscle except in the very early stages of development. Moog (1944) reported that this tissue lost practically all its acid phosphatase activity during differentiation from the mesoderm. Rossi *et al.* (1953, 1954) found a similar picture in fetal human muscle, since they were able to obtain a positive reaction only in the nuclei of the skeletal muscle of their series of fetuses.

In contrast to the observations mentioned above, Wolf *et al.* (1943) found quite a lot of acid phosphatase activity in adult skeletal muscle of various animals. In the skeletal muscle fibers themselves, there was a moderately intense reaction in the nuclei and sarcolemma, and the cross striations were sometimes visible. In addition, the smooth muscle cells and the nuclei of all cell types in the walls of arteries and veins contained acid phosphatase. There was also a strong reaction for this enzyme in the axons, the nuclei and cytoplasm of Schwann cells and in the endoneurial cells of peripheral nerves (Figs. 16 & 17).

The observations of the present authors (Beckett and Bourne, 1958) on fetal goat and fetal human muscle are much more in tune with those of Wolf *et al.* (1943) than with those of Rossi *et al.* (1953, 1954) or Moog (1944). The overall reaction intensity observed in the course of our work on skeletal muscle was, as with cardiac muscle, variable in the extreme, and was apparently not correlated either with fetal age or with known technical differences. A high reaction intensity was seen in nerve axons and in the perinuclear granules of both the muscle fibers themselves and of the connective tissue cells. There was a large variation in the number of groups of positive perinuclear granules. The variation appeared to be random and not connected with different stages of pre-natal life. In fetal human muscle, the perinuclear granules were far less conspicuous than those in fetal goat muscle. The nucleoli of all structures contained in skeletal muscle of both fetal goat and fetal human possessed a high degree of acid phosphatase activity. In addition, a moderate reaction was observed in the walls of blood vessels, and in the nuclei of both connective tissue and skeletal muscle fibers. Sometimes

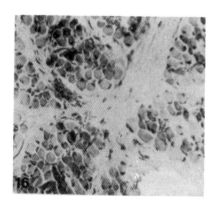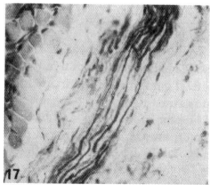

Fig. 16. Acid phosphatase in skeletal muscle of fetus.
Fig. 17. Acid phosphatase, skeletal muscle goat fetus showing positive nerve axons.

acid phosphatase was present in the muscle fibers themselves, and cross striations were then positive. As with a perinuclear granules, the variation in acid phosphatase activity in the muscle fiber seemed to be completely random.

Chodynicki and Kui (1969) found in fetal guinea pigs 12–15 mm long that the morphological differentiation of the fibers was accompanied by increased acid and alkaline phosphatase activities and was also accompanied by increased glycogen in the sarcoplasm of developing myoblasts.

IV. 5-Nucleotidase

The technique for the demonstration of this enzyme in acetone-fixed material was introduced by Gomori (1948), but it has been very rarely used for the study of fetal muscle.

A. 5-Nucleotidase in Cardiac Muscle

McKay *et al.* (1955) noticed that 5-nucleotidase was present in the endothelium of the heart of a 5 mm human embryo. The work of the authors Beckett and Bourne (1958) has to some extent confirmed that of McKay *et al.* and has amplified it. Our observations showed that in fetal muscle there was a specific enzyme that would hydrolyze 5-nucleotide, but not glycerophosphate, at pH 8.25 and that had a distribution totally different from alkaline phosphatase. Long incubation periods (8 or 16 hrs) were required for the demonstration of 5-nucleotidase, but its localization in fetal muscle was precise, with no evidence of diffusion. It must be pointed out, however, that in some other tissues, for instance rat muscle, the incubation period must be considerably shortened in order to attain clearcut localization of enzyme activity.

After 8 hrs incubation, the hearts of the $2\frac{1}{2}$ inch CR goat fetuses used in our series showed 5-nucleotidase activity in the endocardium and in auricular muscle. The intimas of the larger blood vessels (i.e., those larger than capillaries) and the walls of the aorta, where this structure was present in the section, also gave a positive reaction. In all of these sites there was high enzyme activity in the nuclei and moderate activity in the cytoplasm. There was no 5-nucleotidase present in ventricular muscle. If the incubation period was extended to 16 hrs,

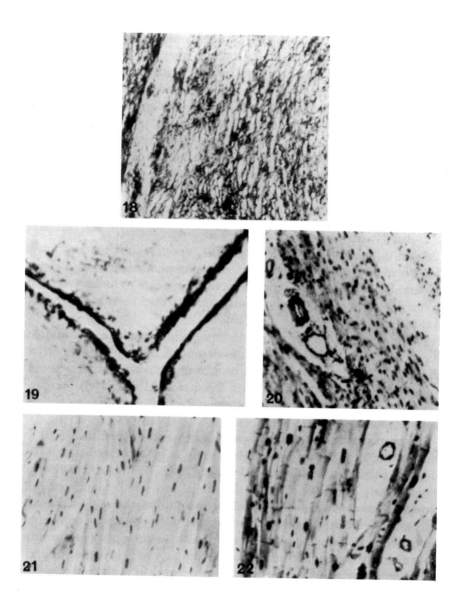

Fig. 18. 5-Nucleotidase in aortic wall of goat fetus.

Fig. 19. A weak 5-nucleotidase reaction in fetal goat cardiac muscle. The enzyme activity is to the endothelial lining of the heart.

the reaction intensity became greater and the nuclei of the ventricular muscle adjacent to the endocardium appeared to contain 5-nucleotidase. Whether the positive reaction in ventricular nuclei was due to diffusion or to the existence of 5-nucleotidase in them is difficult to decide.

In the 3-month human fetal heart, after 8 hr incubation with substrate, 5-nucleotidase activity was almost entirely located in the endocardium and a little of the underlying tissue and also in the heart valves. As in goat, the reaction was primarily nuclear, and there was no activity in ventricular muscle. After longer periods of incubation, 5-nucleotidase activity was visible in the nuclei of the walls of the smaller blood vessels and in auricular muscle (Figs. 18–22).

The degree of enzyme activity displayed in specimens of ventricular muscle taken from older goat fetuses was very variable. In the sarcoplasm and myofibrils of both cardiac muscle fibers and Purkinje fibers, the reaction intensity varied from nil to very intense. Most 5-nucleotidase activity was observed in the endocradium and in the nuclei and intimas of the arteries and veins of the myocardium. In addition, there was much enzyme activity in the nuclei of both capillaries and loose connective tissue and also in the nucleoli of the cardiac muscle fibers. Both adult goat cardiac muscle and that obtained from an 18 inch CR (term) fetus showed a slight reaction in the sarcoplasm and myofibrils and a strong positive reaction in intercalated discs. A similar distribution was seen in human adult cardiac muscle, but with one difference. The nucleoli of the latter tissue contain less 5-nucleotidase activity compared with their nuclei than do the nucleoli of goat cardiac muscle nuclei. The nuclei are therefore more conspicuous in the goat.

B. 5-Nucleotidase in Skeletal Muscle

For the discussion of this subject, the authors will again have to rely only on their own observations on goat and human fetal muscle, since

Fig. 20. A strong 5-nucleotidase reaction in fetal goat cardiac muscle. Enzyme activity is present in the cardiac muscle fibers themselves, as well as in the walls of blood vessels, etc.

Fig. 21. 5-Nucleotidase activity in adult goat cardiac muscle. Note the positive nuclei and intercalated discs.

Fig. 22. 5-Nucleotidase activity in adult human cardiac muscle. The distribution is similar to that seen in the goat. The technique was carried out at pH 7.8 for this tissue.

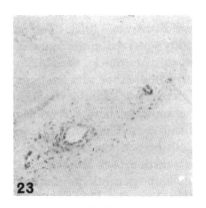 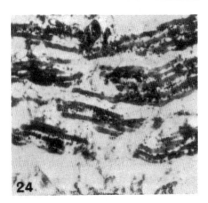

Fig. 23. A weak 5-nucleotidase reaction in fetal goat skeletal muscle. Enzyme activity is limited to blood vessel walls.

Fig. 24. A strong 5-nucleotidase reaction in fetal goat skeletal muscle. There is a great deal of activity in the muscle fibers themselves as well as in blood vessel walls and nerve fibers.

there appears to be no reference to other work in the literature. The distribution of 5-nucleotidase in skeletal muscle was very similar to that in cardiac muscle and was equally variable. Most 5-nucleotidase activity in fetal muscle was present both in connective tissue fibers and nuclei of the arteries and veins. Equally high enzyme activity was present in tendon fibroblasts and in the nuclei and fibers of connective tissue situated between the muscle fibers. In addition, in the goat but not in the human, there was a strong reaction in the axons and connective tissue sheaths of nerves present.

The overall reaction intensity in fetal goat muscle was extremely variable, and this was reflected in the extent to which the muscle fibers themselves displayed enzyme activity. Although in muscle fibers enzyme activity varied from nil to very intense, there appeared to be no correlation between the reaction intensity and the age of the fetus from which the muscle was taken. Adult goat and human muscle showed either no 5-nucleotidase activity at all, or a reaction that was limited to the walls of blood vessels of larger size than capillaries (see Figs. 23 and 24).

V. ATPase

Klein (1966) found that in the developing myocardium of the chicken embryo, a biochemical analysis of the ATPase in the isolated cardiac muscle nuclei paralleled very well the histochemical localization of the enzyme in the nuclei and that the Mg^{2+}-activated ATP phosphophydrolase predominated over the Ca^{2+}-activated enzyme.

In skeletal muscle, Arese and Rinando (1964) found that the ATPase levels in the chick embryo showed a continuous increase with development and that the increase was especially great during late embryonic development. They obtained similar results with creatine and creatine phosphate. Trayer and Perry (1966) found that in rabbit, guinea pig, and chick embryos, all fetal myosins were low in ATPase activity—in the early fetuses the level was only 30% of the adult, but like other workers, they found a progressive increase of the activity of this enzyme with increasing embryonic age. Obinata and his colleagues (1966) found thin filaments of myosin 60–80 Å in width in the myoblasts of chick embryos only 3 days old that showed ATPase activity.

VI. Sulfhydryl and Amino Groups

The histochemical procedures for demonstration of these two types of protein-bound groupings (Volume III, Chapter 9) were carried out by the present authors on one or two specimens of fetal goat and human skeletal muscle. Although the number of specimens used was so small that the results can only be regarded as tentative, it may be worth while recording that there appeared to be a consistently lower concentration of these groups in fetal than in adult muscle.

VII. Glycogen and Glycolytic Enzymes

Rinando and his colleagues (1965) studied the presence of glycogen and a variety of glycogen-related enzymes in the chick embryo from the sixth to the twenty-first day of incubation. In the earlier stages, the muscle demonstrated a very high glycogenesis and a decreased glycogenolysis which was demonstrated by the accumulation of glycogen. During development, there was a striking increase in the enzyme UDPG glycogen glycosyl-transferase, which by the time of hatching had reached levels twice that of the mature chicken. Phosphoglucomutase remained at a low level throughout incubation of the embryo, and these levels were much below those shown by the adult chicken.

Between birth and 18 weeks of age in mammals, Mann and Salafsky (1970) found a marked elevation in glycolytic enzyme activity in the anterior tibialis muscle, whereas the oxidative enzyme activity showed very little change from birth to 18 weeks, and glycolytic activity and oxidative enzyme activity in the soleus muscle were unchanged in that same period. Some of these animals had one limb immobilized from birth for 18 weeks, and this produced atrophy in the two muscles. The authors found that the glycolytic enzymes and the oxidative enzymes in the legs that had casts were generally unchanged, but the contralateral anterior tibialis had a higher glycolytic activity than the normal muscles, a result which is not easily explained.

Various authors have studied the distribution of glycogen in embryonic muscle. Rodriquez and Rebollo (1965) studied glycogen in the chick embryo and found glycogen was present in the early myoblast stage. The largest amount, however, was found when the myotubes were formed, and the glycogen was arranged in thick aggregations along the axis of the tubes. It apparently was greater in amount in those fibers that were derived from secondary myotubes, or in other words, those that produced the future white fibers.

Bocek and Beatty (1967) studied the glycogen turnover in the muscles of the fetus of a Rhesus monkey which was only 90 days old. This is actually a little more than half the period of gestation, and they found that the glycogen could readily be mobilized from the fetal muscle and also from the neonatal muscle in both aerobic conditions and in periods of restricted oxygen availability. They point out that the Rhesus fetus near term shows a suprisingly increased tolerance to a degree of anoxia that would kill an older animal, and they believe this is due to the additional stores of glycogen that exist in the muscles of the fetus.

Yakota (1969) studied distribution of glycogen in the pectoralis major of the albino rat. He found that up to 15 days of gestation, there was very little glycogen in the myoblasts. In the period of preparatory differentiation, which extends from 17 to 22 days, he found that the glycogen had begun to increase, starting at about 17 days and reaching its highest concentration at 22 days. During the period of final differentiation, which was from birth to 10 days, there was a marked reduction in glycogen, together with the establishment of well defined cross striations in the developing muscle fibers and a regular arrangement in the myofibrils. In the adult rats, the author found the most glycogen present in the interfibrillar spaces that were closely associated with the cytoplasmic reticula.

A few people have studied the distribution of acid mucopolysaccharide granules in developing muscle. Vinogradov (1964), studying the muscles of a 16–17-day-old rat embryo, found small granules of this material in the striated muscle fibers, although its appearance was preceded by the appearance of glycogen which appeared at a much earlier stage. The acid mucopolysaccharide granules were found in the connective tissue in the muscle and also in the central region of the myotubes. Rodriguez and Rebollo (1965) also looked at acid mucopolysaccharides, this time in the chick embryo, and found that the granules were present in the periphery of the fibers, and that they corresponded to the region of the sarcolemma. Carinci and Manzoli (1964) made a study of the nuchal muscle in the chick embryo at 72 and 96 hr and before hatching; they found a striking hypertrophy of the fibers of this particular muscle with some interesting histochemical changes. The fibers changed from 9.8 μ at 15 days to 20.5 μ in diameter on the twenty-first day of incubation, and they found that the perimysium and endomysium changed from a loose fibrillar sort of mesenchyme structure into a gelatinous material containing delicate interlacing fibrils of collagen. This material was PAS positive and was metachromic, and the significance of its formation is not obvious.

VIII. Lipids

In adult animals, the muscle fibers vary in the amount of fat and lipid they contain. The dark fibers usually contain a lot of lipids, and Rebolla and Piantelli (1965) showed that in the developing muscles of the chick, lipids were present. In myoblasts derived from the myotome, there were no lipids, but myoblasts that developed from mesenchyme cells showed abundant lipid granules. Myotubes were also positive, not only for phospholipids, but also for fatty acids and neutral fats and also unsaturated lipids, but they contained no cholesterol or cholesterol esters. The authors were able to distinguish between three types of fibers at quite an early stage by the distribution of lipids within them. The red fibers contained the most lipid. They found that the adult lipid pattern in a chick was reached 1 month after hatching.

A limited number of studies have been made on the nucleic acids in developing chick muscle. Gordon and his colleagues (1966) studied DNA in the quadriceps of the rat and found that it rose to a fixed level in the 90-day-old animal. The sarcoplasmic and fibrillar proteins reached the same level in a 140 days. Growth appears to be associated with an increase in nuclei together with hypertrophy of the muscle fibers during the first 90 days. O'Neill and Strohman (1969), in a study of embryonic muscle in myotubes, showed that there was an extensive decrease in DNA polymerase.

IX. Summary

Succinic dehydrogenase activity in fetal cardiac muscle develops from moderate to strong during the course of prenatal life, but never reaches levels seen in adult muscle. In the adult, the intercalated disks are negative and the Purkinje fibers show a positive reaction in their nuclei and in their peripheral myofibrils.

The level of succinic dehydrogenase activity in skeletal muscle is considerably lower than that observed in cardiac muscle, and it increases during fetal life. The difference between fibers having a low succinic dehydrogenase activity and those having a higher activity becomes apparent before birth.

There is probably no simple esterase or lipase activity in fetal cardiac or skeletal muscle.

In the rat, cholinesterase does not appear in end plate areas in appreciable amounts until 16 days (21–22 days in full term), whereas in the goat and human it is present much earlier and in fact is there by the time that about one-third of the total prenatal life has passed. It is probable that the development of the subneural apparatus takes place much more rapidly in the goat than in the human after the first appearance of cholinesterase activity.

In the goat, cholinesterase at musculotendinous junctions appears later than that in end plate areas. This enzyme is sometimes also present in nerve axons in the fetal goat, but not in the fetal human.

Except at very early stages of development, alkaline phosphatase activity is almost entirely limited to the endothelium of capillaries and larger blood vessels in both cardiac and skeletal muscle.

Acid phosphatase has a more widespread distribution than the phosphatase demonstrable at alkaline pH. In both cardiac and skeletal muscle, most activity is centered on the nucleoli and perinuclear granules. There is rather less activity in nuclei and less still in the striated muscle fibers themselves and the smooth muscle cells of blood vessels. The overall reaction intensity of this acid phosphatase is extremely variable, but it is not certain that this is due to technical difficulties. Acid phosphatase activity is sometimes present in peripheral nerves of skeletal muscle.

There appears to be a specific enzyme in cardiac and skeletal muscle capable of hydrolyzing 5-nucleotide. This enzyme is particularly associated with nuclei of all types and with membranes, It is therefore limited in distribution to the walls of blood vessels of larger caliber than capillaries and to the endocardium when there is little total activity. When there is more activity present, it is also situated in the nuclei of connective tissue and sometimes in the substance of muscle fibers. On the whole, there is more 5-nucleotidase in fetal than in adult striated muscle.

Fetal myosin has a low ATPase activity, being only one-third that of adult muscle, but the activity increases progressively through embryonic life.

REFERENCES

Arese, P., and Rinando, M. T. (1964). *Boll. Soc. Ital. Biol. Sper.* **40**, 117.
Beams, H. W. (1929). *Anat. Rec.* **44**, 237.
Beckett, E. B., and Bourne, G. H. (1957). *J. Neurol., Neurosurg. Psychiat.* **20**, 191.
Beckett, E. B., and Bourne, G. H. (1958). *Acta Anat.* **35**, 224.
Bevelander, G., and Johnson, P. L. (1950). *Anat. Rec.* **108**, 1.
Bocek, R. M., and Beatty, C. H. (1967). *Pediatrics* **40**, 412.
Bourne, G. H. (1953). *Nature (London)* **172**, 588.
Brachet, J. (1940). *Arch. Biol.* **51**, 167.
Brotchi, J., and Mildadenov, S. (1967). *Arch. Int. Physiol. Biochim.* **75**, 543.
Buño, W. (1951). *Anat. Rec.* **111**, 123.
Buño, W. (1954). *Gaz. Med. Port.* **7**, 198.
Buño, W., and Mariño, R. G. (1952). *Acta Anat.* **16**, 85.
Carinci, P., and Manzoli, F. A. (1964). *Boll. Soc. Ital. Biol. Sper.* **40**, 110.
Chodynicki, S., and Kui, A. (1969). *Folia Morphol. (Warsaw)* **28**, 395.
Cleland, E. C., and Slater, K. W. (1953). *Quart. J. Microsc. Sci.* **94**, 329.
Cöers, C. (1955). *Acta Neurol. Psychiat. Belg.* **55**, 741.
Couteaux, R. (1953). *C. R. Soc. Biol.* **147**, 1974.
Diculesco, I., Onicesco, D., and Mischin, L. (1964). *J. Histochem. Cytochem.* **12**, 145.
Dorn, A. (1968). *Acta. Histochem.* **31**, 14.
Dubowitz, V. (1965a). *J. Neurol., Neurosurg. Psychiat.* **28**, 516.
Dubowtiz, V. (1965b). *J. Neurol., Neurosurg. Psychiat.* **28**, 519.
East, E. W. (1931). *Anat. Rec.* **50**, 201.
Eastlick, H. L. (1937). *J. Morphol.* **61**, 399.
Elftman, H., and Copenhaver, W. M. (1947). *Anat. Rec.* **97**, 385.
Filogamo, G., and Gabella, G. (1967). *Arch. Biol.* **78**, 1.
Gerebtzoff, M. A. (1956). *Extr. Ann. Histochim.* **1**, 26.
Gerebtzoff, M. A., Philippot, E., and Dallemagne, M. J., (1954). *Acta Anat.* **20**, 237.
Germino, N. L., Wahrmann, J. P., and Dalbora, H. (1966). *Acta Anat.* **64**, 282.
Gomori, G. (1948). *Proc. Soc. Exp. Biol. Med.* **68**, 354.
Gomori, G. (1950). *Amer. J. Clin. Pathol.* **20**, 665.

Goodwin, B. C., and Sizer, I. W. (1965). Develop. Biol. 11, 136.
Gordon, E. E., Kowalski, K., and Fritts, M. (1966). Amer. J. Physiol. 210, 1033.
Horowitz, N. H. (1942). J. Dent. Res. 22, 519.
Kabat, E. A., and Furth, J. (1941). Amer. J. Pathol. 17, 303.
Kamieniecka, Z. (1968). J. Neurol. Sci. 7, 319.
Khera, K. S., and Lohin, Q. N. (1965). J. Histochem. Cytochem. 13, 559.
Kisch, B. (1941). Exp. Med. Surg. 9, 333.
Klein, R. L. (1966). J. Histochem. Cytochem. 14, 669.
Kupler, C., and Koelle, G. (1951). J. Exp. Zool. 116, 397.
Lorch, I. J. (1947). Quart. J. Microsc. Sci. 88, 67.
MacDougald, T. J. (1936). Z. Zellforsch. Mikrosk. Anat. 24, 399.
McKay, D. G., Adams, Jr., E. C., Hertig, A. T., and Danziger, S. (1955). Anat. Rec. 122, 125.
Mann, W. S., and Salafsky, B. J. (1970). J. Physiol. (London) 208, 33.
Moog, F. (1944). Biol. Bull. 86, 51.
Nystrom, B. (1968). Acta. Neurol. Scand. 44, 405.
Obinata, T., Yomamoto, M., and Maruyama, K. (1966). Develop. Biol. 14, 192.
O'Neill, M., and Strohman, R. C. (1969). J. Cell. Physiol. 73, 61.
Pritchard, J. J. (1952). J. Anat. 86, 259.
Rebollo, M. A., and Piantelli, A. (1965). Acta Neurol. Latinoamer. 10, 181.
Rinando, M. T., Guinta, C., Bozzi, M. E., and Bruno, R. (1965). Enzymologia 46, 321.
Rodriguez, M. M., and Rebollo, M. A. (1965). Acta Neurol. Latinoamer. 10, 123.
Rosa, C. G., and Velardo, J. T. (1954). J. Histochem. Cytochem. 2, 110.
Rossi, F., Pescetto, G., and Reale, E. (1951). Z. Anat. Entwicklungsgesch. 115, 500.
Rossi, F., Pescetto, G., and Reale, E. (1952). Ext. C. R. Ass. Anat., 39th Reunion p. 744
Rossi, F., Pescetto, G., and Reale, E. (1953). Z. Anat. Entwicklungsgesch. 117, 36.
Rossi, F., Reale, E., and Pescetto, G. (1954). Ext. C. R. Ass. Anat. pp. 1–102.
Shen, S. C. (1949). Anat. Rec. 105, 489.
Sippel, T. O. (1954). J. Exp. Zool. 126, 205.
Straus, W. L., Jr. (1939). Anat. Rec. 3, Suppl. 109, 50.
Trayer, I. P., and Perry, S. V. (1966). Biochem. Z. 345, 87.
Vinogradov, V. V. (1964). Biol. Histochem. Cytochem. 3, 169.
Wirsen, C., and Larsson, K. S. (1964). J. Embryol. Exp. Morphol. 12, 759.
Wolf, A., Kabat, E. A., and Newman, W. (1943). Amer. J. Pathol. 19, 423.
Yakota, S. (1969). Biol. Anat. Jap. 45, 309.
Zorzoli, A., and Stowell, R. E. (1947). Anat. Rec. 97, 495.

5

POSTEMBRYONIC GROWTH AND DIFFERENTIATION OF STRIATED MUSCLE

GEOFFREY GOLDSPINK

The last decade has seen a considerable increase in our knowledge of how muscle tissue differentiates and grows, both during the embryonic stages and the postnatal stages of development. This chapter is concerned mainly with the changes that take place after the birth of the animal. However, these changes are in many cases a continuation of events that have already commenced in the embryo. Similarly, in the muscles of some species of animals, the processes that are normally regarded as embryonic differentiation are seen to continue after birth. In other words, the sequence of events during the postnatal period varies according to the rate of development of the animal and its state of maturity at birth, and therefore it is not always easy to make the distinction between what is considered to be embryonic development and what is considered to be postembryonic development. However, in the author's opinion it is not important to make this distinction. It is more important to attempt to establish a coherent theme for muscle growth in general, while still bearing in mind that the temporal sequence of the various processes may differ from species to species.

One reason for the increased interest in muscle development over the past few years is that it affords a good example of tissue differentiation and growth, in that this tissue is apparently programmed to produce considerable quantities of unique kinds of proteins, the properties of which have been extensively studied. Muscle, therefore, provides the cellular and molecular biologist who is interested in cellular differentiation and growth with an excellent system for study.

The study of muscle development is also of considerable importance in medicine and in agriculture. From the medical point of view it is important that we gain knowledge of muscle differentiation and growth in order to understand the nature of the derangement of the normal growth processes that occur with muscular dystrophy and related diseases. As striated muscle is the most abundant tissue in the body, it is also important that we understand the way in which it develops in order to appreciate the growth of the body as a whole. Of course, this does not only apply to the human being; it also applies to animals, particularly the commercial meat-producing animals. In most countries, striated muscle provides a substantial proportion of the protein in the diet. One of the most serious problems regarding the food supply for an increasing world population is relative scarcity of protein, particularly a protein source such as meat, with its high content of essential amino acids. From the agricultural standpoint, it is very desirable, therefore, that we study the way in which muscle grows and the factors that influence its growth. When this is fully understood it may be possible to increase the efficiency and the rate at which meat can be produced.

In recent years, most of the relevant studies have been carried out at the cellular and molecular level and therefore the emphasis of this chapter has been placed on the postnatal changes that occur at this level of organization. In addition to attempting to review the present day situation, the author has also taken the liberty of drawing the reader's attention to some of the more outstanding problems that still need to be solved in order to provide the complete picture of this subject.

I. Changes in Muscle Length during Postnatal Growth

A. Sarcomere Length and Number

The limbs of most species of animals approximately double in length during postnatal growth. The increase in limb length is of course accompanied by an increase in the individual muscles of the limb. However, as pointed out by Meara (1947), the fibers of the muscle may not increase in length to the same extent as the belly of the muscle. In some muscles, the fibers run obliquely to the long axis of the muscle, and therefore some of the increase in length of the muscle is attributable to the increase in the girth of the fibers. Even in muscles where the fibers do run parallel to the long axis, there seems to be a rearrangement of the fibers during growth. In these muscles the tendons appear to extend further into the muscle as growth proceeds, so that the myotendon junctions become more staggered. Nevertheless, the fibers of most muscles do lengthen considerably during postnatal growth. This increase in length is associated mainly with an increase in the number of sarcomeres in series along the myofibrils (Goldspink, 1968; Williams and Goldspink, 1971). As well as the increase in sarcomere number, there may also be an increase in the length of the individual sarcomeres in both invertebrate (Aronson, 1961; Shafiq, 1963; Auber, 1965) and vertebrate muscles (Goldspink, 1968; Walker, 1970). Some workers (Fischman, 1967; Goldspink, 1968) have suggested that the A and I filaments may be assembled by the interaction of the myosin cross bridges with the actin filaments. If this is the case, then the new sarcomeres would presumably be assembled at the sarcomere length that gives the maximum interaction of the myosin and actin. Goldspink (1968) found that the sarcomere length increased from 2.3 μ in newborn mouse muscle to 2.8 μ in the adult, with the muscle at its maximum *in vivo* length. Although in some invertebrates there is evidence that the filaments change length during growth (Aronson, 1961), in vertebrates the increase in sarcomere length does not involve a change in filament lengths. The changes in sarcomere length during differentiation and growth are therefore presumably due to the difference in the rate of muscle growth and the rate of production of sarcomeres. The sarcomere length changes may therefore vary between different species and strains of animals according to their rate of growth. The pulling out of the sarcomeres is apparently not uniform along the muscle fiber length, as the terminal sarcomeres are invariably shorter than those in the middle of the fibers. What is certain is that for the muscles studied, the main contributing factor to the increase in muscle length is the increase in the number of sarcomeres in series along the length of the fibers. The number of sarcomeres in series has been counted in individually teased fibers from the soleus and biceps brachii muscle of mice at different ages (Williams and Goldspink, 1971). During growth, the number of sarcomeres in the mouse soleus was found to increase from about 700 to 2200, with most of the increase occurring before the animals were 3 weeks old (Fig. 1).

The way in which the myofibrils increase in length is a fascinating problem. This must of course involve the addition of new sarcomeres to the existing myofibrils; however, the point or points at which the

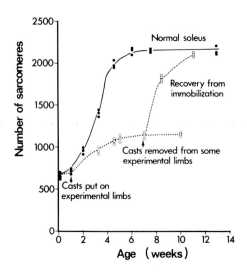

Fig. 1. Graph showing the number of sarcomeres along the length of soleus muscle fibers from mice of different ages; ● = normal soleus muscles; ○ = soleus muscles immobilized by means of plaster casts in the shortened position; □ = soleus muscles that are recovering after a period of immobilization. From Williams and Goldspink (1971).

sarcomeres are added to the myofibril has been a matter for debate for some time. Some authors (Ruska and Edwards, 1957; Schmalbruch, 1968) have suggested that the myofibrils grow interstitially; in other words, new sarcomeres are added to the myofibrils at points along their length. The evidence for this theory is based on the fact that the sarcomeres of adjacent myofibrils are often out of register because of slight differences in sarcomere length. Therefore, for a given length of muscle fiber, some of the myofibrils will have an additional sarcomere, and this is taken as evidence that a sarcomere has been inserted. However, in order to insert new sarcomeres in this way, it would be necessary not only for the myofibril to divide transversely, but it would also involve considerable modification of the sarcoplasmic reticulum and transverse tubular system. Other workers (Heidenhain, 1913; Holtzer et al., 1957; Kitiyakara and Angevine, 1963; Ishikawa, 1965; Griffin et al., 1971; MacKay et al., 1969) have suggested that the lengthening process entails the serial addition of sarcomeres onto the ends of the existing myofibrils. Electron microscope studies of the end regions of growing muscle fibers show numerous ribosome formations and myofilaments that are not properly organized into myofibrils, which suggest that this region is active in protein synthesis and myofibril assembly (Ishikawa, 1965; Mackay et al., 1969; Williams and Goldspink, 1971). The muscle–tendon junction is characterized by peripheral clefts and fingerlike invaginations into the ends of the muscle fibers. Muir (1961) and Ishikawa (1965) suggest hypothetical schemes whereby these sarcolemmal clefts provide regions where myofilaments might be added to the ends of myofibrils without the latter having to relinquish their attachment to the sarcolemma.

More direct evidence for serial addition of sarcomeres has recently been presented (Griffin et al., 1971; Williams and Goldspink, 1971). These workers injected tritiated adenosine into growing mice in an attempt to label newly formed sarcomeres. Adenosine is known to be incorporated into the structural ADP of the actin filaments and ribosomal RNA. The soluble nucleotides are removed by glycerol extraction. Autoradiography was carried out on single teased fibers and longitudinal sections (Fig. 2). Also, some muscles were sectioned transversely on a cryostat, and the amount of label in different parts of the muscle was determined by scintillation counting. The data from the autoradiography and scintillation counting showed that most of the label was incorporated into the end regions of the muscle fibers, thus demonstrating that these regions are more active in the synthesis of actin and ribosomes. This, in turn, strongly suggests that the ends of the fibers are the regions of longitudinal growth and that new sarcomeres are most probably

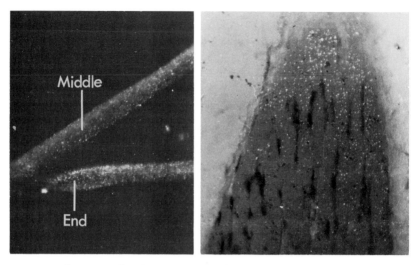

Fig. 2. Radioautographs of a single fiber (*left*) and a longitudinal section (*right*) taken from the muscles of young mice injected with tritiated adenosine for a period of 3–4 days. Note that the adenosine has been incorporated mainly into the ends of the muscle fibers, indicating that this is the region of longitudinal growth. Photographs were taken using Leitz ultrapack optics; therefore, the silver grains appear as white spots. From Griffin *et al.* (1971) and Williams and Goldspink (1971), respectively.

added serially to the ends of the existing myofibrils. This also seems to fit in with the fact that the terminal sarcomeres of the myofibrils are invariably shorter than those in the middle. These are presumably the ones that were laid down last and are not yet fully functional (Goldspink, 1968).

The molecular mechanism of the synthesis and assembly of new sarcomeres is not known; however, certain physiological experiments have shed light onto the sort of factors that may be involved in stimulating muscle fibers to produce more sarcomeres. Surgical modification of the distance that the muscle has to contract is known to radically affect the longitudinal growth of the fiber (Crawford, 1954, 1961; Alder *et al.*, 1959). Williams and Goldspink (1971) have shown that immobilization of limbs of growing mice using plaster casts results in a considerable reduction in the number of sarcomeres in series and therefore in shorter muscle fibers. However, immobilization of adult muscles in their extended position results in an increase in the number of sarcomeres in series along the fibers (Tabary *et al.* 1972; Williams and Goldspink, unpublished work).

The work of Alder *et al.* (1959) has shown that the nervous system has an important influence on longitudinal growth. This is also borne out by the studies of Tabary *et al.* (1971) on the longitudinal growth of muscle in children afflicted with cerebral palsy. In these children, the defect in the central nervous system in some way prevents the muscle fibers producing enough sarcomeres for them to attain the correct length. The inference that one may draw from these physiological and clinical studies is that the number of sarcomeres along the muscle fibers is adjusted according to the functional length of the muscle. Further studies are obviously necessary to elucidate the link between the mechanical event of contraction and the biochemical processes involved in the synthesis and assembly of sarcomeres.

B. Other Changes Associated with Increase in Muscle Fiber Length

Not only does the number of sarcomeres have to increase during growth, but the other components of the muscle have also to increase in number or size. No information seems to be available on how some parts, for example the sarcolemma of the fibers increases with growth; however something is known about the increase in the muscle fiber nuclei. Several workers (Montgomery, 1962; Enesco and Puddy, 1964; Cheek, 1968; Winick and Noble, 1966; Widdowson, 1968) have shown that the total DNA content and number of nuclei increase during muscle growth. Williams and Goldspink (1971) have counted the total number of fiber nuclei in individual teased muscle fibers, while Moss (1968a) has counted nuclei in sections in muscle of different ages. They found that the fiber nuclei continue to increase in number beyond the stage at which there was no further increase in muscle fiber length. They also found that the larger diameter fibers possessed more nuclei. Therefore, it must be concluded that the increase in the number of nuceli is associated with an increase in the girth of the fibers as well as the increase in length. The increase in the number of fiber nuclei does not apparently result from mitosis of the existing nuclei. Instead, certain cells known as satellite cells fuse with the fibers and thus donate nuclei to the growing fibers (Mauro, 1961; Shafiq *et al.*, 1968; Moss and Leblond, 1970; Mauro *et al.*, 1970.) These cells can be seen to be associated with the muscle fibers, particularly in young muscles, and they actually lie under the basal membrane of the fibers. The percentage of satellite cell nuclei with respect to muscle fiber nuclei has been shown in the subclavius muscle of the rat to decline by about 8 times between birth and maturity (Allbrook *et al.*, 1971). Kityiakara and Angevine (1963) have carried out experiments using radioactively labeled thymidine and found that there were more labeled nuclei at the ends of the fibers. This seems to suggest that the satellite cells provide the necessary additional nuclei to allow the muscle fibers to increase in length as well as girth.

II. Changes in Muscle Girth during Postnatal Growth

A. *Muscle Fiber Number*

Some confusion has arisen in the past over the question of whether or not the number of muscle cells increases during postembryonic growth. The confusion is eliminated if the definition of what is meant by a muscle cell is made clear. Cheek (1968) used DNA estimations as an index of cellular development in an extensive study of human muscle growth. During postnatal growth, he found a fourteenfold increase in the number of muscle cell nuclei and interpreted this as representing an increase in muscle cell number, as his concept of a muscle cell is the nucleus together with the cytoplasm over which the nucleus has jurisdiction. However, to most people, the muscle cell is the muscle fiber with its many nuclei. If one takes this latter definition, then it is apparent from the literature (Morpurgo, 1895; MacCallum, 1898; McMeekan, 1940a; Le Gros Clark, 1958; Ham and Leeson, 1961; Adams *et al.*, 1962; Goldspink, 1962a; Chiakulas and Pauly, 1965; Rowe and Goldspink, 1969a) that the number of muscle cells (muscle fibers) does not increase during postnatal growth, except perhaps for a short period after birth. Whether or not there is any postnatal increase at all appears to depend on the state of maturity of the animal at birth. A modest postnatal increase has been reported for rat and mouse muscle (Morpurgo, 1895; Chiakulas and Pauly, 1965; Goldspink, 1962a) and for human muscle (Montgomery, 1962). A very considerable postnatal increase in fiber number has, however, been reported as occurring in muscles of the marsupial *Setonix brachyurus* (the Australian quokka) by Bridge and Allbrook (1970). This marsupial is born in a very immature condition, and it seems that the first stage of muscle growth involves the rapid development of a small number of fibers, just enough to enable the newborn animal to climb to its mother's pouch. The full complement of fibers in the muscles is then attained at about 100 days after birth. The conclusion that can be drawn is that the increase in fiber number, which may occur shortly after birth, should really be regarded as an extension of the embryonic differentiation of the tissue.

There is apparently no difference in the total number of fibers between the same anatomical muscles of males and females (Rowe and Goldspink, 1969a). However, there are considerable differences between different genetic strains and, of course, between different species. Although small mammals tend to have small fibers and large mammals to have larger fibers, the differences in muscle size between species are attributable mainly to differences in fiber number. Black-Schaffer *et al.* (1965) carried out an interesting study on cardiac muscle for which the same situation presumably holds true. They concluded that a rat heart weighing 1.4 gm contains about 9×10^7 fibers, while the heart of a blue whale weighing 2.9×10^5 gm contains 2×10^{13} fibers, an increase in fiber number almost proportional to the weight difference. From the physiological standpoint, it is not feasible to have fibers developing beyond a certain size, because the distance to the center of the fiber would be too great to allow adequate diffusion of oxygen and possibly too great for the transmission of the impulse down the T system to the centermost myofibrils (Adrian *et al.*, 1969). During the evolution of the larger animals, it was therefore obviously necessary for the fiber number to increase rather than the fiber size.

A modification of fiber number has also apparently resulted from the artificial selection of animals. Staun (1963) estimated the number of fibers in the muscles of different breeds of pig and found that fiber number was characteristic for a particular breed. This finding has also been verified for different strains of inbred mice by Luff and Goldspink (1970b). These workers (Luff and Goldspink, 1967) have also counted the number of fibers in the muscles of mice that were originally from the same strains but which had subsequently been selected for largeness and smallness over several generations. They concluded that the difference in muscle size between the large line and the small line could mainly be attributed to difference in the total number of fibers in the muscle and not to a difference in fiber size.

The fact that fiber number becomes fixed before birth or shortly after birth and that it can apparently only normally be altered by natural or artificial selection, suggests that it is under direct or indirect genetic control. However, it may be possible to modify the fiber complement by experimental means, particularly if the experiments are carried out on the fetus or the neonatal animal during the time that the myoblasts are still proliferating. For instance, a slight modification in the rate of mitosis would presumably result in a considerable alteration in fiber number. The cellular mechanisms involved in producing a certain complement of fibers for a given muscle are unfortunately not understood at the present time.

B. *Changes in Muscle Fiber Size during Normal Growth*

It has long been recognized that muscle fibers increase in size very considerably during the postnatal growth of the animal. As stated above, the consensus of opinion is that the postembryonic increase in the girth of the muscle is almost entirely due to the hypertrophy (or further growth) of the existing fibers, not to hyperplasia.

Several workers (Morpurgo, 1895; Hammond and Appleton, 1932; Joubert, 1956; Bowden and Goyer, 1960; Chiakulas and Pauly, 1965; Bridge and Allbrook, 1970) have measured the overall increase in fiber size during growth. For instance, it is well known that the mean fiber diameter is greater in the male than in the female (Bowman, 1840; Schwalbe and Mayeda, 1891; Hammond and Appleton, 1932; Rowe and Goldspink, 1969a). However, few workers have attempted to study the way in which the fibers increase in size or the ultrastructural and biochemical changes that accompany the postnatal growth of the fibers. As pointed out by several authors (Morpurgo, 1895, Rowe and Goldspink, 1969a; Shear, 1969), it is necessary to choose a muscle with a suitable fiber arrangement and to pay particular attention to the method of fixation (Goldspink, 1961) in order to obtain accurate measurements of fiber size. All too often, general histological fixatives have been used for studies of fiber size and insufficient attention has been paid to the choice of muscle for this type of investigation.

The way individual muscle fibers grow is of particular interest. Studies by Goldspink (1962a), Goldspink and Rowe (1968), and Rowe and Goldspink (1969a) on rodent muscles indicate that the muscle fibers grow in a discontinuous way rather than in a gradual and continuous manner. Shortly after birth all the fibers are about the same size, 15–20 μ in diameter. In some muscles, such as the soleus and extensor digitorum longus, the fibers stay at this basic level of development throughout the life of the animal. However, in other muscles, such as the biceps brachii, some of the fibers undergo further growth or hypertrophy to a size of about 40 μ in diameter. The conversion of small fibers into larger fibers can sometimes be detected during growth by plotting fiber size distributions from muscles at different ages (Fig. 3). In the mouse soleus and extensor digitorum longus muscles, the modes and shapes of the distribution plots do not change to any great extent during growth. However, in the mouse biceps brachii and anterior tibialis muscle, the appearance of the large fibers is discernible by 3 weeks of age, when a fiber size peak appears at about 40 μ. As the animal grows, the number of fibers at this peak increases. In the mature biceps brachii this popula-

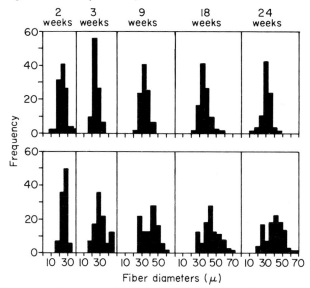

Fig. 3. Frequency histograms showing the distribution of fiber sizes in the soleus (*top*) and biceps brachii (*bottom*) muscles from mice of different ages. Each histogram is for a 100 fiber sample taken at random. From Rowe and Goldspink (1969a).

tion of large fibers is approximately equal to the number in the small phase. As mentioned below, the proportions of fibers in the large or small phase can be altered considerably by exercise or changing the level of nutrition in the animal. The physiological reason for the conversion of the small phase fibers into large phase fibers becomes apparent when the ultrastructure of fiber in these two phases of development is examined (Goldspink, 1970) (Fig. 4). The large phase fibers possess larger myofibrils and a greater number of myofibrils. Indeed, the increase in girth of the fiber can be explained almost entirely by the increase in the number and size of the myofibrils. It must be stated that the biceps brachii and anterior tibialis muscles of the mouse are somewhat unusual in that the muscle fiber growth in this animal is very rapid. In larger muscles, which grow at a slower rate, the discontinuous growth of the fibers apparently is not usually detectable by constructing distribution histograms. In these muscles, the fibers probably undergo several increases in size over a relatively long period of time before they attain their ultimate size. This is contrasted with the situation in the mouse muscle where the fibers undergo only one postembryonic size increase to reach their ultimate size, and this occurs in a relatively short time. A general scheme for the different stages of muscle fiber development is presented in Fig. 5.

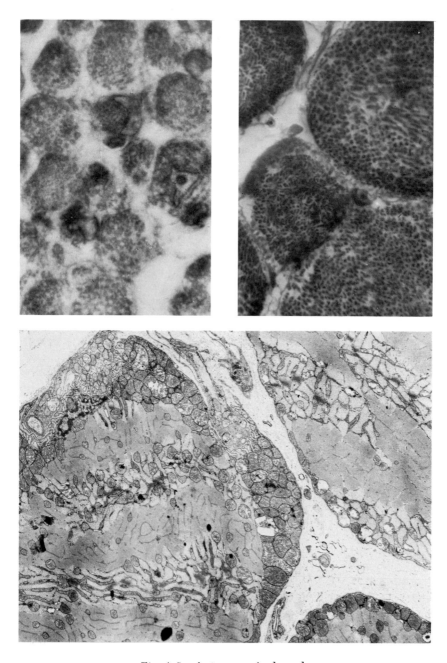

Fig. 4. See facing page for legend.

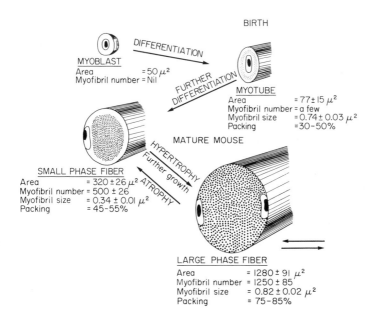

Fig. 5. The different phases of muscle fiber development during the postnatal growth of the mouse. The composition of the mouse biceps brachii muscle is 40% small and 60% large phase fibers, whereas in the soleus muscle all the fibers stay in the small phase. The last set of arrows indicates the further increase in fiber size that may occur in the muscles of larger animals.

The stimulus that causes a fiber to undergo further growth or hypertrophy is presumably the intensity of the work load to which the fiber is subjected. As the body weight of the animal increases during growth, the work load on most of the skeletal muscles must also be considerably increased. It seems reasonable to suggest that when the work load threshold for an individual fiber is exceeded, this in some way triggers the production of more myofibrils and causes the fiber to increase in size. This idea is supported by the information on the effect of exercise

Fig. 4. (*Top*) Light photomicrographs of biceps brachii muscle from a newborn (*left*) and mature (*right*) mouse. In the mature muscle, one small phase fiber is surrounded by several large phase fibers (×1000). (*Bottom*) Electron micrograph of small and large phase fibers in the biceps brachii muscle of a mature mouse. Note the difference in the arrangement of the myofibrils and mitochondria. The myofibrils in the small fiber tend to be more irregular in cross section and the mitochondria are more peripherally situated than in the large fiber (×4000). [From *J. Cell Sci.* G. Goldspink, 9, 123 (1971).]

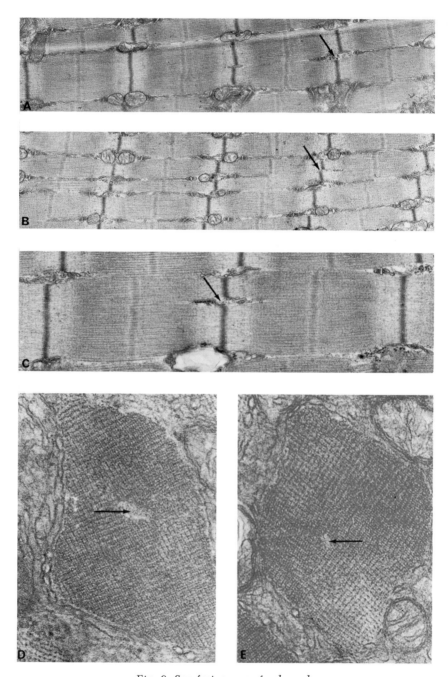

Fig. 6. See facing page for legend.

and other methods of increasing the work load on muscles discussed in Section VI.

III. Ultrastructural Changes Associated with Postnatal Development of Muscle Fibers

Electron microscope studies of embryonic development of muscle tissue are discussed in Chapter 3 of this volume. Therefore, the ultrastructural changes described here relate only to the further development of the muscle fibers that normally occurs during the postnatal growth of the animal. One of the most striking changes that occurs during this period is the increase in the amount of myofibrillar material in the fibers. Recently, the author (Goldspink, 1970) reported that during the postnatal growth of the mouse biceps brachii muscle, the number of myofibrils within an individual muscle fiber may increase as much as fifteenfold. Evidence obtained by examining muscle fibers at different stages of growth strongly suggests that this proliferation of myofibrils is the result of longitudinal splitting of the existing myofibrils once they attain a certain size. The fact that myofibrils do split was recorded many years earlier by Maurer (1894) and Heidenhain (1913). Although Heidenhain suggested that this was the method by which the myofibrils increased in number, he did not, of course, have the advantage of the electron microscope, and therefore no supporting evidence was presented. The recent evidence, however (Goldspink, 1970), includes the observations that the myofibrils in some fibers are about twice the size of the myofibrils in other fibers of the same muscle. This observation has also been very briefly mentioned by Pellegrino and Franzini-Armstrong (1969). The myofibrils of fibers are not usually round but are polygonal in cross section with straight sides. There is also a much higher incidence of apparent splitting in large and intermediate-sized fibers, whereas no splitting is seen in the small-phase fibers. These observations all indicate that longitudinal splitting of the myofibrils is the means by which they proliferate during growth (Fig. 6).

Fig. 6. (A–B) Electron micrographs of longitudinal sections showing myofibrils from the biceps brachii muscle of the mouse that are in the process of splitting (×12,500). (C) Electron micrograph of a myofibril from the biceps brachii muscle of the mouse that has apparently just commenced to split, as at this stage only one Z disc has divided. Note the presence of the sarcotubular systems in the fork of the splits (×25,000). (D–E) Electron micrographs of transverse sections of the biceps brachii muscle of the mouse showing two Z discs that have apparently just commenced to split. The splits are seen to begin as small holes in the center of each Z disc (×65,000). From *J. Cell Sci.* Goldspink (1971).

The possible mechanical reason for the splitting has been investigated (Goldspink, 1971). In mouse muscle fixed in different states of contraction, relaxation and stretch, the peripheral actin filaments of the myofibrils can be seen to run slightly obliquely to the Z disc axis. This is true for all sarcomere lengths except in the very stretched state when there is no overlap of the actin and myosin filaments. When tension is developed by two adjacent sarcomeres the oblique pull on the actin filaments will produce a stress in the center of the Z disc (Fig. 7). When the myofibrils attain a certain thickness the tension developed will be sufficient to tear a small hole in the center of the disc and the rip would then extend to the edge of the Z disc with the direction of the lattice weave. Z discs with holes in their centers have been frequently observed and these are believed to represent the first stage of the splitting process (Fig. 6). Also, it seems likely from the work

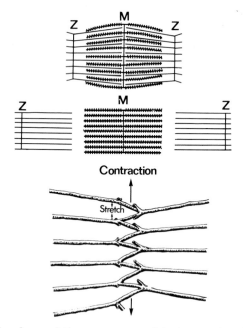

Fig. 7. (A) The shape of the sarcomere and lattice spacing during contraction (*top*) and stretch (*bottom*). Not drawn to scale. (B) The effect of the oblique pull of the actin filaments on the Z disc. When the muscle contracts, the lateral force due to the oblique pull of the actin filaments causes the Z disc lattice to open out. When the muscle is stretched, the pull of the actin filaments is less oblique and the Z disc lattice tends to close up due to its branched structure (Knappeis and Carlson, 1963; Reedy, 1964). When the myofibrils attain a certain size this lateral force is believed to cause the Z disc to rip and the myofibril to split longitudinally (Goldspink, 1969, 1971).

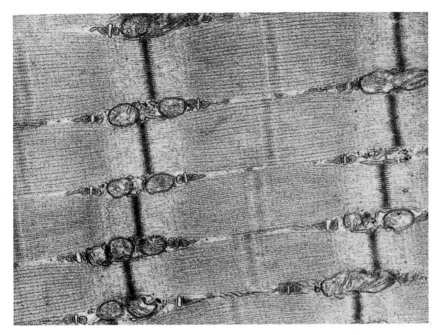

Fig. 8. Electron micrograph of a longitudinal section from a biceps brachii muscle fixed during contraction; sarcomere length = 2.0 μm. Note the oblique pull of the peripheral filaments on the bowing of the myosin filaments $\times 35,000$ if reproduced at this magnification.

of Rowe and Morton (1971) that there are frequently defects in the Z disc lattice, and in that case, one would expect that the rip would commence at the site of the defect. The reason the more peripheral actin filaments run obliquely from the Z disc to the A band is believed to be due to a slight mismatch between the square or slightly rhombic lattice of the Z disc and the hexagonal lattice of the A band. Pringle (1968) has shown that it is theoretically possible to transform the rhombic lattice into the hexagonal lattice with an equidistant displacement of the actin filaments. For a perfect transformation, the lattice dimensions are critical; the ratio of the Z and M lattice spacings should be 1:1.51. However, in all but the very stretched myofibrils, the ratio is nearer to 1:2.0 (Z:M), which means that the peripheral actin filaments are always displaced more than the central ones as they run from the Z disc into the A band. The situation is exaggerated during isotonic contractions of the sarcomeres because the myosin filament lattice expands more than the Z line lattice, although this is compensated for, to some extent, by the bowing of the myosin filaments (Fig. 8). It seems, therefore, that in fast vertebrate muscles there is this built in mechanism for ensuring that the myofilamental mass is subdivided into smaller units. If the myofilamental mass were built up as one uninterrupted mass it would present problems as far as the activation of the contractile apparatus was concerned. However, the splitting mechanism allows the sarcoplasmic reticulum and the transverse tubular system to invade the myofilamental mass and to develop at about the same rate as the myofibril content of the fibers (Goldspink, 1971).

It is well known that the myofibrils of slow muscles tend to be less discrete and more irregular in cross section than those of fast muscles. The different appearances of the myofibrils in these two kinds of muscles have been termed *"Felderstruktur"* in the case of slow muscle and *"Fibrillenstruktur"* in the case of fast muscle (Krügar, 1950; Hess, 1961). In the chicken, the *Felderstruktur* appearance of the slow fibers is known to develop during growth (Shear and Goldspink, 1970b, 1971) and it is believed to be the result of the less frequent and incomplete splitting of the myofibrils. To explain this, it has been suggested (Goldspink, 1971; Shear and Goldspink, 1971) that the rate at which tension developed by the myofibrils is perhaps more important than the amount of tension developed, as far as the ripping of the Z discs is concerned. In the chicken posterior latissimus dorsi (fast muscles) the rate of tension development increases so that it is about three to five times faster than that of the adjacent anterior latissimus dorsi (slow muscle). It is also known that the fast posterior latissimus dorsi develops more tension per unit cross sectional area than the slow anterior muscle. In the slow muscle, an extensive development of sarcoplasmic reticulum is not required, and therefore less subdivision of the myofibril mass is necessary.

It seems that the method of myofibril proliferation during growth involves an increase in the number of actin and myosin filaments in the myofibril and the subsequent splitting of the myofibril once it attains a certain size. This method of myofibril production may of course be different in many respects from that by which myofibrils are first produced during early embryonic development. The way in which more actin and myosin subunits are added to the myofibrils is not known. However, recent radioautography in conjection with electron microscopy have indicated that the newly formed contractile proteins are added to the outside of the myofibrils (Morkin, 1970). This certainly does seem reasonable, as it is difficult to imagine the new filaments being produced or added on at the center of the filament lattice. In embryonic muscle, free whole filaments are often observed in the cytoplasm of the myotubes (Allen and Pepe, 1965; Fischman, 1967; Kelly, 1969),

although some of these filaments apparently do not participate in myofibrillogenesis (Ishikawa *et al.*, 1968; Kelly, 1969). However, in growing
muscle, it is unlikely that whole filaments are added onto the existing
myofibrils, because free filaments are not usually observed in postembryonic muscle except at the end regions of the fibers. The other reason
this is unlikely is that the myofibrils in differentiated muscle are enclosed
by the sarcoplasmic reticulum and it is difficult to see how the filaments,
which are known to be semirigid, could get through this in order to
be connected to the myofibril. It seems more feasible that the myofibrils
are built up by the addition of individual protein subunits and that
the subunits, once they have been synthesized by the ribosomes, will
attach themselves to the nearest "receptive" myofibril. The self assembly
of myosin and actin molecules has been shown *in vitro* by Huxley
(1963), and it seems probable that this process operates *in vivo*. Auber
(1965) has studied the addition of filaments in developing insect flight
muscle and noted that myosin filaments on the periphery of the myofibrils are often smaller in diameter, indicating that the construction
of these peripheral filaments is incomplete. Auber's observation, therefore, lends support to the idea of the *in situ* assembly of the
myofilaments.

As the myofibrils increase in number, concomitant changes must occur
in the sarcoplasmic reticulum and transverse tubular systems. The early
postnatal development of these systems has been studied by several
workers (S. M. Walker and Schrodt, 1968; Schiaffino and Margreth,
1969; Yokota, 1969; Edge, 1970). In rat muscle, the sarcoplasmic reticulum undergoes some striking changes in the first 2 weeks after birth.
The sarcoplasmic reticulum in the newborn rat consists of a few irregularly oriented tubules which do not completely surround the myofibrils.
At this stage, there are also very few contacts between the sarcoplasmic
reticulum and the transverse tubular systems. During the first 2 weeks
after birth, the sarcoplasmic reticulum develops into its mature form
of well-organized tubules, which envelop the myofibrils, and the transverse tubular system, which arises from the plasma membrane, penetrates deep into the muscle fibers, and makes contact with the sarcoplasmic reticulum to form triads level with the edge of the A band. The
number of triads observed in sections during this period increases very
considerably during the first few weeks after birth. The transverse tubular system and the sarcoplasmic reticulum must presumably continue to
extend as long as the fiber is producing myofibrils. In the case of the
transverse tubular system, it is not known if this involves the production
of new tubules from the plasma membrane surface or whether it involves
the lengthening and more extensive branching of tubules. The extent to

which the sarcoplasmic reticulum develops varies from muscle to muscle. In fast muscle fibers, the sarcoplasmic reticulum tends to be more extensive and more highly organised than in slow muscle (Shafiq *et al.*, 1968; Shear and Goldspink, 1971; Pellegrino and Franzini-Armstrong, 1963; Gauthier and Padykula, 1966); however, it is not yet known just how the sarcoplasmic reticulum is fabricated by the muscle fiber or what controls its development in the different kinds of fibers.

IV. Histochemical and Biochemical Changes Associated with Muscle Growth

A. *Metabolic Changes*

It is well established that mature skeletal muscle is not made up of a homogeneous population of fibers. The constituent fibers of the same muscle may differ in size (Goldspink, 1962a, 1964; Rowe and Goldspink, 1969a), in their histochemical characteristics (Padykula, 1952; Ogata, 1958; Beckett and Bourne, 1960; Dubowitz and Pearse, 1960), in their structure (Gauthier, 1970; Shafiq *et al.*, 1969a), and in their physiology (Close, 1967). These differences between fibers are not usually apparent in the muscles of the embryo or neonatal animal (Dubowitz, 1965; Nyström, 1968; Close, 1964), but they become apparent as the animal grows.

The reader interested in more complete information concerning the differences between fibers both of the same muscle and of different muscles is referred to some recent reviews on the subject by Yellin and Guth (1970), Beatty and Bocek (1970), and Padykula and Gauthier (1966) and to Chapter 5 of Volume IV. In this chapter, only the enzyme changes associated with growth will be discussed. The histochemistry of developing muscle has been described by several workers (Beckett and Bourne, 1960; Dubowitz, 1963, 1965; Cosmos 1966; Fenichel, 1963, 1966; Wirsén and Larsson, 1964; Germino *et al.*, 1965; Nyström, 1966, 1968; Beatty *et al.*, 1967; Karpati and Engel, 1967; Goldspink, 1969; Goldspink and Waterson, 1971). What emerges from this work is that the postnatal pattern of enzyme development differs from species to species according to the relative state of maturity of the animal at birth. In animals such as the rat and mouse, which are relatively immature at birth, the differences between the fibers are not very apparent in the neonatal animal (Dubowitz, 1965). This is also true for the cat, except that some of the fibers do stain more intensively for ATPase activity (Nyström, 1968). However, in the guinea pig and human infant, differences between the fibers are already discernible at birth (Dubowitz, 1965; Fenichel, 1966).

Although the timing is different for muscles in different species, the general sequence of events appears to be similar. In the early fetus, the levels of oxidative and other enzymes in the myotubes are low, but as development proceeds some myotubes/fibers develop more rapidly than others with respect both to size and enzyme levels. At this stage, the larger myotube/fibers stain more darkly for succinic dehydrogenase (Goldspink, 1969). However, the slower developing fibers tend to catch up so that just before or just after birth (depending on the species) all the fibers are more or less uniform. This first stage really represents the completion of the embryonic differentiation during which time the myotubes are developing into small but proper muscle fibers. This is then followed by the further growth of some of the fibers and the emergence of enzymic differences between the fibers. The larger fibers that appear are low in oxidative enzymes but tend to develop higher concentrations of phosphorylase and enzymes associated with glycolysis (Fig. 9).

Dubowitz and Pearse (1960) named the small red fibers Type I and the large white fibers Type II; Stein and Padykula (1962) classified muscle fibers into types A, B, and C. This latter method is essentially the same as that of Dubowitz and Pearse, except that they claim to distinguish an intermediate type of fiber. Other classifications have been suggested, including one that claims to distinguish eight different types of fiber (Romanul, 1964). The problems associated with putting fibers into different categories have been discussed recently by Yellin and Guth (1970). As far as growth is concerned, it is more instructive to consider how and why the difference between individual muscle fibers arises during the development and growth of the animal rather than to discuss the pros and cons of the different classifications. Some histochemists have extrapolated from data obtained for separate fast and slow muscles and suggested that the small red fibers are physiologically slow fibers (Jinnai, 1960; James 1968; Beatty and Bocek, 1970). However, Edgerton and Simpson (1969) have recently suggested that the intermediate fibers are the slow fibers. Yet again, if one looks at the information presented by Hall-Craggs (1968) for the rabbit laryngeal muscles, the conclusion is that the small red fibers are the fast fibers. There is, however, one muscle that is known to have only fast motor units, namely the extensor digitorum longus of the rat (Close, 1967), and yet this muscle is made up of a histochemically mixed population of fibers. Therefore, one must conclude that there is no hard and fast

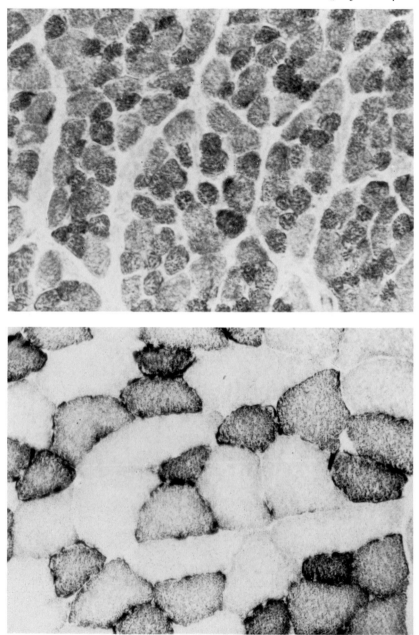

Fig. 9. Muscle fibers in the biceps brachii muscle of a very young mouse (*top*) and a mature mouse (*bottom*) stained for succinic dehydrogenase activity. Note that in the young muscle the staining of the fibers is more uniform, but in the adult muscle the small fibers stain very much more darkly ($\times 500$).

relationship between the histochemical properties of the fiber and its speed of contraction, perhaps with the exception of myosin ATPase staining. In the author's opinion the relationship is an indirect one in that the metabolism of the fibers is adapted to the activity of the muscle and the role it has to play. It is, of course, true that tonic muscles are in general composed of slow-contracting fibers, and because these muscles are involved in maintaining the posture of the animal, they are engaged in almost continuous activity and therefore tend to have high levels of oxidative enzymes. However, there are some muscles that are made up of fast-contracting fibers which are also engaged in almost continuous activity. A good example of such a muscle is the bat coracothyroid muscle, which is part of the bat's sonic radar system that causes the emission of the high frequency sound. This muscle, although it is a fast muscle, possesses many small fibers that are rich in mitochondria and oxidative enzymes (Revel, 1962).

It seems, therefore, that changes in the activity patterns of the muscles during growth may be very important in determining the metabolism of the individual fibers. However, there is still the problem as to why some of the fibers are different from others of the same muscle. It may be that in some muscles the fibers differentiate into two or more different physiological types with different rates of shortening and with different phasic and tonic functions. Whether or not the muscles of an animal contain both fast and slow fibers seems to be related to the body size of the animal. In small rodents such as the mouse, the most muscles appear to be homogeneously fast. However, with increasing mass, there appears to be more division of labor between the fibers into either phasic or tonic, and therefore these muscles appear to be innervated by both fast and slow motor units (Davies and Gunn, 1971). However, there is an additional possibility that some of the histochemical differences that arise are due to the fact that the fibers do not all grow to the same extent. It has been shown that in some muscles there is an inverse relationship between the size of the fiber and its succinic dehydrogenase concentration. The succinic dehydrogenase content of individual fibers has been measured cytophotometrically in both frog muscle (Glasz-Moerts *et al.*, 1968) and mouse muscle (Goldspink, 1969). In both cases there was a strong correlation between fiber size and enzyme concentration. This has also been found to be true for the soleus, extensor digitorum longus, and biceps branchii muscle of the rat (Goldspink and Waterson, 1971). It seems that during the postnatal growth of the fibers in some muscles at least, the mitochondria and oxidative enzyme systems do not keep pace with the production of myofibrils and therefore they become, as it were, diluted out. As a consequence, the fiber tends to rely more on glycolysis for its energy supply, and this results in the increased phosphorylase and glycolytic enzyme levels typical of these fibers. We are left, therefore, with the conclusion that the histochemical difference between fibers may arise during growth because of the differentiation of fibers, within the same muscle, which have different physiological roles (phasic and tonic) and hence different metabolic needs, and they may also arise because of the different rate of growth of the individual fibers in the muscle which results in changes in the metabolism of the faster growing fibers.

Some biochemical studies of enzyme activity during growth have been carried out recently by Perry and his co-workers (Trayer and Perry, 1966; Kendrick-Jones and Perry, 1967; Holland and Perry, 1969). Biochemical methods, although they do not give information about the metabolism of individual muscle fibers, do have the advantage that they are more easily quantitated than cytochemical methods. This group of workers has studied the rate of development of various enzymes including myosin ATPase, the Mg^{2+}-activated ATPase of the mitochondria and sarcoplasmic reticulum, creatine kinase, and 5'-AMP deaminase. The interesting points that emerge from these studies are that the postnatal rate of development of the enzyme systems differs from muscle to muscle and from species to species. Again, this depends on the relative maturity of the particular muscle compared with the rest of the musculature and the state of maturity of the animal at birth. The postnatal development of these enzyme systems can also be related to the activity of the animal. This has also been shown to be true with respect to succinic dehydrogenase (Goldspink, 1962b). This enzyme increases markedly at the time of weaning when the animals begin to move about in search of solid food. Similar changes in the cholinesterase activity of the muscle have been reported by Guth *et al.* (1966). Although the sole plate cholinesterase does not change, the background cholinesterase of the fibers increases considerably, presumably as a result of increased activity of the animal. In addition to changes in the total concentrations of different enzymes, there are also changes in the molecular species of some of the enzymes. Dawson *et al.*, (1964) have shown that change in the isoenzyme composition of lactic dehydrogenase takes place during the postnatal development of striated muscle. This enzyme is made up of four subunits which may be of two different kinds. In heart muscle and embryonic striated muscles, the H subunit is predominant. However, as growth proceeds, the number of M subunits increases at a very striking rate (see Fig. 10), so that in the adult gastrocnemius muscle of the rabbit, the lactic acid dehydro-

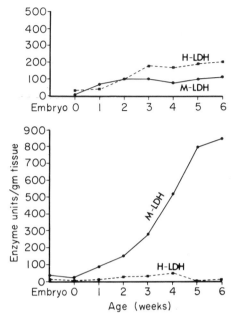

Fig. 10. Changes in enzyme units per gram of tissue (fresh weight) of H type lactic dehydrogenase and M type lactic dehydrogenase in the soleus (*top*) and the gastrocnemius (*bottom*) muscles of the rabbit. From *Science,* D. M. Dawson *et al.* (1964).

genase is almost exclusively made up of the M type. In the red muscles, the production of M type subunits is not so marked, and even in the adult, the H type still tends to be predominant. This is a very good example of metabolic adaptation of the fibers during growth. The higher K_m value of the M type means that it is capable of converting more pyruvate to lactate in a given time. Therefore, this type of subunit is required, when the animal begins to run about and use its phasic muscles, to deal with large and sporadic accumulation of pyruvate in the muscles. However, in muscles that do not engage in vigorous sporadic activity, it is an advantage to retain the H type of subunits, as it is more active at the lower concentrations of pyruvate.

The metabolic changes observed during growth can occur either by the evolutionary programming of the DNA of the presumptive myoblasts and other cells or as an adaptive response to the change in the activity pattern of the growing animal. The nature of the mechanism actually involved in bringing about the different metabolic changes is a subject for more intensive study.

B. Changes in Structure and Rate of Synthesis of Contractile Proteins during Growth

The contractile proteins change both qualitatively and quantitatively during postnatal growth. In most animals, there is a significant increase in the percentage dry weight and percentage nitrogen content of the muscle just after birth (Lawrie, 1961; N. B. Schwartz, 1961; Goldspink, 1962b; Holland, 1968). This increase in dry weight and nitrogen content is associated mainly with an increase in the myofibrillar content of the muscle fibers. In the rat the actin concentration and, therefore, presumably the other contractile proteins double within the first week after birth (Uchino and Tsuboi, 1970). After this early postnatal period, the total amount of actin in the muscles continues to increase but it increases at about the same rate as the other protein fractions. In cattle also, the concentration of myofibrils continues to increase for a time after birth, attaining the adult value at about 7 months (Lawrie, 1961).

In addition to these quantitative changes in the myofibrillar proteins, there are suggestions that the structure of the proteins is altering during this early postnatal growth period. The characteristics of the myosin molecules change in several ways. There is evidence that the specific activity of the myosin ATPase (see Fig. 11) increases and that the inhibitory effect of EGTA on the myofibrillar ATPase system also increases during this period (Bárány et al., 1965; Trayer and Perry, 1966, Holland, 1968). Recently, however, Stracher and Dow, (1970) has disputed the finding that ATPase activity of myosin increases during development. Using a different extraction procedure for the myosin, which apparently protects the low molecular weight components, they claim to be able to obtain high specific activties for the ATPase of very young muscle. The low molecular weight components are believed to be either the enzymatic site or in some way involved in the ATPase activity of the myosin molecule. However, the apparent discrepancy may well be due to a species difference, as Perrie et al. used rabbit muscle and Stracher and Dow used embryonic chick muscle. Under the extraction conditions used by Perrie et al. (1969), the myosin from 1-day-old rabbits contains four components which have a similar electrophoretic mobility as those from adult myosin, although two of the components appear to be present in smaller amounts. Whether the ATPase activity and the low molecular weight components of myosin really do change during postnatal growth is debatable; however, there is more convincing evidence that the amino acid composition of the myosin does change. Such differences, particularly in the 3-methylhistidine con-

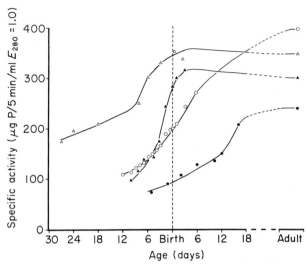

Fig. 11. Specific ATPase activity of purified myosin isolated from mixed skeletal muscle at different stages of development. Conditions of assay: 5 mM ATP, 5 mM calcium chloride, 50 mM tris hydrochloride, pH 8.2, 9.2 M potassium chloride. Adult values (averages of at least three preparations) indicated by the appropriate symbols (*top right*); ○ = rabbit; △ = guinea pig; ▲ = fowl; ● = rat. From Trayer and Perry (1966).

tent of the myosin, have been reported (Trayer *et al.*, 1968). This is supported to some extent by the finding that the immunochemical response of the myosin changes during development, although its molecular weight and hydrodynamic parameters are apparently unchanged (Trayer and Perry, 1968). It has been suggested by Perry (1970) that myosin may exist in the form of two (or more) isoenzymes with different ATPase activity, one being the fetal type and the other the adult type. As the animal develops, the fetal type is replaced by the adult type, although the ratio of fetal to adult would vary from muscle to muscle, depending on whether the muscle was fast or slow. This is a very interesting suggestion, although more work is obviously necessary in order to substantiate it.

Although it is relatively easy to measure the rate of accumulation of the different contractile proteins during growth, it is much more difficult to estimate their rates of synthesis and turnover. There are several problems associated with the measurement of the rates of synthesis *in vivo* using labeled amino acids as precursor. These include the fact that the various proteins have different rates of turnover and the fact that the size of the amino acid pools, the permeability of the plasma membrane, and the blood supply to different muscles at different ages may also differ considerably. However, there is some evidence based on incorporation studies that red muscles are more active in synthesizing proteins than white muscles (Goldberg, 1967). Red muscle is known to contain a higher concentration of RNA (Goldberg, 1967), and certainly it may be expected that because of their usual tonic function these muscles may have a greater turnover of certain enzymes. However, in contrast to this, Wool *et al.* (1968) were able to find no difference in the ability of ribosomes prepared from red and white muscle to synthesize protein. Therefore, it seems that the rate of incorporation of amino acids into the proteins of red and white muscle may be due to other factors, such as the difference in the rate of uptake of the amino acids into the fibers (Dreyfus, 1967; Goldberg, 1967) and differences in the rate of blood flow through the muscles (D. J. Reiss *et al.*, 1967).

In spite of the difficulties experienced in studying protein synthesis *in vivo*, our knowledge of protein synthesis in muscle has expanded considerably over the last five years or so. This subject has been recently reviewed by Young (1970), and therefore only a brief discussion of the most salient finding is given here. Several workers have noted that the total RNA and the ribosomal RNA in muscle decreases during postnatal growth (Devi *et al.*, 1963; Florini and Breuer, 1965; Breuer and Florini, 1965). As well as the decrease in the total ribosome content, the percentage of the ribosomes in polyribosome aggregations also decreases during growth (Breuer and Florini, 1965; Srivastava, 1969). Heywood *et al.* (1967) have shown, using embryonic chick muscle, that myosin is synthesized by very large polyribosomes made up of 55–65 individual ribosomes. The other contractile proteins are synthesized on polyribosomes varying in size between 15–25 ribosomes and 5–8 ribosomes. Tropomyosin, for example, has been shown to be synthesized by the latter category of polyribosomes (Heywood and Rich, 1968). This means that the different contractile proteins are synthesized independently and not produced from the one polycistronic strand of mRNA. This in turn means that the rate of synthesis of the different proteins may be independent of one another and may be controlled by separate mechanisms. The decrease in the polyribosome concentration during growth is probably due mainly to a simple diluting of the polyribosomes by the rapidly accumulating myofibrils. Polyribosome aggregates are certainly much more frequently observed with the electron microscope in embryonic or neonatal muscles than in adult tissue. In developing tissue, the polyribosomes are frequently seen in coiled configurations (Allen and Pepe, 1965; Fischman, 1967), and they often seem to be actually coiled around individual myosin filaments (Larson *et al.*, 1969;

Williams and Goldspink, 1971). In adult muscle, the ribosomes are not so obvious, because although the fibers probably have nearly the same total number of ribosomes (Breuer and Florini, 1965), they are distributed over a greater area, and there is also a tendency for them to be hidden by the mass of myofibrils and glycogen granules. Certainly the ribosomes involved in myofibril synthesis in adult muscle appear to be very closely associated with the myofibrils (Iyengar and Goldspink, 1971; Griffin and Goldspink, 1971). Although the concentration of ribosomes decreases with age, it is probably the production of mRNA that becomes the limiting factor in myofibrillar production. The decrease in the percentage of ribosomes in polyribosomes that occurs during growth is a reflection of the reduction in mRNA synthesis. This has been verified by Srivastava (1969), who demonstrated that the protein synthetic activity of muscle ribosome preparations could be stimulated by synthetic mRNA and that the extent to which they were stimulated depended on the age of the muscle. It is difficult to obtain information about mRNA in muscle or any other tissue, because it represents only a small percentage of the total RNA and it is readily subjected to hydrolysis by the ribonuclease of the tissue. In spite of these difficulties, Heywood and Nwagwu (1968) have apparently been successful in isolating and partly characterizing mRNA for myosin. This species of mRNA has a molecular weight of about 1.7×10^6 and contains about 6000 nucleotides.

The use of cell-free systems to study the mechanism of protein synthesis during growth has also been investigated recently by several other workers (Breuer and Florini, 1965; Srivastava and Chaudhary, 1969; Chen and Young, 1968; Wool et al., 1968; Iyengar and Goldspink, 1971). Breuer and Florini (1965) found that the ability of the cell-free systems in synthesizing proteins was relatable to their polyribosome content; in other words, their messenger RNA content. The polyribosome content of the total ribosome fraction was found to be 83% in the muscle of young rats and 72% in the muscle of older rats. Similar results have also been obtained by Batelle and Florini (1971) for chick muscle (see Fig. 12). Recently, Iyengar and Goldspink (1971) have found that polyribosome preparations from young muscles synthesize a much higher percentage of myofibrillar proteins than those from adult muscles. Therefore, it seems that not only does the quantity of mRNA change, but also the relative proportions of the different species mRNA also change during growth. These studies tend to suggest that the control of growth is at the transcriptional level rather than the translational level. However, more research is required in order to elucidate the exact nature of the control. Striated muscle offers many advantages for studies on protein

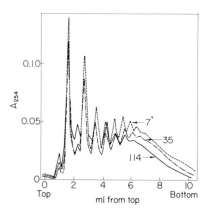

Fig. 12. Sucrose gradient profiles and activity of ribsome preparations from chicks at 7, 35, and 114 days of age. The gradient profiles shown above demonstrate the reduction in the polyribosome content of the muscle with age. The largest peak is that of the monoribosomes, the other major peaks are of polyribosomes of different sizes. The concentrations of the ribosomes at the different levels of the tube following centrifugation in a sucrose gradient were obtained by measuring the absorbance at 254 mμ. The protein-synthesizing activity of the ribosome preparations is shown in the table below. From Battelle and Florini (1971).

Age (days)	Incorporation[a] (DPM/μg ribosomes incubated)
7	400
35	150
114	130

[a] Ribosomes were isolated in parallel using a modification of the procedure of Ionasescu *et al.* [*Arch. Neurol.* **23**, 512 (1970)] and the activity of ribosome preparations was determined as the transfer of ^{3}H from [^{3}H]leucyl-tRNA into protein as described in Table II of Florini and Beuer [*Biochemistry* **4**, 253 (1965)]. Incubations were for 15 min at 37°C; incorporation was linear with time and with ribosome concentration. Data are averages of five assays at three different ribosome concentrations.

synthesis because it produces considerable quantities of unique types of proteins. It is very likely, therefore, that this problem will attract the attention of many more biochemists and cell biologists, with the result that we shall arrive at a more complete understanding of the molecular processes involved in protein synthesis and the control of protein synthesis during growth.

V. Physiological Changes during Postnatal Development

A. *Relative Amounts and Distributions of Water and Electrolytes and Resting Membrane Potentials at Different Stages of Postnatal Growth*

As early as 1857, von Bezold noted that the content of water, salt, and organic matter of biological tissues was dependent on the age of the animal. Some years later Moulton (1923) introduced the term "chemical maturity," which referred to the point during growth at which the concentrations of water, protein, and salt became constant, if expressed on a fat-free basis.

The first detailed analysis of electrolyte change in muscle during development was by Yannet and Darrow (1938) on the limb muscles of cats. They compared the chemical composition of the muscles from a group of younger animals with that of a group of older animals. From their data, it was not possible to observe any progressive changes with growth. However, they concluded that the concentration of sodium and chloride decreased and that the concentration of nitrogen (protein), phosphate, and potassium increased. They also assumed that the chloride was confined to the extracellular spaces throughout growth and that the decrease in this ion meant a decrease in the relative size of the extracellular compartments. Hines and Knowlton (1939) studied the changes in water and chloride during postnatal growth in the rat gastrocnemius muscle and found a decrease in both chloride concentration and the percentage water content of the muscle. The decrease in water content, they suggested, was due to a decrease in the water content of the muscle fibers rather than a decrease in the extracellular fluid. They also concluded that this muscle becomes chemically mature at about 90 days after birth. Barlow and Manery (1954) have studied the changes in the pectoral muscle of chickens after hatching and demonstrated that the young chick muscles had a higher concentration of sodium, chloride, and water and less potassium than the adult muscle. From this they concluded that growth involved simply a displacement of the extracellular phase, which decreased in size but maintained a constant composition throughout the growth period. Another quite extensive study was carried out by Dickerson and Widdowson (1960) on the developmental changes in human and pig muscle. They found a decrease in calcium as well as sodium, chloride, and water and an increase in potassium, phosphorus, and nitrogen. From these results they also concluded that the extracellular fluid decreased throughout growth.

In recent years, the changes in the electrolyte concentration and the decrease in water content of the muscle during growth have been confirmed by several workers (Reimold, 1962; Novikova, 1964; Vernadakis and Woodbury, 1964; Luff and Goldspink, 1970a). The rate at which the electrolyte concentrations change during growth varies between the different muscles in the same animal (Flear *et al.*, 1965; Sréter and Woo, 1963; Luff and Goldspink, 1970a) and between the same anatomical muscles in different species (Kerpel-Fornius *et al.*, 1964). However, these differences may again be due to the relative state of maturity of the animal or the particular muscle at birth. However, it is known that slow muscles in the mature animal do have a slightly different electrolyte balance (Sréter and Woo, 1963; Luff and Goldspink, 1970a).

Although there is no great dispute over the electrolyte changes that occur during postnatal growth, the change in the dimensions of the extracellular spaces during growth, in the author's opinion, is still an open question. Using the chloride space method, several workers have concluded that the extracellular compartments decrease in relative size during post embryonic growth and that this is the reason for the decrease in percentage water content of the muscle (Barlow and Manery, 1954; Dickerson and Widdowson, 1960). Other workers have employed chemical tracers such as inulin, which are believed to be capable of entering the extracellular compartment but not the muscle fibers, and have arrived at the same conclusion (Burr and McLennan, 1960; Kobayashi and Yonemura, 1967). However, the evidence for the decrease in extracellular space in growing muscle has been disputed by Goldspink (1966), Hazlewood and Nichols (1969), and Hazlewood and Ginski (1968). Although the extracellular fluid composition does not apparently change much during growth because the serum electrolytes concentrations are virtually unchanged (Luff and Goldspink, 1970a), the electrolyte concentrations within the fibers almost certainly change during growth because the resting membrane potentials change (Fudel-Osipova and Martynenko, 1965; Hazlewood and Nichols, 1966, 1967; Novikova, 1964; Harris and Luff, 1970). It seems most likely that the membrane permeability changes during the prenatal and postnatal period. This suggestion is supported by the fact that the values for extracellular space using the chemical methods often exceed the values for the total muscle water (Goldspink, 1962b; Hazlewood and Ginski, 1968), and calculations using even the smallest recorded values for the apparent extracellular space lead to negative values for intracellular concentrations of sodium (Hazlewood and Ginski, 1969; Hazlewood and Nichols, 1969). The conclusion at the present time must be that during the growth period, the partition of the electrolytes between the inside and outside of the fiber changes in addition to their relative concentrations in the muscle as a whole. This then invalidates the use of chloride space and other chemical methods for the estimation of extracellular space during development because these methods rely on a constant permeability or total impermeability of the muscle fibers, as the case may be. Also, it seems reasonable to expect the percentage water content of the fibers to decrease with growth, especially during the early phase of postnatal development, because of the considerable increase in the amount of insoluble myofibrillar proteins in the fibers. Therefore, it does not seem necessary to postulate a decrease in the extracellular fluid in order to explain the decrease in percentage water content of the muscle. This view is supported by results of a recent study (Goldspink, unpublished work) in which the percentage of extra fiber space was measured in unfixed cryostat sections of muscles from mice of different ages. It was found that after one week or so there was no further change in the percentage of extra fiber space with growth. The change that occurs during the first week or so is regarded as being part of the process of differentiation of the tissues, which, in the mouse, continues beyond birth.

One way to obtain information about the changes in membrane permeability and the distribution of ions during growth is to measure the membrane potentials and other bioelectric characteristics. Several workers have investigated the changes in the resting membrane potential during growth; Novikova (1964) and Fudel-Osipova and Martynenko (1965) have both reported an increase in the mean resting membrane potential from birth to maturity in the skeletal muscles of the rat. Hazlewood and Nichols (1967; 1968) made measurements on muscle from two groups of rats of different ages and found higher values in the older group. However, in a more recent publication, they report that the mature values are reached at 30 days after birth (Hazlewood and Nichols, 1969). F. Schwartz and Wichan (1966) showed a progressive increase in the mean resting membrane potential from 67.9 mV in the 22-day-old animal to 90.3 mV by the age of 80 days. Karzel (1968), working on breast muscle in the chicken, found a progressive increase from 27 mV in the newly hatched chick to 85 mV by 16 days and 93 mV at 60 days of age. Harris and Luff (1970) measured the resting membrane potentials in the mouse soleus (slow) muscle and the extensor digitorum longus (fast) muscle during growth. In both muscles, the potentials increased by about 20 mV between the ages of 1 week and 16 weeks; however, most of the increase took place in the first 3 weeks after birth. These workers also found that in young animals the resting membrane potentials of the slow muscle fibers were slightly higher than those in the fast muscle, but in the mature animal the situation was reversed. This latter finding is in agreement with work by Yonemura (1967), who also found that the rat soleus had a lower mean resting membrane potential than the extensor digitorum longus. It appears, therefore, that not only does the resting membrane potential change during growth, but that the mature values differ from muscle to muscle. The reason for the change in potential is not known; however, it seems likely that it is connected with the increased efficiency of the sodium pump and other ion pumping systems.

B. Changes in Contractile Properties of Muscles during Postembryonic Growth

Perhaps the two most obvious manifestations of muscle development after birth are the increase in size and the increase in contractile strength (maximum isometric tension). Although the increase in size has been quite extensively studied, the increase in contractile strength has received surprisingly little attention. N. B. Schwartz (1961) reported that the contractile strength of the combined gastrocnemius and soleus muscles of the rat increased in proportion to their wet weight during the period from weaning to maturity. However, if the strength (maximum isometric tension) was expressed on a dry weight or nitrogen basis, the younger muscles developed relatively more tension than the older muscles. Schwartz also found no change in the length–tension relationship of the muscle during the postweaning period. Alder *et al.* (1958) found that the length–tension curve for the anterior tibialis muscle of newborn rabbits was similar to that of the adults. However, they did find some difference in that the tension exerted by this muscle in full dorsiflexion was relatively smaller in the young animal. Goldspink (1968) found that the length–tension curve for the newborn mouse biceps brachii was different from that of the adult in that optimum length was greater that the maximum *in vivo* length. This can be explained as being due to the shorter sarcomere length (see p. 181). However, this difference was not apparent in slightly older mice. The shape of the curve was also different for the young mouse muscles in that the tension dropped more rapidly on each side of the tension obtained at each side of the optimum resting length. In this respect, the newborn mouse biceps brachii muscle is similar to the newborn rabbit muscle (Alder *et al.*,

1958). Recently, McComas (personal communication) has found that the length–tension curve for the human extensor digitorum brevis muscle stimulated *in situ* changes considerably during growth and that the total tension output of this muscle shows a marked increase at puberty. The change in the shape of the length–tension curve during the postnatal period is probably due to a change in the architecture of the muscle; for example, the development of a more staggered arrangement of the fibers (see p. 181). Recently, Shear and Goldspink (1971) measured the increase in contractile strength of the chicken anterior (slow tonic) and posterior (fast twitch) latissimus dorsi muscles. They found that although the fibers of the slow muscle increased in size and myofibril content more than those of the fast muscle, the fast muscle was able to develop more contractile tension per gram than the slow muscle (both muscles are approximately the same length). This was true for all stages of growth except for the prehatching period. The increase in muscle strength during growth is no doubt related to the number of myosin cross bridges in parallel. As the myofibrillar content of the muscle fibers increases during growth, then this will be accompanied by a corresponding increase in the contractile strength of the muscle. However, as suggested by the work of Shear and Goldspink (1971), it may be that the ability of the cross bridges to develop tension may differ from muscle to muscle and it may even change during growth, particularly during the early postnatal stages. This aspect is certainly worthy of further investigation.

Unlike the situation regarding the contractile strength of the muscle, the changes in the speed of contraction during growth have now been quite extensively studied. In some mammals, all limb muscles are uniformly slow at birth, whereas in adult animals they are differentiated into fast muscle and slow muscle (Denny-Brown, 1929; Buller *et al.*, 1960a,b; Buller and Lewis, 1965a; Close, 1964). These two kinds of muscles can be distinguished on the basis of their single-twitch contraction times. These two types of muscles are found in mammals and are referred to as fast and slow twitch muscles. In other vertebrates, there is a different type of slow muscle, an example of which is the anterior latissimus dorsi of the chicken, and this is referred to as a true slow or slow tonic muscle. The postnatal differentiation of fast and slow twitch muscles has been studied in the cat by Buller *et al.* (1960a,b). These workers found that both types of muscles were slow at birth. After birth, the speed of contraction of fast muscles increased considerably, while that of the slow muscles stayed more or less the same. However, surgical isolation of the lumbosacral spinal cord from incoming impulses was found to affect the slow muscle, so that its contraction times approached that of the fast muscle. This finding suggested that the differentiation of these muscles into fast and slow was under the influence of the nervous system. Approximately two years later, Close (1964) published the results of a study on the postnatal development of the soleus (slow) and extensor digitorum longus (fast) muscles of the rat (Fig. 13). He found that at birth the force:velocity properties of the muscles were virtually identical; thereafter, the intrinsic speed of shortening, or in other words the rate of shortening per sarcomere, increased in the extensor digitorum longus muscle by about three times, whereas the soleus muscle underwent little or no change. Gutmann and Hanzlikova (1966)

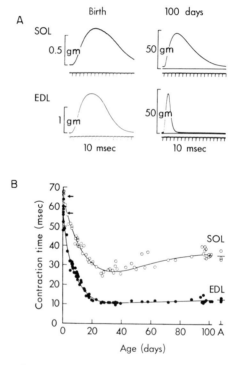

Fig. 13. Contractile properties of the extensor digitorum longus (EDL) and soleus (SOL) muscles of female rats at different stages of development. Representative records of isometric twitches at birth and 100 days are shown in A, and the time courses of change in isometric contraction time are shown in B. Force:velocity properties of muscles from newborn and 35-day-old animals are given in C. In these plots the speed of shortening of a sarcomere is plotted against the load expressed as a percentage of the maximum isometric tension $\%P_0$. The time courses of developmental change in intrinsic speed of shortening (maximum speed when the load is zero) are shown in D; each point represents one muscle. All measurements were made with the muscles at 35°–36°C. From *J. Physiol.* (*London*) Close (1964).

have also studied the differentiation of the same muscles in the rat. However, they studied the change in the responses of these muscles to various drugs such as caffeine and acetylcholine. They found that in the newborn animal, both muscles responded similarly; whereas the sensitivity of the fast muscle began to decline during the early stages of postnatal development. The development of the contractile properties of tonic type slow muscles and fast twitch muscles have been studied in the chicken by Gutmann and Syrovy (1967) and by Shear and Gold-spink (1971). In contrast with the fast and slow twitch muscles, the fast and tonic muscles do show distinct differences in prehatched and newly hatched chicks.

The reason for the change in muscle contraction speed in some muscles but not others during development is not known for certain, but there are two possibilities—either the myosin ATPase is changed as suggested by the work of Trayer and Perry (1967) and Bárány *et al.* (1965) (p. 204) or it may be connected with the more extensive development of the sarcoplasmic reticulum and transverse tubular system (p. 197). However, as Sexton and Gerstein (1967) have shown that there is still a big difference in the rate of shortening of the fibers from fast and slow muscles after glycerol extraction and activation with ATP, it would seem that the former suggestion is the more likely. The physiological reason for the postnatal differentiation of muscles with different rates of shortening becomes apparent when the energetics of these muscles is studied. Recently, Goldspink *et al.* (1970a,b) and Awan and Goldspink (1970) have shown that the slow muscles use much less energy than the fast muscles when maintaining tension. In contrast, the fast muscles use less energy than the slow muscles when they are required to shorten and do external work. It seems that over the course of evolution muscles have become adapted for the kind of role they have to play. It is probably that originally all muscles were slow but some muscles over the course of very many years have developed into fast muscles. The increase in speed of contraction, which takes place during postembryonic growth, may be regarded as an example of ontogeny recapitulating phylogeny.

VI. Factors That Influence Muscle Growth

A. Effects of Exercise and Increased Work Load on Muscle Fiber Growth

Hypertrophy of striated muscle fibers as a result of exercise has been an accepted fact for many years (Morpurgo, 1895, 1897; Siebert, 1928;

Thörner, 1934; Hoffman, 1947). Morpurgo attributed fiber hypertrophy to an increase in the sarcoplasm rather than the myofibrils. However, more recent biochemical studies (Helander, 1961) and cytological studies on striated muscle (Goldspink, 1964, 1965, 1970) and on cardiac muscle (Molbert and Jijima, 1959; Richter and Kellner, 1963) clearly indicate that hypertrophy is usually associated with a large increase in the myofibrillar material of the fiber. However, this does not preclude the possibility that under certain conditions of exercise some hypertrophy of the fibers may be due partly or wholly to an increase in the mitochondrial and sarcoplasmic proteins (Gordon *et al.,* 1967).

It is well known from observations on athletes that exercise of endurance leaves the size of the musculature relatively unchanged, although it is believed to produce biochemical changes that lead to greater stamina. On the other hand, exercises of high intensity such as weight lifting are known to be much more effective in producing muscle fiber hypertrophy. An attempt should always be made therefore to make clear what kind of exercise is being employed; whether it is of high intensity and short duration, low intensity and of long duration or of moderate intensity and moderate duration. When this is done it is easier to understand the nature of the adaptive changes resulting from the particular exercise.

Siebert (1928) exercised dogs by running them on an endless belt and found that running at relatively high speed produced hypertrophy but low speed running had no effect on fiber size. M. G. Walker (1966) exercised mice in the same manner but arrived at a different and at first sight a rather surprising conclusion, that over the range of speeds used, the duration of the exercise was apparently more important than the intensity of the exercise. In the author's opinion both the intensity and duration are important, as there is probably an inverse relationship between the minimum intensity and the minimum duration. In other words, the higher the intensity the shorter the duration of exercise required to induce the hypertrophy. When more data are available it may be possible to construct strength–duration plots for the hypertrophy thresholds of different muscles.

One of the best examples of intensive exercise is weight lifting involving relatively heavy weights. The effect of this type of exercise has been studied using laboratory animals by Goldspink (1964) and Howells and Goldspink (1971). In this work, the animals were required to pull down a food basket that was attached to a pulley system. A known weight was attached at the other end of the pulley system, and thus the amount of work performed by the animals in obtaining their food could be estimated. This type of exercise was found to be very effective in producing fiber hypertrophy. Increases of about 50% in the mean fiber cross-sectional area of the fibers in the mouse and hamster biceps brachii muscles were obtained. However, the increase in mean fiber size was the result of a certain percentage of fibers undergoing the increase in size from the small phase (20 μ diameter) to the large phase (40 μ diameter) and not to a general increase in size of all the fibers (see Fig. 14).

Using forced and voluntary exercise, Carrow *et al.* (1967) observed that in rats there was a much greater increase in cross-sectional area of the smaller red fibers than the larger white fibers. Unfortunately, these authors apparently did not investigate the possibilities of the conversion of one type or size of fiber into the other type or size of fiber. Edgerton *et al.* (1970) have carried out a similar study and concluded that the small red fibers are used preferentially during treadmill running. It may be that the red and white fibers in the rat muscles studied by these workers are physiologically discrete types with different kinds of innervation and are therefore not interconvertible. Certainly, the response of mixed muscles to exercise of various kinds needs to be studied further, because it is likely to be more complex than in the case of the more homogeneous muscles of the mouse. However, before such studies can be carried out in a meaningful way, something must be known about the physiology of the muscles. It must be known to what

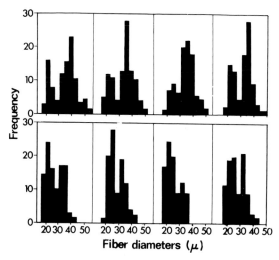

Fig. 14. Distribution histograms of fiber diameters in the biceps brachii muscles from exercised mice (*top*) and from control mice (*bottom*). Each histogram is for a 100 fiber sample taken at random. From Goldspink (1964).

extent they are mixed in terms of fast and slow motor units. The response of different physiological types of fibers has, however, been studied in different muscles that are made up entirely of one type of fiber, as they apparently are in the mouse muscle (Goldspink, 1964). As mentioned below, it is known that muscle fibers from both separate mammalian fast twitch and slow twitch muscles are capable of undergoing hypertrophy (Rowe and Goldspink, 1968). Also, M. G. Walker (1971) has recently examined the response of the fibers of the red and white muscles in marine fish. He found that if fish were repeatedly swum at low speeds, only the fibers of the red muscles underwent hypertrophy. At higher speeds when the white muscles were providing the necessary propulsion power, it was the fibers of these muscles that underwent hypertrophy.

Other methods of increasing the work load on muscle fibers have been used by several workers. These methods usually involve the surgical removal of part of the muscle and/or the removal of a synergetic muscle (Jewell and Zaimis, 1954; Crawford, 1961; Goldberg, 1967; Rowe and Goldspink, 1968; Lesch et al., 1968). Goldberg (1967) carried out tenotomy of the gastrocnemius muscle of normal and hypophysectomized rats and showed that growth hormone was not required for work-induced hypertrophy. Rowe and Goldspink (1968) using essentially the same procedure found that the mouse soleus muscle, which normally has a unimodal distribution of fiber sizes, became bimodal. In this case, the increased work load had not caused all the fibers to increase in size, but instead it caused a certain percentage of them to increase in cross-sectional area by three or four times. These workers also found that there was an increase in the number of fibers in the experimental muscles. Other workers have also recorded the formation of new fibers in greatly overloaded muscles, apparently resulting from splitting of the existing fibers (Van Linge, 1962; Rowe and Goldspink, 1968; Edgerton, 1970; Hall-Craggs, 1970). However, it is not really known whether this is some form of regeneration of damaged fibers or a normal adaptive mechanism. The crucial point that needs to be proven is whether or not these new fibers or offshoots become established as fully functional units. Using the technique in which part of the muscle is removed, it has been shown that both fast and slow muscle fibers are capable of extensive hypertrophy (Rowe and Goldspink, 1968), and it would seem, if anything, that when the work load is artificially increased the small red fibers of the slow muscle are more capable of increasing in size than the large white fibers.

From the studies described above, it is apparent that exercise of an intensive nature will produce muscle fiber hypertrophy. It is also known that this type of hypertrophy is associated with an increase in the myofibril content (Goldspink, 1964) and in the contractile strength of the muscle but not in its intrinsic speed of contraction (M. G. Walker, 1967; Binkhorst, 1969. However, less information is available about how muscles adapt to exercise of low intensity and long duration. It is known that running at relatively low speeds and swimming will induce an increase in the levels of certain enzymes (Gordon *et al.*, 1966, 1967; Jeffress *et al.*, 1968; Hensel and Hildebrandt, 1964; Hearn and Wainio, 1956, 1957) and an increase in the amount of mitochondrial protein (Holloszy, 1967). It is known that moderate exercise also increases the amount of myoglobin (Lawrie, 1953; Pattengale and Holloszy, 1967) and the number of blood capillaries (Carrow *et al.*, 1967; Edgerton *et al.*, 1969; Edgerton and Barnard, 1970).

The different response of the muscle to different types of exercise poses the question of the nature of the link or feedback mechanism between the mechanical event and the biochemical processes involved in the synthesis of myofibrillar proteins or the other type of proteins. Only very recently has it been possible to study cell growth at the molecular level, and already several interesting findings have emerged concerning muscle fiber hypertrophy. Goldberg (1968, 1969) has shown that the incorporation of leucine-^{14}C into both sarcoplasmic and myofibrillar protein is enhanced and the rate of degradation of these proteins reduced during work-induced hypertrophy. Hamosh *et al.* (1967) found that the RNA concentration increased during hypertrophy and that cell free systems prepared from muscles undergoing hypertrophy possess a greater ability to synthesize proteins. These workers found that L-phenylalanine was incorporated into protein at a faster rate by the microsome fractions prepared from hypertrophied muscle both in the absence and in the presence of artificial RNA (poly U). They also found an increase in the RNA content of the microsome fractions. These two findings indicate that the enhanced ability to synthesize proteins during hypertrophy is due to an increase in ribosomes as well as RNA.

B. Effects of Immobilization and Inactivity on Muscle Fiber Growth

Little attention has been directed to the problem of disuse of muscles in which the nerve, tendons, and blood supply remain intact. It is, of course, quite difficult to prevent an animal from using its muscles without resorting to cutting the tendons or nerve supply. One approach is to encase the limb in a plaster cast. Although this method does not prevent the animal from developing isometric tension it does reduce the activity of the muscle quite considerably. Helander (1957) showed that encasing the hind limb of a rabbit in a plaster cast resulted in the considerable atrophy of the calf muscles. This atrophy was associated with a considerable decrease in muscle fiber diameter and reduction in the number of blood capillaries. Also, both the total and the percentage of contractile proteins in the muscle decreased very markedly during disuse (Helander, 1957). Therefore, to all intents and purposes, simple disuse appears to have the exact opposite effects to exercise. An alternative approach has been used by Wells (1969) who produced immobilization of soleus and anterior tibialis muscles by rigidly fixing the ankle joint and found that muscle weights were reduced, but the tetanic tension showed an even greater reduction. Wells found no change in the speed of contraction of the muscles, although apparently fixation of both the ankle and the knee joint does result in an increase in speed of the muscles (Fischbach and Robbins, 1969). As already mentioned in Section I, immobilization of the muscle during growth has a very marked effect on the longitudinal growth of the muscles. Immobilization with the muscle in the shortened or the extended state causes a reduction in the number of sarcomeres in series along the fiber and hence a reduction in muscle fiber length.

Another method of reducing the work load of the fibers is to cut the tendon of the muscle. This has been shown to result in considerable atrophy of the fibers; however, the rate at which they undergo atrophy varies from muscle to muscle. The slow soleus muscle, for instance, is known to undergo much greater atrophy than fast muscles (McMinn and Vrbová 1962, 1964; Bergman and Afifi, 1969). In the electron microscope, fibers of tenotomized muscles show localized degenerative processes similar to those resulting from denervation. The diminution of muscle fiber size is accompanied by a decrease in the diameter of the myofibrils (Wechsler, 1966, Bergman and Afifi, 1969). Tenotomy has also been found to produce an increase in the number of ring fibers in the muscle (Vecchi, 1927; Goerttler, 1935; Morris, 1959; Hikida and Bock, 1970). The atrophy of the fibers in the soleus has been shown to be also associated with central core and nemaline lesions in which there is degeneration of the mitochondria and changes in the Z line (Shafiq *et al.*, 1969b; Resnick *et al.*, 1968). The fact that tenotomy affects the structural integrity of slow muscles more than fast muscles is most probably due to the differences in innervation of these two types of muscles. Connecting with this, it has also been shown that speed of contraction of the slow muscles is increased as a result of tenotomy (Vrbová, 1963; Buller and Lewis, 1965b). It seems that the slow muscles, because they are engaged in a postural function, are more dependent on the reflex pathways. Section of the tendon no doubt results in a considerable change in the sensory output of the muscle, and this in some way causes the degeneration of the muscle fibers. In this respect, tenotomy is more akin to denervation than to simple disuse atrophy.

C. Effect of Nerve Supply on Growth and Development of Muscle Fibers

Over the past decade, there has been a growing awareness of the influence that the nervous system exerts on muscle fiber development. It has long been recognized that denervation of muscles leads to atrophy and eventually the degeneration of the muscle fibers. However, it was not clear whether this was due to the change in the activity of the muscle or whether the nerve produces a substance or in some other way imparts stability to the muscle fibers. More recent work suggests that both effects are important. The earlier work on the effect of denervation has been reviewed by Tower (1939). Also, Guth (1968) has published a comprehensive review on the trophic influence of nerve on muscle. Therefore, the intention here is not to review the entire literature on this subject but to discuss only that work pertinent to muscle growth.

The structural changes that are associated with denervation are different from those resulting from simple disuse in that many of the fibers as well as undergoing atrophy, actually degenerate, and this results in a reduction in the number of fibers in the muscle (Gutmann and Zelena, 1962). These changes in the muscle fibers are accompanied by a reduction in the size of the myofibrils (Wechsler and Hager, 1960; Pellegrino and Franzini-Armstrong, 1963; Muscatello *et al.*, 1965) and an increase in the number of nuclei. Also, the nuclei of the muscle fibers, as well as increasing in number, become more rounded in shape and their chromatin material appears to be more dispersed (Tower, 1935). Biochemical studies have shown that following denervation, myofibrillar proteins are lost at a greater rate than the sarcoplasmic or connective tissue proteins (Fischer and Ramsey, 1946; Fischer 1948; Schmidt and Schlief, 1956; Helander, 1957). Attempts have been made to study the rate of synthesis of proteins *in vivo* using radioactively labeled amino acids, but the results of *in vivo* incorporation of amino acids into denervated and normal muscle proteins are always difficult to interpret because of differences in the pool size and the permeability of the muscle to the different amino acids (Goldberg and Goodman, 1969). In order to overcome some of these difficulties, isolated muscles have been used.

Studies on incubated frog muscles have shown that denervation of the preparations resulted in a decreased incorporation of valine-^{14}C into the myofibrillar proteins (Margreth *et al.*, 1966) but an increased incorporation into the sarcoplasmic proteins (Muscatello *et al.*, 1965). It seems, therefore, that the denervation selectively suppresses the synthesis of the myofibrillar proteins while it accelerates the synthesis of sarcoplasmic proteins. In addition to altering the rate of synthesis, there is also evidence (Goldberg, 1969) that the rate of breakdown of the proteins is increased. This seems to tie in with the fact that the size of the lysosomes is greater in denervated muscle (Pellegrino and Franzini-Armstrong, 1963) and cathepsin activity levels are increased (Hajek *et al.*, 1964; Weinstock and Lukacus, 1965; Weinstock, 1966).

The rapidity of the weight loss following denervation varies between different muscle of the same animal and between the same anatomical muscles in different species. In general, the rate of weight loss can be related to the metabolic rate of the tissue (Knowlton and Hines, 1936). Also, the level at which the nerve supply is sectioned affects the rate of the decrease in muscle weight. Transection of the nerve near to the muscle produces a more rapid change (Gutmann, 1970), which suggests that some sort of substance does pass down the axon. There is one muscle, the chicken anterior latissimus dorsi, which was alleged to undergo hypertrophy following denervation. However, Hikida (1970) has shown that this sort of hypertrophy is associated with an increase in the extracellular compartments while the fibers themselves decrease in size. In other muscles, there is a transient hypertrophy of the fibers shortly following the section of the nerve (Sola and Martin, 1953; Stewart, 1955; Manchester and Harris, 1968; Guth, 1968), but this may be due to a high frequency of action potentials originating from the cut ends of the axons.

In addition to stabilizing the highly differentiated state of the muscle fibers, the nervous system also determines some of their characteristics. This was really first demonstrated by the experiments of Buller *et al.* (1960a,b), who crossed the nerve supply of a fast muscle and a slow muscle. Under these circumstances, the contraction speed of the muscles changes to almost that of the muscles whose nerve supply they receive. This clearly demonstrates that the rate of contraction of the muscles is in some part determined by the type of innervation of the muscle fibers. The change in the speed of contraction following cross innervation has been shown to be associated with a change in the myosin ATPase of the muscles (Mommaerts *et al.*, 1969). Other studies have shown that cross innervation of fast and slow muscles is also accompanied by changes in the levels of oxidative and glycolytic enzymes (Prewitt and Salafsky, 1967; Romanul and Van der Meulen, 1966; Robbins *et al.*, 1969; Guth *et al.*, 1968), myoglobin (McPherson and Tokunage, 1967), glycogen, and potassium (Drahota and Gutmann, 1963). However, it is not known whether these other biochemical changes are the direct result of the cross innervation or whether they are just part of the metabolic readjustments that would be necessary following the change in contraction speed and the change in level of activity of the fibers. However, what is certain is that the type of innervation the muscle fibers receive has a considerable effect on the way they develop and that this will be better understood when we know the nature of this "trophic influence." Also, a recent study by McComas *et al.* (1970) on the number of functional motor units in dystrophic muscle has indicated that neurogenic nature of this disease. Therefore, from the medical standpoint, it is very important that the influences of the nerve on muscle development be more fully understood. The derangements of muscle fiber growth due to dystrophy and other diseases are discussed in Chapter 6 of Volume IV of this treatise.

D. *Effect of Hormones on Muscle Fiber Growth*

Over the years there has accumulated a vast amount of information on the different effects of various hormones on the musculature and other tissue of the body. The intention here is not to review the vast plethora of endocrine literature, but only to mention studies that are strictly relevant to muscle growth, and in particular to discuss those studies that give some insight into the cellular events that are involved in fiber growth.

Androgens may be presumed to have a direct, or at least an indirect influence on the development of the musculature, because the muscles of the males are almost invariably larger than those in the female. As mentioned above, this sex difference in mouse muscles (Goldspink and Rowe, 1968) is attributable to a greater increase in diameter of the fiber, as there is no difference in the number of fibers in the male and female. Whether this applies to species other than rats and mice is not known. It is, however, in keeping with the findings of Venable (1966a), that replacement therapy using testosterone following castration caused a considerable hypertrophy of the fibers in the rat anus retractor muscles. Venable states that the hypertrophy of the fibers was the result of an increase in myofibrillar material in the fibers. There was, however, no change in the number of fibers either during the replacement period or during the atrophy of the muscle following castration.

The degree of responsiveness to androgens of different muscles varies enormously; Kochakian *et al.* (1961) found that the temporal and masseter muscles of the guinea pig are very sensitive to castration and subsequent replacement therapy. The development of these muscles in the guinea pig are, it seems, secondary sexual characteristics of that particular species. In the rat, the response of the muscles to androgens (Kochakian, 1966) are more uniform, with, of course, the exception of the levator ani muscles (Venable, 1966a,b).

There seems to be little doubt that androgens increase the rate of protein synthesis in most muscles and thus influence the rate and extent of the development of the fibers. Several workers (Novak, 1957; De Loecker, 1965; Kochakian, 1966; Buresova *et al.*, 1969) have all reported an increased incorporation of labeled amino acids into muscle protein following administration of androgen to both intact and castrated animals. Breuer and Florini (1965) found that ribosomes obtained from castrated animals are less active in synthesizing protein than those from intact animals. These workers also reported (Florini and Breuer, 1966) that the ability of the ribosomes to synthesize proteins can be modified by using a combination of testosterone and growth hormone. They concluded that one of the important factors in the increase in protein synthesis by hormones was the increase in mRNA production by the chromosomes. The evidence for this is based on the finding that muscle RNA polymerase is increased and that the stimulant effect of the hormone on protein synthesis is prevented by pretreatment with actinomysin D.

The effects of estrogens on muscle fiber growth is less well known. It is not known, for instance, whether estrogen retards muscle growth in the female or whether the female muscles are smaller because of their lower androgen levels. Administration of a single high dose of estrogen into male rats during the first 5 days after birth results in a considerably reduced overall growth rate (Ošťádolová *et al.*, 1969), and presumably this also has a marked effect on muscle development. However, this rather drastic type of treatment may in some way damage the pituitary and affect the production and release of growth hormone. Therefore, there is a need for more information concerning the effects of estrogens at more physiological dose levels.

A considerable number of studies have, however, been carried out on the effects of growth hormone on muscle development and it is apparent from these that this hormone can increase the rate of protein synthesis and hence increase muscle bulk. Bigland and Jehring (1952) reported that although growth hormone increased muscle size, it did not increase the contractile strength of the muscle. Scow and Hagen (1965) reported that the increase in muscle bulk resulting from growth hormone treatment did not involve any change in the composition of the muscle. That the increase in muscle mass is associated with the increased protein synthesis, has been shown by amino acid incorporation studies using both *in vivo* (Friedberg and Greenberg, 1948; Lee and Williams, 1952) and *in vitro* systems (Kostyo and Knobil, 1959; Manchester and Young, 1959; Reiss and Kipnis, 1959). Florini and Breuer (1966) have investigated the nature of the increased synthesis using polyribosome preparations. They found that ribosome synthetic activity is increased after only a single injection of growth hormone, and this preceded the increase in RNA polymerase activity. This is in harmony with the finding that the enhancement of protein synthesis by growth hormone does not initially seem to be associated with an increased RNA synthesis (Martin and Young, 1965; K. G. Dawson *et al.*, 1966). The mode and site of growth hormone action is uncertain (Manchester, 1970); however, it seems likely that the main effect is to increase the uptake of amino acids into the muscle cell. In the same way, other hormones may be implicated, particularly insulin, as this hormone is also known to exert some control over the entry of amino acids across the plasma membrane.

E. Effect of Level of Nutrition on Growth of Muscle Fibers

It is well known that to reduce the food intake of an animal or human being may lead to a considerable reduction in muscle mass (Waterlow, 1956; Allison *et al.*, 1962; Hagan and Scow, 1957; Widdowson *et al.*, 1960; Kerpel-Fronius an Frank, 1949). In fact, excluding fat, muscle seems to be the tissue that is most susceptible to starvation or semistarvation. The decrease in muscle bulk has been shown to be associated with a decrease in mean fiber diameter in cattle (Robertson and Baker, 1933), sheep (Joubert, 1956), pigs (McMeekan, 1940b), the laboratory mouse (Goldspink, 1964, 1965; Rowe, 1968a), and in humans (Vincent and Rademacker, 1959; Montgomery, 1962). The way in which the mean fiber diameter decreases has been studied in some detail by Goldspink (1964) and Rowe (1968a). These workers found that in those muscles of the mouse that are normally composed of large and small fibers, the effect of starvation was to reduce the number of large phase fibers in the muscle so that the fiber size distribution plots tended to become unimodal again. Goldspink (1965) has also shown that the decrease in fiber size is the result of a reduction in the number of myofibrils in the fiber and that this accounts for the decrease in the contractile strength which is normally associated with starvation. The way in which the myofibrils are removed from the fibers is not known, but Bird *et al.* (1968) found increased levels of cathepsins after a period of 5 days of reduced food intake. Wechsler (1964) claimed that the myofibrils at the periphery of the fiber are the first to be removed.

According to the work of Hagan and Scow (1957), Dickerson and McCance (1964), and Montgomery *et al.* (1964) short-term starvation causes a greater reduction in the sarcoplasmic proteins than in the myofibrillar proteins, while those of the connective tissue are relatively unaffected. However, more prolonged malnutrition does not affect the ratio of myofibrillar to sarcoplasmic proteins (Widdowson *et al.*, 1960; Waterlow and Stephen, 1966). There appears to be contradictory evidence concerning possible changes in muscle DNA content and number of fiber nuclei during restricted feeding. Dickerson and McCance (1964) have reported a reduction in the DNA content of undernourished pigs. They also found that when the pigs were rehabilitated on high-level nutrition, the DNA content increased considerably. A decrease in the DNA content of undernourished rats has also been reported by Elliott and Cheek (1968). However, Mendes and Waterlow (1958) were unable to detect any change in DNA content of rat muscle following starvation. The situation is clarified to some extent by the more recent work of Moss (1968a,b), who found that in chicken muscle the total amount of DNA is decreased by restricted feeding although the DNA per gram of muscle remains the same as in well fed chickens. During a period of rapid starvation in which the chickens were totally deprived of food, the DNA (number of nuclei) was not affected, although the muscle bulk still decreased, and therefore the relative amount of DNA was actually greater in the starved muscles. It therefore seems that different results may be obtained depending on whether the experiments involve a rapid or more prolonged reduction in the food intake of the animal.

Fibers of different types of muscles are known to be affected by starvation to different extents. Rowe (1968a) found that the fibers of white, fast muscles of the laboratory mouse exhibited a decrease in size, while those of the red, slow (soleus) muscle remained virtually unchanged. Tasker and Tulpule (1964) found that the levels of glycolytic enzymes in white muscle fiber were not influenced by starvation, but both ATPase and creatine phosphokinase were reduced in both white and red muscles. On the other hand, cytospectrophotometric measurements of the succinic dehydrogenase concentrations in individual fibers by Goldspink and Waterson (1971) have shown that a reduced food intake causes a reduction in the activity of this enzyme in the white fibers but not in the red fibers. The explanation for the difference in response may lie in the kind of innervation of the fiber or the richness of the vascular supply of the muscle. However, it may be at a more biochemical level in that the reduced supply of fuel and raw material could be expected to affect the anabolic and catabolic pathways in different ways in these metabolically different types of fibers.

The biochemical control of protein metabolism is not fully understood especially in relation to starvation. However, it appears that not only is the rate of protein degradation increased during starvation (Bird *et al.*, 1968), but the rate of synthesis of new protein is decreased (Waterlow and Stephen, 1967, 1968). This latter finding may be the result of shortage of one or more essential amino acids in the muscle fibers (Thompson *et al.*, 1950) or due to a reduction in the protein synthesizing capacity of the polyribosomes as suggested by the experiments of Young and Alexis (1968). In any event, the way that muscle protein metabolism is modified during periods of restricted food intake is a very interesting and important problem which is certainly worthy of further study.

VII. Conclusions

The postembryonic growth of muscle may be conveniently divided into two stages—the early postembryonic period and the period following this in which there is further growth of the fibers. During the early period, the myotubes develop into proper muscle fibers and, as well as increasing in girth, the fibers increase very considerably in length. The early postnatal period also sees the development of the mature physiological and biochemical characteristics of the muscle. The myofibrillar proteins increase very rapidly so that at the end of this period the mature proportions of the myofibrillar and other proteins are reached. The concentrations of ions inside and outside the muscle cell change so that the mature values for the resting potentials are gradually attained. In muscles destined to become fast muscles, the speed of contraction increases to that of the mature muscle. Changes in the metabolism of the fibers, as reflected by changes in enzyme activities, are also evident. The early period may, in some ways, be regarded as the postnatal extension of embryonic differentiation process. However, the situation is not as straightforward as this, because after birth the activity of the animal is drastically altered. This in itself would be expected to produce changes in addition to those which are programmed for during the differentiation process.

Once the postembryonic differentiation of the muscle is complete, the fibers of most muscles still continue to grow. Although striated muscle is a highly differentiated and highly specialized tissue, it is remarkably adaptable. The further growth of the fibers may be regarded as an adaptation to the greater work load on the fibers resulting from the change in activity and increased body mass of the animal. This adaptation may take the form not only of changes in size and myofibrillar content of the fibers, but it may also involve further changes in their metabolism.

Striated muscle makes up about half of the bulk of the body, and it is therefore important for the survival of the animal that its most abundant tissue should be adaptable. The extent of the adaptability is well demonstrated by the exercise experiments and reduced food intake studies described in Section IV. Of course, in the wild, the animal may well experience considerable variations in both of these parameters.

The real heart of the problem of muscle development, or for that matter the differentiation and growth of any tissue, is the control of the synthesis of the various proteins. Present day knowledge of molecular biology is such that we know how the amino acid sequence for specific proteins is coded for in the DNA of the cell. We also know something of how this information is relayed to the ribosomes where these proteins are made. However, unfortunately, little information is available about how the synthesis of these proteins is switched on and switched off and indeed what factors control their overall rate of synthesis. Once we understand the control mechanisms for protein synthesis we should be able to get a clear picture of how muscle differentiates and grows both before birth and during postembryonic growth.

REFERENCES

Adams, R. D., Denny-Brown, D., and Pearson, C. M. (1962). "Diseases of Muscle," 2nd ed. Harper, New York.
Adrian, R. H., Constantin, L. L., and Peachy, L. D. (1969). *J. Physiol.* (*London*) **204**, 231.
Alder, A. B., Crawford, G. N. C., and Edwards, R. G. (1958). *Proc. Roy. Soc.*, *Ser. B* **148**, 207.
Alder, A. B., Crawford, G. N. C., and Edwards, R. G. (1959). *Proc. Roy. Soc.*, *Ser. B* **150**, 554.
Allbrook, D. B., Han, M. F., and Hellmuth, A. E. (1971). *Pathology* **3**, 233.
Allen, E. R., and Pepe, F. A. (1965). *Amer. J. Anat.* **116**, 343.
Allison, J. B., Wannemacher, R. W., Banks, W. L., Wunner, W. H., and Gomez-Brenes, R. A. (1962). *J. Nutr.* **78**, 333.
Aronson, J. (1961). *J. Biophys. Biochem. Cytol* **11**, 147.
Auber, M. J. (1965). *C. R. Acad. Sci.* **261**, 4845.

Awan, M. Z., and Goldspink, G. (1970). *Biochim. Biophys. Acta* **216,** 229.
Bárányi, M., Tucci, A. F., Bárány, K., Volpe, A., and Reckard, T. (1965). *Arch. Biochem. Biophys.* **111,** 727.
Barlow, J. S., and Manery, J. F. (1954). *J. Cell. Comp. Physiol.* **43,** 165.
Battelle, B. A. M., and Florini, J. R. (1972). In press.
Beatty, C. H., and Bocek, R. M. (1970). *In* "The Physiology and Biochemistry of Muscle as a Food," pp. 155–191. Univ. of Wisconsin Press, Madison.
Beatty, C. H., Basinger, G. M., and Bocek, R. M. (1967). *J. Histochem. Cytochem.* **15,** 93.
Beckett, E. B., and Bourne, G. H. (1960). *In* "The Structure and Function of Muscle" (G. H. Bourne, ed.), 1st ed., Vol. 1, pp. 89–110. Academic Press, New York.
Bergman, R. A., and Afifi, A. K. (1969). *Johns Hopkins. Med. J.* **124,** 119.
Bigland, B., and Jehring, B. (1952). *J. Physiol. (London)* **116,** 129.
Binkhorst, R. A. (1969). *Pfluegers Arch.* **309,** 193.
Bird, J. W. C., Berg, T., and Leathem, J. H. (1968). *Proc. Soc. Exp. Biol. Med.* **127,** 182.
Black-Schaffer, B., Grinstead, C. E., and Braunstein, J. N. (1965). *Circ. Res.* **16,** 383.
Bowden, D. H., and Goyer, R. A. (1960). *AMA Arch. Pathol.* **69,** 188.
Bowman, W. (1840). *Phil. Trans. Roy. Soc. London* **130,** 457.
Breuer, C. B., and Florini, J. R. (1965). *Biochemistry* **4,** 1544.
Bridge, D. T., and Allbrook, D. (1970). *J. Anat.* **106,** 285.
Buller, A. J., and Lewis, D. M. (1965a). *J. Physiol. (London)* **176,** 355.
Buller, A. J., and Lewis, D. M. (1965b). *J. Physiol. (London)* **178,** 343.
Buller, A. J., Eccles, J. C., and Eccles, R. M. (1960a). *J. Physiol. (London)* **150,** 399.
Buller, A. J., Eccles, J. C., and Eccles, R. M. (1960b). *J. Physiol. (London)* **150,** 417.
Buresova, M., Gutmann, E., and Klicpera, M. (1969). *Physiol. Bohemoslov.* **18,** 137.
Burr, L. H., and McLennan, H. (1960). *Can. J. Biochem. Physiol.* **38,** 829.
Carrow, R. E., Brown, R. E., and Van Huss, W. D. (1967). *Anat. Rec.* **159,** 33.
Cheek, D. B., ed. (1968). "Human Growth." Lea & Febiger, Philadelphia, Pennsylvania.
Chen, S. C., and Young, V. R. (1968). *Biochem. J.* **106,** 61.
Chiakulas, J. J., and Pauly, J. E. (1965). *Anat. Rec.* **152,** 55.
Close, R. (1964). *J. Physiol. (London)* **173,** 74.
Close, R. (1967). *Excerpta Med. Found. Int. Congr. Ser.* **147,** 142.
Cosmos, E. (1966). *Develop. Biol.* **13,** 163.
Crawford, G. N. C. (1954). *J. Bone Joint Surg.* **36,** 294.
Crawford, G. N. C. (1961). *Proc. Roy. Soc. Ser. B* **154,** 134.
Davies, A. S., and Gunn, H. M. (1971). *J. Anat.* **110,** 169.
Dawson, D. M., Goodfriend, T. L., and Kaplan, N. O. (1964) *Science* **143,** 929.
Dawson, D. M., and Romanul, F. C. A. (1964). *Arch. Neurol. (Chicago)* **11,** 369.
Dawson, K. G., Patey, P., Rubenstein, D., and Beck, J. C. (1966). *Mol. Pharmacol.* **2,** 269.
De Loecker, W. (1965). *Arch. Int. Pharmacodyn. Ther.* **153,** 69.
Denny-Brown, D. (1929). *Proc. Roy. Soc., Ser. B* **104,** 371.

Devi, A., Mukundan, M. A., Srivastava, U., and Sarkar, N. K. (1963). *Exp. Cell Res.* 32, 242.

Dickerson, J. W. T., and McCance, R. A. (1964). *Clin. Sci.* 27, 123.

Dickerson, J. W. T., and Widdowson, E. M. (1960). *Biochem. J.* 74, 247.

Drahota, Z., and Gutmann, E. (1963). *Physiol. Bohemoslov.* 12, 339.

Dreyfus, J. C. (1967). *Rev. Fr. Etud. Clin. Biol.* 12, 343.

Dubowitz, V. (1963). *Nature (London)* 197, 1215.

Dubowitz, V. (1965). *J. Neurol., Neurosurg. Psychiat.* [N.S.] 28, 516.

Dubowitz, V., and Pearse, A. G. (1960). *Nature (London)* 185, 701.

Edge, M. B. (1970). *Develop. Biol.* 23, 634–650.

Edgerton, V. R. (1970). *Amer. J. Anat.* 127, 81.

Edgerton, V. R., and Barnard, R. J. (1970). *Experientia* 26, 1222.

Edgerton, V. R., and Simpson, D. R. (1969). *J. Histochem. Cytochem.* 17, 828.

Edgerton, V. R., Gerehman, L., and Carrow, R. (1969). *Exp. Neurol.* 24, 110.

Edgerton, V. R., Simpson, D. R., Barnard, R. J., and Peter, J. B. (1970). *Nature (London)* 225, 866.

Elliott, D. A., and Cheek, D. B. (1968). *In* "Human Growth" (D. B. Cheek, ed.), p. 326. Lea & Febiger, Philadelphia, Pennsylvania.

Enesco, M., and Puddy, D. (1964). *Amer. J. Anat.* 114, 238.

Fenichel, G. M. (1963). *Neurology* 13, 219.

Fenichel, G. M. (1966). *Neurology* 16, 741.

Fischer, E. (1948). *Arch. Phys. Med. Rehabil.* 29, 291.

Fischer, E., and Ramsey, R. W. (1946). *Amer. J. Physiol.* 145, 571.

Fischbach, G. D., and Robbins, N. (1969). *J. Physiol. (London)* 210, 305.

Fischman, D. A. (1967). *J. Cell Biol.* 32, 557.

Flear, C. T. G., Carpenter, R. G., and Florence, I. (1965). *J. Clin. Pathol.* 18, 74.

Florini, J. R., and Breuer, C. B. (1965). *Biochemistry* 4, 253.

Florini, J. R., and Breuer, C. B. (1966). *Biochemistry* 5, 1870.

Friedberg, F., and Greenberg, D. M. (1948). *Arch. Biochem.* 17, 193.

Fudel-Osipova, S. I., and Martynenko, O. A. (1965). *Biofizika* 10, 796.

Gauthier, G. F. (1970). *In* "The Biochemistry and Physiology of Muscle as a Food," pp. 103–130. Univ. of Wisconsin Press, Madison.

Gauthier, G. F., and Padykula, H. (1966). *J. Cell Biol.* 28, 333.

Germino, N. L., D'Albora, H., and Wahrmann, J. P. (1965). *Acta Anat.* 62, 434.

Glasz-Moerts, M. J., Diegenbach, P. C., and Van der Steth, A. (1968). *Biol. Med. Sci.* 71, 377.

Goerttler, K. (1935). *Verh. Deut. Orthop. Ges.* 30, 34.

Goldberg, A. L. (1967). *Amer. J. Physiol.* 213, 1193.

Goldberg, A. L. (1968). *J. Cell Biol.* 36, 653.

Goldberg, A. L. (1969). *J. Biol. Chem.* 244, 3217.

Goldberg, A. L., and Goodman, H. M. (1969). *Amer. J. Physiol.* 216, 1116.

Goldspink, G. (1961). *Nature (London)* 192, 1305.

Goldspink, G. (1962a). *Proc. Roy. Irish Acad, Sect. B* [3] 62, 135.

Goldspink, G. (1962b). *Comp. Biochem. Physiol.* 7, 157.

Goldspink, G. (1964). *J. Cell. Comp. Physiol.* 63, 209.

Goldspink, G. (1965). *Amer. J. Physiol.* 209, 100.

Goldspink, G. (1966). *Can. J. Physiol.* 44, 765.

Goldspink, G. (1968). *J. Cell Sci.* 3, 539.

Goldspink, G. (1969). *Life Sci.* 8, 791.

Goldspink, G. (1970). *J. Cell Sci.* **6**, 593.
Goldspink, G. (1971). *J. Cell Sci.* **9**, 751.
Goldspink, G., and Rowe, R. W. D. (1968). *Proc. Roy. Irish Acad. Sect. B* [3] **66**, 85.
Goldspink, G., and Waterson, S. E. (1971). *Acta Histochem.* **40**, 16.
Goldspink, G., Larson, R. E., and Davies, R. E. (1970a). *Z. Vergl. Physiol.* **66**, 379.
Goldspink, G., Larson, R. E., and Davies, R. E. (1970b). *Z. Vergl. Physiol.* **66**, 389.
Gordon, E. E., Kowalski, K., and Fritts, M. (1966). *Amer. J. Physiol.* **210**, 1033.
Gordon, E. E., Kowalski, K., and Fritts, M. (1967). *J. Amer. Med. Ass.* **199**, 103.
Griffin, G. E., and Goldspink, G. (1971). In preparation.
Griffin, G. E., Williams, P., and Goldspink, G. (1971). *Nature (London)* **232**, 28.
Guth, L. (1968). *Physiol. Rev.* **48**, 645.
Guth, L., Brown, W. C., and Ziemnowicz, J. D. (1966). *Amer. J. Physiol.* **211**, 1113.
Guth, L., Watson, P. K., and Brown, W. C. (1968). *Exp. Neurol.* **20**, 52.
Gutmann, E. (1970). *Curr. Res. Neurosci.* **10**, 54.
Gutmann, E., and Hanzlikova, V. (1966). *Physiol. Bohemoslov.* **16**, 244.
Gutmann, E., and Syrovy, I. (1967). *Physiol. Bohemoslov.* **16**, 232.
Gutmann, E., and Zalena, J. (1962). In "The Denervated Muscle" (E. Gutmann, ed.), p. 57. Czech. Acad. Sci., Prague.
Hagan, S. N., and Scow, R. O. (1957). *Amer. J. Physiol.* **188**, 91.
Hajek, I., Gutmann, E., and Syrovy, I. (1964). *Physiol. Bohemoslov.* **13**, 32.
Hall-Craggs, E. C. B. (1968). *J. Anat.* **102**, 241.
Hall-Craggs, E. C. B. (1970). *J. Anat.* **107**, 459.
Ham, A. W., and Leeson, T. W. (1961). "Histology," 4th ed. Lippincott, Philadelphia, Pennsylvania.
Hammond, J., and Appleton, A. B. (1932). "Growth and Development of Mutton Qualities in the Sheep," Oliver & Boyd, Edinburgh.
Hamosh, M., Lesch, M., Baron, J., and Kaufman, S. (1967). *Science* **157**, 935.
Harris, J. B., and Luff, A. R. (1970). *Comp. Biochem. Physiol.* **33**, 923.
Hazlewood, C. F., and Ginski, J. M. (1968). *Amer. J. Phys. Med.* **47**, 87.
Hazlewood, C. F., and Ginski, J. M. (1969). *Johns Hopkins Med. J.* **124**, 132.
Hazlewood, C. F., and Nichols, B. L. (1966). *Fed. Proc., Fed. Amer. Soc. Exp. Biol.* **25**, 289.
Hazlewood, C. F., and Nichols, B. L. (1967). *Nature (London)* **213**, 935.
Hazlewood, C. F., and Nichols, B. L. (1968). *Johns Hopkins Med. J.* **123**, 198.
Hazlewood, C. F., and Nichols, B. L. (1969). *Johns Hopkins Med. J.* **125**, 119.
Hearn, G. R., and Wainio, W. W. (1956). *Amer. J. Physiol.* **185**, 348.
Hearn, G. R., and Wainio, W. W. (1957). *Amer. J. Physiol.* **190**, 206.
Heidenhain, M. (1913). *Arch. Mikrosk. Anat. Entwicklungsmech.* **83**, 427.
Helander, E. A. S. (1957). *Acta Physiol. Scand.* **41**, Suppl. 141.
Helander, E. A. S. (1961). *Biochem. J.* **78**, 478.
Hensel, H., and Hildebrandt, G. (1964). In "Handbook of Physiology" (Amer. Physiol. Soc., J. Field, ed.), Sect. 4, p. 73. Williams & Wilkins, Baltimore, Maryland.
Hess, A. (1961). *J. Physiol. (London)* **157**, 221.

Heywood, S. M. (1969). *Biochemistry* 8, 3839.

Heywood, S. M., and Nwagwu, M. (1968). *Proc. Nat. Acad. Sci. U.S.* 60, 229.

Heywood, S. M., and Rich, A. (1968). *Proc. Nat. Acad. Sci. U.S.* 59, 590.

Heywood, S. M., Dowben, R. M., and Rich, A. (1967). *Proc. Nat. Acad. Sci. U.S.* 57, 1002.

Hikida, R. S. (1970). *Amer. Zool.* 10, 709.'

Hikida, R. S., and Bock, W. J. (1970). *J. Exp. Zool.* 175, 343.

Hines, M., and Knowlton, G. C. (1939). *Proc. Soc. Exp. Biol. Med.* 42, 133.

Hoffman, A. (1947). *Anat. Anz.* 96, 191.

Holland, D. L. (1968). Ph.D. Thesis, University of Birmingham, England.

Holland, D. L., and Perry, S. V. (1969). *Biochem. J.* 114, 161.

Holloszy, J. O. (1967). *J. Biol. Chem.* 242, 2278.

Holtzer, H., Marshall, J., and Finck, H. (1957). *J. Biophys. Biochem. Cytol.* 3, 705.

Howells, K., and Goldspink, G. (1971). In preparation for publication.

Huxley, H. E. (1963). *J. Mol. Biol.* 7, 281.

Ishikawa, H. (1965). *Arch. Histol.* 25, 275.

Ishikawa, H., Bischoff, R., and Holtzer, H. (1968). *J. Cell Biol.* 38, 538.

Iyengar, M. R., and Goldspink, G. (1971). Unpublished findings.

James, N. T. (1968). *Nature (London)* 219, 1174.

Jeffress, R. N., Peter, J. B., and Lam, D. R. (1968). *Life Sci.* 7, 957.

Jewell, P. A., and Zaimis, E. J. (1954). *J. Physiol. (London)* 124, 429.

Jinnai, D. (1960). *Acta Med. Okayama* 14, 359.

Joubert, D. M. (1956). *J. Agr. Sci.* 47, 59.

Karpati, G. O., and Engel, W. K. (1967). *Nature (London)* 215, 1509.

Karzel, K. (1968). *J. Physiol. (London)* 196, 86p.

Kelly, D. E. (1969). *Anat. Rec.* 163, 403.

Kendrick-Jones, J., and Perry, S. V. (1967). *Biochem. J.* 103, 207.

Kerpel-Fronius, E., and Frank, K. (1949). *Ann. Paediat.* 173, 321.

Kerpel-Fronius, E., Nagy, L., and Magyarka, B. (1964). *Biol. Neonatorum* 6, 177.

Kitiyakara, A., and Angevine, D. M. (1963). *Develop. Biol.* 8, 322.

Knappeis, G. C., and Carlson, F. (1963). *J. Cell Biol.* 13, 323.

Knowlton, G. C., and Hines, H. M. (1936). *Proc. Soc. Exp. Biol. Med.* 35, 394.

Kobayashi, N., and Yonemura, K. (1967). *Jap. J. Physiol.* 17, 698.

Kochakian, C. D. (1966). *In* "Physiology and Biochemistry of Muscle as a Food," pp. 81–112. Univ. of Wisconsin Press, Madison.

Kochakian, C. D., Tanaka, R., and Hill, J. (1961). *Amer. J. Phsiol.* 201, 1068.

Kostyo, J. L., and Knobil, E. (1959). *Endocrinology* 65, 395.

Krügar, P. (1950). *Zool. Anz.* 145, 445.

Larson, P. F., Hudgson, P., and Walton, J. N. (1969). *Nature (London)* 222, 1169.

Lawrie, R. A. (1953). *Nature (London)* 171, 1069.

Lawrie, R. A. (1961). *J. Agr. Sci.* 56, 249.

Lee, N. D., and Williams, R. H. (1952). *Endocrinology* 51, 451.

Le Gros Clark, W. E. (1958). "The Tissues of the Body." Oxford Univ. Press (Clarendon), London and New York.

Lesch, M., Parmley, W. W., Hamosh, M., Kaufman, S., and Sonnenblick, E. H. (1968). *Amer. J. Physiol.* 214, 685.

Luff, A. R., and Goldspink, G. (1967). *Life Sci.* 6, 1821.

Luff, A. R., and Goldspink, G. (1970a). *Comp. Biochem. Physiol.* 32, 581.

Luff, A. R., and Goldspink, G. (1970b). *J. Anim. Sci.* 30, 891.
MacCallum, J. B. (1898). *Bull Johns Hopkins Hosp.* 9, 208.
McComas, A. J., Sica, R. E. P., and Curries, S. (1970). *Nature (London)* 226, 1263.
MacKay, B., Harrop, T. J., and Muir, A. R. (1969). *Acta Anat.* 73, 588.
McMeekan, C. P. (1940a). *J. Agr. Sci.* 30, 276.
McMeekan, C. P. (1940b). *J. Agr. Sci.* 30, 276, 387, and 511.
McMeekan, C. P. (1940c). *J. Agr. Sci.* 31, 1.
McMinn, R. M. H., and Vrbová, G. (1962). *Nature (London)* 195, 502.
McMinn, R. M. H., and Vrbová, G. (1964). *Quart. J. Exp. Physiol.* 49, 424.
McPherson, A., and Tokunage, J. (1967). *J. Physiol. (London)* 188, 121.
Manchester, K. L. (1970). *Mammalian Protein Metab.* 4, 229–281.
Manchester, K. L., and Harris, E. J. (1968). *Biochem. J.* 108, 177.
Manchester, K. L., and Young, F. G. (1959). *Biochem. J.* 72, 136.
Margreth, A., Novello, F., and Aloisi, M. (1966). *Exp. Cell Res.* 41, 666.
Martin, T. E., and Young, F. G. (1965). *Nature (London)* 208, 684.
Maurer, F. (1894). *Morphol. Jahrb.* 21, 371.
Mauro, A. (1961). *J. Biophys. Biochem. Cytol.* 9, 493.
Mauro, A., Shafiq, S. A., and Milhorat, A. T. (1970). "Regeneration of Striated Muscle and Myogenesis." Excerpta Med. Found., Amsterdam.
Meara, P. J. (1947). *Onderstepoort J. Vet. Sci. Anim. Ind.* 21, 329.
Mendes, C. B., and Waterlow, J. C. (1958). *Brit. J. Nutr.* 12, 74.
Molbert, E., and Jijima, S. (1959). *Verh. Deut. Ges. Pathol.* 42, 349.
Mommaerts, W. F. H. M., Buller, A. J., and Seraydarian, K. (1969). *Proc. Nat. Acad. Sci. U.S.* 64, 128.
Montgomery, R. D. (1962). *Nature (London)* 195, 194.
Montgomery, R. D., Dickerson, J. W., and McCance, R. A. (1964). *Brit. J. Nutr.* 18, 587.
Morkin, E. (1970). *Science* 167, 1499.
Morpurgo, B. (1895). *Arch. Sci. Med.* 19, 327.
Morpurgo, B. (1897). *Virchows Arch. Path. Anat. Physiol.* 150, 522.
Morris, R. W. (1959). *AMA Arch. Pathol.* 68, 438.
Moss, F. P. (1968a). *Amer. J. Anat.* 122, 555.
Moss, F. P. (1968b). *Amer. J. Anat.* 122, 565.
Moss, F. P., and Leblond, C. P. (1970). *J. Cell Biol.* 44, 459.
Moulton, C. R. (1923). *J. Biol. Chem.* 57, 79.
Muir, A. R. (1961). In "Electron Microscopy in Anatomy" (Boyd, J. D. *et al.*, eds.), pp. 267–277. Arnold, London.
Muscatello, V., Margreth, A., and Aloisi, M. (1965). *J. Cell Biol.* 27, 1.
Novak, A. (1957). *Amer. J. Physiol.* 191, 306.
Novikova, A. I. (1964). *Sechanov. Physiol. J. USSR* 50, 626.
Nyström, B. (1966). *Nature (London)* 212, 954.
Nyström, B. (1968). *Acta Neurol. Scand.* 44, 405.
Ogata, T. (1958). *Acta Med. Okayama* 12, 216.
Ošťádalová, I., Babieký, A., Kolář, J., Vyhnánek, L., and Pařízek, J. (1969). *Physiol Bohemoslov.* 18, 271.
Padykula, H. A. (1952). *Amer. J. Anat.* 91, 107.
Padykula, H. A., and Gauthier, G. F. (1966). Excerpta Medica International Congress Series No. 147.
Pattengale, P. K., and Holloszy, J. O. (1967). *Amer. J. Physiol.* 213, 783.
Pellegrino, C., and Franzini-Armstrong, C. (1963). *J. Cell Biol.* 17, 327.

Pellegrino, C., and Franzini-Armstrong, C. (1969). *Cytol. Rev.* 7, 139.
Perrie, W. T., Perry, S. V., and Stone, D. (1969). *Biochem. J.* 113, 288.
Perry, S. V. (1970). In "The Physiology and Biochemistry of Muscle as a Food," pp. 537–553. Univ. of Wisconsin Press, Madison.
Prewitt, M. A., and Salafsky, B. (1967). *Amer. J. Physiol.* 213, 295.
Pringle, J. W. S. (1968). *Symp. Soc. Exp. Biol.* XXII, 67.
Reedy, M. D. (1964). *Proc. Roy. Soc.* B,106, 458.
Reimold, V. E. (1962). *Monatsschr. Kinderheilk.* 15, 330.
Reiss, D. J., Wooten, G. F., and Hollenberg, M. (1967). *Amer. J. Physiol.* 213, 592.
Reiss, E., and Kipnis, D. M. (1959). *J. Lab. Clin. Med.* 54, 937.
Resnick, J. S., Engel, W. K., and Nelson, P. G. (1968). *Neurology* 18, 737.
Revel, J. P. (1962). *J. Cell Biol.* 12, 571.
Richter, G. W., and Kellner, A. (1963). *J. Cell Biol.* 18, 195.
Robbins, N., Karpati, G., and Engel, W. K. (1969). *Arch. Neural. (Chicago)* 20, 318.
Robertson, J. D., and Baker, D. D. (1933). *Mo., Agr. Exp. Sta., Res. Bull.* 200.
Romanul, F. C. A. (1964). *Arch. Neurol. (Chicago)* 11, 355.
Romanul, F. C. A., and Van der Meulen, J. P. (1966). *Nature (London)* 212, 1369.
Rowe, R. W. D. (1968a). *J. Exp. Zool.* 167, 353.
Rowe, R. W. D. (1968b). *J. Exp. Zool.* 169, 59.
Rowe, R. W. D., and Goldspink, G. (1968). *Anat. Rec.* 161, 69.
Rowe, R. W. D., and Goldspink, G. (1969a). *J. Anat.* 104, 519.
Rowe, R. W. D., and Goldspink, G. (1969b). *J. Anat.* 104, 531.
Rowe, R. W. D., and Morton, D. J. (1971). *J. Cell Sci.* 9, 139.
Ruska, H., and Edwards, G. A. (1957). *Growth* 21, 73.
Schiaffino, S., and Margreth, A. (1969). *J. Cell Biol.* 41, 855.
Schmalbruch, H. Z. (1968). *Z. Mikrosk-Anat. Forsch.* 79, 493.
Schmidt, C. G., and Schlief, H. (1956). *Z. Gesamte Exp. Med.* 127, 53.
Schwalbe, G., and Mayeda, R. (1891). *Z. Biol. (Munich)* 27, 482.
Schwartz, F., and Wichan, I. (1966). *Acta Biol. Med. Ger.* 17, 96.
Schwartz, N. B. (1961). *Amer. J. Physiol.* 201, 164.
Scow, R. O., and Hagan, S. N. (1965). *Endocrinology* 77, 852.
Sexton, A. W., and Gerstein, J. W. (1967). *Science* 157, 199.
Shafiq, S. A. (1963). *J. Cell Biol.* 17, 363.
Shafiq, S. A., Gorychi, M. A., and Mauro, A. (1968). *J. Anat.* 103, 135.
Shafiq, S. A., Gorychi, M. A., and Milhorat, A. T. (1969a). *J. Anat.* 104, 281.
Shafiq, S. A., Goyrchi, M. A., Steven, A., Asiedu, and Milhorat, A. T. (1969b). *Arch. Neurol. (Chicago)* 20, 625.
Shear, C. R. (1969). Ph.D. Thesis, University of Columbia, New York.
Shear, C. R., and Goldspink, G. (1970a). *Anat. Rec.* 169, 426.
Shear, C. R., and Goldspink, G. (1970b). *Amer. Zool.* 10, 557.
Shear, C. R., and Goldspink, G. (1971). *J. Morphol.* 135, 351.
Siebert, W. W. (1928). *Z. Klin. Med.* 109, 360.
Sola, O. M., and Martin, A. W. (1953). *Amer. J. Physiol.* 172, 324.
Sréter, F. A., and Woo, G. (1963). *Amer. J. Physiol.* 205, 1290.
Srivastava, U. (1969). *Arch. Biochem. Biophys.* 130, 129.

Srivastava, U., and Chaudhary, K. D. (1969). *Can. J. Biochem.* **47**, 231.
Staun, H. (1963). *Acta Agr. Scand.* **13**, 293.
Stein, J. M., and Padykula, H. A. (1962). *Amer. J. Anat.* **110**, 103.
Stewart, D. M. (1955). *Biochem. J.* **59**, 553.
Stracher, A., and Dow, J. (1970). *Proc. Int. Congr. Biochem., 8th, 1969* Abst., p. 28.
Tabary, J. C., Goldspink, G., Tardieu, C., Lombard, M., Tardieu, G. and Chigot, P., (1971). *Rev. Chir. Orthop. Réparatrice de l'Appareil Moteur (Paris)* **57**, 463.
Tabary, J. C., Tabary, C., Tardieu, C., Tardieu G., and Goldspink G. (1972). *J. Physiol.* **224**, 231.
Tasker, K., and Tulpule, P. G. (1964). *Biochem. J.* **92**, 391.
Thompson, H. T., Schurr, P. E., Henderson, L. M., and Elvehjem, C. A. (1950). *J. Biol. Chem.* **182**, 47.
Thörner, S. H. (1934). *Arbeitsphysiologie* **8**, 359.
Tower, S. S. (1935). *Amer. J. Anat.* **56**, 1.
Tower, S. S. (1939). *Physiol. Rev.* **19**, 1.
Trayer, I. P., and Perry, S. V. (1966). *Biochem. Z.* **345**, 87.
Trayer, I. P., Harris, C. I., and Perry, S. V. (1968). *Nature (London)* **217**, 452.
Uchino, J., and Tsuboi, K. K. (1970). *Amer. J. Physiol.* **219**, 154.
Van Linge, B. (1962). *J. Bone Joint Surg., Brit. Vol.* **44**, 711.
Vecchi, G. (1927). *Arch. Sci. Med.* **50**, 377.
Venable, J. H. (1966a). *Amer. J. Anat.* **119**, 263, also 271.
Venable, J. H. (1966b). *Amer. J. Anat.* **119**, 271.
Vernadakis, A., and Woodbury, D. J. (1964). *Amer. J. Physiol.* **206**, 1365.
Vincent, M., and Rademacker, M. A. (1959). *Amer. J. Trop. Med. Hyg.* **8**, 511.
von Bezold, A. (1857). *Z. Wiss. Zool.* **8**, 487.
Vrbová, G. (1963). *J. Physiol. (London)* **166**, 241.
Walker, M. G. (1966). *J. Cell. Comp. Physiol.* **19**, 791.
Walker, M. G. (1967). *Experientia* **24**, 360.
Walker, M. G. (1970). *J. Cons., Cons. Perm. Int. Explor. Mer.* **33**, 228.
Walker, M. G. (1971). *J. Cons., Cons. Perm. Int. Explor. Mer.* **33**, 421.
Walker, S. M., and Schrodt, G. R. (1968). *J. Cell Biol.* **37**, 564.
Waterlow, J. C. (1956). *West Indian Med. J.* **5**, 167.
Waterlow, J. C., and Stephen, J. M. L. (1966). *Brit. J. Nutr.* **20**, 461.
Waterlow, J. C., and Stephen, J. M. L. (1967). *Clin. Sci.* **33**, 489.
Waterlow, J. C., and Stephen, J. M. L. (1968). *Clin. Sci.* **35**, 287.
Wechsler, W. (1964). *Naturwissenschaften* **51**, 91.
Wechsler, W. (1966). *In* "Methods and Achievements in Experimental Biology" (E. Bajusz and G. Jasmin, eds.) pp. 411–438. Karger, Basel.
Wechsler, W., and Hager, H. (1960). *Naturwissenschaften* **47**, 185.
Weinstock, I. M. (1966). *Ann. N.Y. Acad. Sci.* **138**, 199.
Weinstock, I. M., and Lukacs, M. (1965). *Enzymol. Biol. Clin.* **5**, 89.
Wells, J. B. (1969). *Exp. Neurol.* **24**, 514.
Widdowson, E. M. (1968). *In* "Growth and Development of Mammals" (G. A. Lodge and G. M. Lamming, Eds.), p. 224. Plenum Press, New York.
Widdowson, E. M, Dickerson, J. W. T., and McCance, R. A. (1960). *Brit. J. Nutr.* **14**, 457.
Williams, P. E., and Goldspink, G. (1972). *J. Cell Sci.* (in press).
Winick, M., and Noble, A. (1966). *J. Nutr.* **89**, 300.

Wirsén, C., and Larsson, K. S. (1964). *J. Embryol. Exp. Morphol.* **12**, 759.
Wool, I. G., Stirewalt, W. S., Kurihara, K., Low, R. B., Bailey, P., and Oyer, D. (1968). *Recent Progr. Horm. Res.* **24**, 139.
Yannet, H., and Darrow, D. C. (1938). *J. Biol. Chem.* **123**, 293.
Yellin, H., and Guth, L. G. (1970). *Exp. Neurol.* **26**, 424.
Yokota, S., (1969), Okajimas Fol. Anat. Jap. **45**, 309.
Yonemura, K. (1967). *Jap. J. Physiol.* **17**, 708.
Young, V. R. (1970). *Mammalian Protein Metab.* **4**, 586–728.
Young, V. R., and Alexis, S. D. (1968). *J. Nutr.* **96**, 255.

SKELETAL MUSCLE IN CULTURE

MARGARET R. MURRAY

I. Introduction

From the earliest days of tissue culture, muscle attracted the attention of investigators, who lost no time in examining the *in vitro* behavior of the major tissue types. Against the new background of movement, where tissues normally fixed (or "noble") took on a plastic and migratory habit, all three kinds of muscle were conspicuous by their demonstrated ability to contract autonomously and rhythmically in isolation from the nervous and other controls exerted by the whole organism. In the Lewises' culture medium (which consisted of Locke solution plus 0.5% dextrose and 10% chicken bouillon) contraction was especially rife. It is sobering to read the early papers on muscle by these superlative observers (W. H. Lewis and Lewis, 1917; M. R. Lewis, 1920) and to reflect on how little proportionately has been added since, notwithstanding recent significant advances. Problems of development and of structure in relation to function occupied the early workers; tissue culture seemed to have been especially designed for the investigation of these problems, since it threw open the living somatic cell to microscopic observation—serially and under conditions of experiment. The *in vitro* observations of Burrows (1912), Lake (1915–1916), and others contributed substantially to the myogenic theory of heart action; it was early shown also that skeletal muscle is intrinsically capable of rhythmic contraction in the total absence of innervation (M. R. Lewis, 1915), and that the striated types of muscle operate generically like smooth muscle in being able to contract before light microscopic evidence of cross-striation exists in the living cell. Much of this early work was reviewed and discussed by Levi in his monograph of 1934, and a more recent, selective, historical resume was assembled by the writer in 1965. Furthermore, many long standing controversies have been rendered obsolete by the electron microscope and by developments in molecular biology, hence they need not be recounted here. Areas will be selected, however, in which tissue culture actually or potentially offers some special insight into muscle problems through its ability to separate the parts of a system and maintain them in health and isolation. Writers of other chapters in this treatise—Fischman (development), Couteaux, Nachmansohn (neuromyal junctions), Field, Pierce (regeneration), and Basmajian (electromyography), to name a few—will provide the proper systematic expositions of these general topics. This chapter will deal with several areas of muscle research to which tissue culture has made substantive or especially interesting contributions, with only brief reference to the respective masses of background knowledge.

II. Histogenesis (Embryonal)

Ten years ago, the problem of sarcoblastic histogenesis in the embryo was still in dispute. During the ensuing decade, a concerted attack from many quarters, mounted with especial tenacity by Konigsberg and Holtzer and their associates, has established the concept of fusion as the means by which the multinucleate fiber is produced. Hypotheses of myogenesis have been tested experimentally chiefly by *in vitro* methods, although Mintz and Baker (1967) have shown by a very ingenious procedure that fusion occurs similarly *in vivo* (p. 249).

To avoid confusion, we shall define our terms as follows:

Myoblast—a proliferating *uni*nuclear[1] myogenic cell, in the stem-line of fiber-formation.

Myocyte—a nonproliferating uninuclear myogenic cell (which may be synthesizing myosin).

Myotube—a myogenic cell with two or more nuclei; may be multinucleate, and may grade into the sarcoblast.

Sarcoblast—a multinucleate protomuscle fiber in which alignment of myofibrils and emergence of other characteristic structures have begun.

A. Induction

To a considerable degree tissue culture has supplanted the earlier explantation of critical regions of one embryo to another as a means of analyzing the conditions governing induction of new cell types among an early embryonic population whose genetic equipment is presumably uniform. Current thinking tends to express the problem in terms of regional differential release of factors responsible for gene activation or repression. Reduced to lowest terms, a search has been directed toward understanding how simple factors such as local change in ion concentration or pH can alter energy-yielding systems and gene function so as to induce tissue differentiation at the onset of development (Barth and Barth, 1963, 1969). These workers observed that naturally occurring cations such as Ca^{2+}, Mg^{2+}, and Li^+ applied to frog gastrulae at concentrations well below the osmotic shock level will induce both ganglionic and motor nerve patterns in small cultures of presumptive epidermis (peripheral to the neural plate area); anions—Cl^-, Br^-, SO_4^{2-}—were found to be irrelevant. Further study revealed that the neural inductions initiated by Li^+ and other cations were dependent upon an immediate sudden rise in the local concentration of endogenous Na^+ ion, to the equivalent of 88 mM NaCl. With Na^+ at lower concentrations, other tissue types could be induced in this system, e.g., muscle at 44 mM sodium; but once differentiation took place in any direction it could

---

[1] This prefix is preferred here in order to form a compound word that is consistently Latin in derivation and one that may be appropriately used in context with "multinuclear," also Latin.

not be modified by Na⁺. Since sodium involvement is known to characterize transport at many levels—tissue, cellular, nuclear, and subcellular—the working hypothesis was presented that some inductors merely increase the permeability of cell membranes, and the ions in the medium complete the induction—cations possibly combining with nucleic acid phosphate groups, with the consequent freeing of DNA or RNA from proteins.

Histone fractions from calf thymus applied to chick embryos explanted at definitive streak-head fold stages 4–6 (Hamburger and Hamilton, 1951), have a morphostatic effect on central nervous system and brain development comparable to that of actinomycin D, which inhibits RNA synthesis (Malpoix and Emelinckx, 1967). But if they are applied at stages 8–10 (4–10 somites) the result is suppression of the somitic axis (Sherbet, 1966). Buckingham and Herrmann (1967), seeking to produce abnormalities in cultures of developing chick muscle with the teratogen 3-acetylpyridine, could do so only at stage 15 (24–27 somites, limb primordia).

Regarding induction of definitive muscle by other tissue types that are contiguous *in situ,* it seems to have been demonstrated by Avery *et al.* (1956) that spinal cord influences the growth, morphogenesis, and differentiation of somitic mesoderm into skeletal muscle. This influence is not due to the cartilage-inducing factor, since both dorsal and ventral halves of the cord are active; spinal ganglion cells and peripheral nerves are inactive (Holtzer, 1961). The presence of spinal cord in culture greatly accelerates the growth of chick somites isolated at stages 14–24 (21 somites to tip of tail somites) (cf. Section III,B). Chick limb buds (containing skeletal primorida) will develop musculature from mesenchyme if explanted at stages 20–23 (3–4 days *in ovo*) (van Weel, 1948). At stage 24, somites alone will differentiate into muscle.

Cell dissociation and reaggregation techniques introduced by Moscona (1952) extended the reach and precision of culture methods in experimental embryology and laid the groundwork for more effective studies of muscle histogenesis. Using mixed aggregates of myogenic and chondrogenic cells which had been separated by trypsinization from chick embryos of stage 16 (26–28 somites) up to 5 days, Moscona (1956) found that in the presence of chondroblasts (which *in vivo* differentiate slightly before myoblasts), the presumptive myoblasts must be in the field in very large amounts if they were to attain any advanced differentiation. In these aggregates, the myoblasts formed the outer sheath of the cluster, with the chondroblasts segregated in the center. However, in dispersed-cell cultures from embryos as early as 4 days, presumptive myoblasts explanted alone are capable of differentiating outside the tissue framework (Moscona, 1952). In monolayer cultures, Shimada (1968) reported that differentiation of dissociated myogenic cells from older embryos (12 days) was delayed or suppressed by the presence of other types of cells—cartilage, liver, lung, heart, and even spinal cord. However, if a Millipore filter was interposed between the myogenic and the heterotypic cells, maintaining a distance of 25 μ, no inhibition of differentiation occurred. Nameroff and Holtzer (1969) described a "contact-mediated" suppression of myogenesis, which was found to be reversible under certain circumstances: When myogenic cells from trypsinized breast muscle of 11-day chick embryos were plated onto nonproliferating layers of heterotypic or homotypic cells, they attached to these substrates but did not divide. If removed and subcultured, these uninuclear cells formed multinuclear myotubes; they would also differentiate on a Millipore filter over the cellular substrate or in a "hole" cut in the substrate, as well as on a killed substrate if it were fixed by other means than glutaraldehyde. Holtzer attributed this suppression of muscle differentiation to cell–cell or cell–substrate contacts, which primarily inhibit mitosis in myocyte precursors (see Sections B and C below).

Using cloning techniques to produce cultures derived from single myogenic cells, Konigsberg and Hauschka (1966) investigated the relatively exacting requirements for myogenesis of such cell strains. It had been shown earlier by Konigsberg (1963) that colonial derivatives of freshly cloned individual cells from trypsinized leg muscle of 11–12-day chick embryos could be induced to form myotubes 24 hr sooner if offered "conditioned" medium, than in freshly prepared medium. The more favorable medium was conditioned by the growth in it of cells dispersed from whole muscle, which included fibroblasts as well as myoblasts. It was then found that a film of reconstituted collagen from rat tail tendon, if used as substrate would completely replace the requirement of cloned muscle cells for conditioned medium in myogenesis (Hauschka and Konigsberg, 1966). This finding seemed to imply that myogenic cells require for their development the presence of a metabolic product from an associated cell type, the fibroblast, and it was remarked that development of a number of other cell types, especially epithelial, had been shown to be dependent on the close proximity of connective tissue elements; therefore, collagen may play a common role in a variety of differentiative events. In his Ph.D. thesis, Hauschka (1966) demonstrated that soluble, newly synthesized collagen was present in conditioned medium containing [^{14}C]proline, and that this—in the typical proline–hydroxyproline ratio—was strongly adsorbed to the surface of petri plates when they were pretreated with the medium before being seeded with myogenic cells. This observation seems to identify the precise molecule responsible for mediating the fibroblast–myoblast interaction that provides the proper substrate for differentiation of myogenic cells isolated at this relatively late stage of embryonic development. In a later paper, Hauschka (1968) indicated a continued dependence of myoblasts upon collagen products for *maintenance* of the capacity to differentiate further.

B. Derivation of the Sarcoblast

One of the more concrete mysteries of development formerly resided in the means by which transition is effected from uninuclear myoblast to multinucleate sarcoblast. Observers had long agreed that mitoses are frequent in the uninuclear myogenic cells and absent from the developing sarcoblasts, while at the same time the number of nuclei within these latter is undergoing rapid increase. Advocates of direct cleavage (or amitosis) of myotubal nuclei as an explanation for this anomaly never succeeded in adducing quantitatively sufficient evidence, while those favoring the theory that the multinucleate plasmodium is produced by fusion of uninuclear cells in the myogenic stem line have in the past decade offered a multitude of mutually confirmatory observations based on a variety of disciplines, coincidently casting light on some fundamental problems of cytodifferentiation.

W. H. Lewis and Lewis (1917) observed neither mitosis nor amitosis in their multinucleate sarcoblasts, and they questioned whether the apparent amitotic stages seen occasionally in fixed preparations represented the typical mode of nuclear replication. They noted a marked tendency for anastomosis and fusion between separate elongating muscle buds following chance contact. Also on the side of the angels, but much later, was Wilde (1958, 1959), who dissociated 5-day chick somites and 13-day mouse embryo muscle and mixed the cells, whose nuclei were distinguishable on a species basis when stained. Chimeric multinucleate muscle units were formed, which contracted; and staining appeared to show that nuclei of both species were incorporated in them, presumably by reaggregation and fusion of the dissociated uninuclear elements. Holtzer *et al.* (1958) cultured suspended cells from 6–9-day chick myotubes, and observed that this population of myogenic cells reformed myotubes within 36 hours, though mitoses were not observed in any but the uninuclear cells. They inferred from these results that the formation of myotubes was accomplished through fusion.

In 1960, the problem was approached through the administration of nucleic acid inhibitors and tracers. Konigsberg *et al.* found that concentrations of nitrogen mustard that specifically inhibit DNA synthesis had no effect on the development of multinuclearity in the myotubes which formed rapidly in monolayer cultures of embryonic chick muscle. Tritiated thymidine, applied briefly to uninuclear myogenic cells by Stockdale and Holtzer (1960), was picked up 4 days later among the nuclei that populated newly formed myotubes; but when [³H]thymidine was applied to cultures already in the myotube stage, no uptake occurred in those (Fig. 1). In the same year Capers made time-lapse, phase contrast observations of myoblast fusion. Neither mitosis, amitosis nor nuclear budding was observed within the formed myotubes; only transient folds and clefts in nuclei gave the appearance of amitotic stages (Fig. 2). Cooper and Konigsberg (1961a) made similar observations with time-lapse cinemicrography, recording also examples of fusion between uninuclear and multinuclear cells and lateral fusion of myotubes. And Konigsberg (1961a) sagely remarked that myoblast proliferation is self-limiting in this system, hence the high degree of differentiation that can be attained in monolayer cultures of myogenic cells. In the same year, Konigsberg (1961b) introduced the cloning method of Puck into the study of histogenesis, at first producing 10% myogenic colonies derived from single cells; other colonies were "fibroblast-like" in habit.

In a detailed study of muscle differentiation *in vitro*, Engel and Horvath (1960) showed by Coons' fluorescence technique that antigenic myosin was to be found in uninuclear muscle cells as well as in myotubes, appearing first in the form of long, thin, unstriated myofibrils. With further differentiation it appeared only in the A bands (and H bands). Shortly thereafter, Stockdale and Holtzer (1961) published an elegant analysis of proliferation–differentiation relationships in myogenesis based on fluorescein-labeled antibodies to the muscle proteins myosin, meromyosin, and actin, in combination with [³H]thymidine introduced as a measure of DNA synthesis. They found that differentiating myocytes synthesizing the muscle proteins did not concurrently synthesize DNA, and myoblasts that synthesized DNA did not concurrently synthesize the contractile proteins. Such DNA-labeled uninuclear cells could, however, be incorporated later, in random order, into myotubes, presumably after having passed through at least one division. This theme was further elaborated by Okazaki and Holtzer in 1965 (1965b). Apparently independently, Coleman *et al.* reported some of the same observations in 1966. Incorporating the idea of DNA inhibitors, Stockdale *et al.* (1964) administered the thymidine analog 5-bromodeoxyuridine (BUdR) to cultures of dividing myogenic cells. Such cells subsequently failed to participate in fusion and lagged in myosin synthesis, though they continued to synthesize various species of molecules required for

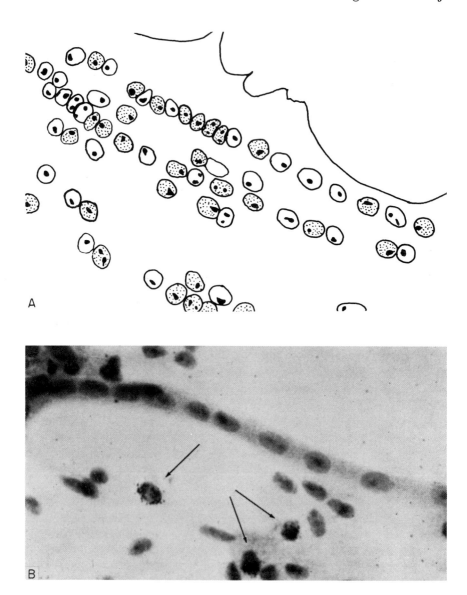

Fig. 1. (A) A tracing of a radioautograph of fixed and stained myotubes from a 4-day-old culture of embryonic chick muscle. Tritiated thymidine was added for 30 min when the culture was 18 hr *in vitro*, and only uninuclear cells were present. The stippled nuclei in this 4-day fusion product are labeled with the radioactive compound. (B) A similar 4-day-old culture; [³H]thymidine was added for 30 min and culture was fixed immediately. Labeled compound is seen only in uninuclear cells, not in myotubes. (After Stockdale and Holtzer, 1961.)

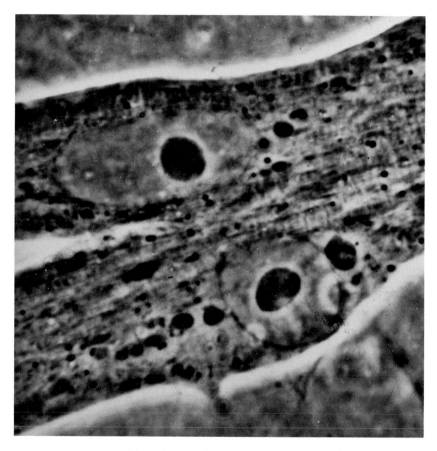

Fig. 2. Fusing sarcoblasts from 9-day culture of 13-day chick embryo skeletal muscle. Banding is irregular; differentiating bundles of myofibrillae are not coordinated. (Photomicrograph with phase contrast by Dr. Charles Capers.)

cell multiplication. If the analog was administered to myocytes that had ceased proliferating, myogenesis was not inhibited. An equimolar concentration of thymidine applied simultaneously to proliferating cells protected against the analog and permitted myogenesis: "interference with myogenesis by the analog may be due to production of an altered DNA." Bischoff and Holtzer (1970) refined these observations somewhat by synchronizing the myogenic cells in 5-fluorodeoxyuridine, to block DNA synthesis and collect myoblasts in the G_1 phase before applying BUdR. They confirmed the earlier observations linking differentiation and capacity for fusion and separating these functions from proliferative DNA synthesis. They had noted also (1968a) that cell strain specificity was retained through many generations in BUdR. Coleman *et al.* (1969) confirmed the Holtzer group's conclusions with a clonal study employing BUdR. Fischbach (1971) introduced a convenient method of eliminating both fibroblasts and early myoblasts from the arena in which dissociated muscle tissue is redifferentiating. Cytosine arabinoside at 10^{-5} or 10^{-6} M is selectively toxic to cells in mitosis; if this is administered when fusion is well advanced, the culture will be left composed almost entirely of myocytes and myotubes.

Bassleer (1962) summarizing his own independent observations with time-lapse cinemicrography and [^{3}H]thymidine autoradioautography (against a careful historical discussion), strongly supported the fusion theory, with the additional cytophotometric observation that sarcoblastic nuclei contained only diploid amounts of DNA: "Augmentation of the numbers of nuclei in myotubes occurs essentially by fusion; but the capacity to fuse must await a certain stage of development, in which the state of their DNA plays a crucial role." Strehler *et al.* (1963) verified and extended Bassleer's quantitative DNA data by automatic recording of absorption spectra of individual nuclei in cell suspensions from 11–12-day embryonic chick leg muscle. Uninuclear cells were of two ploidy classes, exemplified by prophase (tetraploid) and telophase single (diploid) nuclei; myotube nuclei were diploid only. A thoughtful "clonal analysis of myogenesis" at this juncture was published by Konigsberg (1963).

Yaffe and Feldman (1964) entered the field with suppression of RNA synthesis by actinomycin D. Nondifferentiated myogenic cells (myoblasts) were highly susceptible to this antibiotic and died within 24 hr. Differentiated cells (myocytes and myotubes) were relatively resistant, though the agent did penetrate and suppress RNA synthesis. Inhibitors of protein synthesis (e.g., puromycin) abolished the differential effect and damaged all cell types at once. Synthesis of proteins, therefore, continues during suppression of DNA-dependent RNA synthesis. According to Yaffe and Fuchs (1967), [^{3}H]uridine is taken up preferentially by myoblasts in the proliferative stage; [^{3}H]leucine by myotube and sarcoblast nuclei. Actinomycin D reverses these labeling differentials. However, at high dosage actinomycin-D displays an acute cytotoxicity which is tissue-type or cell-strain dependent, according to Sawicki and Godman (1971). This cannot be accounted for by differences in its effect on RNA, DNA or protein synthesis or by overall loss of preformed RNA.

By using mixed cultures in which one species component was labeled with [^{3}H]thymidine, Yaffe and Feldman (1965) observed the formation of hybrid multinucleate sarcoblasts from myoblasts of heterospecific origin, confirming and extending Wilde's earlier report of fusion between species. In the same series of experiments, Yaffe and Feldman found tissue recognition among disaggregated cells taking precedence over species recognition. When skeletal muscle cells of different species were mixed, together with heart or kidney cells, a selective segregation of the skeletal muscle cells occurred, although they were generically different, while homospecific but heterotypic cells were excluded. Skeletal and cardiac myoblasts from the same species would not aggregate or fuse: "cell to cell recognition mechanisms operate at the level of aggregation preceding the actual fusion process." A summary of his laboratory's work on cellular aspects of muscle differentiation is given by Yaffe (1969). Shimada observed in 1968 that myogenesis was delayed in the presence of both heterotypic and heterospecific cells, the degree of retardation being inversely proportional to the concentration of alien cells added in the system. Interposition of a 25 μ Millipore filter between myoblasts and heterotypic cells prevented this inhibition—which he suggests was of mechanical origin, based on the interposition of the wandering heterotypic cells between fusable myocytes. A recognition factor, which is disturbed by disruption of normal RNA and protein synthesis, was proposed by Maslow (1969).

Pursuing a related line of inquiry, Okazaki and Holtzer (1966) also had concluded that activities leading to fusion and myotube formation required recognition between homotypic cell surfaces, since heterotypic cells (connective tissue, chondroblasts, etc.) with labeled nuclei were never incorporated into myotubes. An analysis of prefusion mitotic events, with the aid of colchicine to produce a reversible metaphase arrest, showed that uninucleate cells in S (DNA synthesis), G_2 (post synthetic growth), and M (mitosis) stages could not fuse; hence it was deduced that they must fuse in G_1 (early postmitotic) stage; it followed, therefore, that cell surface alterations are dependent upon the position of the myogenic cell in the division cycle ($M \rightarrow G_1 \rightarrow S \rightarrow G_2$). According to Bischoff and Holtzer (1969), myoblasts withdraw from DNA synthesis before translation for myosin occurs and at the same time begin to acquire a cell surface compatible with fusion. A minimum of 5–8 hr elapses between the end of the differentiatives mitosis (also called "critical" or "quantal") and completion of fusion in the ensuing G_1. A variable number of proliferative mitoses and rounds of DNA synthesis may occur in myoblasts, but a quantal or differentiative mitosis yielding competent myocytes is obligatory before fusion can take place. It was noted by Holtzer *et al.* (1969) that Sendai virus, which alters the cell surface, can induce mass fusion of uninuclear myogenic cells, but the multinuclear units thus produced are nonviable. "This may mean that precursor myogenic cells precociously fused cannot function properly in a myotube: they have failed to undergo that 'quantal' mitosis." (See also Section VI,A.) Holtzer and Bischoff in a review (1970) of mitosis in relation to myogenesis, discuss in greater depth these concepts of the quantal mitosis, and of the action of BUdR in the (reversible) repression of cytodifferentiation. The theory of quantal and proliferative cell cycles in differentiation is also reviewed by Holtzer (1970).

Yaffe (1971) observed that myoblasts in normal medium undergo changes occupying about 52 hr between plating and onset of rapid fusion. When cells cultured for different periods are mixed, fusion takes place only between ripened myocytes. Ripening time can be controlled considerably by the composition of the medium: change from an inhibitory to a permissive medium will induce onset of very intense cell fusion within ~18 hr. Characterization of the active factors in medium was, however, elusive. O'Neill and Stockdale (1972) question some aspects of Holtzer's theory of the quantal cell cycle. Their results indicate that a prolonged G_1 period is not necessary before the onset of fusion; it can be as little as 3 hr, i.e., cells can fuse early or late in G_1. The timing can be altered by manipulating the medium; by this means a fusable myocyte may even be induced to undergo still another cell cycle.

The *in vitro* evolution of the fusion theory need not be pursued further, except to record direct observations (1971) on fine structural changes which accompany the fusion of myocytes. In Shimada's monolayer cultures of trypsinized chick embryo thigh muscle (12 days), the adhesion and fusion of myocytes to one another, and of myocyte to myotube and myotube to myotube, occurred between 4 and 8 days *in vitro* (Fig. 3). At 4 days, junctional thickenings of two adjacent myocyte sarcolemmas (fasciae adherentes) (Fig. 4) were seen. Fingerlike projections from myocyte into caveolae of the myotube were also seen. At 6 days, the myocyte cell border was partially disrupted, allowing cytoplasmic continuity with the adjacent myotube. A slow diffusion of myocyte contents into the myotube took place after complete fusion. Membrane specializations do not persist in the sarcoblast as they do in cohering cardiac cells. Myofibrils were not visible with the electron microscope in fusing uninuclear cells. Cross-striated myofibrils were seen in sarcoblasts after 7 days; at 10 days *in vitro*, they filled the cytoplasmic space and spontaneous contractions were occurring. Lipton and Konigsberg (1971) report in detail electron microscopic observations of fusing myocytes. Their uninuclear cells were selected just before the appearance in the cultures of numerous multinucleate plasmodia, identified as far as possible with phase contrast, marked, embedded *in situ*, and sectioned. Fusion was recognized by the presence of a cytoplasmic

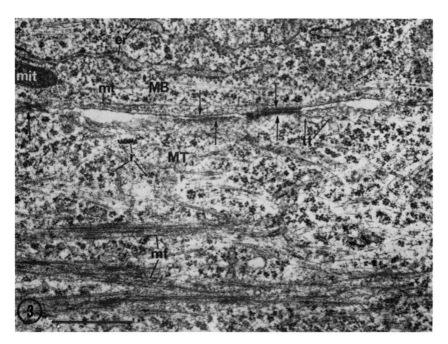

Fig. 3. Electron micrograph of site of attachments of a myocyte (MB) with a myotube (MT) in 4-day culture of embryonic chick muscle. The sarcolemmas generally run parallel. Sarcolemmal thickenings (fasciae adhaerentes) are seen (arrows). Myofibrils (mf), granular endoplasmic reticulum (er), ribosomes (r), T system tubules (tt), microtubules (mt), mitochondria (mit). (After Shimada, 1971.)

bridge between two myocytes running through several adjacent thin sections.

Biochemical confirmation *in vivo* of the histogenetic observations agreed on *in vitro* was provided in an elegant and imaginative experiment by Mintz and Baker (1967) involving hybrid (allophenic) mice created by aggregating cleavage stage blastomers from embryos of different genotypes in which gene control of isocitrate dehydrogenase synthesis existed in different alleles. It was shown that in the mosaic of the descendents from the mixed blastomeres, a uninuclear myogenic cell, of whichever lineage, contained the isozyme of only one locus, as would be expected. However, in multinucleate muscle fibers, an electrophoretic band intermediate in position between those which were contributed by the two original cell strains could be detected, thus indicating that myocytes of both strains had fused to form heterokaryous sarcoblasts.

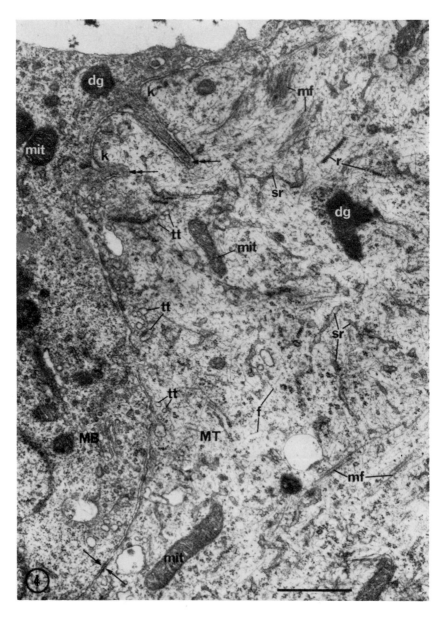

Fig. 4. An electron micrograph similar to that in Fig. 3, showing sarcolemmal thickenings at the area of the opposed membranes. Fingerlike projections of a myocyte membrane (arrows) are seen, which are embedded in caveolae (k) of a myotube. (After Shimada, 1971.)

C. Cytodifferentiation

1. MORPHOLOGY

As the earliest indication of impending myogenesis, Holtfreter (1947) observed an "inherent polarity" in presumptive myoblasts as they developed after isolation from late amphibian gastrulae. These cells adopt a plump spindle shape or a cylindrical modification of it and then stretch out into a long spindle with pseudopods at each end but with a hyaline endplate to distinguish the anterior pole. In their classic paper of 1917, W. H. Lewis and M. R. Lewis observed bipolar cells such as these in the outgrowth area of embryonic chick muscle explants, noting their striking resemblance to myoblasts *in vivo*, and correctly identifying them. In cloned colonies, Konigsberg (1963) recognized constant morphological differences between chick fibroblasts and myoblasts at 18 hr after plating; the myoblast is a relatively refractile cell with a marked bipolar shape and a small ruffled membrane at one pole (Fig. 5A). The development of a muscle colony from a single such myoblast is recorded photographically by Konigsberg in the series of prints reproduced in Figs. 5 and 6.

Although some minor contradictions exist between timing accounts of myofibril appearance, alignment, and banding in embryonic myogenesis, there is general agreement that single unstriated myofilaments, detectable either microscopically or chemically, are first found in myocytes on the verge of fusion, and that parallel alignment and banding begin in the myotube as it is progressing toward sarcoblast status. According to Engel and Horvath (1960), antimyosin fluorescence studies indicate a differentiative sequence as follows: (a) assembly of chemical constituents into unstriated, filamentous structures, (b) arrangement of myofibrillar components into a cross-striated pattern, (c) alignment of the striations of adjacent myofibrils. Moscona (1958) noted banding and twitching "occasionally" in spindle-shaped uninucleate cells of the chick; but Shimada (1971) in electron microscopic observations failed to find myofibrils in most of the fusing chick myocytes, as did Firket (1967). An electron microscopic sampling difficulty, as well as physiological differences attendant upon details of dissociative methods, may have intervened here. Within the differentiating multinucleate fusion product, diffuse masses of myofilaments appear first without organization into bundles (Howarth and Dourmashkin, 1958). A spontaneous spasmodic twitching first occurs [according to Konigsberg (1960)] in individual chick myotubes, during which the contained myofibrils do not

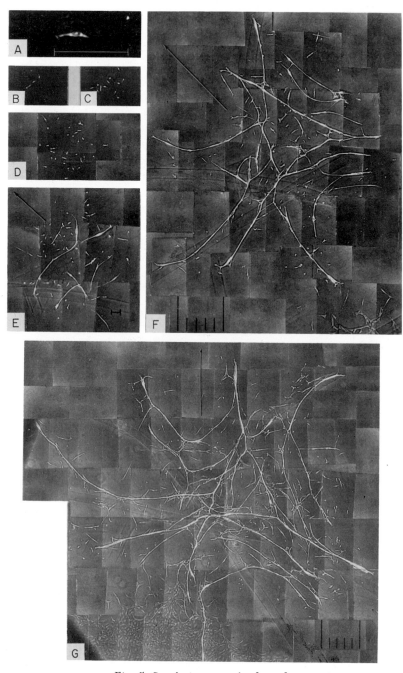

Fig. 5. See facing page for legend.

coordinate. They appear first against the sarcolemma, often paranuclear, and eventually fill the sarcoblast as the nuclei migrate, or are distributed more uniformly through it. Dawkins and Lamont (1971) in a somewhat inconclusive exploration of early myogenesis by immunofluorescent staining with a variety of muscle antisera, suggest that the active antigen in their system, rather than being myosin, is actin or actin-associated.

Several ultrastructural accounts of cytodifferentiation *in vitro* appeared simultaneously in 1967 [cf. Huxley (1966) for *in vivo* fine structure]. Fischman *et al.* (1967) described the condition of myocytes released by trypsin from the comparatively well developed leg muscle of 12-day chick embryos. At first these contained myofibrils which had lost the alignment that obtained in the original sarcoblasts. After 1 day *in vitro* myofibrils were no longer observed, but thin filaments began to appear, and myofibrils were seen anew within 2–3 more days. As fusion took place, the cytoplasmic contents of myotubes were poorly admixed, giving a mosaic appearance. Shimada *et al.* (1967) noted a striking ultrastructural similarity between differentiating muscle *in vitro* and *in vivo*, par-

---

Fig. 5. Photomicrographic record of the development of a muscle colony from a single bipolar cell. The living cells were photographed at an initial magnification of 200 × (A) or 100 × (B–G); bright-medium phase contrast optics were used. (D–G) Composites of several successive overlapping frames covering the progressively greater expanse of the colony on succeeding days. (A) The single cell photographed some time between 18 and 24 hr (day 1) after plating. The nucleus contains one prominent nucleolus. A cluster of highly refractile granules is present in each zone next to the nucleus. Note the ruffled membrane at the tip of the process at left. The scale marker represents 0.1 mm. (B) A colony produced by the cell in A during the first 24 hr of recording (day 2). The scale marker in E pertains to photomicrographs B–E and represents 0.1 mm. (C) The colony on day 3. Three of the cells in the field are rounded and are presumably in an early stage of division. (D) The colony on day 4. (E) The colony on day 6. Long multinuclear myotubes have formed (these were observed 24 hr earlier). The arrows in photomicrographs E, F, and G indicate orientation of the colony with respect to the orientation in G (the orientation was changed to facilitate photographing). The orientation is confirmed by the matching pattern of strain marks in the plastic petri plate. (F) The colony on day 9. The network of myotubes has expanded considerably, but note the presence of single cells among the myotubes. Cells of the leading edge of an invading colony of fibroblast-like cells can be seen at lower right. Each division of the scale represents 0.1 mm. (G) The colony on day 13. The myotubes are longer and more numerous than they are in F. Single cells are still present. Continued proliferation of these single cells is suggested by the rounded appearance of some of them and by associations typical of late anaphase. The invading fibroblastlike cells observed in F are now quite obviously the periphery of a contiguous colony. Compare the cells of the impinging colony with the single cells of the muscle colony. (Scale same as in F.) (After Konigsberg, 1963.)

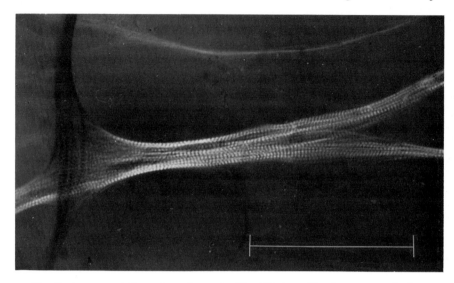

Fig. 6. An area of the colony shown in Fig. 5G (roughly the area at the lower edge of the sixth frame from the right in the fourth row from the bottom). The use of polarizing optics demonstrates the presence of cross-striated myofibrils typical of striated muscle. Fixation: osmium vapor after storage at —20°C in 50% glycerol. Scale marker represents 0.1 mm. (After Konigsberg, 1963.)

ticularly as regards myofibrils and sarcoplasmic reticulum. Firket, working with 12-day chick muscle explants, listed the following ultrastructural aspects of myofibril formation:

1. Uninucleate myogenic cells have a reduced amount of rough ER and a large number of polyribosomes. Some microtubules are present.

2. Differentiation of myofilaments begins only in myotubes. The number of ribosomes decreases while three types of scattered filaments accumulate: (a) microtubules (abundant at first but decreasing), (b) thick filaments 90–120 Å, (c) thin filaments ~50 Å.

3. Thick and thin filaments associate in bundles of myofilaments, where they regularly alternate.

4. Z substance appears in or near the membrane of the extending T system, and Z bodies align themselves on neighboring bundles of myofilaments. In their immediate neighborhood, the thick filaments are replaced by an I band. At the same time, glycogen appears between the bundles, and neighboring bundles join to form a Z line.

A morphological account of differentiation of the sarcoplasmic reticulum and T system with functional interpretations was given by Ezerman and Ishikawa (1967). In 1968, Ishikawa offered extended data on T system formation in early myotubes, stressed his observation of numerous inpocketings of the sarcolemma, and presented schemata leading to the suggestion that these caveolae, or micropinocytotic vesicles, might afford the basis for T system networks (Fig. 7).

A class of filaments of ∼100 Å, intermediate in size between actin and myosin filaments, was described by Ishikawa *et al.* (1968) in cultured sarcoblasts. These are seen scattered through the sarcoplasm at all stages of development and show no obvious association with the myofibrils. They are particularly conspicuous in myotubes fragmented by colchicine and other mitotic inhibitors, and are found in metaphase arrested cells. Similar filaments are present in fibroblasts, chondrocytes, and proliferating myogenic cells. It seems possible that these are the ubiquitous cellular microfilaments which may be sterically interchangeable with microtubules bindable by colchicine and its congeners. In colchicine-treated neurons (R. P. Bunge and M. B. Bunge, 1968, 1969), the normal quota of microtubules disappears, being replaced by masses of such microfilaments, and the cytoplasmic architecture of the soma collapses. Upon withdrawal of the mitotic inhibitor, cytoplasmic organization is restored in reverse order. Bischoff and Holtzer (1968b) discuss the well known disintegrative effect of colchicine upon multinuclear myotubes (described by Godman and Murray, 1953) and report that at very low concentrations (10^{-8} *M*) fusion of myocytes may occur, but these cells fail to elongate. Actin and myosin can be assembled in them, however, to form cross-striated myofibrils. We are probably concerned here with the effects of colchicine and its congeners on the microtubular skeletal framework which myogenic cells possess in common with other types of cells and which, as reported by Firket (1967) and Ishikawa *et al.* (1968) is distinct from the specialized system of contractile proteins characterizing muscle. It is interesting to speculate on possible regroupings of microtubular equipment in the sarcoblast on the basis of Przybylski's observation (1971) that developing skeletal muscle loses its early quota of centrioles and cilia after myotube formation, whereas cardiac muscle retains them throughout life, as it retains its capacity for mitotic division.

2. CYTOCHEMISTRY[2]

According to Moscona (1955), the first cytochemical indication of differentiation in a myogenic stem line (plated from 4-day chick limb bud) is the appearance within presumptive myoblasts of characteristic

---

[2] See also Section V. General Nutrition and Metabolism.

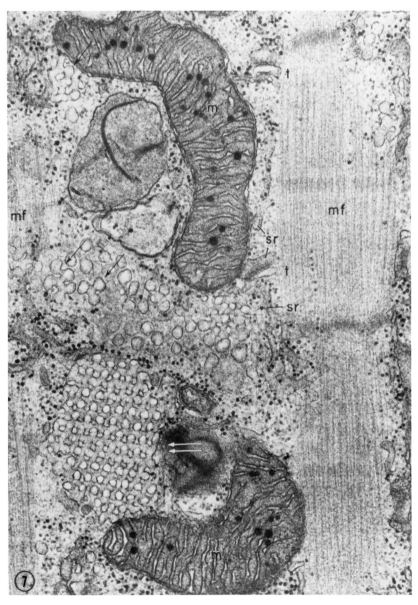

Fig. 7. Longitudinal section of a well-developed sarcoblast. An elaborate membranous network (double arrow) can be seen between the myofibrils (mf). The network appears crystalline. Simpler, and presumably early, configurations of the network also can be found (single arrows). Triadic connections (t) between the T system and sarcoblastic reticulum (sr); mitochondria (m). Sixteen-day culture of 11-day chick embryo pectoral muscle. (After Ishikawa, 1968.)

granules which react with reagents for polysaccharides and nucleoproteins. In this respect, they can be distinguished from accompanying connective tissue cells. At this stage they multiply rapidly by mitosis. After an indeterminate number of divisions, they begin to assume the characteristic spindle shape (p. 251) and to elongate. Concurrently the cytoplasmic granules decrease in size and disappear, while fine fibrils first become discernible.

Succinic dehydrogenase activity is seen in uninuclear myoblasts (Cooper and Konigsberg, 1961b), but greatly increased succinic dehydrogenase activity is associated with multinuclearity. In regions of suspected fusion, reactivity was reduced to the low level of the myoblast. Creatine phosphokinase, a protein concerned with energy transfer but not related directly to the contractile process, can be used (Reporter *et al.,* 1963; Konigsberg, 1964) as an index of differentiation, since creatine phosphokinase activity first makes its appearance concurrently with multinucleation and the enzyme accumulates as the sarcoblast develops. O'Neill and Strohman (1969) reported that a steep decline in DNA polymerase activity closely parallels the time of rapid fusion and formation of myotubes. Remaining activity is almost exclusively associated with uninuclear cells. These findings were confirmed on muscle developing *in vivo* by a report of Stockdale (1970), who found 5-nucleotide phosphatase activity appearing simultaneously with the loss of DNA polymerase.

Ross and his associates have concerned themselves intensively with intranuclear events occurring during differentiation of myoblasts, especially in the first beginnings of fusion. Measurements were made by interference microscopy on the nucleoli of uninuclear myogenic cells and binucleate and multinucleate complexes (Ross, 1964), and these showed that nucleoli in binucleate cells had less than one-half the dry mass and volume of those in uninuclear myoblasts. Serial and time-lapse photomicrographs covering the association and fusion of two myocytes also showed that one or more of these nucleoli shrinks in linear dimensions during the fusion process; after fusion, nucleoli enlarge again. Mean values for dry weight and volume of nucleoli in myotubes were intermediate between those of uni- and binucleate cells. This was interpreted as signifying that myotubes were the product of both recent and less recent fusions (Ross and Jans, 1968a,b). The fusion process (in the mouse) did not take more than an hour, and there were many transitory contacts between cells that did not result in fusion. Optical density measurements indicated that during binary fusion RNA leaves the nucleoli in at least one of the partners at a greater rate than it can be synthesized, and at the same time, it appears in unusually high concentration in that cell's cytoplasm. Electron microscopic studies by Ross *et al.* (1970) on embryonic and regenerating mouse myoblasts confirmed this earlier observation by showing that where two uninucleate cells are fusing and contiguous cell membranes are in a state of dissolution, one of the two (cell A) contains a greater number of ribosomes in the vicinity of the junction than does the other (cell B). Ribosomes A are in polyribosome packets, and thus indicative of active protein synthesis; most ribosomes B aggregate in smaller units. Refractile cytoplasmic organelles (mitochondria, lipochondria, lysosomes, etc.) are also often concentrated at the fusion area in both cells. In primary cultures of differentiated muscle from newborn rats, phospholipid phosphatase was located in lysosomes but not in other organelles (Reporter and Norris, 1969).

Among Engel's early studies localizing the appearance of formed products in isolated differentiating muscle, is a report on glycogen (1961a). By the Schiff–periodic acid method, this was not found in uninuclear myogenic cells, but made its first appearance perinuclearly in the myotube. In the striated sarcoblast stage, glycogen was located as occurring profusely in perinuclear and subsarcolemmal cytoplasm and in transverse bands (cf. Firket, 1967). Cholinesterase (by Koelle and Coers methods) was found diffused through the cytoplasm in all stages of muscle development (Engel, 1961b). In more mature stages acetyl-cholinesterase (AChE) was located along the Z lines. In these completely denervated cultures from 7–13-day chick embryo thigh, he found no motor end plate structures and no enzyme activity at the sarcolemmal level. Wilson and Stinnett (1969) and Wilson (1970) found the three AChE isozymes (that are formed *in situ*) present in myotubes developing *in vitro*. Growth, but not myotube formation, was inhibited in these cultures by malathion (an organophosphorus inhibitor of respiration). AChE activity was much reduced by malaoxon, a product of malathion metabolism with no effect on respiratory rate.

Goodwin and Sizer (1965) examined organ cultures and trypsin-dissociated cultures of chick thigh muscle (explanted from 7 days *in ovo* to hatching), in the presence of cultured lumbosacral spinal cord (from 66-hr and 4-day embryos), which was separated from the muscle cultures by a Millipore filter as well as by considerable amounts of medium. From these observations, they concluded that AChE is myogenic in origin up to the sixteenth day of development *in ovo*, when motor end-plates appear. At this time, the presence *in vitro* of 4-day spinal cord enhanced the uninnervated muscle growth somewhat, but depressed the specific activity of AChE. Addition of acetyl-β-methylcholine, a substrate of AChE, raised the enzyme titer only slightly. Hence, they concluded that substrate induction is not a primary mechanism for AChE control in developing chick muscle, and that the influence of spinal cord is exerted through some other means (see Section IV,C.). Kinetics of ACh receptor production and incorporation into membranes of developing muscle fibers were investigated *in vitro* by Hartzell and Fambrough (1971) by permanently blocking existing receptors with α-Bungarotoxin, then following the reappearance of ACh sensitivity—which paralleled the emergence of the new ACh receptors. The rate of reappearance was depressed by introduction of a protein inhibitor (cycloheximide) and by the ATP inhibitors dinitrophenol and iodoacetate.

Changes in the isozyme patterns of lactic dehydrogenase during development of muscle tissue *in vitro* were noted by Fujisawa (1969) and by Delain (1969a,b). In contrast, the isozymes of aldolase and creatine kinase retain the embryonic character in culture at the same time that lactic dehydrogenase is differentiating from the H type, which predominates in the embryo, to the M type of adult muscle tissue. Emmart *et al.* (1963), in some elegant observations by immunofluorescence on muscle developing in culture, localized the enzyme glyceraldehyde-3-phosphate dehydrogenase in mitochondria and in the A bands of striated fibers. This enzyme is involved in glucose metabolism.

III. Function

A. Denervated or Uninnervated Muscle

1. Spontaneous Rhythmic Contraction

M. R. Lewis reported in 1915 that rhythmical contractions occurred in cultures of embryonic chick muscle entirely free from nervous influences. The same observation has been made by many subsequent workers on a variety of explanted material; Pogogeff and Murray (1946) extended it to adult mammalian muscle. This spontaneous activity is not continuous; rest intervals occur more frequently than in cardiac muscle. The duration of rest periods is variable, as is the rhythm of single fibers and groups of fibers. Excitability seems to vary with the general physiological condition of the culture, which involves developmental status, nutrition, oxygen tension, and accumulation of metabolic wastes, as well as pH, temperature, and electrolyte balance.

A film by W. G. Cooper (1962) shows pulsation of myoblasts. In dispersed cell cultures from chick leg muscle, Konigsberg (1960) observed a spontaneous twitching activity after 7 days *in vitro* which was confined to single myotubes. There was no transfer of impulse even to contiguous cells. Within a single syncytial unit, the twitchings of different fibrils did not necessarily coordinate. Capers (1960) reported observing spontaneous contractions throughout the whole course of differentiation, from myoblast on; contractions become stronger and more regular in later stages. His material came from 13-day chick embryo muscle maintained in Rose chambers.

2. Drug Action

Rhythmically contracting cultures of isolated skeletal muscle afford a valuable source of uninjured fibers with intact sarcolemma, which are at the same time entirely devoid of neural connections and end plates. In such material it should be possible to distinguish directly between general membrane action, chemical transmission at the specialized myoneural junction, and effects of chemical agents on the contractile system itself.

Using embryonic chick material, Sacerdote de Lustig (1942, 1943) reported that physostigmine and low concentrations of acetylcholine activated the automatic contractility of skeletal and smooth muscle, while adrenaline, atropine, and high concentrations of acetylcholine paralyzed it. Several curarizing substances tested on skeletal muscle activated it at low concentrations (10^{-5}) but paralyzed at higher levels. Curare, erythrine, and cobra venom antagonized only the paralyzing action of acetylcholine; strychnine and veratrine could reactivate cultures paralyzed by physostigmin, acetylcholine, erythrine, or curare.

The sporadic and nonuniform nature of contraction in this type of preparation, presents a substantial source of error. In the attempt to arrive at more precise data, an arithmetical system of evaluation involving the averaging of 15 counts was devised by the writer for experimental and control cultures of fetal rat skeletal muscle (Murray, 1960, pp. 128–129). Application of this system led to the conclusion that procaine, veratrine, tubocurarine, and ryanodine are inhibitory in these circumstances (though at very low concentrations tubocurarine may stimulate); ATP and acetylcholine are without effect or inhibitory; and caffeine and mapharsen activate contraction. Of some interest is the indication that in this material entirely devoid of myoneural junctions, the tested curarizing drugs exert a blocking effect (cf. Section III,B,2). Ryanodine, an alkaloid with some curare-like properties in vivo, appears to be membrane-active and to require no nervous intervention. The question still exists whether some or all of these compounds are able to penetrate (via the sarcoplasmic reticulum), or to affect membrane permeability, or to enter by pinocytosis, or whether they act solely at the membrane surface. However, in membraneless Szent-Györgyi models, cocaine and veratrine are without effect, as are ryanodine and caffeine; mapharsen is inhibitory; and ATP reacts directly with the contractile substance.

An uncontrolled factor in all the above tests may be the length of time (or age) *in vitro* of the muscle tissue. In organ cultures of frog muscle developing denervation effects *in vitro*, A. J. Harris and Miledi (1966) found that the muscle membrane sensitivity to local application of acetylcholine became progressively more widespread during a 2 week period after explanation; this may also apply to other membrane-active substances.

B. Innervation in Vitro

1. MORPHOLOGY AND CHEMISTRY

Neurons explanted *in vitro* typically evolve long neurites which make transient "exploratory contacts" with each other and with other types of cells present in the field. The occurrence of such nonspecific cellular contacts should not be confused with organotypic behavior on the part of developing nervous tissues, in which progressive histological and functional maturation culminates in establishment and maintenance of specialized characteristic properties which are unique to the organ, or area of it, from which the tissue originates [see Crain's exposition (1966) of these distinctions]. A case in point is the neuromuscular junction, which has been described as developing *de novo* in culture, in some cases with strong morphological and functional verification, in others with an appearance more indicative of transient, nonspecific contacts between nerve and muscle fiber or of very primitive junctions.

A spate of such reports emerged in 1969. In light microscopic studies with cord and muscle explants placed about a millimeter apart, Nakai (1969) shows bulbous tips of many exploring nerve fibers in contact with sarcoblasts; a few of these, by silver and AChE staining, appear to be primitive junctions. With electron microscopic observations on short-term cultures (about 1 week), James and Tresman (1969) show exploring fibers in nonspecific contacts with myoblasts. Shimada *et al.* 1969b) using suspensions of 12-day chick embryo leg muscle with 6-day ventral spinal cord plated over them and cultured for about 2 weeks, show differentiating myotubes and nerve fibers in contact; two synaptic profiles appear, but no lemmoblast alignment is shown. Veneroni and Murray (1969) planted fragments of chick cord with dissociated muscle elements; the latter soon reconstituted sarcoblastic conformations and were visited by exploring fibers, some of which appeared with the light microscope to establish primitive neuromyal junctions involving highly localized AChE activity coupled with clusters of nuclei of undetermined origin during 3–4 weeks *in vitro*. Fry and Wilson (1970) used the scanning electron microscope to investigate these problems in short-term (about 1 week) cultures of trypsinized cord and muscle, finding anew that many apparent junctions were nonspecific. AChE staining appeared to show both diffuse and localized activity in the absence of nerve contacts, as can be expected in early muscle development *in vitro*.

Cieselski-Treska *et al.* (1970) combined in a Rose-chamber cultures of myoblasts with sensory and motor neurons "isolated by microdissection." Fibrillatory contractions occurred after about a week, then fascicular contractions. There was moderate AChE activity in both neurons and muscle cells—which became more marked at points of contact between them. Two morphological types of contact were observed—one from the cord neurons and another from the dorsal root ganglion. A. J. Harris *et al.* (1971) combined a cloned line of normal myoblasts with a clone of neurons from the omnipresent mouse neuroblastoma, finding that though resting potentials could be recorded from myocytes and early fusion products, action potentials could not be evoked until a greater degree of sarcoblast maturity had been reached; at about this time spontaneous contractions began. The muscle cells were ACh-sensitive at all periods; but sometimes an increased sensitivity was found at the point of contact between myotube and an exploring neoplastic neurite. Functional synaptic transmission, however, was not encountered.

As early as the forties, whole somites with adjacent cord areas (from 4-day chicks, stages 23–24) were explanted and the pattern of their functional development in isolation was recorded in terms of spontaneous activity and electrical excitability (Szepsenwohl, 1947). A morphological study of developing neuromuscular relationships in such organotypic preparations, from 10–16-day mouse embryos, was made in 1968 by Bornstein *et al.* Myelinated axons with synaptic terminals and contracting cross-striated muscle fibers developed and persisted over months in culture. Localized activation of AChE was first suggested after 6 days *in vitro*, concentrations of AChE activity were found in some numbers after 2 weeks, and at 36 days they were very numerous and strongly outlined. Electron micrographs showed the neuromuscular apparatus well developed after 16 days. In these preparations it is likely that some neurites had already infiltrated the muscle anlagen at the time of explantation; however, a high degree of further differentiation undoubtedly took place *in vitro*.

Peterson and Crain (1970a,b) observed the formation of functional neuromuscular junctions between spatially separated explants of fetal rodent cord and muscle, even across species lines (rat and mouse). Appropriate orientation of the respective tissues was found to be important for the establishment of specialized contacts, and the arrival of exploring nerve fibers from the cord explants appeared to enhance muscle differentiation. Partially atrophied muscle explants could be rejuvenated by the introduction of ventral cord explants into the area. This "trophic influence" was especially marked when cortisone was included in the medium. In 5–11-week-old cultures, motor endplates were found with characteristic subneural infoldings, increased sole-plate sarcoplasm and terminal lemmoblats. Synchronized contractions of a network of fibers within the muscle explants could be evoked by selective electrical stimulation of ventral cord or root.

Starting with fetal rodent spinal cord, Crain *et al.* (1970) have been able to effect functional couplings with adult skeletal muscle explants of both rodent and human origin, in cultures maintained from 2–7 weeks *in vitro*. During this time, bundles of neurites resembling a ventral root are evolved from the cord placed <1 mm away and terminate characteristically within the muscle tissue. Although it is well known (e.g., Waser and Nickel, 1969) that end plates retain their structure to a recognizable degree for as much as 2 months after denervation *in vivo*, it is stated (Peterson, 1970) that in these adult muscle fibers the held over end plates are not reinnervated by the fetal neurites, but new sites of neuromuscular junction are established. Biopsied human muscle used for these explants was "dramatically activated" to regenerate by the fetal rodent neurons [see Section IV,C]. More recently this group (Pappas *et al.* 1971) has reported that the fine-structure of the neuromuscular synapses, especially in older cultures, is essentially comparable to that of adult mammalian motor end plates *in situ*. Cord-evoked muscle contractions were demonstrated before fixation of these elegant model preparations for electron microscopy.

In 1970, Fischbach reported functional synaptic contacts developing after 3–4 weeks in cultures of initially dissociated cord and muscle cells from 7- and 11-day chick embryos. It is suggested that these were comparable in differentiation to the preparations of Shimada *et al.* (1969) (see p. 261).

2. ELECTROPHYSIOLOGY

Until recently, electrophysiological studies of skeletal muscle cultures have been quite limited in number. Many workers have made EKG records from tissue cultures but chiefly of contracting heart muscle. A detailed review of this earlier work is given by Crain (1965). Usually, summation effects were obtained with relatively large electrodes (30 μ), as by Szepsenwohl (1946) who recorded from both cardiac and skeletal muscle; in the latter, during spontaneous contraction, a single, quick, bipolar wave was found where a group of fibers had the same frequency, but this was replaced with a slow mono- or bipolar wave when several different rhythms were present. Szepsenwohl also showed (1947) that the threshold for electrical excitation of cultured embryonic skeletal muscle decreased with differentiation *in vitro,* reaching a low level at the onset of spontaneous activity. When he explanted myotomes along with adjacent portions of spinal cord from 4-day chick embryos, he found the transition to low excitation threshold and spontaneous contractions occurring within 24 hr *in vitro,* but the same myotomes explanted without spinal cord required 5–6 days to reach a corresponding stage of functional development. These observations contrast significantly with those on cardiac muscle, which showed low excitation threshold and spontaneous activity at all stages *in vitro.*

In 1954, an intracellular recording technique was applied by Crain to cultured, spontaneously contracting rat muscle fibers, demonstrating membrane resting potentials up to 70 mV and action potentials to 100 mV (approximately normal amplitudes). This array involved a microdissection assembly with oscilloscope hookup (described by Crain, 1956); potassium chloride glass electrodes of 0.5 μ diameter were inserted into neurons or muscle cells visualized with the high-power compound microscope. With such a small electrode, the impaled cells could remain viable for half an hour or more while typical potentials were registered (See Fig. 8 with Crain's description). A characteristic pacemaker prepotential is evident, beginning immediately after each spike and leading to the triggering of the next spike, a fundamental pattern observable also in cultured cardiac muscle and in sensory ganglion cells.

These bioelectric properties of cultured skeletal muscle cells were confirmed and extended in primary explants for 13 day chick embryos by Li *et al.* (1959), who also demonstrated spontaneous rhythmic oscillations of membrane potential in the absence of spike discharges (Fig. 9). When these attained a critical level of depolarization, spikes were engendered. In other experiments a twitch response could be evoked; this consisted of a spike potential followed by a prolonged after-potential that persisted for as long as 2 sec during which rhythmic oscillations occurred (Fig. 10). Li concluded that the spontaneous rhythmic potentials must arise at some point along the muscle fiber itself. He showed in 1960 that electrical activity *in vitro* was synchronous with the fibril-

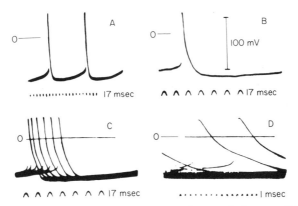

Fig. 8. Membrane resting and action potentials recorded intracellularly from a cultured, spontaneously contracting skeletal muscle fiber. Note the large overshoot and the negative afterpotential (similar in pattern to those of cardiac muscle *in vitro*). The family of action potentials shown in C was obtained by nearly synchronizing the sweep frequency with that of the fiber discharges. Newborn rat muscle 9 days *in vitro*. Recordings were made by Dr. S. M. Crain (1954).

latory movements of the fiber, both spontaneous and evoked, as these were observed microscopically. Spontaneous activity *in vitro* was similar in character to that recorded *in vivo* from denervated muscle.

With the progressive development of techniques for organotypic culture of nervous tissues (Murray, 1965b; Crain, 1966), bioelectric studies

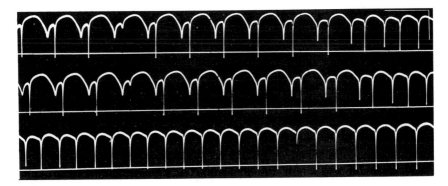

Fig. 9. Intracellular recording from a twitching chick embryo skeletal muscle cell in culture. Note dissociation of spontaneous rhythmic spike discharges and oscillations of potential. Continuous records. First line represents zero potential, vertical bar represents, 50 mV, and horizontal bar represents 200 msec. (After Li *et al.*, 1959.)

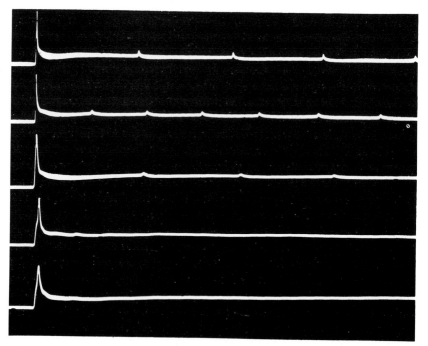

Fig. 10. Twitch responses from cultured chick embryo skeletal muscle cell following repetitive stimulation at frequency of about once every 10 sec. First line represents zero potentials, vertical bar represents 50 mV, and horizontal bar represents 200 msec. (After Li *et al.*, 1959.)

of skeletal muscle innervated *in vitro* have advanced *pari passu*. These have served also to confirm, with functional data, the morphological delineation of specialized junctions established in culture between different regions of the nervous system and between nerve and muscle, thus providing model systems of great value for experiments that can be performed in isolation with microscopic visualization. Electrophysiological procedures applied to cultured cells have likewise undergone refinement, e.g., Crain (1970a).

Thus, it has been possible to extend the early studies of Szepsenwohl on cultured cord with attached myotomes by microelectrophysiological experiments on explants consisting of fetal mouse cord and myotomes as they continued to differentiate *in vitro* during several weeks, eventually evolving myelinated axons from the cord which penetrated bundles of cross-striated muscle fibers developing from somites. During 1–2 months in culture, explants of cord alone evolve a great number of functional synapses (Crain and Peterson, 1964, 1967; Peterson *et al.*, 1965; M. B. Bunge *et al.*, 1967). Where myotomes were explanted along with cord from 12-day fetuses, bioelectric recordings (Crain, 1966) demonstrated that characteristic neuromuscular transmission was present. Following a cord or ventral root stimulus critically localized with extracellular electrode, skeletal muscle responses began, with latencies of at least several milliseconds after the onset of spike barrages in the spinal cord. In simultaneous microscopic observations, synchronized contractions of large groups of muscle fibers (often located at some distance from the cord) could be seen occurring concomitantly with the cord-evoked muscle potentials. Selective stimulation of an attached dorsal root ganglion also might trigger cord activity, which in turn led to muscle contraction (a reflex arc in culture).

Recently, more systematic exploitation of these experimental potentialities has been undertaken by Crain and colleagues, especially with material in which fetal rodent cord is coupled *in vitro* with explants of muscle placed 0.5–1.0 mm away, one tissue having been grown in isolation before being presented with the other (see p. 263). In these preparations *d*-tubocurarine (1–10 μg/ml) selectively blocks neurally evoked muscle contractions in about 1 min, while electric stimuli applied directly to muscle fibers are still effective, and spontaneous asynchronous fibrillations are not altered by cord stimuli nor blocked by curare (cf. p. 260). Eserine administration accelerates recovery of normal function after withdrawal of curare. Strychnine (10 μg/ml) produces characteristic convulsive discharges in the cord and muscle explants. Chronic (1 month) exposure of paired explants of adult muscle and fetal cord to *d*-tubocurarine permits some muscle regeneration, but only partial maturation and innervation. There is a sustained neuromuscular block at the higher curare levels (about 100 μg/ml). Hemicholinium-3 applied at 5 μg/ml interferes with ventral root development, and regenerative activity is aborted (Crain and Peterson, 1971). For more detailed accounts of bioelectric findings in this type of model, Crain (1970b) and Crain *et al.* (1970) should be consulted.

Apparently it is possible to develop primitive synaptic contacts which may be functional in relatively short-term cultures of initially dissociated cord and muscle cells from chick embryos (Fischbach, 1970). During 2 or more weeks of differentiation, myogenesis proceeded as described by others (Shimada *et al.*, 1969a,b), and spinal cord neurons plated over the muscle developed long processes, some of which appeared microscopically to make contacts with other neurons and muscle fibers.

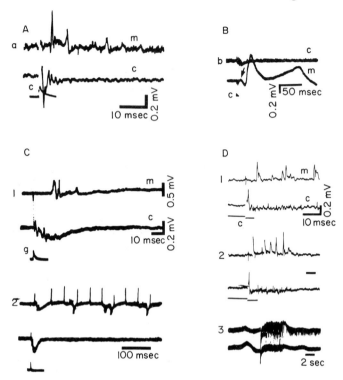

Fig. 11. Bioelectric records of neuromuscular transmission between separate explants of fetal rodent spinal cord and skeletal muscle in long-term cultures. (A)
Simultaneous recordings of responses evoked in explants of mouse spinal cord
(c = second sweep) and muscle (m = first sweep) by stimulus to nearby cord
site (c = third sweep). (Muscle was added to 4-day cord culture and maintained
for 4 weeks *in vitro*.) Muscle action potential (and contraction) arises shortly after
end of cord response and additional muscle potentials occur during the following
40 msec. (B) Similar discharges evoked in explants of rat spinal cord (c = first
sweep) and muscle (m) by stimulus to nearby cord site (c). (Muscle tissue was
added to 4-day cord culture and maintained for 5 weeks *in vitro*.) Rapid synchronous
twitch of many muscle fibers occurs concomitant with long-duration negativity following brief muscle action potential (at arrow). Note that sweep rate is much slower
than in (A). (C) 1. Similar responses evoked in explants of rat spinal cord (c)
and muscle (m), but stimulus is now located in dorsal root ganglion (g) attached
to cord. (The cord tissue was added to the muscle explant at 3 weeks *in vitro*
and then cultured for 6 more weeks.) Latency of muscle action potential is now
much longer than with direct stimulus to ventral cord cf. (C_1) and (A)]. 2. Single
stimulus to dorsal root ganglion evokes, at times, long-lasting repetitive discharges
(and contractions) in muscle following primary evoked cord and muscle responses
[note slower sweep rate in (C_2)]. (D) 1 and 2. Similar responses evoked in
explants of mouse spinal cord (c) and rat muscle (m) by stimulus to nearby
cord site (c). (The cord tissue was added to the muscle explant at 2 months
in vitro and then cultured for 2 more months.) Note repetitive bursts of muscle

Intracellular electrodes to the muscle cells registered action potentials approaching normal amplitude; membrane response was obtained from iontophoretically applied acetylcholine. In timing and pattern, spontaneous and evoked potentials recorded from some neurons and apparent neuromyal contacts resembled responses of spinal cord and neuromuscular junctions *in situ* (cf. pp. 261–263). In later papers (Kano *et al.*, and Kano and Shimada, 1971) these observations have been elaborated further. In 12–13-day cultures, end-plate potentials were evoked in originally monolayered muscle cells by stimulation of spinal-cord fragments 250–500 μm away. Recording was by intracellular microelectrodes in the muscle-fibers. Tubocurarine and eserine produced the expected results; administration of tetrodotoxin blocked end-plate potentials. Two different patterns observed in end-plate potentials were thought to delineate fast and slow muscle reactions. A thoughtful review of embryonal formation of neuromuscular junctions was published by Shimada and Kano (1971).

Most recently, Robbins and Yonezawa (1971a,b) have reported critical observations of the onset of chemical transmission at developing neuromyal junctions, placing it at about the time when nerve–muscle contacts are first visible by light microscopy (4–10 days *in vitro*). With conventional recording techniques applied to juxtaposed explants of rat cord and muscle, they found that chemical transmission from its onset was discrete and short lived, qualitatively similar to the adult, and was in some cases established in immature noncontracting myotubes. Excitatory junctional potentials could be recorded when the resting potential was 12 mV or more and the myotube contained three or more nuclei. Contractility and functional transmission developed independently. The transmittor in early junctions was assumed to be acetylcholine, since *d*-tubocurarine completely eliminated excitatory junctional potentials. Innervation appeared to be related to cross-striation, since denervation *in vitro* led to loss of striation. But ACh was believed ruled out as a trophic agent by the findings of Crain and Peterson (1971) on chronic exposure of cord-muscle cultures to anticholinergic agents [see pp. 267, 279, 281].

---

spikes concomitant with cord afterdischarge. 3. After introduction of strychnine (10 μg/ml), high-frequency bursts begin to occur spontaneously and synchronously in both cord and muscle explants. Note repetitive discharge continuing in muscle long after end of cord barrage. Recordings were made with chloridized silver electrodes via saline-filled pipettes with 3–5 μ tips; stimuli (0.1–0.2 msec duration) were applied locally through pairs of saline-filled pipettes with 10 μ tips. (After Peterson and Crain, 1970b.)

IV. Histogenetic Aspects of Regeneration

Since the general subject of regeneration is covered systematically elsewhere in these volumes (Hudgson and Field, Volume II, Chapter 6) only some aspects to which data from tissue culture seem especially germane, will be considered here.

A. Modulation Range

In contrast to tissue *determination*, the progressive irreversible restriction of cellular potency with divergence of tissue type—which normally occurs during early development, a less persistent and less far reaching change in cellular form and function is recognized and is known as *modulation*. This term encompasses the reversible changes, in response to external stimuli that may be assumed by cells of a given tissue type once this type is determined (Weiss, 1950). In modulation capacity, skeletal muscle is particularly versatile, having a morphological range that extends from giant multinucleate forms to macrophage or histiocyte-like forms. The series of morphological changes normally undergone by embryonic muscle, from spindle-shaped myoblasts to plasmodial structure and to cross-striated sarcoblast, have been recounted above. They are not necessarily permanent; differentiation may regress as well as progress *in vitro*, and even cross-banding may be unstable and may appear and disappear more than once in the same cell (Veneroni and Murray, 1969).

When adult rat or human muscle is explanted (Pogogeff and Murray, 1946), a great variety of modulated forms may be observed in primary cultures (Fig. 12, 2–6)—round, spindle, strap, rhomboid, trapezoid, ribbon, and even stellate or irregular shapes, which may vary greatly in size. After a lag period of 1–3 weeks, these wander or grow out from explants containing fibers in varying degrees of injury. This outgrowth, therefore, contains retrogressive forms mingled with differentiating forms. Cultures of uninnervated adult rat muscle were maintained continuously *in situ* on coverslips by these authors for as long as a year and a half, during which time they displayed both differentiation and dedifferentiation, according to area and conditions. Contraction was present until the observations ended, but not continuously in the same groups of cells.

In a paper on the stability of the differentiated state in clonal lines of skeletal muscle, Hauschka (1968) shows graphically some differences in morphology which appeared and were perpetuated in several sub-

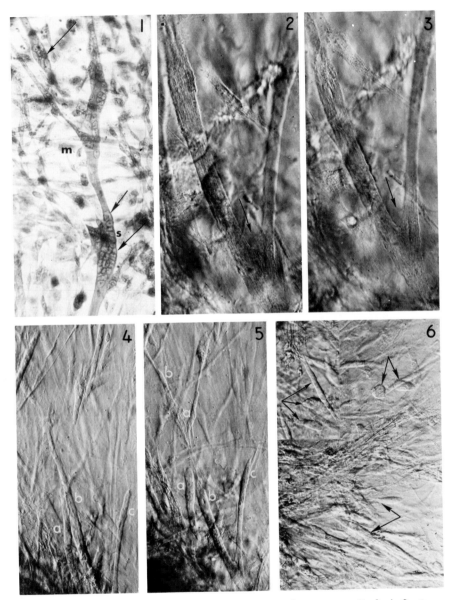

Fig. 12. (1) Embryonic muscle, 18-day rat fetus, 3 days *in vitro*. Zenker's fixative, haematoxylin stain. Uninuclear myoblast (m), strap-form myotube(s), with crowded nuclei (arrows). (2 and 3) Living adult human muscle, the same culture at 20 and 21 days *in vitro*. Note longitudinally oriented myofibrillar bundles. Arrow points to region of apparent fusion between two sarcoblasts. (4 and 5) Another adult human muscle culture, 27 and 30 days *in vitro*. Separation of distal tips and their migration away from parent fibers a and b. (6) Typical viable, differentiating migrant forms (arrows), a little precocious at 16 days *in vitro*.

clones. The subclones responded differently also to the presence of collagen as substrate (Section II,A). Richler and Yaffe (1970) report the establishment by selective serial passages, of six independently isolated myogenic cell lines, which manifested distinct differences in cell morphology and extent of differentiation over periods of months. Though the establishment of a cell line *ipso facto* involves heritable alterations from the primary tissue of explantation, these appear to be based on modifications of the genome that are separate from the restrictions of potency that normally occur in earlier development. Yaffe (1968) produced two such modified lines by brief exposures to a carcinogen (methylcholanthrene). The term "myofibromyoblast" is suggested by Lappano-Coletta (1970) to describe the basic cell type that is produced in primary cultures of skeletal muscle fibers dedifferentiating *in vitro*. Such uninuclear cells are said to be derived directly from the reorganization of cytoplasmic contents within the explanted fully differentiated adult muscle fibers, and the origin of their plasma membranes is subsarcolemmal. During culture, repeated cyclic alternation between myogenesis and dedifferentiation are observed with the myofibromyoblast taking on one or the other form, depending on the developmental stage. Such modulations as are depicted here may afford take-off points for the heritable variations reported in cultures of dissociated muscle fibers by Hauschka and by Yaffe. For a discussion in depth of cell culture in relation to somatic variation, M. Harris (1964, especially Chapter III) should be consulted.

B. Fiber Reformation

The pathological anatomists of the late nineteenth century were aware of the considerable capacity of mammalian skeletal muscle to regenerate after necrotizing injury where the stromal pattern was retained, as in typhoid fever. However, because of its abortive or slight regenerative performance after traumatic wounds and burns, the idea later became widespread that muscle has little inherent growth capacity once it has reached the differentiated state, and medical students were thus indoctrinated. By now, the wheel has come full circle; even the observations described above on reconstitution of dissociated embryonic muscle (see Section II,B and C) should leave no doubt that mammalian skeletal muscle has a substantial potential for repair. It is probable that the actual extent of regeneration in any given case is conditioned more by extrinsic factors governing stroma, scar formation, revascularization, and reinnervation than by any intrinsic limitation of growth capacity in the adult.

When muscle injury occurs *in situ,* the local inflammatory response coupled with the modulatory changes involved in the breakdown and reconstitution of these highly organized syncytial units produce a histological chaos, which has given rise to a plethora of theory seeking to clarify the cellular mechanics of regeneration. The transformation of mature fibers to unicellular forms such as histiocytes, fibrocytes, and fat cells has been proposed, as well as their disintegration into myoblasts which then redifferentiate in the embryonal manner; the intervention of satellite cells is now invoked, budding from original fibers has likewise been described, as has the reversion of the mature muscle to an undifferentiated embryonal blastema. Some ambiguity seems to have resulted from the equating of findings in higher and lower vertebrates, whose overall regenerative capacities differ, and from differences in the degree of maturity of the regenerating fibers under observation. Now that the sequence of events in the normal histogenesis of embryonal muscle is understood, there is no need to labor earlier theories of regeneration. Nevertheless, since the take-off point is different, the behavior of mature fibers in response to injury affords some areas of difference from the developmental course, which may profitably be discussed. Since studies on the behavior of muscle in tissue culture have the advantage of eliminating systemic factors that lead to confusion or indirectly affect regenerative growth, a number of investigators have approached regenerative problems by that means; among the more recent are Capers, Cox, Firket, Ross, Stockdale, and their respective colleagues. A useful review correlating regenerative data *in vivo* and *in vitro* was published (1966) by Betz *et al.*

For functional regeneration *in situ,* it is of course necessary to have the injured or destroyed fibers replaced within the general framework of the whole muscle and to have them adequately innervated. Only the initial stages of replacement and the local cellular aspects of injury (devoid of systemic intervention) have been profitably studied *in vitro.* It is now clear that two main types of behavior occur at the cellular level—(1) budding from original fibers; and (2) breakdown of some fibers into uninuclear units, with subsequent reconstitution of myotube.

1. BUDDING (CONTINUOUS MODE)

As early as 1917, the Lewises described growth (i.e., elongation) taking place from the sectioned ends of sarcoblasts in explant cultures of 11-day chick embryo muscle. These buds were less differentiated than the parent fibers; there was also a marked tendency for anastomoses and fusion between them, following chance contact. Branches with one,

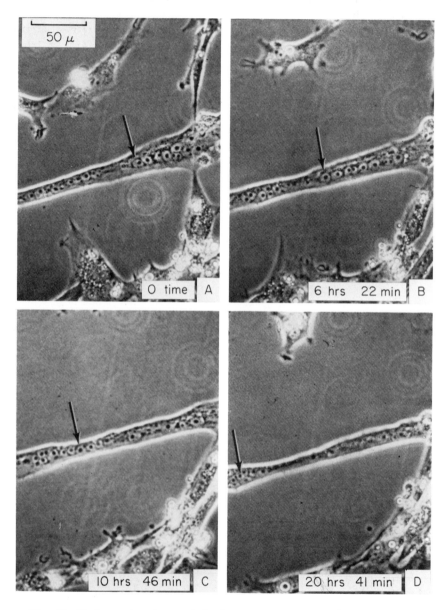

Fig. 13. Nuclear migration typical of young regenerating fibers can be seen by following the location of the arrow in frames A–D, selected from a time-lapse film covering 26 hr and 34 min. These events took place before the appearance of myofibrils visible by phase contrast at this magnification. (After Capers, 1960.)

two, or several nuclei might also split off from the new growths and wander freely. Some buds seemed to consist of chains of myoblasts which tended to break apart from one another. Primitive cross-striations were also seen in uninuclear myogenic cells. The Lewises believed that sarcoblasts could be reformed from the fusion of these isolated cells. Capers (1960) explanted somewhat more advanced embryonic muscle (13-day chick) in Rose chambers, where the tissue after having been teased remained undisturbed for as long as 4 months, and its behavior could be recorded with phase contrast, time-lapse cinemicrography. Though myocyte fusion did take place in these circumstances, the majority of fibers were formed as a result of the repair and growth of damaged but viable fibers in the original explant. Though no nuclear division was noted in the sarcoblasts, there was much nuclear translocation from one centrally situated nuclear aggregation to another (Fig. 13), especially to the growing tip. Lash *et al.* (1957) reported that mobilization of nuclei at some distance behind a wound in adult anterior tibialis fibers was responsible for the central alignment of nuclei within the regenerating fiber elongations. Resnik and Firket (1964) observed in regenerating fibers of 12-day embryonic chick that AChE first appeared near the nuclei in buds at some distance from their source in normal fibers, and some AChE-positive sarcoblasts were entirely separate from the explant. Budding, as well as release of uninuclear muscle cells were observed by Pogogeff and Murray (1946) in explant cultures of adult muscle. As in the material of Capers, little or no mitosis of free, uninuclear myogenic cells was recorded, and none was seen in multinuclear forms (Figs. 12, 14, 15).

2. LIBERATION AND SUBSEQUENT FUSION OF MYOBLASTS (DISCONTINUOUS MODE)

Where sarcoblasts but not mature adult muscle are cultured, and where formed fibers are dissociated and the uninuclear products cultured as monolayers, regeneration in birds and mammals seems to follow substantially the embryonal mode of histogenesis, with division and fusion of myogenic cells playing a major role. When regenerating muscle in the adult lizard tail is removed and cultured thus, a similar sequence is noted (Cox and Simpson, 1970). Matagne-Dhoossche (1969) trypsinized rat striated muscle elements that had differentiated in culture and reseeded them, with the same results. However, when muscle injured by excision is replanted under conditions that do not stimulate rapid growth (e.g. Capers, 1960; Pogogeff and Murray, 1946), budding appears to predominate, and the number of uninuclear myogenic cells is minimal.

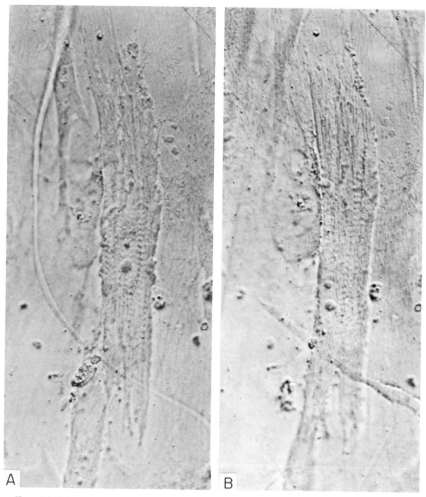

Fig. 14. Living regenerative uninuclear form in a culture of adult human skeletal muscle (A) at 30 days, (B) at 31 days, *in vitro*. Many myofibrillar bundles show primitive banding.

It is probable that the constitution of the media, as well as physical and mechanical factors, all of which differ importantly in these two culture situations, influence the outcome. Simpson and Cox (1967), culturing the dissociation products of regenerating adult lizard muscle, report that myoblast proliferation and fusion can be selectively controlled by altering the culture medium: in Ham's medium (1963) for diploid cell lines, myoblasts remain as single cells for months, but if

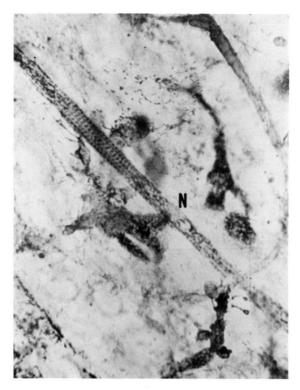

Fig. 15. Adult rat leg muscle, 41 days *in vitro*. Fixed in formol–Locke solution, Polak's silver nitrate impregnation for mitochondria and fibrillar structures. In this regenerative sarcoblast, note banding in one rather circumscribed region; mitochondria are longitudinally aligned above and below a nucleus (N). Note also polymorphic ameboid cells nearby.

Ham's medium is replaced by Eagle's basal medium (1956, 1960) plus glutamine, fusion and myotube formation take place. In monolayer cultures, the effects of dispersing agents (such as trypsin or colchicine) on myocyte membranes may be prolonged, and the fusion that would normally occur quickly between the uninuclear products of traumatic fiber breakdown in explant cultures may be retarded, as described by Ross *et al.* (1970).

But how are we to account for the multiplication of nuclei which even before their aggregation into myotubes have become permanently diploid? Does their dissociation from the fiber as units surrounded by a shell of the original cytoplasm serve to rejuvenate them metabolically or is the dogma at fault? Thanks to the studies of Mauro (1961) and

subsequent workers on the muscle satellite cell, there is a ram in the
thicket; and an assistant ram has been let loose recently by Stockdale
(1971).

The satellite is a small, elongate uninuclear cell which is found at
the surface, outside the plasmalemma but inside the basement membrane
of adult skeletal muscle fibers. In growing animals (e.g., rats) a significant number of dividing nuclei is found within the basement membrane
of the muscle fibers, and it has been suggested that satellite cells might
be the source; Mauro hypothesized that adult satellite cells might represent dormant myoblasts (a stem line) held over from embryonic life.
In recent experiments in situ, Resnik (1969a) and Moss and Leblond
(1970) have investigated with [³H]thymidine the status of DNA-pro-
ducing cells in regenerating and in growing muscle, respectively. In
both cases, they concluded that the tritium-incorporating nuclei belonged
to satellite cells. Moss and Leblond go so far as to say on the basis
of the changing position of the tracer that satellite cells divide and
are then incorporated into the fiber as true muscle nuclei. According
to Resnik's (1969b) electron microscope study of regeneration in situ
(in adult rabbits and mice), myoblasts are pinched off the muscle sarcoplasm as uninuclear cells that pass through a satellite cell stage. The
plasma membranes that isolate these formerly aggregated myonuclei
and their surrounding envelopes of cytoplasm are formed by the progressive coalescence of vesicles and clefts—probably in connection with the
extracellular space, and presumably in contact with the transverse tubular system. The uninuclear cells after being isolated multiply by mitosis.
Hess and Rosner (1970) describe the evolution in situ of satellite cells
by budding from denervated muscle of adult guinea-pigs. Holtzer in
recent reviews of his work (e.g., 1970) incorporates the satellite-cell
lore: having observed that some myoblasts in culture may remain in the
G_1 phase for many days, he postulates that they can hold over in vivo
for many years, synthesizing neither myosin nor DNA.

Stockdale (1971), ignoring the satellite cell, has shown that under
"repair" conditions (initiated by ultraviolet irradiation of differentiated
muscle cultures), DNA synthesis can be resumed in the nuclei of
myotubes even though they may have abandoned this function in the
normal course of development. It is not proved by these experiments
whether their incorporation of [³H]thymidine represents a return to
the DNA metabolism that characterizes the myoblast, or whether a special and different type of DNA polymerase remains available during
differentiated existence to be drawn upon for repair. Stockdale had previously shown in vivo (1970) that the development of nucleotide phospha-
tases and loss of DNA polymerase occur abruptly at the time myoblasts

cease DNA synthesis, corroborating the report of O'Neill and Strohman (1969) noted on p. 257.

C. Innervation—Trophic Aspects

Much has been said and little proven about "trophic" effects of muscle innervation *in vivo*, largely because it is difficult to distinguish between intrinsic influence of the nerve presence *in situ* and the well-established effects of function induced by innervation upon muscle maintenance.

A beginning seems to have been made toward separating these entities in organized nerve–muscle cultures. It is well known that differentiated fetal muscle isolated for long periods *in vitro* gradually becomes somewhat atrophied, even though it continues in some fibrillatory function. Adult muscle isolated by explantation, especially human muscle, only very rarely contracts at all (Pogogeff and Murray, 1946; Peterson and Crain, 1970a). The introduction of spinal cord into such cultures even after a month *in vitro* leads to gradual increase in size, number, and cross-banding of the fibers. In the case of adult human and rodent muscle, a dynamic regeneration process is initiated by the arrival of exploring fibers whether or not neuromyal junctions have been formed (Fig. 16) (Peterson and Crain, 1970a,b). When optimally oriented cord–ganglion fragments are presented to atrophied muscle as much as two months after its original explanation, neurites spread over the muscle surface within 2 days, and significant morphological changes can be observed in the muscle within 4 days. Within a week, contractions may begin, although functional innervation, according to these investigators, does not take place in less than 3 weeks. Particularly striking is the reversal of cortisone effects upon muscle development by the invasion of ventral root nerve fibers. Though differentiation of noninnervated muscle is repressed severely by the presence of cortisone in its medium so that myocyte fusion may be completely eliminated, introduction of properly oriented spinal cord will induce myotubes, fibrillations, cross-banding, and eventually synchronous contractions (in that order) under the same culture conditions. Cortisone has a favorable, rather than an inhibitory effect upon cord maturation (Murray and Peterson, 1965).

The above results obtained with explant or organotypic cultures contrast with those reported by Shimada (1968) for combinations of dissociated embryonic myogenic cells and other tissue types, including spinal cord (see p. 241) and with those of Fischbach (1970) using similar monolayer culture procedures (see p. 263). Robbins and Yonezawa (p. 269) note that since, in their own preparations, bare nerve endings

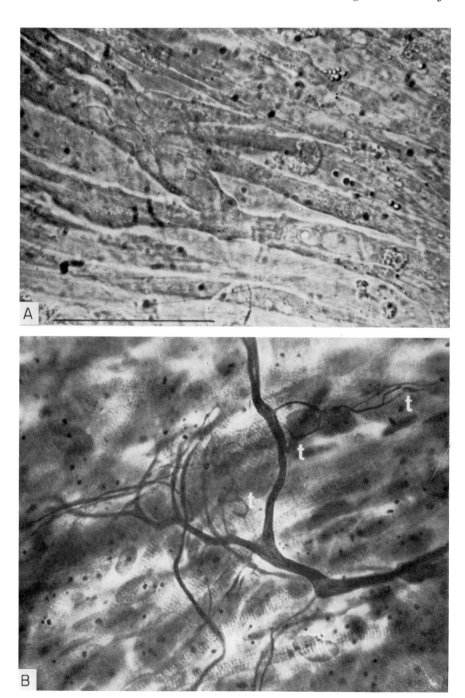

Fig. 16. See facing page for legend.

without any lemmal investiture are functionally active at a stage when Peterson and Crain (1970a) first notice neural enhancement of muscle development, the release of neurotransmitter, or some concomitant event, could be responsible for those trophic effects. However, ACh does not seem to fulfil the requirements. Fambrough (1970) suggests that the regulatory effects of motoneurons on ACh sensitivity in muscle may be produced by restriction, on the part of the neuron, of gene activity in the muscle fibers. Inhibitors of RNA and protein synthesis prevent development of ACh sensitivity, but do not affect it once it is established.

An approach to nerve trophic function by Lentz (1971) utilizes adult amphibian muscle in organ culture and assays its muscle cholinesterase activity in response to explanted sensory ganglia, spinal cord, and liver, also to nerve homogenates, separated from the muscle by a Millipore filter. The presence of fresh sensory ganglia slowed the normal loss of cholinesterase activity in the excised muscle and especially served to maintain morphology and enzyme activity at endplates. Spinal cord was not as effective as ganglion. Boiled ganglia and other tissues were ineffective. Lentz concludes that the trophic effect is mediated by a diffusible chemical substance produced by nerves.

Gutmann *et al.* (1969) produced a correlated study of avian fast and slow muscle (posterior versus anterior latissimus dorsi) development *in vivo*, with their latent periods *in vitro*. The two types of muscle showed no significant growth differences in culture, although *in vivo* the slow muscle (LDA) develops faster. When muscle regeneration was induced *in situ*, the LDA showed faster repair also. Neural regulation *in situ* is invoked.

A test isolating the effects of function was devised by Nakai (1965), who imposed continuous tendinous stress upon developing chick intercostal muscle explanted in its cartilaginous milieu. Unstretched preparations showed eventual degeneration of muscle and other tissues except cartilage, while the muscle subjected to stress continued to develop and maintain its differentiated state. Growth, however, was about half the normal in extent.

Fig. 16. Atrophy of fetal rat muscle explants grown in isolation, and recovery after coupling with cord.   (A) Atrophied muscle, 8 weeks in culture. Myotubes show abnormal swellings, and central nuclei vary greatly in size; cross-banding is absent.   (B) Month-old atrophied rat muscle 5 weeks after addition of mouse cord explant. Note arborization of broad nerve fiber which spreads over muscle tissue; several simple terminals (t) can be seen, and well developed cross-banding is evident.   Holmes' silver impregnation method was used. Scale: 50 μ. (After Peterson and Crain, 1970b.)

V. General Nutrition and Metabolism

Nutrition in cultured cells is far from being an exact science, if only because the conditions of culture (e.g., dispersed cells, monolayers, primary cultures, organotypic cultures) themselves impose a wide range of circumstances governing survival and growth. Chemically defined media are sufficient, with small biological supplements, for certain cell lines, but are seriously inadequate for differentiating primary cultures. Established cell lines are *ipso facto* modified genetically from their tissue of origin. Consequently, caution must be exercised in extrapolating specific findings from particular cell types, even though the parent cell may have been isolated by cloning. Further insight into these general problems may be sought in M. Harris (1964, Chapter 5).

From early times, "fibroblasts" derived from cardiac tissue have been employed in quantitative studies dealing with general concepts of growth rates and growth factors, and strains of fibroblast-like cells have similarly been derived from skeletal muscle (see Section IV,A). In an early attempt to distinguish among grossly modified cell strains from various parent tissues, Parker (1933) reported a characteristic growth rate in standard biological media for skeletal muscle, which differed from those for other rapidly multiplying cells of fibroblastic morphology derived from other mesenchymal tissues such as heart and bone. M. Harris and Kutsky (1958), using freshly isolated monolayers from chick skeletal muscle, which presumably included a fibroblastic component, found that logarithmic growth was supported by supplementation of defined media with the nucleoprotein fraction from adult chicken spleen. The specific action of nucleoprotein fraction from other adult organs was not so high, and embryonic nucleoprotein fraction was only weakly stimulating. Growth stimulation was found to be as great with the protein subfraction as with complete nucleoprotein fraction, and it was suggested that the nucleic acid acted as a stabilizing factor. M. Harris later reported (1959) on an essential growth factor for skeletal muscle tissue that was present in serum dialyzate used as a supplement to Medium 199. In this setting, proteose peptone 0.1% could be substituted for the serum dialyzate. "Growth," in all these investigations, referred to cell multiplication and migration, not to tissue differentiation; and the cell strain was selected on the basis of convenience.

Rinaldini (1959) published in great detail a method developed at the Strangeways Laboratory for the dissociation and quantitative cultivation of embryonic cardiac and skeletal muscle cells. Lucy and Rinaldini (1959) employed these procedures in studies of amino acid utilization by dispersed cell cultures of pectoral muscles from 13-day chick embryos cultivated in defined media. They found that leucine was utilized in large amounts (assayed in terms of medium depletion), while glutamic acid was accumulated in the culture medium. Aspartic acid was utilized at the greatest rate by the least differentiated cultures. A high utilization of leucine and isoleucine relative to that of tyrosine and phenylalanine reflected the proportions in which those amino acids occur in the muscle protein; otherwise, there was no correlation in these developing tissues between muscle protein constitution and the pattern of amino acid utilization from the media supplied to the cultures. Studies on the synthesis of nonessential amino acids from ^{14}C-labeled glucose in natural (biological) media indicated that alanine, aspartic acid, glutamic acid, serine, glycine, and proline could be synthesized by skeletal muscle *in vitro* [Rinaldini (1958) reviewed these and other data]. Granick and Granick (1971) describe a morphological response in chick myoblasts which is peculiar to lack of arginine in the medium. This "nucleolar necklace" formation is abolished by restoration of arginine. They suggest that the nucleolar aberration may be due to depletion of some rapidly turning over, arginine-rich proteins that normally attach to ribosomal RNA precursor molecules.

Reference has been made above (p. 241) to the dependence of avian skeletal myoblasts on collagen products for further differentiation. Where monolayer cultures are constituted of dispersed cells from whole muscle, the quota of connective tissue cells that is always present in such an assortment probably provides the proline–hydroxyproline supplement needed by the myoblasts. But when such a population is subjected to clonal analysis, the isolated pure lines of muscle cells require the presence of collagen or its precursors in some form (e.g., a reconstituted collagen substrate). Population density constitutes an important factor in nutrition; the single cloned cell is initially at a great disadvantage. White and Hauschka (1971) report a still unidentified new factor in conditioned medium which greatly enhances the level of differentiation of cloned muscle colonies. This acts in addition to exogenous collagen (which continues to be required as substrate); it is produced by chick muscle precursor cells within their initial 24 hr in confluent monolayer culture. Dissociated cells from early developmental stages (7½ days *in ovo*) respond more strikingly to this factor than do later ones (12 days). It is thought to be a macromolecular contribution to the medium by the monolayered cells.

According to Okazaki and Holtzer (1965b), variations in rate or amount of myotube formation can be induced by "trivial" changes in culture conditions: cells grown on glass in a 2:3 ratio of horse serum to Eagle's medium (which contains neither proline nor hydroxyproline) proliferate to form a thick, multilayered mat, but do not differentiate into definitive myoblasts or form lasting myotubes. But cells grown on glass at a population density of 2×10^{-6} per milliliter in a medium with proportions of 8 parts Eagle's MEM, 1 part horse serum, and 1 part embryo extract fuse rapidly and form cross-striations as well as multiplying rapidly. Cox (1968) also ascribes a constant high relationship of multiplication and fusion in clone-type density cultures to a permissive medium consisting of Eagle's MEM with glutamine, 10% serum, 4–5% embryonic extract, plus small amounts of sodium bicarbonate, and antibiotic mixture. Ham's (1963) mammalian cell medium, with the same supplements, is not effective for these dissociated cells, which come from regenerating reptilian tail muscle. Shainberg et al. (1969) report controlling avian histogenesis by reductions of Ca^{2+} concentration in a medium consisting otherwise of Eagle's MEM, with 10% horse serum and 1% chick embryo extract. Cultures receiving 1400 μM calcium chloride (the concentration normally present) produced many fusing myotubes, while those receiving 270 μM or less showed only myoblasts at 72 hr. This inhibition of fusion was quickly reversed by the addition of calcium chloride in the normal amount. Thymidine and uridine incorporation were not affected by low Ca^{2+}, and leucine only a little; but creatine phosphokinase activity, involved in the differentiated contractile activity of muscle, was correspondingly reduced by low calcium chloride and restored by normal calcium chloride.

Antimycin A, an inhibitor of electron transport, also inhibits oxidative phosphorylation and respiration in isolated mitochondria. Konigsberg (1964) first observed that it has a selective destructive effect on developing skeletal muscle syncytia, and Reporter and Ebert (1965) undertook to investigate its mechanism and site of action at this stage of myogenesis. In the course of their study, an inhibitor of the limited destructive action by antimycin A was found in an aqueous extract of chicken liver mitochondria. The active factor appears to be a protein of $\sim 10,000$ molecular weight with a 410 mμ absorption peak. It is active in low concentration (~ 0.035 μg/ml) but is not effective against oligomycin. The factor is antigenic and occurs normally in the early chick embryo. It cannot yet be identified with known mitochondrial enzymes, and its mode of interference with antimycin A is still unresolved.

There is general agreement that endocrine products influence the overall course of muscle development, especially after myotube formation, when proteins other than nucleoproteins begin to be synthesized in large amounts. De la Haba et al. (1966, 1968) report that in a culture situation that requires a serum component added to Eagle's medium in order to promote fusion of myocytes, insulin and somatotrophin can replace serum. However, myotubes thus induced do not develop striations or contract, and they die in 3 days. Full differentiation can be brought about by the timely restoration of serum. Glycogen synthetase, which normally reaches a high titer in the myotube stage, remains low in cultures supplied with Eagle's MEM alone and in cultures developing myotubes under the influence of insulin and somatotropin only. Addition of whole serum restores the normal quota of glycogen synthetase; a nonprotein, low molecular weight serum factor also will effect this, but it will not alone induce striation. More serum factors are being sought which may operate in successive differentiative stages. Perhaps one or a series of hormones may be rate-controlling, by analogy with the apparent function of hydrocortisone in the biosynthesis of glutamine synthetase in developing retina (Alescio and Moscona, 1969).

VI. Pathology

A. Neoplasia

1. Spontaneous Neoplasia

Morphological exuberance may ensue when a form normally as highly organized as the skeletal muscle fiber escapes from restraining influences and yet remains viable and reproductive, as in neoplasia. Spontaneous rhabdomyosarcomas explanted *in vitro* recapitulate as a group the modulative repertory of regenerating and developing muscle in culture, but rarely reach the end point of cross-striation. Round forms, spindle forms, fibroblasts, sarcoblastic ribbons that may or may not develop myofibrillae, and giant multinucleate forms can be identified, though not all of these need appear in a single tumor. In addition to these conventional modulants, exceedingly bizarre forms are sometimes seen in very malignant tumors (Figs. 17–19) which yet in their texture and outline generically resemble some of the cultured normal cells. An experienced observer of the modulation range that characterizes muscle *in vitro* can put this knowledge to use in human tumor diagnosis, as Timofeevskii (1946–1947) and Murray and Stout (1958) have done in the classification of tumors from a variety of tissue origins. For human rhabdomyosarcoma, the differential diagnosis usually lies between this and liposarcoma or fibrosarcoma, neither of whose normal parent tissues run a comparable gamut of modulation.

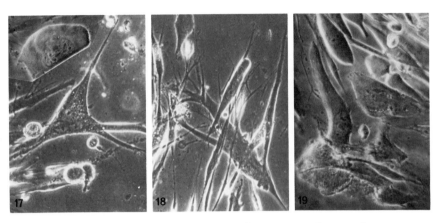

Figs. 17–19. Human rhabdomyosarcoma, 3 days *in vitro*. Living, phase contrast. Note extreme pleomorphism, with bizarre forms.

Hiramoto *et al.* (1961) undertook to explain by immunohistochemical studies the fact that some human rhabdomyosarcomas contain myosin while others do not. The latter include connective tissue components that are not observed in the myosin-containing tumors. This type of "rhabdomyosarcoma" (according to the authors) either is not derived from muscle or produces too little myosin to be picked up by their paired fluorescent label treatment, while producing enough connective tissue components to be readily demonstrated. Normal skeletal muscle also grown *in vitro* was observed actually to make such a histochemical shift during cultivation. (Section IV,A) (Yaffe, 1968). Holtzer (1970) remarks that in embryonic development myoglobin synthesis seems to stand in the same relation as myosin to the mitotic cycle: myoglobin is not synthesized by replicating myoblasts but is seen only in postmitotic myotubes. In papers dealing with stable cell lines derived from two human rhabdomyosarcomas McAllister *et al.* (1969, 1971) report that the progeny of one tumor (RD) but not the other, contained myoglobin. Both lines resembled the original tumors morphologically; they showed no myofibrils or virus particles. Chromosome counts ranged from 45 to 170. A cell-line from the RD tumor, when inoculated into fetuses of pregnant cats, produced in the kittens disseminated rhabdomyosarcomas that were histologically similar to the parent tumor, and contained myoglobin.

A transplantable rat rhabdomyosarcoma was used recently by Barendsen and colleagues in an analysis of radiation (X-ray and neutron) effects on cell proliferation in this type of neoplasm. Survival curves were derived by direct cloning of cells which had been irradiated *in vivo* and *in vitro*. The cultured cells were irradiated in equilibrium with air or under hypoxic conditions. As expected, the mean lethal dose for cells in air was substantially less than for hypoxic cultures. There was always a wide deviation in susceptibility of individual cells, especially in tumor populations irradiated *in situ*, so that nests of viable cells survived high doses 300 kV X rays and 15 MeV neutrons, and resumed proliferation (Hermens and Barendsen, 1969; Barendsen and Broerse, 1969). Effects of ^{60}Co radiation on mitotic activity of normal myoblasts and fibroblasts in culture were described by Pagani and Lullini in a brief report (1966).

2. Experimental Neoplasia

Considerable attention has been devoted to the muscle tumors induced in the rat by implanted metals, especially nickel. Basrur and Gilman approached the subject with direct observation, first of two cell strains derived from such tumors (1963), and second (1967) by exposing normal muscle cells and tumor cells to nickel sulfide in culture. The two strains derived from rat rhabdomyosarcomas had different histories over a 1 year period. The first remained predominantly diploid, evolved myotubes, and eventually ceased to produce tumors on inoculation into rats. The second became predominantly heterploid and failed to exhibit any differentiation, producing only uninuclear myoblasts. When normal muscle cultures were exposed to nickel sulfide, their myoblasts were greatly modified in morphology and in DNA synthesis; older cultures which contained myotubes were less severely affected; cultures from the nickel-induced tumors were apparently unaffected. Swierenga and Basrur (1968) reported a marked decrease in mitotic index and the induction of abnormal mitotic figures in normal muscle cells exposed to nickel sulfide. Further cytological and ultrastructural observations of changes in rat embryo myoblasts exposed to this carcinogen are reported by Sykes and Basrur (1971).

Corbeil (1967) observed that mechanisms of multinucleation as carried out by cells explanted from the nickel-induced tumors were similar to those from normal newborn muscle. However, in the strap cell stage, the rate and extent of differentiation of the tumor tissue diverged from the normal. He observed in 1969 the accepted fusion type of multinucleation in monolayers but reported occasional mitoses in strap cells of both normal and neoplastic origin, asserting, therefore, that numerical increase within the myotube *can* be accomplished by mitosis. Fluorescent antibody technique showed that the cultured tumor cells from several different rhabdomyosarcomas possessed four to five tumor-specific antigens located in the cytoplasm, while at the same time having lost one or two muscle-specific antigens (Corbeil, 1968).

A study at subcellular level of the mechanism by which nickel and other metals, including cobalt, induce muscle tumors was undertaken by Heath and Webb (1967), largely *in vivo*. They found that metallic Co powder, implanted in rat muscle, slowly dissolved and disappeared from the injection site, with an accompaniment of redox cycles involving certain organic molecules. When administered as cobalt chloride to normal myoblasts in culture, it produced cytological changes that were similar to those seen in the vicinity of Co implants *in vivo*. The metallic powder, incubated up to a month with horse serum, was less toxic for myoblasts in culture than the equivalent ionic amount presented as cobalt chloride. The Co^{2+}–serum complex, in which the metal is bound to globulins and albumins, was thought to be adsorbed at the surface of the myoblast and to enter the cell by endocytosis. Digestion of the carrier proteins there by lysosomal proteinase could then liberate Co^{2+} and initiate changes in redox potentials, thus inducing the cancerous transformation more gradually (Health *et al.*, 1969).

Although investigations of viral oncogenesis are legion, few of these have been concerned with muscle transformation as such. Oppenheim *et al.* (1968) studying a mouse sarcoma virus that produces rhabdomyosarcoma in mouse, rat, and hamster found it showing the same morphology and antigenicity as the murine leukemogenic viruses. *In vitro* it appears to be a defective virus, i.e., multiplying only in the presence of a leukemogenic "helper" virus.

Mention has been made above (p. 247) of the application of Sendai virus to differentiating muscle cultures and its effects upon the surface of uninuclear cells. Kaign *et al.* (1966), using clonal analysis in the manner of Konigsberg, showed that skeletal myoblasts and fibroblasts from the chick embryo are equally susceptible to Rous sarcoma virus, not only in the realm of morphological transformation, but also in ability to produce infectious virus. In 1968, this group (Lee *et al.*) applied Rous sarcoma virus in culture to induce [³H]thymidine incorporation into multinucleated myotubes, but could not ascertain whether the viral effect on DNA synthesis was exerted only on the fusing myoblasts or directly on the sarcoblast. As in the experiments of Holtzer *et al.* (1969), the primary aim of these studies was to unravel the skein of muscle differentiation.

Explanted muscle from newborn mice, maintained under relatively anaerobic conditions is susceptible to coxsackie A-4 viruses, which produce a slow cytopathogenic effect, combined with reproduction of virus particles. Newly produced virus particles can be passaged in muscle cultures and retain their pathogenicity for mice. Immune anti-A-4 serum delays the appearance of the CPE in infected cultures (Kaňtoch and Siemiňska, 1965). Infection of fetal mouse muscle cells in culture with coxsackievirus A13 is followed by replication of the agent in both myoblasts and myotubes: newly synthesized A13 antigen is demonstrated by immunofluorescent assay. Susceptibility to virus decreased after fusion, also if fusion was inhibited. When this inhibition was reversed, susceptibility reverted to the degree found in early differentiation. Primary cultures were, however, resistant to poliovirus T_1. "Surface changes on cells may alter or efface specific virus receptors" (Goldberg and Cromwell, 1971).

In an effort to study the appearance of T antigen (which is not related to production of virus particles, and whose biological function is unknown), Fogel and Defendi (1967) infected muscle cultures from various mammalian species with oncogenic DNA viruses (simian virus 40 and polyoma). T antigen appeared in the myotubes of young, differentiating cultures, but fluorescent antibody techniques very rarely revealed *viral* antigen in them. Uninuclear myogenic cells were as susceptible to infection as embryonic fibroblasts from the same species, including the human. Myoblasts infected before fusion were capable of forming myotubes in which many or all of the nuclei contained T antigen. Fifty to 100% of myotubes containing at least one infected myoblast showed DNA synthesis, which was completely repressed, as expected, in the noninfected myotubes. Fully formed sarcoblasts, however, could not be infected. When the nuclei of SV40 transformed myoblasts are fused with those of nontransformed cells, T antigen appears in the latter, according to Steplewski *et al.* (1968). Its appearance *de novo* also occurs in the presence of 5-fluorodeoxyuridine, which suppresses DNA synthesis. Transmission of T antigen is inhibited, however, by actinomycin D and by cycloheximide; therefore, it is dependent upon new RNA and protein synthesis. Here as elsewhere, unique histogenetic attributes of developing muscle have afforded material for crucial experiments in another field.

B. Dystrophy

As a more basic extension of investigations performed *in vivo* on the muscular dystrophies, some study of dystrophic muscle both human and animal has been pursued in culture. Generally it may be said that behavior of the "diseased" muscle repeatedly shows points of difference in comparison with normal muscle when isolated *in vitro*, thus lending color to the belief that the progressive dystrophy arises from conditions intrinsic to the muscle itself. The situation *in vitro* is complicated by the action of many factors that might be deemed irrelevant—such as variations in the growth media and culture methods used (Sections IV,B and V), differences in performance between normal muscle tissues from various locations *in situ* (Pogogeff and Murray, 1950), etc. Nevertheless, there is sufficient agreement among observors to warrant some discussion here.

1. HUMAN DISEASE

An early report on human dystrophic muscle cultured by Geiger and Garvin (1957) indicated a shorter latent period and relative lack of fibrillatory contraction in cells from the disease biopsy as compared to the normal biopsy. Herrmann *et al.* (1960) found that muscle explanted from patients with muscular dystrophy did not produce cross-striae, but were not significantly different in other parameters from explanted normal muscle. They reported a good deal of behavioral inconsistency, which they regarded as irrelevant to the pathology of the tissue, in both the human cultures and comparative explants from 14-day old genetically dystrophic chicks. Goyle *et al.* (1967, 1968), using normal and dystrophic muscle biopsies from human adults, remarked on greater outgrowth with smaller and more pleomorphic myoblasts in the muscular dystrophy cultures, but more myotubes and multinucleate formations in the normal. They observed cross-striations in both after 1 month *in vitro*. Bateson (1968) observed no cross-striations in muscular dystrophy cultures; Skeate *et al.* (1969) report striations not varying in incidence from the normal. The latter group, in a more detailed study, have reproduced their original results (Bishop *et al.*, 1971). A preliminary survey by Kakulas *et al.* (1968) included cultures of normal human fetal and adult skeletal muscle compared to overt Duchenne (sex-linked) muscular dystrophy and its heterozygous carriers, as well as Becker pseudohypertrophic muscular dystrophy. In the latter, they reported indolent but bizarre growth; in the two Duchenne groups, the appearance of a few myotubes, some cross-striated; in general, the diseased muscle showed disordered maturation and occurrence of irregular multipolar cells. *In vivo* evidence that substantial denervation occurs before muscular disability is established has led McComas *et al.* (1970) to regard Duchenne muscular dystrophy as at least in part a neuropathy. Albuquerque and Warnick (1971) suggest that a decrease in sodium permeability is in part responsible for the observed alterations in electrical properties of the nerve-terminal and postsynaptic muscle membrane at the neuromyal junction.

2. Hereditary Dystrophies

Genetically based muscular dystrophies in animals of course allow more systematic study; for this, dystrophic chicks and mice have generally been used. After further work, Herrmann (1964) and Herrmann *et al.* (1964) became convinced that differences in growth pattern of cultures from normal and homozygous dystrophic avian muscle do exist and are highly reproducible. Pectoralis muscle of 11-day chicks (which becomes progressively dystrophic in the diseased birds) was evaluated comparatively as to degree of histological deterioration in the original explant and extent of cell spreading. Up to the eighteenth day *in ovo,* there is no difference in these respects between normal and dystrophic muscle; both respond to culture conditions by blurring of topography and spreading of cells. But shortly after hatching, the normal muscles lose this capacity, while the dystrophic do not, and those consistently showed the greatest amount and most extreme forms of histological retrogression and spreading. Herrmann points out that there are distinct, successive, periods during ontogeny in which muscle tissue becomes dependent upon extracellular factors of, for example, neuronal and hormonal nature, for continued typical development, and that dystrophic symptoms also become manifest during distinct stages of maturation. Since the same growth medium (usually including serum and/or embryonic extract) is supplied to both controls and experimentals, perhaps the dystrophic cells are unable to utilize the hormonal or other inducing substances that are offered to them (see p. 284). Askanas *et al.* (1971) report some different patterns of growth in normal vs. dystrophic breast muscle dissociated and explanted from embryonic chicks at 12 and 15 days, respectively, *in ovo.*

In several reports by Pearce (1963a,b) of comparative studies on normal and dystrophic mice, it was shown that explants from dystrophic mice produced a relatively satisfactory migration of myoblasts as compared to normal adult mouse muscle (similar to the behavior characterized by Herrmann, above, as "spreading"). Having observed areas of spontaneous regeneration within dystrophic muscles, Pearce hypothesized that these supplied the source of the outgrowth in culture, and ingeniously proceeded, therefore, to traumatize the normal muscle 2–5 days before explanting it; this gave an abundant outgrowth of myoblasts, being generally more productive than the dystrophic muscle. Study of the Bar Harbor mutant mice (dy dy) was continued by Ross who re-

ported (1965) that myoblasts from their tissue *in vitro* very rarely underwent fusion, and this only in explants from young mice. These potentially
dystrophic myoblasts, even when explanted at less than 4 months post
partum, had an abnormally high proportion of nucleoli with very low
dry mass. In mice about 7 months old, in which the disease had progressed further, more than 50% of the nucleoli contained less than 0.75
pg solid material. Cytoplasmic RNA in dystrophic myoblasts which had
lost the capacity to fuse also was significantly lower than in myoblasts
from normal mice [see p. 257]. Ross and Jans (1968b) extended these
observations, finding that as the disease advanced, the differences in
cytoplasmic RNA became more marked, concluding that by this time
most of the dystrophic myoblasts were lacking the necessary RNA to
trigger the differentiation which precedes and follows fusion. Some confirmation of this hypothesis is contributed by Little and Meyer (1970)
in the finding that skeletal muscle from dystrophic mice contains severalfold higher than normal levels of free alkaline RNase II activity and
none of the free RNase inhibitor that is normally present.

A chapter on tissue culture in muscle disease by Ross and Hudgson
(1969) includes a review of literature on myogenesis, a critique of methods and merits of distinguishing myoblasts from associated fibroblasts
in mammals, and an account of muscle tissue culture in relation to
pathology. They recount normal or abortive regenerative processes in
cultures of pretraumatized normal and pathological muscle which closely
parallel the processes of regeneration *in situ* in relation to morphology,
chemistry, and behavior of myoblasts. But though normal skeletal muscle
in situ responds to trauma (of mechanical injury or virus infection)
by regenerating effectively and often completely and typically, dystrophic muscle, while retaining some capacity to regenerate, exhibits
in vivo even more abnormal patterns than are reflected in the tissue
culture preparations, where myoblasts rarely achieve fusion. The authors
suggest that in early stages of the disease the explants contain both
normal muscle and dystrophic cells attempting to regenerate, hence the
mixed bag of responses.

C. Other Diseases

Studies of polymyositis *in vitro* seem to be identified mainly with
Kakulas, who has been concerned with an inquiry into a possible im-
munological etiology for this disease. Two papers in 1966 recount the
destruction of muscle cultures from 19–20-day fetal rats by sensitized
lymphoid cells from young adult rats. The adults were injected with

rabbit muscle and adjuvant, and their lymph nodes were excised after 3–6 weeks, when symptoms of muscle disease had appeared. Ninety-five percent "destruction" followed exposure of cultures to the muscle-sensitized lymphocytes; and 17%, 15%, and 20%, respectively, followed exposure to control cells from untreated, adjuvant-treated, and kidney-immunized adults. Destruction of embryonic bovine muscle cells in culture following their exposure to rabbit antiserum produced by the microsome fraction of calf kidney cells was described in some detail by Landshütz (1961). The antiserum heated to 50°C for 30 min produced no effect, but whole serum plus complement caused violent contraction and death of fibers. Serum diluted 1:600 plus complement produced contraction only of myofibrillae.

Bateson (1968), Skeate *et al.* (1969), and Ross *et al.* (1970) have described the morphological parameter of cultured tissues from polymyositis patients, along with other myopathies, including neurogenic atrophies. Most interesting morphological and physiological descriptions of a hereditary progressive neurogenic muscular atrophy in the mouse were given by Duchen *et al.* (1965, 1966), but to the writer's knowledge this material has not yet been studied in culture, although recent advances in method (i.e., development of neuromyal junctions) have now made such a study eminently feasible.

ACKNOWLEDGMENTS

This chapter was prepared during the writer's tenure of Research Career Award 5-KO6-GM-15372 from the U.S. Department of Health, Education, and Welfare, to which grateful acknowledgement is made. Sincere thanks also are offered to colleagues here and in other institutions of learning who have generously permitted the reproduction of their published and unpublished material.

REFERENCES

Albuquerque, E. X., and Warnick, J. E. (1971). *Science* 172, 1260.
Alescio, T., and Moscona, A. A. (1969). *Biochem. Biophys. Res. Commun.* 34, 176.
Askanas, V., Shafiq, S. A., and Milhorat, A. T. (1971). *Arch. Neurol.* 24, 259.
Avery, G., Chow, M., and Holtzer, H. (1956). *J. Exp. Zool.* 132, 409.
Barendsen, G. W., and Broerse, J. J. (1969). *Eur. J. Cancer* 5, 373.
Barth, L. G., and Barth, L. J. (1963). *Biol. Bull.* 124, 125.
Barth, L. G., and Barth, L. J. (1969). *Develop. Biol.* 20, 236.
Basrur, P. K., and Gilman, J. P. W. (1963). *J. Nat. Cancer Inst.* 30, 163.
Basrur, P. K., and Gilman, J. P. (1967). *Cancer Res.* 27, 1168.
Bassleer, R. (1962). *Z. Anat. Entwicklungsgesch.* 123, 184.
Bateson, R. G. (1968). Ph.D. Thesis, University of London.

Betz, E. H., Firket, H., and Reznik, M. (1966). *Int. Rev. Cytol.* 19, 203–227.
Bischoff, R., and Holtzer, H. (1968a). *Amer. Zool.* 8, 784.
Bischoff, R., and Holtzer, H. (1968b). *J. Cell Biol.* 36, 111.
Bischoff, R., and Holtzer, H. (1969). *J. Cell Biol.* 41, 188.
Bischoff, R., and Holtzer, H. (1970). *J. Cell Biol.* 44, 134.
Bishop, A., Gallup, B., Skeate, Y., and Dubowitz, V. (1971). *J. Neurol. Sci.* 13, 333.
Bornstein, M. B., Iwanami, H., Lehrer, G. M., and Breitbart, L. (1968). *Z. Zellforsch. Mikrosk. Anat.* 92, 197.
Buckingham, B. J., and Herrmann, H. (1967). *J. Embryol. Exp. Morphol.* 17, 239.
Bunge, M. B., Bunge, R. P., and Peterson, E. R. (1967). *Brain Res.* 6, 728.
Bunge, R. P., and Bunge, M. B. (1968). *Anat. Rec.* 160, 323.
Bunge, R. P., and Bunge, M. B. (1969). *J. Neuropathol. Exp. Neurol.* 28, 169.
Burrows, M. T. (1912). *Muenchen. Med. Wochenschr.* 59, 1473.
Capers, C. R. (1960). *J. Biophys. Biochem. Cytol.* 7, 559.
Cieselski-Treska, J., Hermetet, J. C., Sensenbrenner, M., and Mandel, P. (1970). *J. Physiol.* (*Paris*) 62, Suppl. 2, 260.
Coleman, J. R., Coleman, A. W., and Roy, H. (1966). *Amer. Zool.* 6, 557.
Coleman, J. R., Coleman, A. W., and Hartline, E. J. H. (1969). *Develop. Biol.* 19, 527.
Cooper, W. G., and Konigsberg, I. R. (1961a). *Anat. Rec.* 140, 195.
Cooper, W. G., and Konigsberg, I. R. (1961b). *Exp. Cell Res.* 23, 576.
Corbeil, L. B. (1967). *Cancer* (*New York*) 20, 572.
Corbeil, L. B. (1968). *Cancer* (*New York*) 21, 184.
Corbeil, L. B. (1969). *Life Sci.* 8, 651.
Cox, P. G. (1968). *J. Morphol.* 126, 1.
Cox, P. G., and Simpson, S. B., Jr. (1970). *Develop. Biol.* 23, 433.
Crain, S. M. (1954). *Diss. Abstr.* 10, 785 (Univ. Microfilms, Ann Arbor, Michigan)
Crain, S. M. (1956). *J. Comp. Neurol.* 104, 285.
Crain, S. M. (1965). *In* "Cells and Tissues in Culture" (E. N. Willmer, ed.), Vol. 2, pp. 335–339 and 345–347. Academic Press, New York.
Crain, S. M. (1966). *Int. Rev. Neurobiol.* 9, 1–43.
Crain, S. M. (1970a). *J. Cell Biol.* 47, 43a.
Crain, S. M. (1970b). *J. Exp. Zool.* 173, 353.
Crain, S. M., Alfei, L., and Peterson, E. R. (1970). *J. Neurobiol.* 1, 471.
Crain, S. M., and Peterson, E. R. (1964). *J. Cell. Comp. Physiol.* 64, 1.
Crain, S. M., and Peterson, E. R. (1967). *Brain Res.* 6, 750.
Crain, S. M., and Peterson, E. R. (1971). *In Vitro* 6, 37.
Dawkins, R. L., and Lamont, M. (1971). *Exp. Cell Res.* 67, 1.
De la Haba, G., Cooper, G. W., and Elting, V. (1966). *Proc. Nat. Acad. Sci. U.S.* 56, 1719.
De la Haba, G., Cooper, G. W., and Elting, V. (1968). *J. Cell. Physiol.* 72, 21.
Delain, D. (1969a). *C. R. Acad. Sci., Ser. D* 268, 2515.
Delain, D. (1969b). *C. R. Soc. Biol.* 162, 1920.
Duchen, L. W., Falconer, D. S., and Strich, S. J. (1965). *J. Physiol.* (*London*) 183, 53.
Duchen, L. W., Searle, A. G., and Strich, S. J. (1966). *J. Physiol.* (*London*) 189, 4.
Eagle, H. (1956). *Science* 122, 501.

Eagle, H. (1960). *Proc. Nat. Acad. Sci. U.S.* **46**, 427.

Emmart, E. W., Komenz, D. R., and Miquel, J. (1963). *J. Histochem. Cytochem.* **11**, 207.

Engel, W. K. (1961a). *J. Histochem. Cytochem.* **9**, 38.

Engel, W. K. (1961b). *J. Histochem. Cytochem.* **9**, 66.

Engel, W. K., and Horvath, B. (1960). *J. Exp. Zool.* **144**, 209.

Ezerman, E. B., and Ishikawa, H. (1967). *J. Cell Biol.* **35**, 405.

Fambrough, D. M. (1970). *Science* **168**, 372.

Firket, H. (1967). *Z. Zellforsch. mikrosk. Anat.* **78**, 313.

Fischbach, G. D. (1970). *Science* **169**, 1331.

Fischbach, G. D. (1971). Soc. for Neuroscience 1st Annual Meeting, Washington, 1971, Abstracts, p. 161.

Fischman, D. A., Shimada, Y., and Moscona, A. A. (1967). *J. Cell Biol.* **35**, 39A.

Fogel, M., and Defendi, V. (1967). *Proc. Nat. Acad. Sci. U.S.* **58**, 967.

Fry, D. M., and Wilson, B. W. (1970). *J. Cell Biol.* **47**, 66a.

Fujisawa, H (1969). *Embryologia* **10**, 256.

Geiger, R. S., and Garvin, J. S. (1957). *J. Neuropathol. Exp. Neurol.* **16**, 532.

Godman, G. C., and Murray, M. R. (1953). *Proc. Soc. Exp. Biol. Med.* **84**, 668.

Goldberg, R., and Cromwell, R. L. (1971). *J. Virol.* **7**, 759.

Goodwin, B. C., and Sizer, I. W. (1965). *Develop. Biol.* **11**, 136.

Goyle, R. S., Kalra, S. L., and Singh, B. (1967). *Neurology (India)* **15**, 149.

Goyle, R. S., Kalra, S. L., and Singh, B. (1968). *Neurology (India)* **16**, 87.

Granick, S., and Granick, D. (1971). *J. Cell Biol.* **51**, 636.

Gutmann, T., Hanzlíková, V., and Holečková, E. (1969). *Exp. Cell Res.* **56**, 33.

Ham, R. G. (1963). *Exp. Cell Res.* **29**, 515.

Hamburger, V., and Hamilton, H. (1951). *J. Morphol.* **88**, 49.

Harris, A. J., Heineman, S., Schubert, D., and Tarakis, H. (1971). *Nature (London)* **231**, 296.

Harris, A. J., and Miledi, R. (1966). *Nature (London)* **209**, 716.

Harris, M. (1959). *Proc. Soc. Exp. Biol. Med.* **102**, 468.

Harris, M. (1964). "Cell Culture and Somatic Variation." Holt, New York.

Harris, M., and Kutsky, R. J. (1958). *Cancer Res.* **18**, 585.

Hartzell, H. C., and Fambrough, D. M. (1971). Society for Neuroscience 1st Annual Meeting, Washington 1971. Abstr. p. 161.

Hauschka, S. D. (1966). Ph.D. Thesis, Johns Hopkins University, Baltimore, Maryland.

Hauschka, S. D. (1968). In "Results and Problems in Cell Differentiation" (H. Ursprung, ed.), Vol. 1, pp. 37–57. Springer Publ., New York.

Hauschka, S. D., and Konigsberg, I. R. (1966). *Proc. Nat. Acad. Sci. U.S.* **55**, 119.

Heath, J. C., and Webb, M. (1967). *Brit. J. Cancer* **21**, 768.

Heath, J. C., Webb, M., and Caffrey, M. (1969). *Brit. J. Cancer* **23**, 153.

Hermens, A. F., and Barendsen, G. W. (1969). *Eur. J. Cancer* **5**, 173.

Herrmann, H. (1964). *Tex. Rep. Biol. Med.* **22**, 911.

Herrmann, H., Konigsberg, U. R., and Robinson, G. (1960). *Proc. Soc. Exp. Biol. Med.* **105**, 217.

Herrmann, H., Klein, N. W., and Albers, G. (1964). *J. Cell Biol.* **22**, 391.

Hess, A., and Rosner, S. (1970). *Amer. J. Anat.* **129**, 21.

Hiramoto, R., Cairns, J., and Pressman, D. (1961). *J. Nat. Cancer Inst.* **27**, 937.

Holtfreter, J. (1947). *J. Morphol.* **80**, 57.

Holtzer, H. (1961). *In* "Molecular and Cellular Synthesis" (D. Rudnick, ed.), pp. 35–87. Ronald Press, New York.

Holtzer, H. (1970). *In* "Control Mechanisms in the Expression of Cellular Phenotypes" (H. A. Padykula, ed.) (*Symp. Int. Soc. Cell Biol.* 9, 69–88.) Academic Press, New York.

Holtzer, H., Abbott, J., and Lash, J. (1958). *Anat. Rec.* 131, 567.

Holtzer, H., and Bischoff, R. (1970). *In* "The Physiology and Biochemistry of Muscle as a Food" (E. J. Briskey, R. G. Cassens, and B. B. Marsh, eds.), Vol. 2, pp. 29–52. Univ. of Wisconsin Press, Madison.

Holtzer, H., Bischoff, R., and Chacko, S. (1969). *In* "Cellular Recognition" (R. T. Smith and R. A. Good, eds.), pp. 19–25. Appleton, New York.

Howarth, S., and Dourmashkin, R. A. (1958). *Exp. Cell Res.* 15, 613.

Huxley, H. E. (1966). *Harvey Lect.* 60, 85–118.

Ishikawa, H. (1968). *J. Cell Biol.* 38, 51.

Ishikawa, H., Bischoff, R., and Holtzer, H. (1968). *J. Cell Biol.* 38, 538.

James, D. W., and Tresman, R. L. (1969). *Z. Zellforsch. Mikrosk. Anat.* 100, 126.

Kaighn, M. E., Ebert, J. D., and Stott, P. M. (1966). *Proc. Nat. Acad. Sci. U.S.* 56, 133.

Kakulas, B. A. (1966a). *Nature (London)* 210, 1115.

Kakulas, B. A. (1966b). *J. Pathol. Bacteriol.* 91, 495.

Kakulas, B. A., Papadimitriou, J. M., Knight, J. O., and Mastaglia, F. L. (1968). *Proc. Aust. Ass. Neurol.* 5, 79.

Kano, M., Ishikawa, K., and Shimada, Y. (1971). *Brain Res.* 25, 216.

Kano, M., and Shimada, Y. (1971). *Brain Res.* 27, 402.

Kańtoch, M., and Siemińska, A. (1965). *Arch. Immunol. Ther. Exp.* 13, 413.

Konigsberg, I. R. (1960). *Exp. Cell Res.* 21, 414.

Konigsberg, I. R. (1961a). *Circulation* 24, 447.

Konigsberg, I. R. (1961b). *Proc. Nat. Acad. Sci. U.S.* 47, 1808.

Konigsberg, I. R. (1963). *Science* 140, 1273.

Konigsberg, I. R. (1964). *Carnegie Inst. Wash., Yearb.* 63, 516.

Konigsberg, I. R., and Hauschka, S. D. (1966). *In* "Reproduction: Molecular, Subcellular and Cellular" (M. Locke, ed.), pp. 243–286. Academic Press, New York.

Konigsberg, I. R., McElvain, N., Tootle, M., and Herrmann, H. (1960). *J. Biophys. Biochem. Cytol.* 8, 333.

Lake, N. C. (1915–1916). *J. Physiol. (London)* 50, 364.

Landschütz, C. (1961). *Z. Naturforsch. B* 16, 769.

Lappano-Colletta, E. R. (1970). *J. Cell Biol.* 47, 116a.

Lash, J. W., Holtzer, H., and Swift, H. (1957). *Anat. Rec.* 128, 679.

Lee, H. H., Kaighn, M. E., and Ebert, J. D. (1968). *Int. J. Cancer* 3, 126.

Lentz, T. L. (1971). *Science* 171, 187.

Levi, G. (1934). *Ergeb. Anat. Entwicklungsgesch.* 31, 125.

Lewis, M. R. (1915). *Amer. J. Physiol.* 38, 153.

Lewis, M. R. (1920). *Contrib. Embryol. Carnegie Inst.* 9, No. 35, 191.

Lewis, W. H., and Lewis, M. R. (1917). *Amer. J. Anat.* 22, 169.

Li, C.-L. (1960). *Science* 132, 1889.

Li, C.-L., Engel, K., and Klatzo, I. (1959). *J. Cell Comp. Physiol.* 53, 421.

Lipton, B. H., and Konigsberg, I. R. (1971). *Anat. Rec.* 169, 368.

Little, B. W., and Meyer, W. L. (1970). *Science* 170, 747.

Lucy, J. A., and Rinaldini, L. M. (1959). *Exp. Cell Res.* **17**, 385.
McAllister, R. M., Melnyk, J., Finklestein, J. Z., Adams, Jr., E. C., and Gardner, M. B. (1969). *Cancer (New York)* **24**, 520.
McAllister, R. M., Nelson-Rees, W. A., Johnson, E. Y., Rongey, R. W., and Gardner, M. B. (1971). *J. Nat. Cancer Inst.* **47**, 603.
McComas, A. J., Sica, R. E. P., and Currie, S. (1970). *Nature (London)* **226**, 1263.
Malpoix, P., and Emelinckx, A. (1967). *J. Embryol. Exp. Morphol.* **18**, 143.
Maslow, D. E. (1969). *Exp. Cell Res.* **54**, 381.
Matagne-Dhoosche, F. (1969). *C. R. Soc. Biol.* **162**, 1860.
Mauro, A. (1961). *J. Biophys. Biochem. Cytol.* **9**, 493.
Mintz, B., and Baker, W. W. (1967). *Proc. Nat. Acad. Sci. U.S.* **58**, 592.
Moscona, A. (1952). *J. Anat.* **86**, 287.
Moscona, A. (1955). *Exp. Cell Res.* **9**, 377.
Moscona, A. (1956). *Proc. Soc. Exp. Biol. Med.* **92**, 410.
Moscona, A. (1958). *In* "Cytodifferentiation" (D. Rudnick, ed.), pp. 49–51. Univ. of Chicago Press, Chicago.
Moss, F. P., and Leblond, C. P. (1970). *J. Cell Biol.* **44**, 459.
Murray, M. R. (1960). *In* "The Structure and Function of Muscle" (G. H. Bourne, ed.), Vol. 1, pp. 111–136. Academic Press, New York.
Murray, M. R. (1965a). *In* "Cells and Tissues in Culture" (E. N. Willmer, ed.), Vol. 2, pp. 311–372. Academic Press, New York.
Murray, M. R. (1965b). *In* "Cells and Tissues in Culture" (E. N. Willmer, ed.), Vol. 2, pp. 373–455. Academic Press, New York.
Murray, M. R., and Peterson, E. R. (1965). *Excerpta Med. Found. Congr. Ser.* **94**, E-221-222.
Murray, M. R., and Stout, A. P. (1958). *In* "Treatment of Cancer and Allied Diseases" (G. T. Pack and I. M. Ariel, eds.), Vol. 1, pp. 124–142. Harper (Hoeber), New York.
Nakai, J. (1965). *Exp. Cell Res.* **40**, 307.
Nakai, J. (1969). *J. Exp. Zool.* **170**, 85.
Nameroff, M., and Holtzer, H. (1969). *Develop. Biol.* **19**, 380.
Okazaki, K., and Holtzer, H. (1965a). *J. Cell Biol.* **27**, 75a.
Okazaki, K., and Holtzer, H. (1965b). *J. Histochem. Cytochem.* **13**, 726.
Okazaki, K., and Holtzer, H. (1966). *Proc. Nat. Acad. Sci. U.S.* **56**, 1484.
O'Neill, M. C., and Stockdale, F. E. (1972). *J. Cell Biol.* **52**, 52.
O'Neill, M., and Strohman, R. (1969). *J. Cell. Physiol.* **73**, 61.
Oppenheim, S., Varet, B., and Le Clerc, J.-C. (1968). *Exp. Anim.* **1**, 95.
Pagani, P. A., and Lullini, I. G. (1966). *Boll. Soc. Ital. Biol. Sper.* **42**, 145.
Pappas, G. D., Peterson, E. R., Masurovsky, E. B., and Crain, S. M. (1971). *New York Acad. Sci. Ann.* **183**, 33.
Parker, R. C. (1933). *J. Exp. Med.* **58**, 401.
Pearce, G. W. (1963a). *In* "Muscular Dystrophy in Man and Animals" (G. H. Bourne and N. Golarz, eds.), p. 178. Hafner, New York.
Pearce, G. W. (1963b). *In* "Research in Muscular Dystrophy," pp. 75–85. Pitman, London.
Peterson, E. R. (1970). Personal communication.
Peterson, E. R., and Crain, S. M. (1970a). *Proc. Int. Congr. Neuropathol., 6th, 1970* pp. 734–735.
Peterson, E. R., and Crain, S. M. (1970b). *Z. Zellforsch. Mikrosk. Anat.* **106**, 1.

Peterson, E. R., Crain, S. M., and Murray, M. R. (1965). Z. *Zellforsch. Mikrosk. Anat.* **66**, 130.
Pogogeff, I. A., and Murray, M. A. (1950). Unpublished observations.
Pogogeff, I. A., and Murray, M. R. (1946). *Anat. Rec.* **95**, 321.
Przybylski, R. J. (1971). *J. Cell Biol.* **49**, 215.
Reporter, M. C., and Ebert, J. D. (1965). *Develop. Biol.* **12**, 154.
Reporter, M. C., and Norris, G. (1969). *J. Cell Biol.* **43**, 115a.
Reporter, M. C., Konigsberg, I. R., and Strehler, B. L. (1963). *Exp. Cell Res.* **30**, 410.
Reznik, M. (1969a). *J. Cell Biol.* **40**, 568.
Reznik, M. (1969b). *Lab. Invest.* **20**, 353.
Reznik, M., and Firket, H. (1964). *C. R. Soc. Biol.* **158**, 1168.
Richler, C., and Yaffe, G. (1970). *Develop. Biol.* **23**, 1.
Rinaldini, L. M. J. (1958). *Int. Rev. Cytol.* **7**, 587–647.
Rinaldini, L. M. J. (1959). *Exp. Cell Res.* **16**, 477.
Robbins, N., and Yonezawa, T. (1971a). *Science* **172**, 395.
Robbins, N., and Yonezawa, T. (1971b). *J. Genl. Physiol.* **58**, 467.
Ross, K. F. A. (1964). *Quart. J. Microsc. Sci.* **105**, 423.
Ross, K. F. A. (1965). *In* "Research in Muscular Dystrophy," pp. 119–132. Pitman, London.
Ross, K. F. A., and Hudgson, P. (1969). *In* "Disorders of Voluntary Muscle" (J. N. Walton, ed.), 2nd ed., pp. 473–515. Churchill, London.
Ross, K. F. A., and Jans, D. E. (1968a). *In* "Cell Structure and its Interpretation" (S. M. McGee-Russell and K. F. A. Ross, eds.), pp. 275–304. Arnold, London.
Ross, K. F. A., and Jans, D. E. (1968b). *In* "Research in Muscular Dystrophy," pp. 240–254. Pitman, London.
Ross, K. F. A., Jans, D. E., Larson, P. F., Mastaglia, F. L., Parson, R., Fulthorpe, J. J., Jenkison, M., and Walton, J. N. (1970). *Nature (London)* **226**, 545.
Sacerdote de Lustig, E. (1942). *Soc. Argentina Biol. Rev.* **18**, 524.
Sacerdote de Lustig, E. (1943). *Soc. Argentina Biol. Rev.* **19**, 159.
Sawicki, S. G., and Godman, G. C. (1971). *J. Cell Biol.* **50**, 746.
Shainberg, A., Yagil, G., and Yaffe, G. (1969). *Exp. Cell Res.* **58**, 163.
Sherbet, G. V. (1966). *J. Embryol. Exp. Morphol.* **16**, 159.
Shimada, Y. (1968). *Exp. Cell Res.* **51**, 564.
Shimada, Y. (1971). *J. Cell Biol.* **48**, 128.
Shimada, Y., Fischman, D. A., and Moscona, A. A. (1967). *J. Cell Biol.* **35**, 445.
Shimada, Y., Fischman, D. A., and Moscona, A. A. (1969a). *J. Cell Biol.* **43**, 382.
Shimada, Y., Fischman, D. A., and Moscona, A. A. (1969b). *Proc. Nat. Acad. Sci. U.S.* **62**, 715.
Shimada, Y., and Kano, M. (1971). *Arch. Histol. Jap.* **33**, 95.
Simpson, S. B., Jr., and Cox, P. G. (1967). *Science* **157**, 1330.
Skeate, Y., Bishop, A., and Dubowitz, V. (1969). *Cell Tissue Kinet.* **2**, 307.
Steplewski, Z., Knowles, B. B., and Koprowski, H. (1968). *Proc. Nat. Acad. Sci. U.S.* **59**, 769.
Stockdale, F. E. (1970). *Develop. Biol.* **21**, 462.
Stockdale, F. E. (1971). *Science* **171**, 1145.
Stockdale, F. E., and Holtzer, H. (1960). *Anat. Rec.* **138**, 384.
Stockdale, F. E., and Holtzer, H. (1961). *Exp. Cell Res.* **24**, 508.

Stockdale, F. E., Okasaki, K., Nameroff, M., and Holtzer, H. (1964). *Science* **146**, 533.

Strehler, B. L., Konigsberg, I. R., and Kelley, E. T. (1963). *Exp. Cell Res.* **32**, 232.

Swierenga, S. H., and Basrur, P. (1968). *Lab. Invest.* **19**, 663.

Sykes, A. K., and Basrur P. K. (1971). *In Vitro* **6**, 377.

Szepsenwohl, J. (1946). *Anat. Rec.* **95**, 125.

Szepsenwohl, J. (1947). *Anat. Rec.* **98**, 67.

Timofeevskii, A. D. (1946–1947). *Amer. Rev. Sov. Med.* **4**, 106.

van Weel, P. B. (1948). *J. Anat.* **82**, 49.

Veneroni, G., and Murray, M. R. (1969). *J. Embryol. Exp. Morphol.* **21**, 369.

Waser, P. G., and Nickel, E. (1969). *Progr. Brain Res.* **31**, 157.

Weiss, P. (1950). *Quart. Rev. Biol.* **25**, 177.

White, N. K., and Hauschka, S. D. (1971). *Exp. Cell Res.* **67**, 479.

Wilde, C. E., Jr. (1958). *Anat. Rec.* **132**, 517.

Wilde, C. E., Jr. (1959). *In* "Cell, Organism and Milieu" (D. Rudnick, ed.), pp. 3–43. Ronald Press, New York.

Wilson, B. W. (1970). *Proc. Tissue Cult. Ass., 21st Annu. Meet.; In Vitro* **6**, 222.

Wilson, B. W., and Stinnett, H. O. (1969). *Proc. Soc. Exp. Biol. Med.* **130**, 30.

Yaffe, D. (1968). *Proc. Nat. Acad.. Sci. U.S.* **61**, 477.

Yaffe, D. (1969). *Curr. Top. Develop. Biol.* **4**, 37–77.

Yaffe, D. (1971). *Exp. Cell Res.* **66**, 33.

Yaffe, D., and Feldman, M. (1964). *Develop. Biol.* **9**, 347.

Yaffe, D., and Feldman, M. (1965). *Develop. Biol.* **11**, 300.

Yaffe, D., and Fuchs, S. (1967). *Develop. Biol.* **15**, 33.

MOLECULAR BASIS OF CONTRACTION IN CROSS-STRIATED MUSCLES

H. E. HUXLEY

I. Introduction

Muscle presents a very favorable system for the study of the relation between molecular architecture and function, in respect both of the large scale properties of the muscle and of the detailed enzyme mechanisms by which energy is transformed, in a controlled fashion, from the chemical to the mechanical form. Many of the physiological, biochemical and structural properties of cross-striated muscle have now been rather clearly defined, but although a great deal is now known about the general nature of the molecular events during contraction, the mechanism is not likely to be completely understood until the three-dimensional structure of the various proteins molecules involved is known to atomic resolution. Such information must await the crystallization of these proteins and the complete analysis of their structure by X-ray diffraction. However, our present knowledge indicates in some detail a number of the properties of these molecules that are required for their various roles in the contractile system and how their behavior is integrated together in a muscle. This chapter will describe the experimental basis of that knowledge.

II. Structure of Muscle Fibers

A. Introduction

The smallest unit of structure in a muscle that can give a normal physiological response is the single muscle fiber. Each fiber contains an assembly of contractile material enclosed within an electrically polarized membrane. Muscle fibers characteristically have diameters between about 10 μ and 100 μ, and most often between 50 μ and 100 μ, though in some species much larger diameters are found—e.g., the crab *Maia squinado* with diameters up to a millimeter or two (Caldwell and Webster, 1961, 1963) or the barnacle *Balanus nubilus* with diameters up to 2 mm (Hoyle and Smyth, 1963). The lengths of the fibers depend on the lengths and construction of the muscles from which they are derived; commonly, as in muscles such as frog sartorius or rabbit psoas, which are favorable laboratory materials, the fibers are several centimeters in length.

Each fiber is a multinucleate cell, formed during embryonic development by the coalescence of a large number of mononucleated myoblasts. As well as the contractile material, muscle fibers contain the normal constituents of many other cell types, such as mitochondria, ribosomes, storage granules, and glycogen particles. They also contain an elaborate internal membrane system, the sarcoplasmic reticulum, which is concerned with the switching on and off of the contractile mechanism (probably by controlling the level of free calcium in the muscle), and which is described in detail in Volume II, Chapter 9 of this book.

The contractile material itself is contained in structures known as myofibrils (Fig. 1). These are usually 1–2 μ in diameter, and so far as is known, each fibril extends for the whole length of the fiber in which it is contained (though some branching may take place, see, for example, Goldspink, 1970). These myofibrils have a banded appearance along their length and they are usually arranged in the fiber with their band patterns in register. It is this feature which gives rise to the characteristic cross-striated appearance of the muscle fiber as a whole, readily visible in the light microscope.

B. Structure of the Myofibrils

The fibrils of vertebrate striated muscle, when examined in the light microscope or the electron microscope (Figs. 2–4), show a very regular system of transverse bands. These arise from a variation of density along the length of the fibril. Each repeat of the pattern is known as a sarcomere. In vertebrate muscles, sarcomere lengths at the normal resting length in the body are commonly about 2.3–2.8 μ. Shorter resting sarcomere lengths than this have not been reported, but longer sarcomeres occur quite commonly in other species. Thus, for example, in the indirect flight number of Calliphora (Hanson, 1956), sarcomere lengths of about

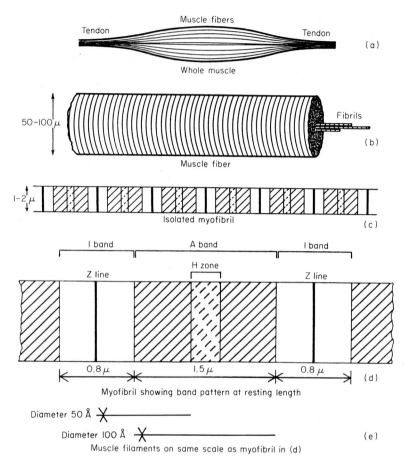

Fig. 1. Dimensions and arrangement of the contractile components in a muscle. The whole muscle (a) is made up of fibers (b) which contain cross-striated myofibrils (c, d). These are constructed of two kinds of protein filaments (e) put together as shown in Fig. 5.

3.3 μ are found; in the extensor carpopodite muscle of the crayfish, the sarcomere length is in the region of 10 μ (Zachau and Zacharova, 1966); in the claw muscle of the crab *Cancer*, Gillis (1969) reports sarcomere lengths of 10–15 μ, while Fahrenbach (1967) finds that the distal, phasic fibers of the accessory flexor muscle of *Cancer magister* have a resting sarcomere length of 4.5 μ; and in the carpopodite flexor of the walking legs of the crab *Portunus depurator*, Franzini-Armstrong (1970) finds three different groups of fibers with sarcomere lengths

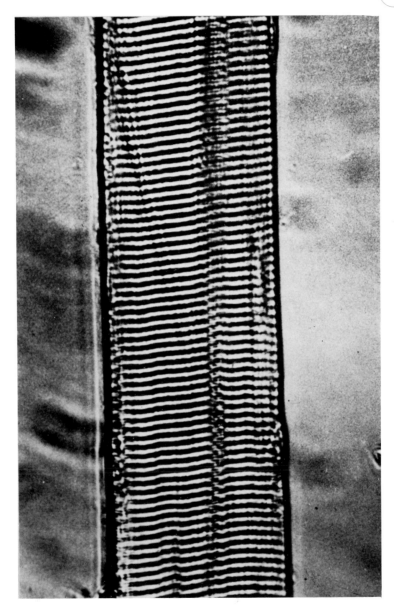

Fig. 2. A single muscle fiber from a rabbit psoas muscle photographed in the phase contrast light microscope. A bands are dark, I bands light. The sarcomere repeat is approximately 2.6 μ, and the diameter of this fiber is approximately 50 μ. (Magnification 940×.)

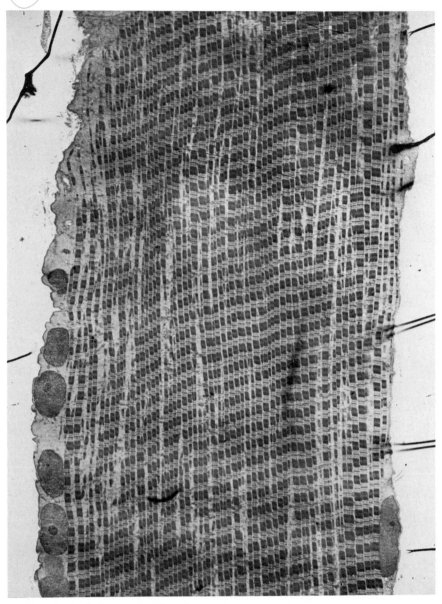

Fig. 3. A single muscle fiber from a human muscle, fixed, stained, embedded in plastic, sectioned longitudinally, and viewed at relatively low magnification in the electron microscope. It can be seen that the striations visible in Fig. 2 arise from the characteristic band pattern of the myofibrils, visible here, with diameters of about 1 μ. The diameter of this fiber in the plane of sectioning is about 55 μ. At the edge of the fiber, several nuclei can be seen just below the surface membrane (sarcolemma). (Magnification 1600×.)

Fig. 4. Single fiber from rabbit psoas muscle showing the repeating band pattern along the myofibrils. Although filaments are visible, the section is too thick for their exact arrangement to be perceived. (Magnification ×15,000.)

approximately 4 μ, 5 μ, and 7 μ. Despite variations in sarcomere length the band patterns of these muscles all share the same main features (Fig. 5). The ends of the sarcomere are defined by narrow dense lines or disks known as the Z lines. Within each sarcomere, we can distinguish a denser, more birefringent zone known as the A (anisotropic) band, which is located symmetrically with respect to the Z lines and is separated from these by zones of lower density and birefringence known as the I (isotropic) bands. In a frog muscle at rest length (sarcomere length approximately 2.3 μ), the A band is about 1.55 μ long, the two half I bands are each about 0.375 μ in length. The central region of the A band in a muscle at this sarcomere length is somewhat less dense than the lateral regions. This central region is known as the H zone, and in the situation already described, its length is approximately 0.35 μ. In many types of muscle, there is a narrow dense line, the M line, in the center of the H zone, about 400–800 Å in width, flanked on either side by a narrow zone of lower density than the rest of the H zone; the width of this whole central zone is about 1500 Å.

There are a number of other very interesting and significant features of the structure which can be seen quite readily, but we will defer description of these until the main underlying elements of the structure which give rise to the band pattern have been described.

C. Arrangement of the Filaments

The band pattern arises because the fibrils are constructed from a long series of partially overlapping arrays of longitudinal protein filaments (Fig. 5). The evidence for this comes from electron microscopy, light microscopy, and X-ray diffraction. Historically, the structure was recognized by the combination of all three techniques, but it will facilitate description and explanation if the electron microscope evidence (some of which was obtained later on) is described first.

When very thin sections of striated muscle are examined in the electron microscope (Fig. 6) (H. E. Huxley, 1957) it can be seen that while only thin filaments (about 50–70 Å diameter) are visible in the parts of the sarcomere corresponding to the I bands, both thin and thicker filaments (diameter 100–120 Å) are visible in the A bands (Fig. 6). When sections are obtained in which the muscle filaments are accurately parallel to the plane of the section (Fig. 7), then it can be seen that each thick filament is continuous from one end of the A band to the other; the thin filaments, on the other hand, do not extend as far as the center of the A band in a muscle at rest length, but terminate

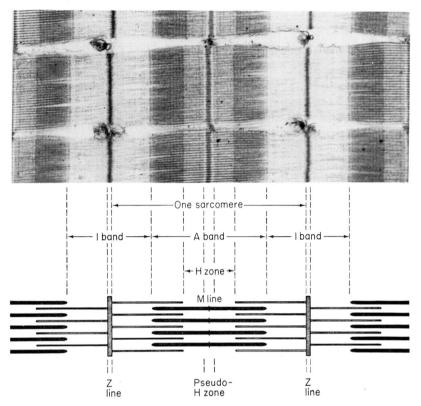

Fig. 5. Longitudinal section of frog sartorius muscle (top) together with diagram showing the overlap of filaments that gives rise to the band pattern. The A band is most dense in its lateral zones where the thick and thin filaments overlap. The central zone of the A band (the H zone) is less dense, since it contains thick filaments only. The I bands are less dense still because they contain only thin filaments. Sarcomere length here is about 2.5 μ.

at a point which corresponds to the boundary of the H zone. All the thick filaments have the same characteristic length (1.55 μ in frog sartorius muscle) and are arranged in register; all the thin filaments also have the same length as each other (about 0.975 μ measured from the center of the Z line) and they too are arranged in register. Consequently, these overlapping arrays of filaments give rise to sharp variations in density along the length of the fibrils, and it is this arrangement that gives rise to the characteristic banded appearance in the manner illustrated in Fig. 5.

When cross sections of striated muscle are examined in the electron microscope, the appearance seen depends on which part of the sarcomere

Fig. 6. Longitudinal section of glycerinated rabbit psoas muscle, cut sufficiently thin so that only single layers of filaments lie within the plane of the section. The thick and thin filaments can be seen clearly under these conditions, thick filaments in the A bands, and thin filaments in the I bands and extending into the A bands. The structure is shown at higher magnification in Fig. 7. (Magnification ×53,000.)

has been cut by the section (H. E. Huxley, 1953b; H. E. Huxley, 1957). If the section has passed through the I band region, only cross sections of thinner (50–70 Å) diameter filaments are seen, spaced out in a rather irregular way (Fig. 8c). If the section has passed near the central region of the A band, within the H zone, then only the thicker filaments are seen, arranged a few hundred Angstrom units apart in a very regular hexagonal lattice (Fig. 8a). Lateral dimensions in embedded muscles tend to be somewhat reduced by dehydration, but X-ray evidence, which we will describe later, shows that in a live frog muscle at rest length, for example, these filaments have an approximately 400 Å center-to-center distance. If the section has passed through the denser lateral parts of the A bands, a double array of filaments is seen, with the thicker filaments at the lattice points of the hexagonal lattice and the thinner filaments lying in between them (Figs. 8b and 9). In vertebrate striated muscle, the thin filaments are located at the trigonal points of the hexagonal lattice (Fig. 10), i.e., symmetrically between three thick filaments. In this type of lattice, there are twice as many thin filaments as thick ones. Some other types of striated muscle have somewhat different arrangements and correspondingly different number ratios.

As a consequence of the great regularity of the lattice, and the fact that this regularity is preserved in fixed and embedded material, it is possible to cut very thin longitudinal sections 100–200 Å in thickness which include only a single layer of filaments, as we have already seen. When the section happens to pass through the lattice in certain directions such a layer displays the arrangement of thick and thin filaments with particular clarity. This can be seen in Fig. 11a. A section parallel to the 11 crystallographic directions of the hexagonal lattice of vertebrate striated muscle (from rabbit in this instance) can contain an alternation of pairs of thin filament lying in between the thicker ones. The partially overlapped arrangement of the filaments can be seen very readily. Other sections that pass through the muscle parallel to the 10 crystallographic directions display a simple alternation of thick and thin filaments (Fig. 11b). However, in general there will be two superposed thin filaments within the thickness of such a section, which can be somewhat thicker than the previous type. If the sections are thicker still, however, thick and thin filaments will be superposed within the section and their pattern of overlap obscured. This is the appearance one generally sees in muscle sections unless special efforts are made to obtain extremely thin ones.

Some other features of the structure may be noted at this point. The thick filaments taper at either end, over the last 1500 Å or so of their length. They often appear somewhat thickened at the very center of

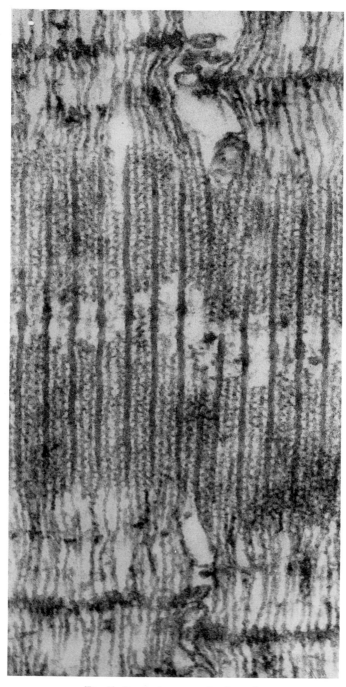

Fig. 7. See facing page for legend.

the A band, and in well preserved material (Fig. 12), cross bridges may be observed between the thick filaments at this point, and these bridges have been studied in some detail by Pepe (1967) and by Knappeis and Carlson (1968). These may occur in 3, 5, or 7 distinct sets, so that they appear as a sequence of regularly spaced lines about 200 Å apart in longitudinal sections (Fig. 13).

However, the most interesting feature visible in thin sections is the presence of cross bridges that extend from the thick filaments to the thin filaments alongside them. These occur at intervals of about 400 Å (as measured in the electron microscope) between a given thick and thin filament. Since there are six thin filaments around each thick one, there must be six cross bridges leaving each thick filament for every 400 Å interval (approximately) of its length. The cross bridges can be seen on the thick filaments on the outside of the fibril (Fig. 14), where there is no thin filament alongside, extending blindly into the sarcoplasm. They are also visible in the H zones of a muscle, where no thin filaments are present. On the other hand, no such projections are seen on the thin filaments in the I bands. It is therefore clear that these cross bridges are a permanent part of the structure of the thick filaments. In the electron microscope, they appear about 50 Å in width; it is not easy to estimate their length because of the sideways shrinkage of the tissue during processing, but it appears to be somewhere in the region of 100 Å.

D. X-Ray Diffraction Evidence Concerning Filament Arrangement

Although our confidence in electron microscope observations is now such that a model of muscle structure based solely on the evidence described above would be considered well established, this was not always the case. In the early days of electron microscopy, it still had to be proved that the structures seen in fixed and stained material did in fact bear a reasonably close relationship to those present in the living

---

Fig. 7. An electron micrograph of a longitudinal section through the sarcomeres of two adjacent myofibrils, such as those shown at lower magnification in Fig. 6. The Z lines bounding the sarcomeres are at the top and bottom of the picture. Two kinds of filaments are visible, thick ones (about 110 Å) in an array confined to the A band, and thin ones (about 50 Å) in two arrays which terminate at the borders of the H zone in the middle of the picture. The two kinds of filaments interdigitate in the A band (except the H zone); the plane of sectioning through the lattice of interdigitating filaments which will produce a longitudinal section like this one is illustrated in Fig. 11b. Cross links between thick and thin filaments are visible. (Magnification ×148,000.)

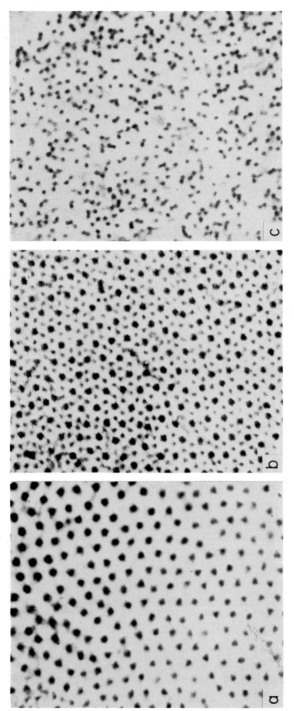

Fig. 8. Electron micrographs of cross sections of rabbit psoas muscle. (a) The simple hexagonal array of thick filaments in the H zone. (b) The double hexagonal array of thick and thin filaments elsewhere in the A band. (c) The thin filament alone in the I Band. (Magnification ×150,000.)

material. Indeed, experience based on muscle structure was one of the strongest pieces of evidence for the validity of electron microscope observations, since independent evidence about the nature of the structure was already available from X-ray diffraction studies (H. E. Huxley, 1951, 1952, 1953a).

It was found that frog sartorius muscles, maintained in Ringer solution so that they would contract perfectly normally, gave a sharp and detailed series of low-angle X-ray reflections. On the equator of the diagram, where the reflections arise from periodicities perpendicular to the long axis of the muscle, a pattern of reflections is seen (Fig. 15) that corresponds to that which would be given by a hexagonal array of long rod-shaped structures spaced about 400 Å apart. Thus, it is quite certain that the hexagonal array of filaments seen in the electron micrographs corresponds closely to the structure actually present in the live muscle.

Furthermore, it was observed that in muscles in rigor the relative intensities of the reflections showed that there was a substantial amount of material present at or near the trigonal position of the lattice. This was interpreted in terms of the presence of a second set of filaments lying in between the hexagonal array of primary filaments. As we will see presently, this interpretation, though correct, is not perfectly straightforward. Nevertheless, the appearance in the electron microscope of thin filaments at the trigonal points of lattice was completely in accord with the X-ray evidence and could therefore be accepted with a high degree of confidence.

E. X-Ray Reflections from the Thick Filaments

Other reflections in the X-ray diagram from live muscles correspond to periodicities in an axial direction and thus show the presence of structural regularities along the length of the filaments. At very low angles, there is a set of reflections with an axial periodicity of 429 Å which are believed to arise from the thicker filaments for reasons discussed below. It was shown by Elliott (1964) that this system of reflections contains both meridional and off-meridional components. In a detailed analysis of the pattern (Fig. 16), H. E. Huxley and Brown (1967) showed that the distribution of X-ray intensity along the layer lines could be accounted for satisfactorily on the basis of a helical arrangement of scattering units. The most prominent feature in this diagram is a strong meridional reflection on the third layer line at 143 Å, showing that the subunit repeat (i.e., the axial translation between the repeating units) is one-third of the helical repeat of 429 Å. Since there is electron

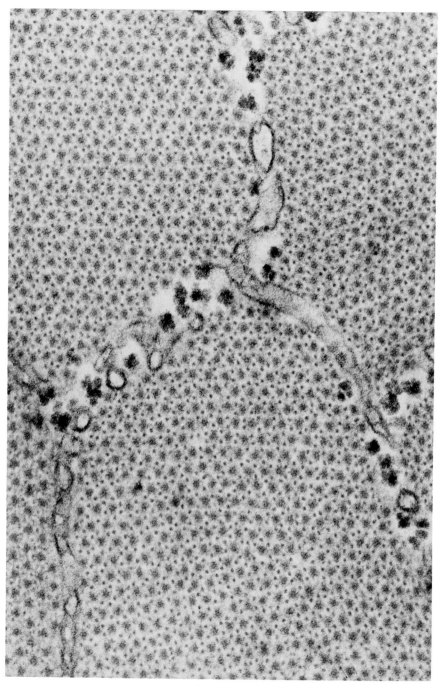

Fig. 9. See facing page for legend.

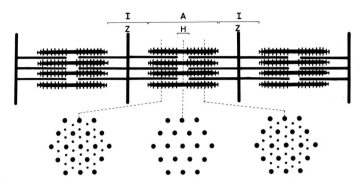

Fig. 10. Structure of striated muscle, showing overlapping arrays of actin- and myosin-containing filaments, the latter with projecting cross bridges on them. For convenience of representation, the structure is drawn with considerable longitudinal foreshortening; with filament diameters and side-spacings as shown, the filament lengths should be about five times the lengths shown.

microscopic evidence that there are six cross bridges on each thick filament for each 400 Å interval (approximately), the model that suggests itself is one in which pairs of cross bridges occur on either side of the filament at 143 Å intervals, with each pair rotated by 120° with respect to its neighbors (Fig. 17). This arrangement gives good agreement with the observed X-ray pattern if the helically arranged scattering units extend outward from a radius of about 60 Å to a radius of about 130 Å and if they are about 50 Å in diameter.

It appears, therefore, that this part of the X-ray diagram is dominated by the contribution from the cross bridges; and indeed this is what we should expect, since they project out into the sarcoplasm from the backbone of the thick filaments; and since their density is greater than that of the sarcoplasm, they produce a variation in density as one moves through the structure in an axial direction; this is what shows up in

Fig. 9. Electron micrograph of cross section of live frog sartorius muscle fixed and processed in conventional manner (glutaraldehyde fixation, osmium tetroxide postfixation, Araldite embedding). The thin actin-containing filaments can be seen at the trigonal points of the hexagonal lattice of thick myosin-containing filaments. The myosin filaments have a considerably larger diameter than the actin filaments and often appear to have a dense central core (probably representing the backbone of the thick filament) and a less dense, sometimes angular, penumbra (probably representing the projected view of all the cross bridges in the section, the helical arrangement of which is not well preserved during the preparative procedures). Rabbit psoas muscle, fixed live, has a similar appearance. (Magnification ×123,000.)

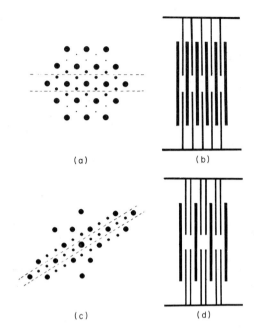

(a)   (b)

(c)   (d)

Fig. 11. (a) End-on view of a double hexagonal array of filaments. The two dotted lines indicate the outline of a longitudinal section, about 250 Å in thickness, parallel to the 10 lattice planes, at the appropriate level to include one layer of primary filaments and two layers of secondary filaments. (b) Expected appearance of longitudinal section cut as indicated in (a). Note simple alternation of primary and secondary filaments. The latter will represent two filaments lying vertically above each other in the section. (c) As in (a), but with dotted lines showing a longitudinal section about 150 Å in thickness, cut parallel to the 11 planes of the lattice. Primary and secondary filaments in this case all lie in the same layer, and two secondary filaments occur between each pair of primary ones. (d) Expected appearance of longitudinal section, cut as in (c). Note characteristic appearance of two secondary filaments between each pair of primary filaments.

the X-ray diagram. The backbone of the thick filaments, on the other hand, appears to consist of rather solidly packed protein and would therefore contribute less strongly to the axial density fluctuations.

One can say, then, that the general features of the X-ray diagram from living muscles agree closely with those that would be given by filaments having the structure seen in the electron micrographs of fixed and embedded material. Thus, those X-ray results provide clear confirmation of the picture of muscle derived from the electron micrographic evidence.

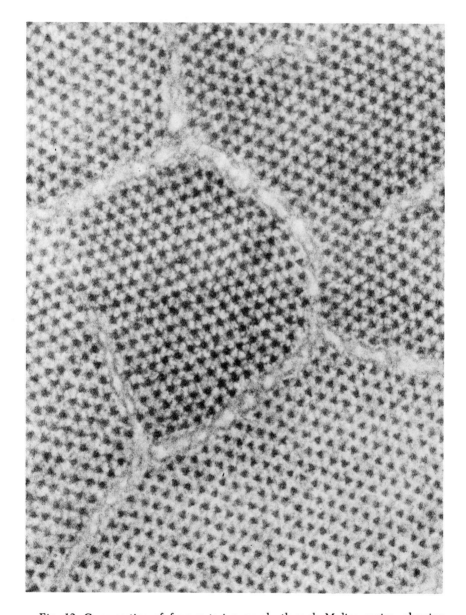

Fig. 12. Cross section of frog sartorius muscle through M line region, showing cross bridges between thick filaments. The triangular appearance of the thick filaments in the region adjacent to the M line itself is readily visible in the fibrils in the lower right and upper left of the picture. (Magnification ×121,500.)

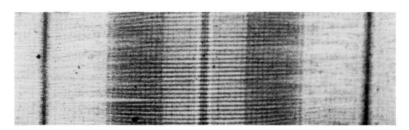

Fig. 13. Longitudinal section of frog sartorius muscle, showing three prominent transverse lines in the M line region, with a slightly less dense zone on either side (not to be confused with the much wider H zone). The narrow lines represent M line cross bridges visible in cross-sectional view in Fig. 12; the narrow light zone is believed to represent the central zone of the thick filaments which lacks myosin cross bridges. (Magnification ×34,200.)

There are some additional features of the X-ray diagram from resting muscles that should be mentioned at this point. The first concerns the shape of the reflections. The meridional reflections are extremely sharp in an axial direction, showing that they arise from structures at least several thousand angstroms in length (H. E. Huxley, 1952; H. E. Huxley

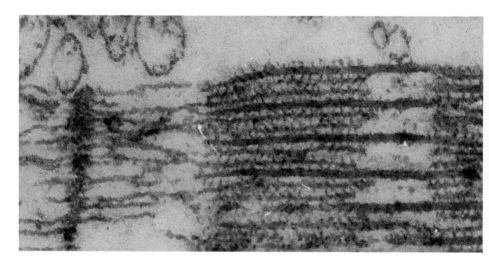

Fig. 14. Very thin longitudinal section of glycerinated rabbit psoas showing cross bridges on the thick myosin containing filaments even on the outside filament which has no thin filaments alongside it. The absence of bridges in the central bare zone of the thick filaments, and the thickening in the M line region—probably representing the residue of the M line bridges degraded by the glycerination procedure —can also be seen. (Magnification ×144,000.)

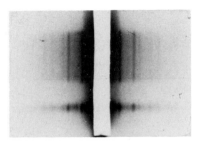

Fig. 15. Low-angle equatorial X-ray diffraction diagram of surviving frog sartorius muscle taken with fine-focus rotating anode X-ray tube, bent quartz crystal focussing monochromator, and a totally reflecting mirror arrangement, which also allowed through some of the beam that had not been reflected by the glass, thereby giving a combined slit (upper recording) and pinhole (lower recording) type pattern. The very strong inner reflection (10 crystallographic planes) and less strong second reflection (11 planes) are visible, together with higher order reflections from the hexagonal lattice. An intermediate reflection may arise from the Z line lattice.

and Brown, 1967), as would be the case if they were derived from regularly repeating cross bridges along the length of the thick filaments. These reflections are also very narrow in a direction at right angles to the meridian. However, the width of the reflections in this direction cannot be used as a measure of the diameter of the filaments, since the filaments are arranged in a very regular lattice, and it is the size of the whole lattice that determines the lateral spread of the reflections. In the patterns from well oriented muscles photographed using a double monochromator type camera, it can be seen from the small lateral spread of the meridional reflections (Fig. 18) that the lateral ordering of the cross bridge lattice extended over several thousand angstrom units. Moreover, sharp sampling of the off-meridional reflections can also often be seen, showing that there is some type of systematic rotational registration of the helical arrays of cross bridges on neighboring filaments. The nature of this lattice sampling can be used as independent evidence to show that the cross bridges are arranged on a threefold screw axis; the original paper should be consulted for details (H. E. Huxley and Brown, 1967).

Thus, the evidence that the 429 Å helical pattern of reflections comes from the thick filaments consists of two main arguments: (1) that the distribution of X-ray reflections along the layer lines is that which would be expected from filaments having a backbone diameter of about 120 Å and cross bridges extending outward from it to a total diameter of about 260 Å; (2) that the lattice sampling effects require that filaments

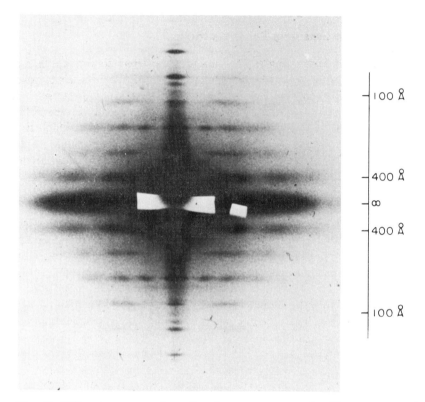

— 100 Å

— 400 Å
— ∞
— 400 Å

— 100 Å

Fig. 16. Diffraction pattern from live frog sartorius muscle; fiber axis vertical. The pattern of layer-line reflections corresponds to a repeat of 429 Å and is believed to arise from the helical arrangement of cross bridges on the myosin filaments. Note strong first layer line and strong meridional reflection on third layer line. The distributions of intensity along the first, second, fourth, and fifth layer lines are all similar to one another. Sampling of the transform is evident on the third layer line.

giving rise to the layer line pattern be located at the lattice points of the hexagonal lattice and have a threefold screw axes. It should also be mentioned that certain forbidden reflections occur in the muscle diagram, i.e., meridional reflections on layer lines other than the third, sixth, ninth, etc., which would not be given if all the material in the thick filaments were arranged with the helical symmetry we have described. This indicates that while most of the structure has helical symmetry, there is present either some distortion, or some other parts of the structure arranged with a different symmetry.

F. X-Ray Reflections from Thin Filaments

At somewhat larger angles than those considered so far, another set of meridional and near-meridional X-ray reflections (the actin reflections) are visible. We will defer detailed discussion of these until the protein composition of the muscle filaments has been discussed and merely mention that they appear to arise from filaments in which there are subunits arranged in helical fashion, with a subunit repeat of 27.3 Å and a helical repeat of 360–370 Å (i.e., the helix is nonintegral). The radius of the helix along which the subunits are places is approximately 25 Å (Selby and Bear, 1956), which would be consistent with a model in which the pattern arose from the thinner filaments. As we will see below, there is very strong evidence to support this view.

It should also be mentioned at this point that there are a number of other meridional reflections in the diagram from live muscles, which arise from repeating periodicities additional to those we have described so far. So as not to present too complex a picture while we are still discussing only the basic structure of the muscle, description of these will be deferred (to p. 354). However, it will be convenient to refer now to a prominent meridional reflection at a spacing of 385 Å, a periodicity which is significantly different from that of both the thick and the thin filament helices but which agrees closely in spacing with a fine axial periodicity which can be discerned in the thin filaments in electron micrographs.

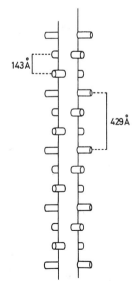

Fig. 17. Schematic diagram showing arrangement of cross bridges on 6/2 helix. Helical repeat is 429 Å, but true meridional repeat (i.e., periodicity of the variation in density of structure projected on to long axis of filaments) is 143 Å.

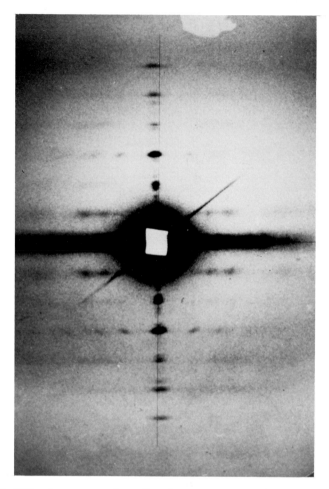

Fig. 18. Diffraction pattern from live frog sartorius muscle; fiber axis vertical. Photographed with double monochromator arrangement. The diagonal streak is an instrumental artifact. Note sampling on layer lines, indicating three-dimensional order present in cross bridge arrangement within the fibrils. The lattice spots present on the first layer line occur on different row lines to those on the equator and on the third layer line, and indicate the presence of a superlattice of spacing $\sqrt{3}$ times that of the standard lattice.

III. Protein Components of the Filaments

A. Introduction

So far, we have described muscle fibrils and filaments in purely structural terms. Now, we must begin to fill in this rather abstract picture

and discuss the location of the various protein molecules from which muscles are constructed. Rabbit skeletal muscles have been employed much more than any others for biochemical studies, for reasons which have been explained very delightfully by A. Szent-Györgyi (1951), and they have also been used extensively in electron microscope and X-ray work and in other studies designed to find out the location of the principal proteins. The main results almost certainly apply to other cross-striated muscles, too, but some caution should be exercised over points of detail.

The two principal proteins, whose presence has been demonstrated in a large variety of muscles, are myosin and actin. In rabbit skeletal muscle, myosin accounts for about 38% of the total protein in the muscle fibers (Hasselbach and Schneider 1951) or about 54% of the total structural protein of the fibrils washed free of soluble proteins. The quantity of actin is less well known; earlier estimates (Hanson and Huxley, 1957; Perry and Corsi, 1958) indicated a value of 20–25% of the fibrillar protein, but later estimates based on the ADP content of muscle fibrils suggest a figure of 18% (Weber *et al.*, 1969), corresponding to about 12% of total muscle protein, or 24 mg actin per gram of muscle (see also p. 354).

The location of these two proteins in the muscle structure can be demonstrated in a very straightforward way (Hanson and Huxley, 1953; Hasselbach, 1953). Myosin can be extracted from these muscles by strong salt solutions (e.g., 0.47 M, 0.16 M phosphate buffer) in the presence of agents that keep myosin and actin dissociated from each other—either the indigenous ATP present in recently dissected muscles, or added pyrophosphate (10 mM) plus magnesium. With short extraction times, not much actin is removed from either fresh muscle or glycerinated material (A. G. Szent-Györgyi *et al.*, 1955). Once extracted, the myosin can be precipitated at low ionic strength and estimated.

When these standard myosin extraction methods are applied to isolated myofibrils observed in the phase contrast light microscope, it can be seen immediately that the dense A bands of the fibrils are dissolved (Fig. 19) leaving behind a ghost fibril consisting of the Z lines and segments of material on either side having approximately the density of the I bands and occupying the region of the sarcomere corresponding to the position of the thin filaments. That is, these "I segments" extend from the Z lines to the position of the edges of the original H-zone. Electron microscopic observations on the extracted fibrils show that the thick filaments have been dissolved and that the thin filaments remain.

Thus, there is very strong qualitative evidence that myosin is contained in the thick filaments and that actin, which is not extracted, is present in that part of the structure which remains, i.e., the thin filaments. There is also strong quantitative evidence to support this.

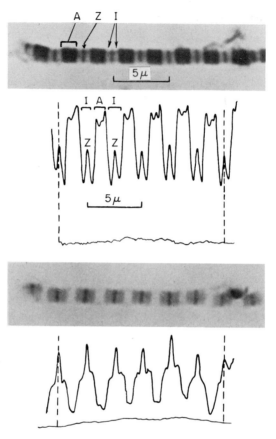

Fig. 19. Photographs taken in an interference microscope (and densitometer tracings of the negatives) of one glycerol-extracted rabbit fibril before and after the
extraction of myosin. The A substance is removed, leaving behind the I substance
and the Z lines. To the eye, there appear to be gaps where the I substance is
discontinuous in the center of the sarcomere (H zone), but the densitometer tracing
demonstrates that there is a small amount of material here. The tracings were
taken along the length of the fibril and show peaks corresponding to Z lines and
A bands, and troughs in the I bands. The heights of such peaks or troughs above
the base line (tracing of background density) are related to the amount of protein
present.

B. Interference Microscope Measurements and Biochemical Estimations

The interference microscope permits measurements of the phase retardation of the light beam as it passes through any chosen part of
the specimen. From such measurements it is possible to calculate the

amount of protein present in the different bands of a muscle fibril, so that the total amount of protein extracted, and the amounts from the different bands, can be determined. These quantities can then be compared with large-scale biochemical analyses of the same material. The results of these experiments are very straightforward and are given in Tables I and II (H. E. Huxley and Hanson, 1957; Hanson

TABLE I

INTERFERENCE MICROSCOPE MEASUREMENTS ON STRIATED MUSCLE

	Quantity (% of total myofibrillar protein)[a]		
Substance	Method 1	Method 2	Method 3
A substance	54.5 ± 2.0	55.3 ± 2.0	49.6–54.0
I substance	36.4 ± 2.0	35.7 ± 2.0	36.5–37.9
Extra material in Z lines	6 ± 1	6 ± 1	—
Material in H zone gap after extraction	3.1 ± 0.3	(3.1 ± 0.3)	3.5–6.9
A substance extracted	54.5 ± 2.0	—	—
I substance extracted	10.0 ± 1.0	—	—
Total extracted	64.5 ± 3	61.5 ± 4	—

[a] The three methods represent different ways of analyzing the experimental results; see original paper for details (H. E. Huxley and Hanson, 1957).

TABLE II

BIOCHEMICAL ANALYSIS OF STRIATED MUSCLE

	Percent of total protein[a]								
	Fresh muscle			Glycerinated muscle			Washed fibrils		
Protein fraction	H-S	S-G	H-H	H-S	S-G	H-H	H-S	S-G	H-H
Soluble	28	(48)	34	—	42	28	—	—	—
Myosin	38	(26)	34	—	29	37	(53)	50	51
X protein[b]	—	(9)	7	—	10	8	—	18	11
Myosin + X protein	—	(35)	41	—	39	45	—	68	62
Actin	14	—	—	—	—	—	(19.5)	—	—
Actin + residue	34	(17)	25	—	19	27	(47)	32	38

[a] Analyses by Hasselbach and Schneider (H-S) (1951), A. G. Szent-Györgyi et al. (S-G) (1955), and Hanson and Huxley (H-H) (1957). Figures in parentheses calculated by H. E. Huxley and Hanson (1957).

[b] Extra protein, extracted at the same time as myosin, but not precipitated at $\mu = 0.04$.

and Huxley, 1957). The extra material of the A band, as measured in the interference microscope, amounts to 50–55% of the total protein in the myofibril. When myosin is extracted from the fibril, all of this material is removed, together with a further 10% of the total myofibrillar protein which is extracted from the I material (i.e., the filaments stretching from the Z lines to the edges of the H zone). Biochemical analysis of the same extraction process shows that a total of about 62.5% of the total myofibrillar protein is removed; 51% of the myofibrillar protein (i.e., about 80% of the extract) can then be precipitated under conditions which are classically used to precipitate myosin, while the remainder (11%) remains in solution. These results show that at least four-fifths of all the myosin must be present in the A bands, and they are in perfect agreement with the hypothesis that all the myosin is located there, in the thick filaments, as envisaged by the model under consideration, and that a small quantity of other proteins are extracted from the I filaments.

The results of various biochemical analyses of muscle are summarized in Table II. Soluble proteins, which no doubt include many of the enzymes involved in the glycolytic and oxidative processes, make up about 34% of the total dry weight of the whole rabbit psoas muscle. The I substance (excluding the extra material at the Z lines) makes up about 36% of the protein of the myofibrils or about 25% of the dry weight of the whole muscle. This is substantially more than the amount of actin that has been identified in muscle extracts and indicates the presence of other protein components in the thin filaments. We will return to this topic later, for it is a very important one.

Other evidence concerning the locating of actin can be obtained by taking advantage of the finding of A. G. Szent-Györgyi (1951a,b) that actin can be extracted from rabbit muscle by 0.6 M potassium iodide, the myosin having previously been extracted by one of the standard methods. When such a potassium iodide solution was applied to myosin-free fibrils under the phase contrast light microscope, it was found that a large part of the residual I segments was extracted.

A different kind of experiment (Hanson and Huxley, 1955) again indicates that the I substance contains actin. It is well known that actin and myosin in vitro form a complex (actomyosin) which contracts when it is treated with ATP; neither protein by itself is contractile (Hayashi et al., 1958). Rabbit fibrils from which the A substance had been removed, and which showed no structural changes in the presence of ATP, were treated with a solution of rabbit myosin under the microscope. The protein was taken up by the I substance, whose optical density increased. When these "reconstituted" fibrils were treated with

ATP, the I substance in each half sarcomere contracted to the adjacent Z line, where a thick contraction band was formed. Similar results were obtained (Hanson, 1956) when insect fibrils, freed from A substance, were treated with rabbit myosin and then with ATP. It is clear that in these experiments, an actomyosin system had been formed from the thin filaments and the added myosin.

C. Additional Evidence Concerning Location of the Muscle Proteins

Following the experiments described above, Perry and Corsi (1958) obtained evidence in a somewhat different way. They found that prolonged extractions of myofibrils in a low ionic strength medium (0.078 M borate buffer, pH 7.1, or 5 mM tris buffer, pH 7.7) dissolved out the material of the Z lines and I bands, leaving behind the A bands. The extract was found to contain actin and tropomyosin, and though the absolute amounts varied with the extraction conditions, the relative amounts of the two proteins tended to be the same. Accordingly, it was suggested that actin and tropomyosin might be associated in the structure of the I filaments. It was also found that virtually the whole of the ATPase of the preparation remained associated with the A band residue, confirming the location of myosin there.

Further evidence still has come from antibody studies. It was shown very clearly by Marshall *et al.* (1959) and by A. G. Szent-Györgyi *et al.* (1964) that antibody against myosin was bound by muscle fibrils in the A band region only, as could be seen by the location of the fluorescent label that had been attached to the antibody. Antibody against actin or tropomyosin, on the other hand, was bound to the part of the sarcomere occupied by the I filaments.

Further evidence can be obtained from observations in the electron microscope by the negative staining technique on separated muscle filaments prepared by physical disruption of the tissue in a medium that dissociates actin and myosin from each other (H. E. Huxley, 1961, 1963). Two types of filament are found, corresponding closely to the thin and thick filaments seen in tissue sections. The thinner filaments, which can often be seen still attached to fragments of the original Z lines, show a characteristic fine structure (Hanson and Lowy, 1963; H. E. Huxley, 1963) in the form of a double helical arrangement of two intertwined chairs of globular subunits, the subunits being about 50 Å apart; the two chains cross each other (when viewed from any particular direction) at intervals of about 360–370 Å. The identical structure is seen in filaments of purified actin.

Purified myosin, on the other hand, will, under suitable ionic conditions, aggregate into synthetic filaments which closely resemble the A band filaments derived from muscle. The synthetic filaments can have similar dimensions to the natural ones (though their lengths are very variable, unlike the natural ones whose lengths are maintained constant by a mechanism which is not understood at present) and show the same characteristic projections or cross bridges; they can even show the same absence of bridges in a short central bare zone of the filament. Thus, the evidence again clearly confirms the location of myosin in the thick filaments and actin in the thin ones, and I think it would be fair to say that this localization can be taken as established beyond all possible doubt.

IV. Changes in the Band Pattern of Muscle during Contraction and Stretch

A. Historical Background

We have seen in the preceding sections that the pattern of striations arises directly from the way in which the contractile proteins of the muscle are organized into the two separate but overlapping arrays of longitudinal filaments. It is therefore apparent that knowledge of the behavior of the pattern of striations during active or passive changes in the length of the muscle will give rather direct information about the behavior of these actin and myosin filaments during these processes. It is a very fortunate state of affairs that what we can see in the light microscope should be so straightforwardly related to what is happening at the level of the molecular filaments. But the light microscope information proved more difficult to gain than might at first be supposed, and the problem took almost 100 years to settle! The observations are beset by optical artifacts that arise because of the large thickness of the fibers in comparison with the lengths of the bands, because of the different refractive indexes of the different bands, because of alterations in band lengths which tend to take place during any fixing, embedding, and sectioning, whether for light or electron microscopy, and because of the high speed at which the changes take place during normal contraction. These factors account for a good deal of the conflict that has existed between different observations of band changes. This history of these conflicts, and the reasons for them, have been described by H. E. Huxley and Hanson (1954) and particularly engagingly by A. F. Huxley (1957) and need not be repeated here.

B. Recent Measurements

It is clear now that there are only two reliable methods at present available for measuring accurately the band lengths in striated muscle using the light microscope. These are either (1) to use whole fibers immersed in a medium whose refractive index matches the average refractive index of the whole fiber, eliminating the cylindrical shape artifact (A. F. Huxley, 1957), and to examine the striations in the fibers in a phase contrast or intereference microscope with a sufficiently small depth of focus to show only a thin optical section of the fiber, or (2) to use isolated myofibrils, so that the thickness of the specimen (1 μ or less) introduces little error into the measurement of band length, which again is made in the phase contrast light microscope. Method (1) has been used by A. F. Huxley and Niedergerke (1954) and method (2) by Dr. Jean Hanson and the present author (H. E. Huxley and Hanson, 1954).

The former authors examined living frog fibers during passive stretch and during isometric and isotonic contractions, both in twitches and in tetani. Hanson and Huxley used isolated rabbit myofibrils, which could be made to contract by irrigating them with solutions of ATP, and whose length could be controlled by a form of micromanipulation. The use of isolated fibrils has the disadvantage that it is one step removed from the living muscle, but it has the advantage that the band changes, in particular those occurring in the H zone, can be seen more clearly than in whole fibers. Moreover, myosin extraction can be performed on the isolated fibrils at any chosen stage in the contraction process, making interpretation of the observations much more straightforward.

The results of the measurements are very simple. Both sets of authors found that over a wide range of muscle lengths the length of the A band does not change perceptibly, either during stretch or during isotonic or isometric contraction. Changes in the length of the sarcomere are accounted for by changes in the length of the I bands alone. Figure 20 illustrates the results obtained on myofibrils.

When the muscle has contracted down to about 60% of its resting length (at rest length the lengths of the A and I bands were 1.5 μ and 0.8 μ, respectively), the I bands disappear and the A bands come into contact with the adjacent Z lines. If contraction proceeds beyond this point (which under physiological conditions it usually does not), then contraction bands form around the Z lines. The constancy of the length of the A band over the working range shows that the thick fila-

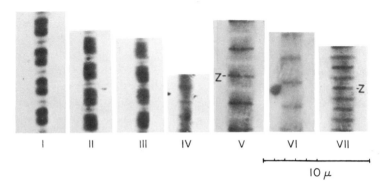

Fig. 20. Photographs showing the changes in band pattern that occur when a fibril contracts. The top row shows intact fibrils, the bottom row shows fibrils from which the A-substance has been extracted at the length shown. The latter pictures show more clearly how the I-substance changes position during contraction and extension. (I)–(IV) One fibril photographed at different stages during contraction in ATP from resting length down to about 50% of that length. The H zone closes up and is replaced by a dense line; when the I bands have disappeared, contraction bands form at the Z lines. (V)–(VII) Three fibrils, similar in length to those above, from which myosin has been extracted. In (VI), the H zone (gap) has disappeared, and in (VII), a dense line has appeared in the center of the sarcomere.

ments that make up the A band must remain at constant length during both stretch and contraction, down to the point at which their ends, for steric reasons, have to crumple up.

The behavior of the secondary filaments can be deduced from the other changes which take place in the band pattern. In isolated myofibrils, the H zones are seen to be much wider in stretched material than in fibrils at rest length (see Fig. 20). During shortening, the H zones become narrower and disappear at about 85% rest length, being replaced on further shortening by a dense line. If myosin extraction is performed on fibrils at the different stages of shortening, it is found that the different lengths of the H zone are duplicated by the different lengths of the gap between the ends of the secondary material, and that in fibrils where the H zone has been replaced by a dark line, this line remains after extraction. Myofibrils in these various conditions are shown in Fig. 20. Down to the point where this line forms, that is, from about 140% rest length (the limit of stretch) to 85% rest length, the distance from the Z lines to the edge of the H zone (or to the edge of the gap after myosin extraction) remains constant. It is apparent from these observations that the secondary filaments, too, do not change their length perceptibly over a considerable range of muscle lengths. Accordingly, the two sets of filaments must slide past each other.

C. Other Observations on Band Lengths

Measurements on isolated myofibrils in the phase contrast light microscope have also been reported by de Villafranca (1957). The results are very similar to those described above; the A band remains apparently constant in length to within ±0.075 μ (i.e., any changes in length are less than the resolving power of the light microscope), over the range of sarcomere lengths used, 2.3–3.3 μ. The I bands were, of course, not constant in length, and the H zones, too, were found to be much longer in stretched muscle, although the observed differences in the latter case were somewhat smaller than would have been anticipated; however, this measurement is a difficult one to make accurately, for the density difference at the boundaries is rather small.

Carlsen and Knappeis (1955) and Knappeis and Carlsen (1956) reported measurements on band lengths in intact fibers in papers that continue the earlier work of Buchtal *et al.* (1936). The results obtained by these workers, who found considerable changes in A band lengths, are not in agreement with the ones we have described above, which were obtained under conditions where optical artifacts were minimized. The optical artifacts which may have been present in the former measurements have been discussed by A. F. Huxley (1957).

It is also of some interest to compare low-power electron micrographs of rest length and of stretched muscle. In order to compare band lengths, it is necessary to cut sections with the microtome knife parallel to the fiber axis, so that any compressive flow of the plastic will not affect the measurements. If this is done, and if the muscle is held at a fixed length during fixation, dehydration, and embedding, it will be seen that the A band length is virtually constant at the different sarcomere lengths (Fig. 21). Moreover, the constancy in the distance from the Z lines to the edges of the H zone may be seen in the same pictures (Fig. 21). These observations cannot, of course, be accepted immediately as independent evidence for these characteristics of the band pattern, for one cannot assume in this type of experiment that no changes have taken place in the relative A and I band lengths during fixation. They do, however, provide perhaps a clearer visualization of the changes that have been seen in the light microscope.

Sjöstrand and Andersson-Cedergren (1957) reported observations of an apparent decrease of both A and I band lengths at sarcomere lengths

Fig. 21. The changes in band pattern at different muscle lengths as seen in the electron microscope (thick sections; oriented for sectioning so as not to foreshorten band lengths.)   (A) I. Stretched muscle showing long I bands and H zone. II. Rest length muscle, showing decrease in length of I bands and H zone, and constancy of length of A band.   (B) I and II. Same pair of preparations showing constancy of distance between Z lines and edge of H zone. (Magnification ×30,000.)

below rest length. These changes were not associated with changes in the pattern of bands such as we have been discussing; and the present author has observed such effects only in specimens that were incorrectly oriented during sectioning, so that compressive flow reduced the apparent band lengths.

The question of changes in filament length under different conditions was also raised again by Carlsen et al. (1961) on the basis of electron microscope observations. The changes they believed they saw were of the order of 10%, i.e., much less than those they reported previously from light microscope studies. Nevertheless, if present, the changes would have been of great significance, and the problem was therefore investigated very carefully by Page (1964; Page and Huxley, 1963). It was found that some changes in filament lengths could indeed be seen in electron microscope specimens under some conditions. However, when measures were taken to minimize any changes in the lengths during the processing for electron microscopy, then the differences between filament lengths under different conditions became very much less, and the small ones that remained could be accounted for perfectly reasonably on the basis of residual changes during fixation, dehydration, and embedding, which varied according to the state of the muscle.

Observations on muscles that had been stimulated electrically, allowed to shorten against a load, and then immersed in the glutaraldehyde fixative while they were still actively contracting at the shortened length, again showed constancy of filament lengths. In this type of experiment, unfortunately, one can only fix the overall length of the muscle, and one cannot tell whether the internal fine structure has actually been fixed in its actively contracting configuration. It is unlikely that it has, in detail anyway. However, it was interesting to find that at sarcomere lengths below that at which the thin filaments would be expected to meet in the center of the A band, a region of double overlap of thin filaments developed in that position (Figs. 22, 23) (H. E. Huxley, 1965). The thin filaments apparently had kept on sliding toward each other, while their ends by-passed each other and still maintained the same filament length. At shorter lengths, contraction bands developed where the thick filaments came up against the Z lines, but shortening continued and the regions of double overlap grew larger still. These findings provide a rather vivid illustration that the underlying mechanism is one which produces a relative sliding movement of the thick and thin filaments past each other, irrespective of the situation at the ends of the filaments.

D. X-Ray Diffraction Observations

The most decisive evidence that the actin- and myosin-containing filaments in muscle always remain constant in length comes from low-angle X-ray diffraction observations on live muscles. As I have already explained, there are two sets of meridional reflections in the X-ray diagrams, which can be identified as arising from the actin and myosin filaments, respectively. One of the earliest observations about the muscle diagram was that the spacings of both these patterns remained unchanged when the length of the muscle (and hence of each sarcomere

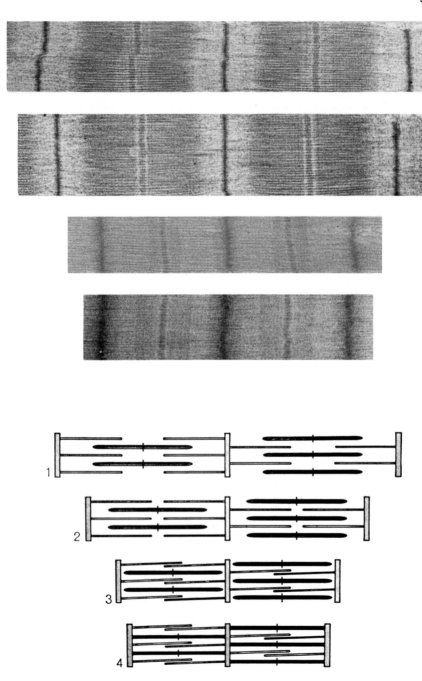

Fig. 22. See facing page for legend.

TABLE III

EFFECT OF CONTRACTION ON SPACINGS OF PRINCIPAL MYOSIN
AND ACTIN REFLECTIONS

Condition of muscle	Myosin spacing (Å)	Actin spacing (Å)
Relaxed	143 ± 0.1	59.25 ± 0.05
Active	144.6 ± 0.1	59.11 ± 0.05
Percentage change	1.15 ± 0.1 (Significant increase on contraction)	0.24 ± 0.1 (No significant change)

[a] The figures are obtained from measurements of 48 separate experiments. In no experiment did the active myosin spacing fail to exceed the relaxed one. The probable errors quoted represent probable error of mean, derived from variation of measured values from their average value.

in it) was changed by as much as 40% by passive stretch. More recently, the X-ray diagrams from muscles during active contractions have been examined (H. E. Huxley *et al.*, 1965; Elliott *et al.*, 1965, 1967; H. E. Huxley and Brown, 1967), and it was found here, too, that the axial periodicities in both the actin and myosin filaments were virtually unchanged (Table III) during both isometric contraction and contraction when shortening had been allowed to take place. Indeed, a small increase, by about 1%, was observed in the periodicity of the myosin filaments.

E. Conclusions

From the evidence described above, it is quite clear that changes in the length of a striated muscle, whether they be passive or active,

Fig. 22. Longitudinal sections of frog sartorius muscle, fixed during isometric contraction at different muscle lengths and examined (at low resolution) in the electron microscope. The change in the appearance of the central region of the A band is particularly noticeable. In the longest sarcomeres, a clear H zone is visible. This closes up as the muscle shortens (leaving the denser M line, and the narrow light zone on either side of it unchanged), and is replaced by a zone denser than the rest of the A band, which increases in width as shortening proceeds. This is believed to arise from a double overlap of the thin actin-containing filaments, as illustrated in the diagrams below. The filaments themselves remain at approximately constant length, until the thick filaments are crumpled up against the Z lines and form contraction bands there.

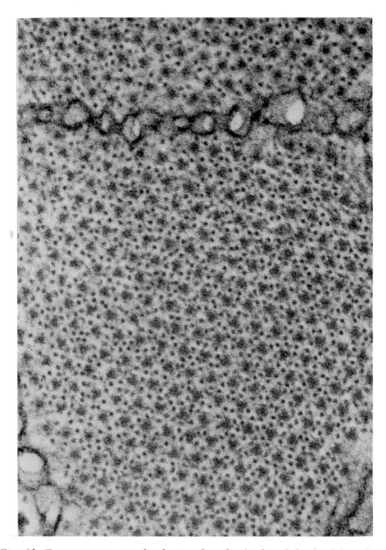

Fig. 23. Frog sartorius muscle shortened under load and fixed while still being stimulated. Cross section through the central region of the sarcomere of a number of shortened fibrils (i.e., through the double overlap region) showing twice the normal number of thin filaments interdigitated between the thick ones.

are brought about by a process in which the two arrays of filaments slide past each other, the lengths of both the thick myosin-containing filaments and the thin actin-containing filaments in the arrays remaining essentially constant.

F. Implications

Since the principal muscle proteins involved in contraction, namely actin and myosin, have been shown to be organized into separate filaments which slide past each other during contraction, it is clear that a relative sliding force must be developed between the filaments as a consequence of the chemical events associated with contraction.

The chemical reaction most closely linked to contraction is the splitting off of the terminal phosphate of adenosine triphosphate (ATP). This was suspected to be the basic energy-yielding reaction for many years, but it was not until 1962 that it was satisfactorily demonstrated to be the case in live muscle by Davies and his colleagues (Cain *et al.*, 1962). It had been shown earlier by Engelhardt and Ljubimova (1939) that the structural protein myosin was also an enzyme, an ATPase. A. Szent-Györgyi (1942) found that artificial fibers prepared from a complex of myosin with a newly discovered structural protein, actin (Straub, 1942), would contract when immersed in a solution of ATP and suitable salts. Subsequent work (described in Volume III, Chapter 8) showed that the splitting of ATP by myosin could be very strongly activated by actin under conditions that favored the association of the two proteins.

These findings provide strong indications that when a muscle is activated, an interaction takes place between myosin molecules in the thick filaments and actin molecules in the thin filaments; that this interaction enables the myosin to split ATP at a rapid rate, thereby providing the energy for contraction; and that this chemical change is somehow or other used to produce a force which causes the filaments to slide past each other.

This model can be described in more concrete terms if we bring into the picture the cross bridges on the thick filaments, visible in the electron microscope, and whose presence is amply confirmed by the low-angle X-ray diffraction observations. It is very natural to suppose that these projections represent the parts of the myosin molecules that interact with actin (H. E. Huxley, 1957), and we will see that there is strong evidence to support this view. Furthermore, given a structure that allows a direct physical contact between myosin and actin (presumably, from a biochemical standpoint, in order to activate the ATP splitting), one might further hypothesize that the cross bridges were directly involved, in a mechanical sense, in developing the relative sliding force between the filaments. Since the likely range of movement of a cross bridge would be small compared to the several thousand angstroms of relative sliding movement between neighboring thick and thin filaments during muscle shortening, it would be reasonable to suppose that the cross bridges function repetitively during contraction, each one going through many cycles of enzymic and structural change as the muscle shortens. In such a model (Fig. 24) the cross bridge would attach to a monomer in the actin filament in one configuration; then a structural change would take place which resulted in the actin filament being drawn along a certain distance. The cross bridge would then release from the actin and return to its starting position. Meanwhile,

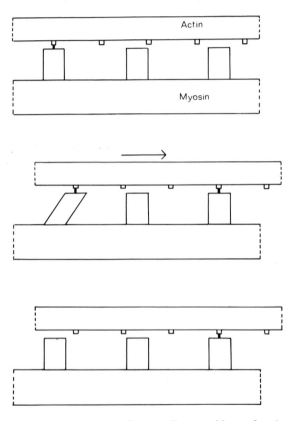

Fig. 24. Diagram showing, very schematically, possible mode of action of cross bridges. A cross bridge attaches to a specific site on the actin filament, then undergoes some configurational change that causes the point of attachment to move closer to the center of the A band, pulling the actin filament along in the required manner. At the end of its working stroke, the bridge detaches and returns to its starting configuration in preparation for another cycle. During each cycle, probably one molecule of ATP is dephosphorylated. Asynchronous attachment of other bridges maintains steady force.

other bridges would have attached, so that a steady force was developed. Each time a cross bridge went through this cycle, one (or perhaps two) molecules of ATP would be split. When activation of the muscle was ended, interaction between cross bridges and actin filaments would cease, ATP splitting would fall to a low level, no sliding force would be developed, and the muscle could be extended again.

The detailed evidence I have described so far has been concerned with the very broad general features of muscle structure and function. That is, I have described the evidence for the sliding mechanism in some detail, but I have so far merely pointed out the existence of the cross bridges and have simply indicated a possible way in which they could be involved in the contraction process. To go into this matter further, we must now consider the structure in greater detail.

V. Molecular Structure of the Filaments

Information from several sources provides evidence about the molecular structure of the muscle filaments. In the first place, one can study the molecular weights and dimensions of the principal muscle proteins in solution. Next, one can study the muscle filaments in the electron microscope, either in sections or by using the negative staining technique on separated filaments, and can compare these observations with analogous ones on synthetic filaments made from purified muscle proteins. Then one can study the muscle structure (and also the structure of certain synthetic filaments) by X-ray diffraction and obtain evidence not only about how the protein molecules pack together (from the low-angle diffraction diagrams), but also about the nature and arrangement of the polypeptide chains within the molecules (from the wide-angle diagrams). The power of all these techniques has undergone continuous development in recent years, and these advances are continuing. A considerable amount of very relevant detail is now known, and it seems likely that more will be forthcoming.

A. The Myosin Filaments

1. Number of Molecules per Filament

It is clear that each thick filament must contain a considerable number of myosin molecules, and we can arrive at an estimate of that number from measurements of the molecular weight of myosin, the amount of myosin present in a given volume of muscle, and the number of thick filaments present in the same volume.

If we consider rabbit psoas muscle at rest length, where the sarcomere length is 2.6 μ and the hexagonal lattice spacing 425 Å, then, if the filament lattice occupied the entire cross section of the muscle, there would be approximately 2.5×10^{14} thick filaments in 1 cm^3 of muscle. A similar volume would contain approximately 76 mg of myosin, if we take the myosin content as 36% of the total protein (Hanson and Huxley, 1957), a protein content of 20% by weight of the whole muscle, and the density of muscle as approximately 1.06. If the molecular weight of myosin is taken to be 475,000 (Lowey *et al.*, 1969), then 76 mg of myosin would represent $(76 \times 10^{-3}N)/(4.75 \times 10^5)$ molecules of myosin, i.e., 9.6×10^{16} molecules (where N = Avagadro's Number, 6×10^{23}). Hence, the number of myosin molecules in each thick filament should be approximately 384.

The number of cross bridges that appear on each thick filament in the electron microscope is approximately 216 (about 18 sets of 6 bridges in each half A band), and a similar figure can be derived from the X-ray diffraction results (which indicate that there are 6 cross bridges in each interval of 429 Å along the thick filaments). Now in the first place, it is of great interest that the number of cross bridges observed should be so similar to the number of myosin molecules calculated to be present in each thick filament from the other data. It indicates that the identification of the cross bridges as representing those parts of each of the myosin molecules which interact with actin as at least in approximate agreement with the known myosin content of muscle. However, we may enquire why the agreement is not more accurate than is actually found. Clearly, there are only a limited number of possibilities. It cannot be because we have neglected the spaces between myofibrils and between muscle fibers, because this would lead to an underestimate of the number of molecules per filament from the biochemical composition data. It might be that each cross bridge in fact represents two myosin molecules, or it might be that for some reason about one-half of the cross bridges are concealed. Alternatively, it might be that there is some other major component present in the thick filaments which has escaped detection because of its similarity to myosin, and which causes the amount of myosin present always to be overestimated biochemically. However, the fact that such a component has eluded discovery for so long indicates that it is not likely to be present in sufficiently large amounts to account for the discrepancy that exists. On the whole, a miscounting of the cross bridges or an erroneous assumption that each one represents only one myosin molecule seems more likely.

At all events, the data are not in conflict with the supposition that each of the visible cross bridges does represent the active part of either one or two molecules, and we must now consider the structure of the myosin molecule in greater detail.

2. STRUCTURE OF THE MYOSIN MOLECULE

Individual myosin molecules can be seen in the electron microscope using the shadow-casting technique (Rice, 1961a,b, 1964; Zobel and Carlson, 1963; H. E. Huxley, 1963). They have the appearance of long rod-shaped structures with a globular region at one end (Fig. 25). The rods are about 20 Å in diameter and 1500 Å in length, and the globular region appears about 50 Å in diameter and 100–200 Å in length. This appearance fits in very well with what was known at that time from other observations on myosin, and subsequent work has confirmed that it is generally correct, although the globular region is now known to consist of two separate subunits, closely similar to each other. Mommaerts (1951) deduced a value of 1500 Å ± 100 Å from light scattering measurements, and Holtzer and Lowey (1956) calculated dimensions of 1650 × 28 Å in a similar way. Later on, Lowey and Cohen (1962) arrived at dimension of 1600 Å × 20 Å by hydrodynamic measurements, and also deduced that there was probably a thickening of the molecule near one end. They calculated a molecular weight of 470,000 ± 25,000.

Very important information has come from studies of the cleavage of myosin into well defined subfragments by brief treatment with proteolytic enzymes (Gergely, 1950, 1951, 1953; Perry, 1951; Mihalyi, 1953; Mihalyi and Szent-Györgyi, 1953a,b; Mueller and Perry, 1961; Kominz *et al.*, 1965; Lowey *et al.*, 1966, 1967, 1969). Tryptic digestion breaks myosin into two large pieces, called heavy meromyosin (HMM) and light meromyosin (LMM) (A. G. Szent-Györgyi, 1953), having molecular weights of 340,000 and 140,000, respectively (Holtzer *et al.*, 1962). HMM carries the biological activity of the myosin, i.e., its ATPase activity and actin-binding ability, and is soluble at all ionic strengths, whereas LMM carries the self-assembly properties, i.e., the ability to aggregate into filaments at physiological ionic strength. In the electron microscope, LMM is seen to be a simple rod-shaped structure, having the same diameter as the rod part of the intact myosin and is probably about 960 Å in length. HMM appears as a double-headed tadpole-shaped structure, i.e., two globules sharing a short tail, probably about 400–500 Å in length. A satisfactory model for myosin simply places these two subfragments end to end, with a short trypsin-sensitive region joining them.

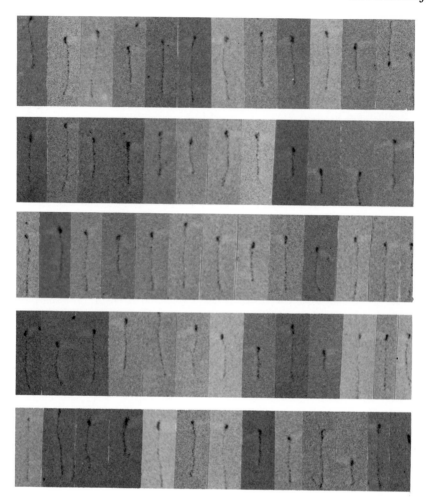

Fig. 25. A number of selected myosin molecules from shadowed preparations demonstrating the existence of the head and tail structure in a large number of instances. The average length of the particles in this particular group is 1670 Å. The globular head region (now known to be a doublet structure) is believed to contain the ATPase and actin binding properties, the linear tail the properties that permit association into filaments. (Magnification ×80,000.)

Using papain as a cleaving agent, Lowey and her co-workers (Lowey et al., 1969) have shown that under suitable conditions, the globular end of myosin can be cleaved away entirely from the rod portion; it was found to consist of two seemingly identical subunits (known as the S1 subfragments). Each of the subunits has a molecular weight of about 117,000 and has actin-activated ATPase activity and actin-bind-

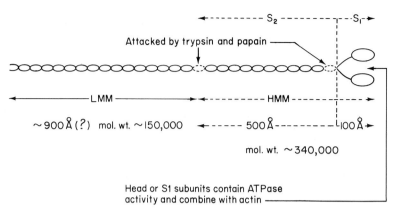

Fig. 26. Diagrammatic representation of myosin molecule (after Lowey *et al.*, 1969).

ing ability, like the parent myosin. Surprisingly, the rod part of HMM (S2) was found to be soluble at all ionic strengths.

It is now recognized (see Volume III, Chapter 7) that the myosin molecule is built out of two probably identical polypeptide chains, wound around each other to form a two-chain α-helix in the long rod part of the molecule, and then continuing on and folding up separately in the two S1 head units (Fig. 26). Optical rotatory dispersion measurements on LMM (A. G. Szent-Györgyi *et al.*, 1960) show that the rod part contains a very high proportion of α-helix, and X-ray diffraction observations both on oriented preparations of myosin (Boehm and Weber, 1932; Astbury and Dickinson, 1935a,b, 1940) and of LMM (A. G. Szent-Györgyi *et al.*, 1960) show an α-helical X-ray diagram. A similar diagram is given by whole muscle, suggesting that the light meromyosin parts of the myosin molecules are arranged with their long axes parallel to the long axis of the muscle. This leads very naturally to the idea that the backbone of the thick filaments is made up of the LMM parts of the myosin molecule, with the HMM parts projecting out sideways at regular intervals to interact with the actin filaments alongside. There is now a great deal of experimental evidence to support this picture.

3. ARRANGEMENT OF MYOSIN MOLECULES IN THE THICK FILAMENTS

We have already seen that the low-angle X-ray diffraction diagrams indicate that the cross bridges on the thick filaments are arranged (approximately) in helical fashion, and we can now interpret this as showing us the arrangement of the HMM part of the myosin molecules. It is clear that this arrangement must be established by the packing of the

LMM units in the backbone of the thick filaments, giving rise to a subunit repeat of 143 Å and a helical repeat of 429 Å. LMM will aggregate by itself into paracrystals with a well defined axial periodicity of about 425 Å (Philpott and Szent-Györgyi, 1954; A. G. Szent-Györgyi *et al.*, 1960) very similar to the 429 Å helical periodicity characteristic of the thick filaments. On some occasions, LMM will aggregate with a strong periodicity one third of the value mentioned above, again having a strong correlation with the subunit repeat in the thick filaments, which is one-third of the helical repeat.

On the other hand, the exact mode of the side-to-side packing of the LMM units is still unclear. As I have already mentioned, the X-ray diagram is dominated by the contribution from the cross bridges, and there is little direct evidence about the internal packing in the backbone. Departures from helical symmetry have already been noted, and it may well be that the backbone packing, which must be influenced by longitudinal overlap of these long molecules, may not be of a simple helical nature. Pepe (1967) has proposed a very interesting model for this packing, but it lies outside the scope of this chapter and the original papers should be consulted for details.

In general then, we would expect the myosin molecules to be placed in pairs, one on either side of a thick filament, and successive pairs to be staggered by 143 Å, and rotated relative to each other so as to place the projecting cross bridges on a 429 Å helix. If the entire rod part of myosin is about 1500 Å in length, then in general, a cross section of the filament will include about 20 molecules, sectioned at various points along their length. The terminal 1500 Å of the filaments would be tapered, as indeed is found to be the case (H. E. Huxley, 1957). The X-ray evidence indicates that in resting muscle, the cross bridges extend out sideways to a radius of about 130 Å from the center of the thick filaments i.e. they extend about 70 Å from the surface of the thick filaments. This indicates that in resting muscle, the linear portion of HMM (i.e. the S2 rod) is likely to be closely applied to backbone.

4. Polarity of the Thick Filaments

Myosin molecules will aggregate to form filaments when the ionic strength of the medium in which they are dissolved is reduced, for example from 0.6 μ to 0.1 μ (Fig. 27). The filaments are of variable length, have diameters up to 100–200 Å and show many short projections on their surfaces when examined in the electron microscope by the negative staining technique; these projections are similar to those seen on natural thick filaments under the same conditions. It is found that in many cases these projections are absent from a short zone about

Fig. 27. Preparation of purified myosin, at ionic strength $\mu = 0.15$, examined by the negative staining technique, showing filamentous aggregates of myosin, of varying lengths up to 2 μ, with diameters similar to those of the natural thick filaments, and like them, having tapered ends. (Magnification $\times 40,000$.)

0.15–0.2 μ long in the center of the filament. Thus, the shorter filaments have a cluster of projections at either end of a bare rod-shaped portion in the middle (Fig. 28), while longer filaments have a bare central region of about the same length but longer lateral portions with projections on them (Fig. 29). Inspection of natural thick filaments reveals that they too have the same structural feature (Fig. 30). Since individual

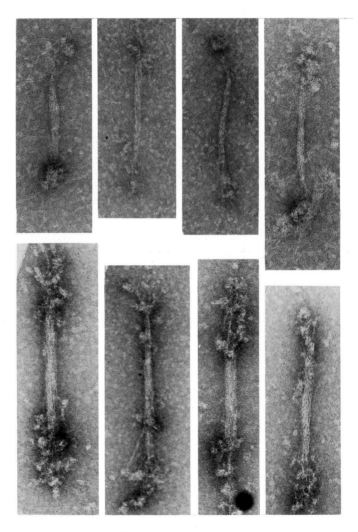

Fig. 28. Synthetic myosin filaments of different lengths ranging from 2500 to 4500 Å, showing characteristic bare central shaft of approximately constant length (1500–2000 Å) together with irregular projections along the rest of the filament. (Magnification ×145,000.)

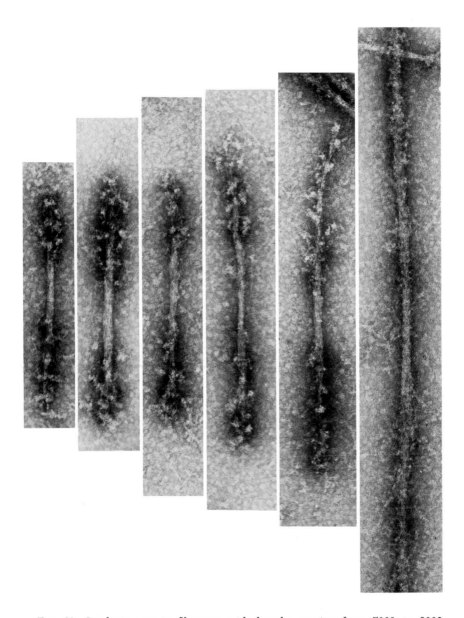

Fig. 29. Synthetic myosin filaments with lengths ranging from 5000 to 9000 Å together with one of length greater than 14,000 Å (1.4 μ). All of them show the same characteristic pattern of a bare central shaft and projections all the way along the rest of the length of the filament. The appearance of the projections is rather variable, perhaps due to clumping together in some cases. (Magnification ×112,000.)

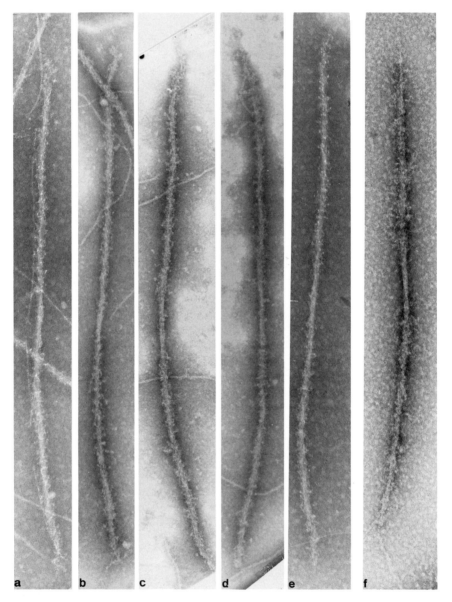

a   b   c   d   e   f

Fig. 30. (a)–(e) Thick filaments from muscle, showing projections all the way along their length except for a short central region about 0.2 μ long, which is comparatively bare but has a slight thickening in the middle. These specimens were fixed in formalin before negative staining, since it was found that the projections were better preserved under these conditions. (f) Synthetic thick filament made from purified myosin. Note close similarity to appearance of natural thick filaments, including the bare central shaft. (Magnification ×84,000.)

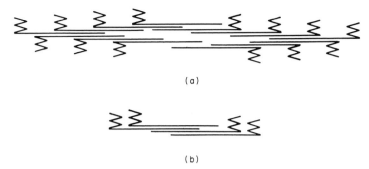

(a)

(b)

Fig. 31. (a) Possible arrangement of myosin molecules, with globular region at one end only, to produce short filaments of type observed, with globular region at either end and straight shaft in center. The polarity of the myosin molecules is simply reversed on either side of the center.   (b) Possible arrangement of same myosin molecules to produce longer filaments in which the straight shaft in the center is still present, but in which a longer region on either side now has globular projections on it. The polarity of the myosin molecules is reversed on either side of the center, but all the molecules on the same side have the same polarity.

myosin molecules have a globular (HMM) region at one end only, it is apparent that such aggregates must be formed by the antiparallel association of myosin molecules (H. E. Huxley, 1963) as illustrated in Fig. 31. The simplest way to account for the appearance observed is in terms of a model in which all the myosin molecules in one half filament are oriented in one direction and all those in the other half of the filament are oriented in the opposite direction; the length of the bridge-free central region is determined by the packing arrangements where antiparallel molecules overlap.

The significance of such an arrangement is that it provides a possible way in which the direction of the sliding force can be specified appropriately. An efficient sliding filament model requires that, in one half A band, all the elements of force generated by the cross bridges add up in the same direction as each other; that direction must be such as to draw the actin filaments toward the center of the A band. In the other half A band, the direction of all the individual elements of force must be reversed. If the direction of the force is determined by the structure of the cross bridge, then reversing the orientation of the myosin molecules, which will reverse the orientation of the cross bridges, will reverse the direction of the force (Fig. 32). Thus, filaments having the characteristic reversal of polarity that seems to be present in both the natural and the synthetic thick filaments, could, on this basis, draw two actin filaments toward each other and hence provide the basis for

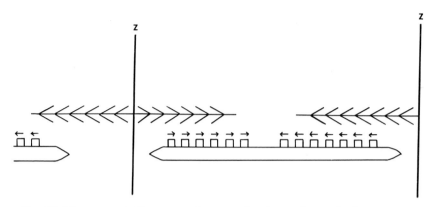

Fig. 32. Diagrammatic illustration of structural polarity of cross bridges on myosin filaments and of molecules of actin in the thin filaments, showing how their interaction could produce sliding forces which would move the actin filaments toward each other in the center of the sarcomere.

a shortening mechanism. It is most interesting that myosin molecules have a built in ability to assemble themselves even *in vitro* into such filaments.

B. The Actin-Containing Filaments

1. ARRANGEMENT OF THE ACTIN MOLECULES

As was mentioned earlier (p. 323), the actin molecules form a two-chain helical structure with a subunit repeat of 54.6 Å along either chain and a helical repeat of 360–370 Å (Fig. 33). Thus, actin forms a nonintegral helix. The I filaments in many muscles are about 1 μ in length (from Z line to H zone), though a small characteristic variation is found between different species (Page, 1966). Each I filament would contain about 360–370 actin monomers. If we take the molecular weight of actin to be 47,600 (Johnson *et al.*, 1967), then we can calculate the actin content of a muscle in much the same way as we calculated the number of myosin molecules per thick filament. A figure of about 26 mg actin per gram of muscle is obtained in this way. This may be compared with values determined by other techniques. Hasselbach and Schneider (1951) found that about 14% of total muscle protein was obtained by their actin extraction procedure; this would correspond to about 28 mg of actin per gram of wet muscle. However, this degree of agreement is almost certainly illusory, since it was not realized at the time of these measurements that actin forms a tight complex with the regulatory protein system in muscle (tropomyosin and troponin) and that some of these proteins may be still present in the extract. Perry and Corsi (1958) made separate estimates of the amounts of actin

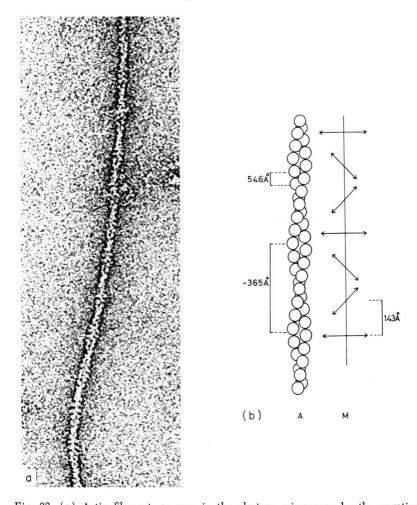

Fig. 33. (a) Actin filament, as seen in the electron microscope by the negative staining technique. The structure consists of twin strings of globular units (G actin molecules) wound round each other in a double helix with a patch of 355–370 Å. The subunits repeat along either strand at a spacing of 54.6 Å. (b) Diagram showing actin double helix, alongside myosin filament on which the position and rotational orientation of the cross bridges is indicated. Noted that the pitches and subunit repeats of the actin and myosin helices are different.

(20–25%) and tropomyosin (11%) as a proportion of total myofibrillar protein, using electrophoretic methods for separation and identification, but as Corsi *et al.* (1966) point out, considerable uncertainties still remain in all these estimates (but see p. 325.)

Interference microscope measurements (p. 326), indicate that the I filaments represent about 25% of the total dry mass of the muscle (or of the order of 50 mg per gram wet muscle), and it is therefore clear that there must be other proteins as well as actin present in them (H. E. Huxley and Hanson, 1957; Hanson and Huxley, 1957; Perry and Corsi, 1958). There is very good evidence that tropomyosin is one of these. Antibodies against tropomyosin, made visible by coupling a fluorescent dye to them, can be seen in the light microscope to bind specifically to the region of the sarcomere occupied by the thin filament (Pepe, 1966), and an increase in the density of the I filaments when they are treated with antibody against tropomyosin can be seen in the electron microscope. Interestingly, such antibody staining does not show up any periodicity in the thin filaments; this indicates that the tropomyosin forms a rather continuous structure along the thin filaments, and that the antitropomyosin binds all the way along it.

Tropomyosin is a fibrous molecule of molecular weight 70,000 and length about 450 Å, containing a very high proportion of α-helix; it consists of two apparently identical polypeptide chains pointing in the same direction (Caspar *et al.*, 1969). In the presence of magnesium ions, tropomyosin forms paracrystals with a very strong 395 Å axial period (Cohen and Longley, 1966; Caspar *et al.*, 1969). This periodicity is somewhat less than the probable length of the molecules, which we might therefore expect to overlap each other longitudinally in the paracrystals.

X-ray diffraction observations on muscle (H. E. Huxley and Brown, 1967) show that there is a meridional reflection with a spacing of 385 Å, which corresponds closely to a periodicity visible in the I filaments which had previously, for various reasons, been thought to be in the region of 400–410 Å. It seems very likely that this represents the tropomyosin period in muscle, though whether the value of 395 Å measured on paracrystals represents a significant difference is not yet clear. If the tropomyosin molecule has the same steric relationship to each of the actin monomers, then it must follow a helical path in the actin filament, perhaps in the two grooves between the subunits as suggested by Hanson and Lowy (1963). Moreover, there must in such a structure be an integral relationship between the actin subunit period and the tropomyosin period, if the latter is attached by specific bonding. If the subunit spacing of the actin monomers is 54.6 Å, then seven of them

(382.2 Å) would correspond quite closely to one tropomyosin period (385 Å ± 2 Å).

Endo and his colleagues (1966) have shown that antibody against troponin (see Volume III, Chapter 7) binds to I filaments with an approximate 400 Å periodicity. It seems very likely that this again corresponds to the periodic repeat of a specific binding site for troponin on the tropomyosin molecules, and a model for the structure suggested by Ebashi is shown in Fig. 34. Hanson (1967, 1968, 1970) has also reported evidence, from the studies which she initiated an actin paracrystals, that troponin is attached with this periodicity.

Recent analysis of the structure of actin filaments by three-dimensional reconstruction of electron micrographs (Moore *et al.*, 1970) gives some information both about the shape of the actin monomers and also about the possible location of the tropomyosin–troponin component (see Figs. 35 and 36). The actin monomers appear to be fairly globular in shape, having dimensions of about 55 Å × 35 Å × 50 Å. In filaments of purified actin, two chains of these monomers are seen on their own, but when thin filaments prepared directly from muscle are used in which the relaxing proteins are probably still present, then additional structures are visible, lying in the grooves of the double helical structure. OBrien *et al* (1971) and Spudich and Huxley (1972) have also found evidence for such a location from electron microscopic studies on actin paracrystals and the subject is in a state of active development.

A noticeable feature of the X-ray diagrams in which the actin reflections are visible is that although sharp reflections are seen close to the meridian out to spacings of about 5 Å or less, there are no reflections visible further away from the meridian than about 30 Å. This shows that although there is a high degree of longitudinal ordering in the actin filaments, the helical ordering, i.e., the ordering dependent on radial or circumferential position, is much less well developed. Expressed

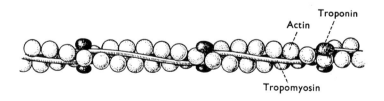

Fig. 34. A model for the fine structure of the thin filament. In this model, it is assumed that two molecules of tropomyosin and troponin exist in each period (cf. Ebashi *et al.*, 1968). The pitch of the double helix in the thin filament formed by the actin molecules is considered to be 360–370 Å, which is slightly shorter than the period due to troponin. (After Ebashi *et al.*, 1969.)

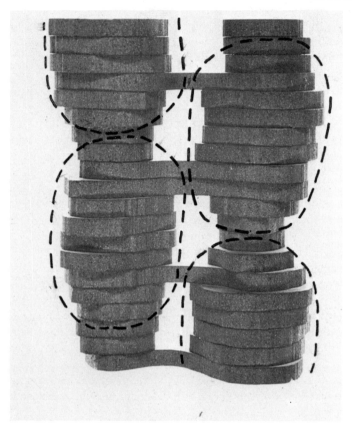

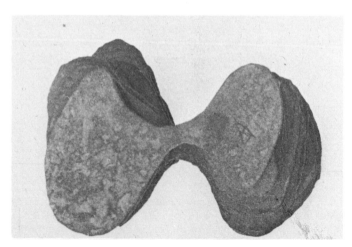

Fig. 35. See facing page for legend.

in simpler terms, one could say that the actin behaved like a twin string of beads which could be twisted easily but not stretched.

2. THE POLARITY OF THE ACTIN FILAMENTS

Although the actin monomers are often portrayed as featureless globular units, they are of course intricately constructed protein molecules with very specific myosin-binding and ATPase-activating sites on their surfaces, to say nothing of the regulatory mechanisms also present. The myosin binding sites on actin presumably have to enter into a very stereospecific combination with the S1 subunits of myosin, whose orientation is, as we have seen, appropriately arranged. Thus, one would anticipate that only actin monomers in the appropriate orientation should be able to interact with the cross bridges along one half of a thick filament, and that the orientation would be reversed at the Z lines to match the reversed orientation of the cross bridges in what corresponds to the opposite half of the thick filament. As Fig. 32 shows, this leads to a reversal of polarity of the actin filaments on either size of the Z lines. Of course, if half the actin monomers in each chain were reversed, then perhaps these would merely be nonfunctional, but the system would be less efficient.

In fact, there is very good evidence that the actin filaments are indeed polar structures, i.e., that all the monomers in both chains are oriented in the same direction, and that this polarity reverses at the Z lines. Actin filaments alone show insufficient detail in the electron microscope for one to discern their polarity reliably, but if they are allowed to combine with myosin, or better still (since the pictures are clearer), with HMM or with S1, then a very clear structural polarity becomes evident; the myosin subunits attach on the outside of the actin double helix, giving a composite helix which displays, in the electron microscope, a very characteristic arrowhead appearance (Fig. 37) (H. E.

Fig. 35. Model of a portion of an F actin filament, built from data obtained by three-dimensional reconstruction. The probable outlines of the G actin subunits are indicated by dotted lines on the longitudinal view. The lower photograph shows an end-on view of the model. The doubling of the thickness of alternate oblique contacts between actin monomers is an artifact of the construction of the model. In reality, symmetry considerations require that all actin monomers form identical relationships with their neighbors. The oblique contacts are 27.5 Å apart, while the sections used to build the model are 5 Å thick. Thus, the peak of the density at the oblique contact points between the two actin monomers falls alternately within one layer and at the junction of two layers of the model. In the latter case, the density within two layers is raised above the chosen cut-off point.

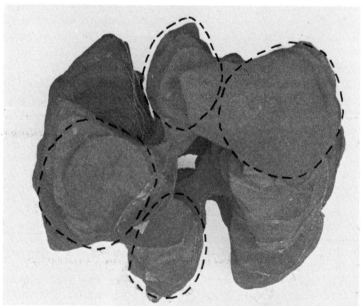

50 Å

Fig. 36. See facing page for legend.

Huxley, 1963). The arrows always point in a constant direction along any given actin filament, and it can be seen that the same subunit and helical repeat are maintained in the composite filament as are present in the underlying actin backbone. It is apparent that the arrowhead appearance arises in some way from the characteristic angle of attachment of the myosin units, and that this must be the same for each actin monomer (otherwise the same periodicities would not be maintained). Accordingly, all the actin monomers must be positioned in the filament with the same structural polarity.

Moreover, it is found that when a group of thin filaments still attached to a fragment of Z line are "decorated" in this manner, all of them can be seen to point in the same direction as their neighbors on the same side of the Z line, with the arrowheads pointing away from the Z-lines (Fig. 38, 39). On the other side of any given Z line, this polarity is reversed. Thus the structure has those features we should expect if the relative sliding force between the filaments is produced by a stereospecific interaction of myosin cross bridges on the thick filaments and actin monomers in the thin filaments.

The preceding argument does not depend on an exact knowledge of the mode of attachment of the myosin subunits to actin; and in fact it was not possible to ascertain this from simple visual inspection of electron micrographs. However, these structural details of the complex have now become available through three-dimensional reconstruction of actin filaments decorated with S1 (Moore *et al.*, 1970). These show the S1 subunits as somewhat elongated and slightly curved structures, about 150 Å × 45 Å × 30 Å, which are attached to the actin monomers, not only tilted relative to the axis of the actin filament by about 45°, but also slewed (i.e., rotated about an axis perpendicular to the filament axis and passing through the S1 unit) by about the same angle. It is the combination of tilting and slewing which produces the arrowhead appearance, and the original paper should be consulted for details.

Tilted cross bridges were also detected in intact insect flight muscle by Reedy *et al.* (1965), both by electron microscopy and X-ray diffraction; the direction of tilt seen in the micrographs of the intact muscle was the same as that expected from the direction of the arrowheads seen on decorated I filaments still attached to Z lines.

Fig. 36. Model of a portion of a thin filament built from data obtained by three-dimensional reconstruction. The probable position of the G actin subunits is shown by the dotted lines enclosing the larger subunits. Additional material present in the model is indicated by the other enclosed areas and is discussed in the text. The upper photograph shows a longitudinal view of the model, the lower one, an end-on view.

Fig. 37. A general field of view of thin filaments decorated with S1 subunits and negatively stained. The individual subunits and their slightly curved appearance can be clearly seen in several places. The variation in the diameter of the complex is caused by anisotropic shrinkage of the layer of stain (over a hole in the carbon film) in which the filaments are embedded. (Magnification ×148,000.)

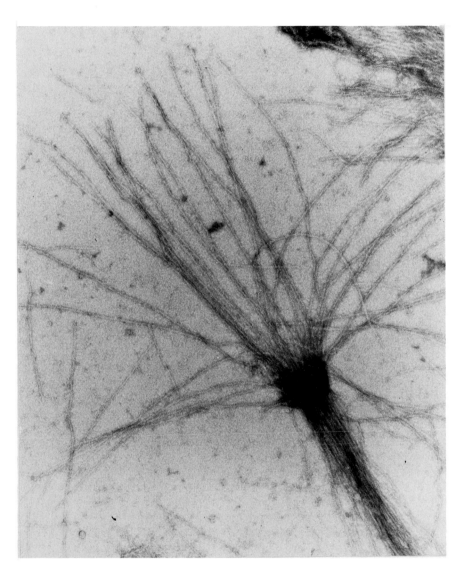

Fig. 38. Isolated fragment of Z line with attached thin filaments (I segment) prepared by mechanical disruption of glycerinated rabbit posoas muscle and viewed in electron microscope using negative staining technique. The characteristic beaded appearance of the actin structure can be seen in the filaments, but insufficient detail is visible to establish their structural polarity. This picture is a control for Fig. 39. (Magnification ×90,000.)

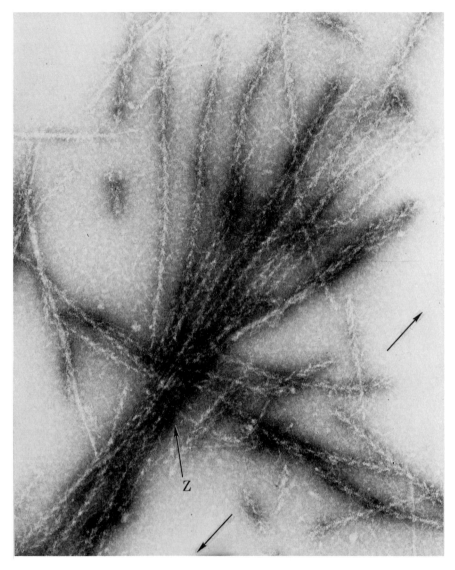

Fig. 39. I segment treated with solution of heavy meromyosin. The thin filaments now show the characteristic polarized structure, and all filaments on the same side of a Z line point in the same direction. Those on the other side of the Z line point in the opposite direction. (Magnification ×96,000.)

The actin filaments are not continuous through the Z line, but are joined there to some other structure which, besides providing mechanical stability to hold them in register, ensures that the required reversal of polarity is observed. Cross sections of the Z line (Knappeis and Carlsen, 1962; Reedy, 1964; Franzini-Armstrong and Porter, 1964) show a characteristic square or tetragonal lattice that is remarkably similar in dimensions to that seen in two-dimensional crystals of tropomyosin (Fig. 40) (H. E. Huxley, 1963). It seems conceivable that the tropomyo-

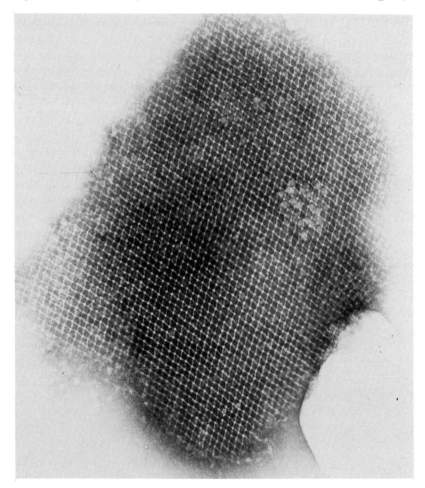

Fig. 40. Lattice structure obtained by homogenization of tropomyosin B crystals, negatively stained with uranyl acetate. Lattice spacing ~200 Å. The woven appearance arises because alternate lattice points along either of the two principal lattice directions are displaced alternately slightly to either side of the mean position. Possibly the lattice points do not all originally lie in one plane, but in two, and the displacement seen is a consequence of the drying down of this structure. The lines joining the lattice points frequently appear double, consisting of two fine filaments measuring about 10–15 Å in diameter and lying about 25 Å apart. (Magnification ×93,000.)

sin in the actin filaments might link with other tropomyosin molecules in a different polymorphic form in the Z lines. However, this hypothesis still remains to be substantiated. In any case, the density of the Z lines and the presence of a considerable amount of additional material, visible in the electron microscope, between the connecting filaments which form the lattice show that there are other components present too. Some evidence has been presented (Briskey *et al.*, 1967; Masaki *et al.*, 1967; Stromer *et al.*, 1969) that the protein α-actinin may be involved and that this or another protein fraction (Masaki *et al.*, 1968; Stromer *et al.*, 1969) may participate in the M line structure which forms a system of connecting cross bridges between the thick filaments in the center of the H zone.

VI. Molecular Changes during Contraction and in Rigor

A. Introduction

While electron microscopy of chemically fixed tissue has been found to give reliable information about the general structural organization of resting muscle, much more difficult problems are encountered when one turns to the molecular changes that may occur during contraction. First, many of the changes are likely to be cyclical ones, taking place on a time scale of the order of milliseconds; chemical fixation appears to be a somewhat slower process. Thus, it seems unlikely that this technique would be able to fix the cross bridges at a clearly defined point of their cycle of action, and indeed, it is quite likely that fixation and the subsequent steps involved in processing tissue for electron microscopy will introduce artifactual changes in the detailed structure of the more labile components of the system. It was of course very gratifying that similar fears about the preservation of fine structure immediately below the level of resolution of the light microscope proved to be unfounded. However, it seems that this was the case because the fixatives were able to cross link the native structure together without altering it very much and because the larger structural entities could maintain their relative positions in the tissue even when water was removed. It seems very unlikely that this will still be the case when we are concerned with the fine details of the structure of the proteins involved, and indeed all the evidence so far indicates that a considerable degradation of the structures does take place.

It is clear, therefore, that we have to have alternate ways of obtaining structural information about contracting muscle, and the technique of

X-ray diffraction seems a very natural choice for such studies. As we have already seen, the X-ray diagrams give information about the regularly arranged protein subunits in both the actin- and myosin-containing filaments, and also about the side-to-side position of the filaments. Moreover, this information can be obtained from intact living muscle, and the passage of the X-ray beam does not affect the tissue in any measurable way. However, there is a very serious difficulty with the technique. It is that the diagrams take a long time to record. Until relatively recently, it took many hours to record even rather low-resolution X-ray pictures from muscle, and even now such diagrams need 10–20 minutes or longer. The former durations are quite long in comparison with the total time for which an isolated muscle can easily be made to contract even when well supplied with oxygenated Ringer solution and glucose; the latter durations are more feasible, if somewhat demanding. The longest total contraction times are obtained when such a muscle is briefly stimulated, for a few seconds or less, and then allowed to recover for a minute or two before the next stimulus; the X-ray diagram is recorded only during the time the muscle is fully active, by using a suitable shutter mechanism. Diagrams showing some of the main features of the contracting pattern have now been obtained, using new types of X-ray tubes and low-angle cameras, and the original papers should be consulted for technical details (Elliott *et al.*, 1967; H. E. Huxley and Brown, 1967).

B. X-Ray Diffraction by Contracting Muscles

It was found by the authors mentioned in the previous paragraph that fully active muscles, which were either contracting isometrically or had been allowed to shorten, still gave clear low-angle X-ray diagrams in both axial and equatorial directions. This result showed that a high proportion of the actin and myosin structure remained organized in regularly arranged filaments and that these structures were therefore still present in active muscle and could be involved in force generation. Such an assumption is of course fundamental to the sliding filament mechanism, and it is important that any such assumptions should be experimentally verified, as this one has been, beyond all possible doubt.

Second, it was found that the meridional spacing of the reflections from both the actin and myosin filaments remained virtually unchanged in the contracting muscle, even when shortening had occurred. In fact, the myosin reflections increased in spacing by about 1%. This showed that the repeating periodicities along the filaments remained essentially

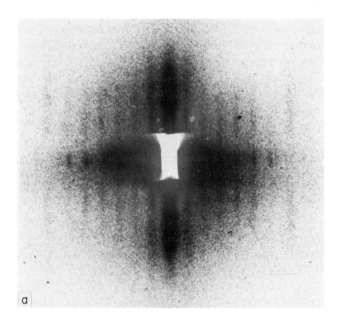

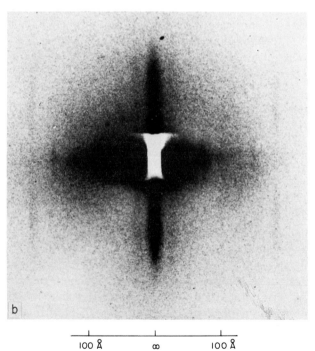

Fig. 41. See facing page for legend.

constant and hence that the filaments did not undergo any uniform shortening along their length. It could not be excluded that some local shortening might occur, nor that some reduction in filament length might be accomplished by limited depolymerization, but clearly, the great bulk of the backbone of the filaments retained their original axial cohesion.

Third, in the case of the actin pattern, all the original reflections are clearly seen in the patterns from contracting muscles—if anything, somewhat stronger than in the resting muscle (perhaps because of an improvement in orientation produced by the tension in the muscle). Thus, it is clear that the great bulk of the whole actin filament structure remains unchanged during contraction (though, as we will see later, there is now some indications of a small charge possibly associated with the troponin–tropomyosin system).

Finally, however, there are very striking changes in the layer line pattern arising from the cross bridges on the myosin filaments. While this pattern remains visible, and no new system of reflections appears, the intensity of all the reflections, especially the off-meridional components, decreases very dramatically in the contracting muscles (Fig. 41). The intensity of the first layer line at 429 Å was found by H. E. Huxley and Brown (1967) to decrease to about one-third of its value in resting muscle, and even that small residual intensity could have been due in part to fatigued fibers (which give a normal resting pattern). On the other hand, the intensity of the 143 Å meridional reflection only decreased to about two-thirds of its original value.

These results showed that in contracting muscle, the cross bridges move from the regular helical pattern found in resting muscle to a much more disordered arrangement. The disorder affects the helical regularity of the bridge pattern more than it does the axial regularity. Thus, there must be some axial movement of the cross bridges and a very considerable amount of radial and/or circumferential movement (H. E. Huxley and Brown, 1967). At the same time, the backbone structure of the thick filaments is relatively unaffected.

This result is perfectly consistent with a mechanism in which the sliding force is generated by movement by the cross bridges, which it would seem are able to search the region around the thick filaments and find suitably disposed actin sites to which to attach and then go

Fig. 41. X-ray diagrams from rest length frog sartorius muscle; fiber axis horizontal. (a) Control picture of muscle in resting state. (b) Picture given by contracting muscle under same conditions and with same total exposure time, but recorded only during fully active state. The system of myosin layer line based on 429 Å disappears, but the 143 and 71.5 Å meridional reflections remain. The 59 Å actin reflection also remains unchanged.

through their force generating cycle of movement. The movement of individual cross bridges would be unsynchronized, as would be expected if the system is to develop a steady average force. While the result does not prove that force is developed in this way, it does show that the cross bridges are not fixed, rigid structures, as they would be in certain other types of contraction mechanism.

These X-ray measurements, then, either confirm various aspects of the sliding filament mechanism or are consistent with it. At present, the practical limitations I have already mentioned make it difficult to search for weaker X-ray reflections in the axial part of the diagram which might give clues about the positions to which the cross bridges move. However, experimental evidence about cross bridge movement can also be obtained from the equatorial reflections, and we will now consider this approach.

C. Changes in Equatorial X-Ray Reflections Associated with Actin–Myosin Interaction

One of the first observations made about the equatorial X-ray reflections from striated muscle was that a very large difference in the relative intensities of the two principal reflections exists in the patterns from live, relaxed muscle as compared with those from muscles in rigor (H. E. Huxley, 1952, 1953a). The two reflections in question are the 10 and 11 reflections from the hexagonal lattice of filaments, which are associated with the two sets of lattice planes shown in Fig. 42. In the live, relaxed muscle, the 10 reflection is stronger than the 11 reflection, whereas in a muscle in rigor (but still held at rest length), the 11 reflection is stronger than the 10 reflection (Fig. 43). It can be seen from Fig. 42 that material at the position of the trigonal points of the lattice (i.e., at the position of the actin filaments) tends to fill up the spaces between 10 type lattice planes if these are drawn through the myosin filaments. On the other hand, material both at the trigonal points and at the hexagonal lattice points lies in the same planes in the 11 directions. Thus, the presence of more material at the position of the actin filaments will have the effect of decreasing the intensity of the 10 reflection and increasing the intensity of the 11 reflection. Accordingly, this must correspond to the change that takes place when a muscle goes into rigor. One can see this in another way, by constructing Fourier projections (i.e., electron density maps) of the end on view of the filament lattice (assuming hexagonal symmetry) for the two states of the muscle (Fig. 44).

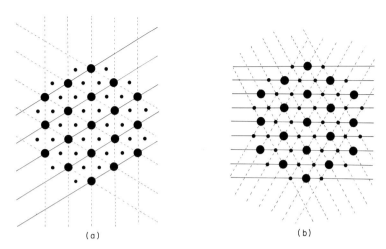

Fig. 42. (a) End-on view of double hexagonal lattice of thick and thin filaments, characteristic of overlap region in A bands of vertebrate striated muscle, showing the three sets of lattice planes in the (10) crystallographic directions. The actin filaments at the trigonal points of the lattice will tend to fill in the space between the dense planes of filaments at the hexagonal lattice points, thereby decreasing the intensity of the (10) X-ray reflections given by the thick filaments on their own. (b) Similar view, showing lattice planes in (11) crystallographic directions In this case, both the actin and myosin filaments lie in the same lattice planes and so their contributions to the intensity of the corresponding X-ray reflection are additive. Hence, as the amount of material at the trigonal points is increased, the intensity of the (10) reflections decreases and that of the (11) reflections increases.

Originally, it was thought that the appearance of extra material at the trigonal positions in the rigor muscle might arise from a more precise ordering of actin filaments when they became attached to the myosin filaments following the loss of all the ATP in the muscle (which was thought, correctly, to act as a dissociating agent in the relaxed muscle). However, it was shown by Elliott *et al.* (1963) that the intensity of the 11 reflections even in resting muscles was sufficiently high to require that all the actin filaments within the overlap zone be already quite precisely positioned at the trigonal points in between the myosin filaments. Thus, the large increase in the relative intensity of the 11 reflection in rigor could not be explained by any further increase in ordering; and so, doubts were expressed about the validity and significance of the original observations.

The situation was clarified when it was shown (H. E. Huxley, 1968) first of all, that a reversal of the relative intensity of the reflections

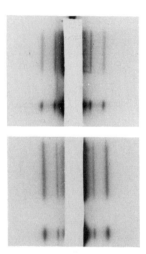

Fig. 43. Low-angle equatorial X-ray patterns from rabbit psoas muscle: (top) live; (bottom) in rigor. The patterns show the 10 and 11 reflections from the hexagonal lattice of myosin and actin filaments. The reversal of the relative intensities of the reflections is believed to be caused by the cross bridges reaching farther out from the myosin filaments in rigor and attaching to the actin filaments at the trigonal positions.

in question really did exist, as between resting and rigor muscles at the same sarcomere lengths, and second, that this reversal could be accounted for in terms of outward cross-bridge movement, in the following way. If, in a resting muscle, the S1 subunits—attached to the LMM backbone of the thick filaments via the S2 linkage (Fig. 45)—are applied rather closely to the surfce of the thick filaments, and if, in rigor, when all the cross bridges are believed to attach to the actin filaments, the S1 subunits hinge outward on the S2 links so that they lie in contact with the actin filaments (Fig. 46), then a very substantial transfer of material will take place, adding to the density around the trigonal points and subtracting from the density around the hexagonal lattice points. It can be shown (see original paper for details) that the observed change in the X-ray reflections can be accounted for very well in this way, on the basis of the known mass of the S1 subunits and of the actin and myosin filaments. It had been noted previously (p. 317) that certain features of the layer line pattern from resting muscle suggested that the cross bridges in resting, relaxed state only extended out to a radius of about 130 Å, some 70 Å short of the surface of the actin filaments, so that observation also fell into place. Somewhat similar findings, on insect flight muscle, have been reported by Miller and Tregear (1970).

However, the real significance of such a structural mechanism is that it offers an explanation of the fact that muscles are able to develop tension over a considerable range of filament side spacings, an observa-

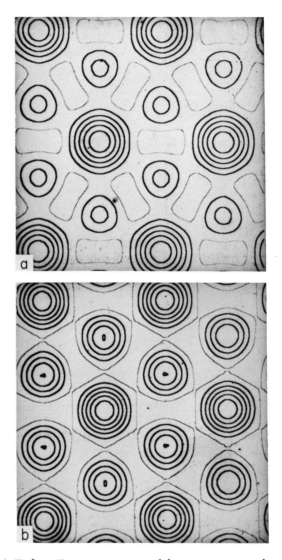

Fig. 44. (a) End-on Fourier projection of live sartorius muscle of frog (using 10 and 11 reflections) showing large peaks of electron density at the lattice points and much smaller ones at the trigonal positions. (b) Corresponding Fourier projection for frog sartorius muscle in rigor, showing much larger peaks of electron density in the trigonal positions.

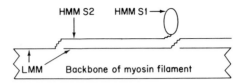

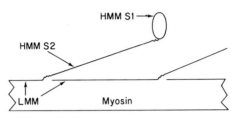

Fig. 45. Suggested behavior of myosin molecules in the thick filaments. The light meromyosin (LMM) part of the molecule is bonded into the backbone of the filament, while the linear portion of the heavy meromyosin (HMM) component can tilt further out from the filament (by bending at the HMM–LMM junction), allowing the globular part of HMM (that is, the S1 fragment) to attach to actin over a range of different side spacings, while maintaining the same orientation.

tion that might otherwise be difficult to explain on the basis of cross bridge action. As can be seen from Fig. 47, the S2 hinge would allow the S1 subunit to attach to actin in the same orientation and go through the same sequence of changes in the effective angle of attachment, over a wide range of myosin–actin distances. Some elements of this mechanism had been suggested by Pepe (1966), on the basis of very interesting antibody staining experiments, which contain, however, what are to the present writer some puzzling features.

Cross bridge behavior of this kind was also predicted from the behavior of the low-angle axial reflections during both rigor and contraction (H. E. Huxley and Brown, 1967), especially from the observation that the 429 Å layer line system could either almost completely disappear, or else (in rigor), change into a new helical pattern, based on the actin helical repeat, and yet leave the 143 Å meridional reflection virtually unaffected. Such behavior was readily explicable if the head of the myosin molecules could undergo substantial radial and circumferential movement with very little change in its axial position, as would be possible if it were attached to a hinge (the S2 subunit) several hundred angstrom units in length. The discovery by Lowey and her coworkers (Lowey *et al.*, 1966, 1967) that this S2 subunit was soluble at all ionic strengths and did not aggregate with LMM, so that it would not be tightly bonded into the backbone of the thick filaments, provided the necessary physicochemical correlation with the structural evidence, which the authors in question soon realized.

The next step in the X-ray experiments was to see whether changes in the relative intensities of the equatorials also took place in actively

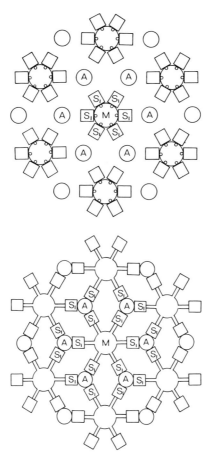

Fig. 46. Cross section through double array of actin and myosin filaments, illustrating possible rearrangement of the cross bridge material, as between relaxed muscle (upper picture) where the S1 units lie close to the backbone of the thick filaments, and rigor (or contracting) muscle, where all (or a portion of) the bridges hinge out sideways bringing the S1 units into direct contact with the actin filaments. The resultant distribution of density in the lattice, so that a greater proportion of it lies near the trigonal positions of the unit cell (i.e., near the actin filaments) would account for the changes observed in the equatorial X-ray diagram.

contracting muscles, as well as in the rigor muscles described above. It is indeed found (Haselgrove, 1970; Haselgrove and Huxley, 1972) that such changes do occur when resting and contracting muscles at the same sarcomere length are compared, but that they are, as expected,

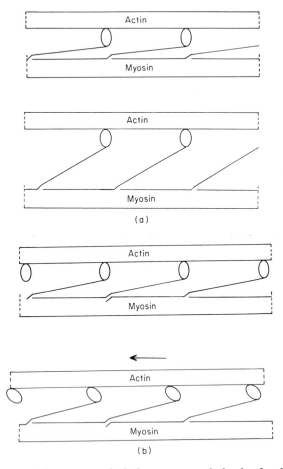

Fig. 47. (a) Possible way in which hinging out of the head subunits of the myosin molecules or the thick filaments could allow attachment of the cross bridges to the actin filaments to take place over a wide range of interfilament spacings, yet always at exactly the same angle, utilizing assumedly flexible linkages at either end of the linear (S2) portion of the molecule. (b) Possible way in which a sliding force between actin and myosin filaments could be developed by a change in the effective angle of attachment of the head subunits to the actin filament. The model assumes flexible linkages between the head subunits and the rest of the myosin molecule, and the existence of a force balance which maintains the filaments at a fixed distance apart for small charges in the extent of overlap.

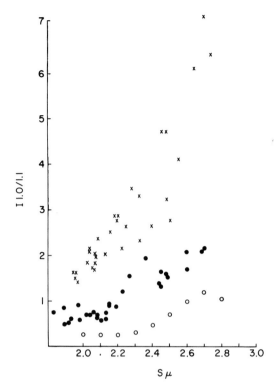

Fig. 48. Graph showing observed ratios between intensities of the (10) and (11) equatorial X-ray reflections as a function of the sarcomere length of the muscle. Key: ✕ = resting muscle, ● = contracting muscle, ○ = muscle in rigor. The ratio decreases—i.e., the relative intensity of the (11) reflection becomes greater—as the sarcomere length decreases (greater overlap, hence greater length of actin filament ordered at trigonal points in A band lattice). The changes in the ratio for different physiological states of the muscle are believed to be associated with lateral cross bridge movements (see text). From Haselgrove (1970).

somewhat smaller in extent than in the rigor muscles (see Fig. 48). The changes could be accounted for if approximately one-half of the cross bridges were leaning out from the thick filaments so that their 'heads' were close to the actin filaments at any one time. In terms of a mechanism in which each cross bridge goes through a cycle of attachment and detachment, this could represent the proportion of cross bridges in the two states, but it does not necessarily follow that all the myosin heads near to the actin filaments have to be attached to them. It is conceivable that the activation mechanism acts in at least two stages, the first stage allowing the cross bridge to move out to the operating position, the next stage, possibly the rate-limiting step, being the actual attachment to the actin filament. This topic represents an area of very active work at present, and much remains to be clarified.

The present general conclusion one can draw from the X-ray results is that the cross bridges appear to have all the necessary properties to act as structural elements that generate a sliding force between filaments whose lateral separation might vary; but that the precise details of their operation is still unclear and may emerge from further experiments employing more intense X-ray sources.

VII. Some Biochemical and Physiological Implications of the Structural Results

The model for muscular contraction put forward in this chapter is a very straightforward one, and its general features are fairly generally accepted. Thus, it can serve as a useful structural background for many of the topics discussed elsewhere in these volumes. It clearly fits in with many different observations; to take a simple example, the inextensibility of muscles in rigor is readily accounted for by the fixed attachment of all the cross bridges to the actin filaments (thereby inhibiting sliding) that we would expect in the absence of ATP; but it would be outside the scope of the present chapter to deal with all of these exhaustively, and many of them will be discussed elsewhere in these volumes. However, there are one or two aspects of the mechanism which should be mentioned here, either because of their fundamental importance, or because they seem particularly relevant to problems that are still unresolved.

A. Activation of Myosin ATPase by Actin

It was observed a considerable time ago that the ATPase activity of purified myosin on its own at neutral pH and in the presence of magnesium ions was relatively low, certainly too low to account for the rate at which energy is released in a muscle, despite the fact that these are the prevailing ionic conditions in the tissue. However, in the presence of actin, under conditions of ionic strength that favor the association of actin and myosin—i.e., in the physiological range—the ATPase activity of the myosin is greatly increased.

In terms of the cross bridge model, this was interpreted (H. E. Huxley,

1960) in the following terms; that when the myosin cross bridge is not attached to the actin filaments, the enzyme site on myosin splits ATP only very slowly, but that when attachment takes place, the splitting is greatly accelerated. It was pointed out that this acceleration could be exerted on any of the steps of the process, i.e., either on the binding step, on the cleavage of the phosphate bond, or on the release of the products ADP and phosphate. Recently, Lymn and Taylor (1970) and Taylor, *et al.* (1970) have described evidence that it is the last step, the release of the products, that seems to be the rate-limiting one. At all events, the significance of the mechanism seems to be that only those cross bridges attached to the thin filaments are able to complete the cycle of splitting of ATP. This clearly provides a mechanism by which the utilization of energy by the muscle can be regulated. If mechanical attachment of the cross bridge to the actin filament is required for the generation of force, then only the cross bridges that can exert a useful mechanical action will be able to split ATP. In other words, ATP will not be split wastefully by unattached bridges. There is good physiological evidence that muscles do indeed use less total energy when shortening against smaller loads (Fenn 1923a,b.)

In an active muscle, the tension varies with the velocity of shortening, being maximal when no shortening takes place (in an isometric contraction) and progressively decreasing as the shortening velocity increases. This can be interpreted in terms of a variation in the proportion of cross links that are attached at any one time at different velocities of sliding. This variation could arise if the rate-limiting step is the attachment of the cross bridges. On such a model, the number of attached, tension-generating cross bridges would always be such as to match the load, and the velocity of shortening for a particular load would settle down at a value which, in the steady state, maintains that number constant and depends on the attachment rate for that particular muscle. Sudden changes in the load, or in the allowed velocity of shortening, would generate transients, and so give useful information about the detailed behavior of the cross bridges. Such an approach has been described by Podolsky and co-workers (1969; Podolsky, 1960; Civan and Podolsky, 1966) and by Armstrong *et al.* (1966).

B. Activation of Contraction by Calcium Ions

The work of Ebashi and his colleagues and of others have given an apparently clear picture of how calcium ions interact with the actin–myosin system in vertebrate striated muscle so as to switch on the splitting of ATP (see Volume III, Chapter 7). It seems that a regulatory protein complex is present in the actin filaments and consists of two components, tropomyosin and troponin, the latter itself also containing more than one component. In the absence of calcium, this regulatory complex prevents actin from activating the myosin ATPase; this inhibition is removed when calcium is supplied, by release from the reticulum. In the absence of the troponin–tropomyosin complex, actin activates myosin ATPase whether calcium is present or not.

If physical combination of actin with myosin is required for activation, then the regulatory complex could work simply by blocking that combination. It seems to be the combination that is blocked, rather than the splitting process itself, since, in a relaxed muscle, not only is the ATP splitting kept at a very low value, but the muscle is readily extensible, and most of the bridges are evidently not attached to the actin filaments. On the other hand, actomyosin and myofibrils seem to stay in the precipitated, contracted state in the absence of both ATP and calcium, even though the regulatory proteins are still present, and HMM and subfragment 1 will both bind to actin filaments containing the regulatory proteins in the absence of ATP and calcium. Thus, it seems that the presence of ATP is required for the inhibition of actin binding to occur. The simplest way for this to happen would be if only a myosin cross bridge carrying ATP (or its split products) was inhibited from combination with actin.

The structural results available so far leave a considerable enigma hanging over the actual mechanism by which this inhibitory effect is exercised. Ebashi and his co-workers (1968) have calculated from their results that one molecule or complex of troponin (molecular weight 70,000) can confer full calcium sensitivity on an amount of the actin–tropomyosin–myosin system containing 6–7 monomers of actin. Electron microscopic studies show that the presence of this amount of troponin gives actin filaments a well marked 400 Å periodicity (Spudich and Huxley, 1972) and indicate, therefore, that the troponin, whatever its molecular weight or subunit composition, cannot be distributed so that it has an identical structural relationship with each actin monomer. It seems more probable that the inhibiting influence is exercised through the tropomyosin component, which at least seems to be evenly distributed along the actin filaments.* It is strange that the main control is not exercised (as far as can be deduced from the biochemical experiments on vertebrate striated muscle) directly on the myosin moiety. In this connection, it should be noted that control via myosin does seem

* For more recent work on this topic see papers by J. C. Haselgrove, by H. E. Huxley, and by A. Weber, in Cold Spring Harbor Symposium on Quantitative Biology Vol. XXXVII, to be published in 1973.

to be exercised in certain molluscan muscles (Kendrick-Jones *et al.*, 1970).

Although the biochemical results on vertebrate striated muscle seem clear cut, some of the results from X-ray diffraction experiments relevant to this issue are by no means equally straightforward. A detailed discussion here would be outside the scope of this article, but in summary, several different types of experiment, especially certain ones by Haselgrove (1970), indicate that movement of the cross bridges, as evidenced by the change in the myosin layer line pattern or a change in the relative intensity of the equatorial reflections, does not seem to depend on the presence of actin alongside them. This suggests that the thick filaments alone may be able to recognize ionic changes in the muscle—say the presence of calcium—and may perhaps function as a kind of safety catch, holding the cross bridges away from the actin filaments until the signal for contraction arrives, so that a double inhibiting mechanism is present. Such an idea is highly speculative, but one should at least note that there exist data that do not readily fit into the simpler forms of the model for activation.

C. Quantitative Considerations of Energy Production

The structural findings described in this chapter make it possible to give approximate estimates of the number of individual reacting sites and of the way that the elements of tension developed and of energy released would add up. Consequently, if one knows the corresponding quantities for the muscle, as a whole, one can calculate something about the contribution made by each cross bridge.

Thus, if we take a figure of 3 kg/cm² on the maximum tension per unit area developed by muscle (e.g., frog sartorius muscle, to which the subsequent figures will apply), then, assuming a thick filament separation of 400 Å, an A filament length of 1.55 μ (with a 0.15 μ bare zone), a helical repeat of 430 Å in the thick filaments with 6 bridges per repeat, the number of cross bridges whose tensions could add up in parallel would be maximally about 6×10^{12} per square centimeter. This would be the number if the entire cross section were occupied by myofibrils with no intervening spaces. Consequently, the force developed at each bridge would be at least 5×10^{-7} dynes. As a comparison, we may note that this is approximately equal to the force exerted by gravity on a water droplet about 5 μ in radius.

When a muscle is shortening, each cross bridge will develop tension as it moves through a certain range of positions. If we take the value already calculated as the average tension per bridge for extremely slow speeds of shortening, we can calculate a figure for the energy release per cross bridge for a given distance of sliding. For 100 Å of sliding, this would be 5×10^{-13} ergs. If we assume a value of 10,000 calories per mole for the free energy of hydrolysis of ATP (this being simply a nominal figure for the purposes of order of magnitude calculations), then the free energy per molecule of ATP is approximately 7×10^{-13} ergs; this would correspond to the energy release needed for a sliding distance of about 140 Å.

I should make clear the significance of this figure. It represents an estimate of the extent of movement of the filaments past each other operating against maximum load for the splitting of one molecule of ATP at each cross bridge. It does not imply that each bridge exerts its average force uniformly for the whole 140 Å distance. It is perfectly possible, for example, that each bridge operates over a change in position of, say, 54 Å, and then remains detached for the rest of the distance, in which case the average force at the bridge during this movement will be approximately 13×10^{-7} dynes and almost 40% of the cross bridges will be attached at any one time. It is essentially a figure showing us how often a cross bridge has to operate when it is working maximally.

In the case of a muscle shortening at 1.6 muscle length per second under half maximal load (say 1.5 kg/cm^2), the energy required per unit volume of muscle for the mechanical work done (ignoring heat production) is 2.4×10^6 ergs per second. Given that the same volume of muscle contains approximately 5×10^{16} cross bridges, then each cross bridge must liberate approximately 5×10^{-11} ergs per second, equivalent on the basis of the figures already used to the splitting of about 70 molecules of ATP per second at each cross bridge.

It is not a straightforward matter to compare such a figure with measured values of ATP splitting in vitro. The relevant quantity would be the actin-activated ATPase of myosin. However, when this quantity is actually measured, it is found to depend (for a given myosin concentration) on the actin concentration (Eisenberg and Moos, 1968; Szent-Kiralyi and Oplatka, 1969), presumably because the rate-limiting factor is the rate of collision of actin and myosin. In a muscle, the structure is no doubt arranged so as to facilitate collision in the correct orientation, and so it would be reasonable to use an ATPase value related to the maximum extrapolated value in vitro (i.e., extrapolated to infinite actin concentration).

From Eisenberg and Moos' values for rabbit HMM (4 μmoles/min per milligram of HMM) and a molecular weight of 300,000, a figure of 20 molecules per second can be calculated. No figures are available for frog myosin, but even allowing for the species difference, the figures indicate no calamitous discrepancy with the maximum values needed *in vivo;* the values obtained *in vitro* in solution may well be somewhat lower than those in the highly organized muscle structure, and future work may establish the rate constants of the various steps involved.

D. Factors Limiting the Maximum Tensions and Maximum Speed of Shortening of Muscles

The velocity of shortening of corresponding muscles from different animal species varies considerably, those from the smaller ones being generally more rapid (see, for example, Close, 1965). Thus, expressed in terms of the velocity of sliding of actin filaments past myosin filaments, a foot muscle from a mouse would give a figure of 30 μ/sec, whereas a cat soleus muscle would give 12 μ/sec.

The practical reason for this variation was pointed out by A. V. Hill some years ago (1950). Essentially, the argument is that the forces required to accelerate and decelerate the limbs of an animal when it is moving as fast as it can are likely in different animals to impose similar stresses (force per unit area) on the various structures involved, if such stress is the limiting factor in speed. This seems to be the case for sprintlike activity, anyway. It can then be shown by a simple dimensional argument that animals of similar construction will all tend to have similar top speeds, irrespective of very large variations in their size (see Table IV). However, to achieve these speeds, the smaller animals have to move their legs to and fro much faster. So their muscles need to have higher shortening velocities; or rather, can have higher velocities without causing damage.

It is apparent, therefore, that the sliding velocity of actin filaments past myosin filaments is a quantity that can be varied within wide limits. In the larger animals, it could in principle have been made much greater, if that had not imposed excessive stresses on the muscles, tendons, joints, and bones. The design problem could in theory have been solved in a different way if the sarcomere length of the muscles had been made greater than it actually is and the sliding velocity increased proportionately, and it is of some interest to explore what the consequences would have been. The velocity of shortening of the muscle (as a percentage of its resting length per second) would have remained constant, and the stresses imposed on the other parts of the structure would have been unchanged. However, because of the longer sarcomeres, the number of cross-bridges between a given thick and thin filament would be greater, and so the tension per unit area would be greater. Consequently, muscles of smaller cross sections could have been used. But

TABLE IV

MAXIMUM SPEEDS (YARDS/SECOND)[a,b]

Man	12 (at about 50 yards in sprint)
Greyhound (55–65 lb.)	18.3 (average over 525 yards)
Whippet (20–21 lb)	16.7 (average over 200 yards)
Horse	
(with rider)	20.7 (gallop: average over 660 yards)
(with vehicle)	15.3 (trotting: average over 1 mile)
Ostrich	25 ($\frac{1}{2}$-mile)
Hare (*L. alleni*)	20
Gazelle	30 (short dash)
Mongolian	20 (10 miles)
	$12\frac{1}{2}$ (1–2 days old)
African	> 25 ($\frac{1}{4}$ mile)
Roan antelope (*Ergoceros*)	> $17\frac{1}{2}$ (short)
American pronghorn antelope	24 (regularly)
	29 (good conditions)
White-tailed deer (*O. virginianus*)	15
American bison	16
Wild donkey (*E. hemionus*)	20 (short)
	15 (16 miles)
Giraffe	15
Black rhinoceros (*Diceros*)	10
Elephant	
Indian	$7\frac{1}{2}$ (short)
African	12 (short)
Coyote	17 (short)
Alaskan wolf	20
Mongolian wolf	$17\frac{1}{2}$
Gobi wolf	18 (short)
Red fox (*Vulpes* 8 lb.)	22
Gray fox (*Urocyon*)	20
Cheetah (hunting leopard)	32–34 (short)

[a] For miles/hour multiply by 2.05.
[b] From Hill (1950).

in practice, it is found that, among vertebrates at least, the A filament lengths are all almost identical, even in muscles with very different speeds of shortening.

This suggests that some other factor limits the tension that can be sustained by a given pair of thick and thin filaments and makes it impractical to take advantage of higher ATPase activities and higher sliding velocities to afford economies in construction of the kind discussed above. The most likely factor is the strength of the thin filaments themselves, or of the structure to which they are attached at the point where their polarity is reversed. It is interesting that in the case of muscles with longer A bands (e.g., arthropods) the number of thin filaments associated with each thick filament is also greater so that the tension present in each is reduced, e.g., in the cockroach (Hagopian, 1966). The actual total tension developed will depend, in addition, on the closeness of the filament packing, which in turn depends on how many cross bridges can be packed in per unit area. Muscles are structures with relatively large cross-sectional areas, which suggests that the size of the basic contractile unit cannot readily be reduced.

The other details of the band pattern—i.e., the I-filament length and the extent of overlap at rest length—are likely to represent a compromise between a number of conflicting requirements. If the I bands were made longer, then a muscle could shorten further before the Z lines came up against the ends of the A band. It could then be arranged to work at a better mechanical advantage (i.e., attached further away from the fulcrum of the movement), thereby requiring a smaller area of muscle for the same couple, providing that the velocity of shortening was increased. On the other hand, if the muscle at rest length was already working with more or less complete overlap between thick and thin filaments, then as soon as it is started to shorten, double overlap would occur in the center of the sarcomere, which might reduce tension, and efficiency too. If the muscle started off with a larger H zone gap, however, then more of the active sites would not be used for part of the time.

Since we have at present very little comparative data on the efficiency of muscles that contract with different velocities of shortening, it is difficult to assess the likely contribution of the different factors involved in this kind of compromise. The present discussion is a very speculative one and intended simply to raise some of the questions, rather than to define the ways in which they might be answered. And as to the question "Why striations?" the best that can be said at present is that it seems to offer the most efficient way of packing the maximum number of active sites into the smallest possible volume with an optimal extent of filament overlap.

VIII. Outstanding Problems and Future Work

What are the next steps that have to be taken to further explore the details of the contraction process? Many of them are obvious enough in principle, though technically difficult in practice. First, more details about the movement of the cross bridges must be discovered, by improved X-ray diffraction experiments and by other methods. Next, we

384   *H. E. Huxley*

have to crystallize the myosin S1 subunit so that a detailed crystallographic analysis can be performed on it. Third, electron microscope
techniques have to be improved so that we can see meaningful fine
detail below the present 20 Å limit. Fourth, much more intense X-ray
sources, perhaps electron storage rings, must be developed so that the
time resolution in the X-ray experiments can match that available in
mechanical, thermal, and chemical measurements. So there is plenty
of work to do for the future, and real hope of finding out more and
more of the molecular details of this remarkable phenomenon.

REFERENCES

Armstrong, C. F., Huxley, A. F., and Julian, F. J. (1966). *J. Physiol. (London)*
    186, 26P.
Astbury, W. T., and Dickinson, S. (1935a). *Nature (London)* **135**, 95.
Astbury, W. T., and Dickinson, S. (1935b). *Nature (London)* **135**, 765.
Astbury, W. T., and Dickinson, S. (1940). *Proc. Roy. Soc., Ser. B* **129**, 307.
Boehm, G., and Weber, H. H. (1932). *Kolloid-Z.* **61**, 269.
Briskey, E. J., Seraydarian, K., and Mommaerts, W. H. E. M. (1967). *Biochim.
    Biophys. Acta* **133**, 424.
Buchthal, F., Knappeis, G. G., and Lindhard, J. (1936). *Skand. Arch. Physiol.* **73**,
    163.
Cain, D. F., Infante, A., and Davies, R. E. (1962). *Nature (London)* **196**, 214.
Caldwell, P. C., and Webster, G. E. (1961). *J. Physiol. (London)* **157**, 36P.
Caldwell, P. C., and Webster, G. E. (1963). *J. Physiol. (London)* **169**, 353.
Carlsen, F., and Knappeis, G. G. (1955). *Exp. Cell Res.* **8**, 329.
Carlsen, F., Knappeis, G. G., and Buchtal, F. (1961). *J. Biophys. Biochem. Cytol.*
    11, 95.
Caspar, D. L. D., Cohen, C., and Longley, W. (1969). *J. Mol. Biol.* **41**, 87.
Civan, M. M., and Podolsky, R. J. (1966). *J. Physiol. (London)* **184**, 511.
Close, R. (1965). *J. Physiol. (London)* **180**, 542.
Cohen, C., and Longley, W. (1966). *Science* **152**, 794.
Corsi, A., Ronchetti, I., and Cigognetti, C. (1966). *Biochem. J.* **100**, 110.
de Villafranca, G. W. (1957). *Exp. Cell Res.* **12**, 410.
Ebashi, S., Kodama, A., and Ebashi, S. (1968). *J. Biochem (Tokyo)* **64**, 465.
Ebashi, S., Endo, M., and Ohtsuki, I. (1969). *Quart. Rev. Biophysics* **2**, 351.
Eisenberg, E., and Moos, C. (1968). *Biochemistry* **7**, 1486.
Elliott, G. F. (1964). *Proc. Roy. Soc., Ser. B* **160**, 467.
Elliott, G. F., Lowy, J., and Worthington, C. R. (1963). *J. Mol. Biol.* **6**, 295.
Elliott, G. F., Lowy, J., and Millman, B. M. (1965). *Nature (London)* **206**, 1357.
Elliott, G. F., Lowy, J., and Millman, B. M. (1967). *J. Mol. Biol.* **25**, 31.
Endo, M., Nonomura, Y., Masaki, T., Ohtsuki, I., and Ebashi, S. (1966). *J. Biochem.
    (Tokyo)* **60**, 605.
Engelhardt, V. A., and Ljubimova, M. N. (1939). *Nature (London)* **144**, 669.
Fahrenbach, W. H. (1967). *J. Cell Biol.* **35**, 69.
Fenn, W. O. (1923a). *J. Physiol. (London)* **58**, 175.
Fenn, W. O. (1923b). *J. Physiol. (London)* **58**, 373.

Franzini-Armstrong, C. (1970). *J. Cell Sci.* **6**, 559.
Franzini-Armstrong, C., and Porter, K. R. (1964). *Z. Zellforsch. Mikrosk. Anat.* **61**, 661.
Gergely, J. (1950). *Fed. Proc., Fed. Amer. Soc. Exp. Biol.* **9**, 176.
Gergely, J. (1951). *Fed. Proc., Fed. Amer. Soc. Exp. Biol.* **10**, 188.
Gergely, J. (1953). *J. Biol. Chem.* **200**, 543.
Gillis, J. M. (1969). *J. Physiol. (London)* **200**, 849.
Goldspink, G. (1970). *J. Cell Sci.* **6**, 593.
Hagopian, M. (1966). *J. Cell Biol.* **28**, 545.
Hanson, J. (1956). *J. Biophys. Biochem. Cytol.* **2**, 691.
Hanson, J., and Huxley, H. E. (1953). *Nature (London)* **172**, 530.
Hanson, J., and Huxley, H. E. (1955). *Symp. Soc. Exp. Biol.* **9**, 228.
Hanson, J., and Huxley, H. E. (1957). *Biochim. Biophys. Acta* **23**, 229.
Hanson, J., and Lowy, J. (1963). *J. Mol. Biol.* **6**, 46.
Hanson, J. (1967) *Nature (London)* **213**, 353.
Hanson, J. (1968). Symposium on muscle (E. Ernst and F. B. Straub, eds.), p. 93. Budapest, Akademiai Kiadó.
Hanson, J. (1970). Personal communication.
Haselgrove, J. (1970). Ph.D. Thesis, University of Cambridge.
Haselgrove, J., and Huxley, H. E. (1972). In preparation.
Hasselbach, W. (1953). *Z. Naturforsch. B* **8**, 449.
Hasselbach, W., and Schneider, G. (1951). *Biochem. Z.* **321**, 461.
Hayashi, T., Rosenbluth, R., Satir, P., and Vozick, M. (1958). *Biochim. Biophys. Acta* **28**, 1.
Hill, A. V. (1950). *Sci. Progr. (London)* **150**, 209.
Holtzer, A., and Lowey, S. (1956). *J. Amer. Chem. Soc.* **78**, 5954.
Holtzer, A., Lowey, S., and Schuster, T. (1962). In "The Molecular Basis of Neoplasia," p. 259. University of Texas Press, Austin, Texas.
Hoyle, G., and Smyth, T. (1963). *Comp. Biochem. Physiol.* **10**, 291.
Huxley, A. F. (1957). *Progr. Biophys. Biophys. Chem.* **7**, 257.
Huxley, A. F., and Niedergerke, R. (1954). *Nature (London)* **173**, 971.
Huxley, H. E. (1951). *Discuss. Faraday Soc.* **11**, 148.
Huxley, H. E. (1952). Ph.D. Thesis, University of Cambridge.
Huxley, H. E. (1953a). *Proc. Roy. Soc. Ser. B* **141**, 59.
Huxley, H. E. (1953b). *Biochim. Biophys. Acta* **12**, 387.
Huxley, H. E. (1957). *J. Biophys. Biochem. Cytol.* **3**, 631.
Huxley, H. E., (1960). In "The Cell" (J. Brachet and A. E. Girsky, eds.). Academic Press, Vol. IV, 365.
Huxley, H. E. (1961). *Circulation* **24**, 328.
Huxley, H. E. (1963). *J. Mol. Biol.* **7**, 281
Huxley, H. E. (1965). *Muscle, Proc. Symp., 1964*, p. 3.
Huxley, H. E. (1968). *J. Mol. Biol.* **37**, 507.
Huxley, H. E., and Brown, W. (1967). *J. Mol. Biol.* **30**, 383.
Huxley, H. E., and Hanson, J. (1954). *Nature (London)* **173**, 973.
Huxley, H. E., and Hanson, J. (1957). *Biochim. Biophys. Acta* **23**, 229.
Huxley, H. E., Brown, W., and Holmes, K. C. (1965). *Nature (London)* **206**, 1358.
Johnson, P., Harris, C. I., and Perry, S. V. (1967). *Biochem. J.* **105**, 361.
Kendrick-Jones, J., Lehman, W., and Szent-Györgyi, A. G. (1970). *J. Mol. Biol.* **54**, 313.

386   *H. E. Huxley*

Knappeis, G. G., and Carlsen, F. (1956). *J. Biophys. Biochem. Cytol.* **2**, 201.
Knappeis, G. G., and Carlsen, F. (1962). *J. Cell Biol.* **13**, 323.
Knappeis, G. G., and Carlsen, F. (1968). *J. Cell Biol.* **38**, 202.
Kominz, D. R., Mitchell, E. R., Nihei, T., and Kay, C. M. (1965). *Biochemistry* **4**, 2373.
Lowey, S., and Cohen, C. (1962). *J. Mol. Biol.* **4**, 293.
Lowey, S., Goldstein, L., and Luck, S. M. (1966). *Biochem. Z.* **345**, 248.
Lowey, S., Goldstein, L., Cohen, C., and Luck, S. M. (1967). *J. Mol. Biol.* **23**, 287.
Lowey, S., Slater, H. S., Weeds, A. G., and Baker, H. (1969). *J. Mol. Biol.* **42**, 1.
Lymn, R. W., and Taylor, E. W. (1970). *Biochemistry* **9**, 2975.
Marshall, J. M., Holtzer, H., Finck, H., and Pepe, F. (1959). *Exp. Cell Res.*, *Suppl.* **7**, 219.
Masaki, T., Endo, M., and Ebashi, S. (1967). *J. Biochem.* (*Tokyo*) **62**, 630.
Masaki, T., Takaiti, O., and Ebashi, S. (1968). *J. Biochem.* (*Tokyo*) **64**, 909.
Mihalyi, E. (1953). *J. Biol. Chem.* **201**, 197.
Mihalyi, E., and Szent-Györgyi, A. G. (1953a). *J. Biol. Chem.* **201**, 189.
Mihalyi, E., and Szent-Györgyi, A. G. (1953b). *J. Biol. Chem.* **201**, 211.
Miller, A., and Tregear, R. T. (1970) *Nature* (*London*) **226**, 1060.
Mommaerts, W. H. F. M. (1951). *J. Biol. Chem.* **188**, 553.
Moore, P. B., Huxley, H. E., and DeRosier, D. J. (1970). *J. Mol. Biol.* **50**, 279.
Mueller, H., and Perry, S. V. (1961). *Biochem. J.* **80**, 217.
O'Brien, E. J., Bennett, P. M., and Hanson, J. (1971). *Phil. Trans. Roy. Soc. London* **B261**, 201.
Page, S. G. (1964). *Proc. Roy. Soc., Ser. B* **160**, 460.
Page, S. G. (1966). Personal communication.
Page, S. G., and Huxley, H. E. (1963). *J. Cell Biol.* **19**, 369.
Pepe, F. A. (1966). *J. Cell Biol.* **28**, 505.
Pepe, F. A. (1967). *J. Mol. Biol.* **27**, 203.
Perry, S. V. (1951). *Biochem. J.* **48**, 257.
Perry, S. V., and Corsi, A. (1958). *Biochem. J.* **68**, 5.
Philpott, D. E., and Szent-Györgyi, A. G. (1954). *Biochim. Biophys. Acta* **15**, 165.
Podolsky, R. J. (1960). *Nature* (*London*) **188**, 666.
Podolsky, R. J., Nolan, A. C., and Zavelev, S. A. (1969). *Proc. Nat. Acad. Sci. U.S.* **64**, 504.
Reedy, M. K. (1964). *Proc. Roy. Soc., Ser. B* **160**, 458.
Reedy, M. K., Holmes, K. C., and Tregear, R. T. (1965). *Nature* (*London*) **207**, 1276.
Rice, R. V. (1961a). *Biochim. Biophys. Acta* **52**, 602.
Rice, R. V. (1961b). *Biochim. Biophys. Acta* **53**, 29.
Rice, R. V. (1964). *In* "Biochemistry of Muscle Contraction" (J. Gergely, ed.), p. 41. Little, Brown, Boston, Massachusetts.
Selby, C. C., and Bear, R. S. (1956). *J. Biophys. Biochem. Cytol.* **2**, 71.
Sjöstrand, F. S., and Andersson-Cedergren, E. (1957). *J. Ultrastruct. Res.* **1**, 74.
Spudich, J., and Huxley, H. E. (1972). In preparation.
Straub, F. B. (1942). *Stud. Inst. Med. Chem. Univ. Szeged* **2**, 3.
Stromer, M. H., Hartshorne, D. J., Mueller, H., and Rice, R. V. (1969). *J. Cell Biol.* **40**, 167.
Szent-Györgyi, A. (1942). *Stud. Inst. Med. Chem. Univ. Szeged* **1**, 17.

Szent-Györgyi, A. (1951). "Chemistry of Muscular Contraction," 2nd rev. ed., p. 10. Academic Press, New York.
Szent-Györgyi, A. G. (1951a). *Arch. Biochem. Biophys.* **31**, 97.
Szent-Györgyi, A. G. (1951b). *J. Biol. Chem.* **192**, 361.
Szent-Györgyi, A. G. (1953). *Arch. Biochem. Biophys.* **42**, 305.
Szent-Györgyi, A. G., Mazia, D., and Szent-Györgyi, A. (1955). *Biochim. Biophys. Acta* **16**, 339.
Szent-Györgyi, A. G., Cohen, C., and Philpott, D. E. (1960). *J. Mol. Biol.* **2**, 133.
Szent-Györgyi, A. G., Holtzer, H., and Johnson, W. H. (1964). *In* "Biochemistry of Muscle Contraction" (J. Gergely, ed.), p. 354. Little, Brown, Boston, Massachusetts.
Szent-Kiralyi, E. M., and Oplatka, A. (1969). *J. Mol. Biol.* **43**, 551.
Taylor, E. W., Lymn, R. W., and Moll, G. (1970). *Biochemistry* **9**, 2984.
Weber, A., Herz, R., and Reiss, I. (1969). *Biochem.* **8**, 2266.
Zachau, J., and Zacharova, D. (1966). *J. Physiol. (London)* **186**, 596.
Zobel, C. R., and Carlson, F. D. (1963). *J. Mol. Biol.* **7**, 78.

8

OBLIQUELY STRIATED MUSCLE

JACK ROSENBLUTH

I. Introduction

Students of human histology are ordinarily taught that muscles fall into two broad categories, smooth and striated. This classification, while correct for vertebrate muscles, really describes two special cases and ignores a wide range of muscle types whose histological and physiological properties are intermediate, but which have been somewhat neglected, presumably because they occur only among invertebrates.

One such type, most simply referred to as "obliquely striated muscle," has been investigated intermittently for over a century and has now been shown to be sufficiently different from classic smooth and cross-striated muscles to justify its separate designation. The characteristic feature of this kind of muscle is a distinctive form of sarcomere whose structural organization underlies the patterns seen in both "helical" and "double obliquely striated" muscles and which has also been found to occur in a number of muscles formerly considered to be "smooth."

The purpose of this chapter is to summarize the evidence from which this basic organization has been inferred and to compare some typical examples of this type of muscle to others with respect to the contractile apparatus, sarcoplasmic reticulum and T system, and innervation, and to consider the functional significance of these structural characteristics. Recent ultrastructural studies of obliquely striated muscles have been carried out mainly on samples from molluscs, (Hanson and Lowy, 1961; Kawaguti, 1962), annelids (Bouligand, 1966, Heumann and Zebe, 1967; Ikemoto, 1963; Kawaguti, 1962; Knapp and Mill, 1971; Mill and Knapp, 1970a and 1970b; Rosenbluth, 1968, 1972; Wissocq, 1970), and nematodes (Hirumi *et al.*, 1971; Hope, 1969; Rosenbluth, 1963, 1965a, 1965b, 1967; Wright, 1966). Since molluscan muscle has been reviewed elsewhere (Hoyle, 1964) it will not be considered in detail here and the examples used in this chapter will be drawn primarily from nematode and annelid muscles.

II. General Organization

When examined in teased preparations, obliquely striated muscles may exhibit a so-called diamond lattice pattern (Schwalbe, 1869) or a helical pattern of bands within the individual fibers. The basis of these band patterns has been a matter of continuing interest and controversy (Bowden, 1958), particularly since the angle formed by the bands changes as the muscle shortens or elongates. One possible explanation for them is that in these muscles the myofilaments are oblique rather than parallel to the axis of the fibers and that the direction of the obliquity is reversed on opposite sides of each fiber. Bundles of myofilaments from the two sides of each fiber when superimposed would then form criss-crossing patterns. This was Marceau's (1909) view, and indeed his calculations were based on the assumption that the contractile elements are obliquely oriented. A second possibility is that the myofilaments are parallel to the fiber axis, and that the oblique pattern derives from the *arrangement* rather than the *orientation* of the myofilaments within the bands. Englemann as early as 1881 supported the latter viewpoint with his polarization studies which showed

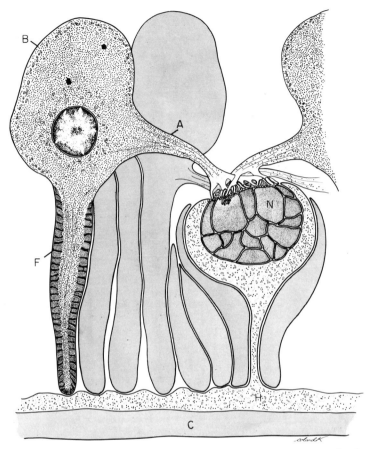

Fig. 1. Diagram of a muscle cell from the body wall of *Ascaris lumbricoides* (transverse section). The cell at the left consists of three parts: a sac-like muscle belly (B), which contains an abundance of glycogen, the flattened muscle fiber (F) which contains the U-shaped contractile material, and the muscle arm (A), or innervation process, which stretches to a nerve cord (N) where a neuromuscular junction is formed; H is the hypodermis and C is the cuticle. From Rosenbluth, 1965b.

that the birefringent material in obliquely striated muscle is roughly axial in orientation and that it remains so despite band angle changes during contraction.

Electron microscopic studies on a variety of such muscles thus far all support Englemann's view by showing myofilaments arranged in staircase patterns in which the thick filaments adjacent to each other in one longitudinal plane are not coextensive, but rather are displaced with respect to each other. As a result, a line drawn through the tips

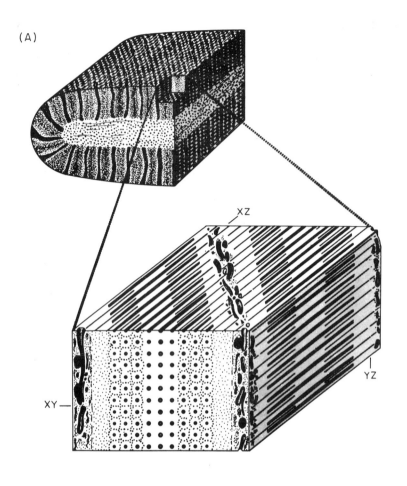

(A)

XZ

YZ

XY

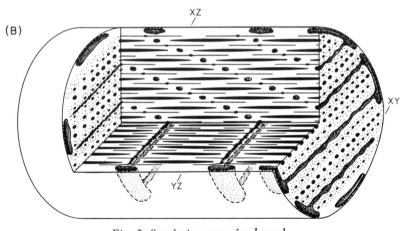

(B)

XZ

XY

YZ

Fig. 2. See facing page for legend.

of the filaments in this plane does not form a 90° angle with the axis of the filaments as it does in cross-striated muscles, but forms instead an acute angle, in some cases as small as 1°; i.e., the striations in this plane, instead of being transverse, are virtually longitudinal with respect to both the fiber and filament axes. The Z line is represented in many obliquely striated muscles by an interrupted chain of "dense bodies" which serve to anchor the thin filaments. Indeed, their role in obliquely striated muscle is one of the principal bases for assuming that the equivalent structures in smooth muscle also correspond to Z lines. A line drawn through the chain of dense bodies parallels the line drawn through the ends of the thick filaments and is usually the most prominent component of the striation pattern. Since some obliquely striated muscles do not have prominent dense bodies, their striation pattern may be almost imperceptible, and the fibers therefore considered to be smooth.

The principle features of this architectural arrangement are summarized in Fig. 2A, which represents part of a muscle fiber from the body wall of *Ascaris lumbricoides* (Fig. 1). The fiber is flattened from side to side and has the general configuration of a ribbon. The flattened sides are oriented radially in the worm, and one of the thin edges rests against the animal's hypodermis peripherally. As in the case of vertebrate smooth muscle, neighboring fibers are not coextensive and the same transverse section will therefore cut through various parts of different fibers from their midportions to their tips. For the sake of description, the three planes shown in Fig. 2A have been designated as follows: (1) that which cuts perpendicularly across the axis of the fiber is the transverse, or XY plane; (2) that which is parallel to the flat sides of the fiber and runs longitudinally is the XZ plane, and (3) the YZ plane, which is also longitudinal but is perpendicular to the XZ plane and to the flat sides of the fibers.

Within the fibers there are both contractile and sarcoplasmic components. The former are disposed in two sheets situated along the flat

Fig. 2. (A) Diagram of an *Ascaris* muscle fiber. The cell has been rotated 90° compared with Fig. 1, and the contractile "U" is therefore lying on its side. The block which has been enlarged has three visible surfaces. The plane facing front is defined as the transverse, or XY, plane. That facing right is the YZ plane, and the plane facing upward is defined as the XZ plane. The oblique bands seen in the XZ plane are in reality almost longitudinal in orientation. From Rosenbluth, 1965a. (B) Equivalent diagram of annelid muscle. In this case the XZ plane is facing front. As the XY plane shows, the two halves of the fiber are in direct contact with no intervening sarcoplasmic core. Here as in (A) the striation angle appears larger in the XZ plane than is actually the case (cf. Fig. 19). From Rosenbluth, 1968.

sides of the fibers and are separated from each other by a layer of sarcoplasm. The contractile material of the two sides is continuous around the edge of the ribbon facing peripherally. Thus, in the XY, or transverse, plane, the contractile apparatus describes a U that has very elongated arms connected by a very sharp bend. A section through the XZ plane may pass through the contractile material on either side of the sarcoplasmic core or through the core itself. A section in the YZ plane will pass through both arms of the U, as well as through the intervening sarcoplasm (except at the bend in the U).

Figure 2B is an equivalent diagram of an annelid obliquely striated muscle showing the same three reference planes. Here there is no sarcoplasmic core separating the two halves of the fiber.

The appearance of the contractile apparatus in these various planes is very characteristic, and in particular, the transverse plane has proven to be the most useful one for analyzing the three-dimensional organization of the muscle (Rosenbluth, 1965a). In this plane, one sees the myofilaments in cross section as large or small dots representing the thick and thin myofilaments (Fig. 2). The filament cross sections are organized into a regular pattern from which it is possible to infer the ratio of thin to thick filament length, the degree of overlap between thick and thin filaments, the lateral spacing of the filaments in different parts of the sarcomere, and the degree of displacement, or stagger, in each of the other two planes.

In the YZ plane, the pattern seen bears a considerable resemblance to that of cross striated muscle. In the XZ plane, however, the bands exhibit the characteristic marked obliquity to the axis of the filaments. These striking variations in form from one plane of section to the next probably account for much of the confusion over this muscle that has existed and for some of the disagreements between different workers about its histological appearance (Roskin, 1925; Plenk, 1924). The variations are easily explained by the three-dimensional models shown in Fig. 2. Here, it is clear that in one longitudinal plane (YZ), the neighboring thick filaments are approximately coextensive; i.e., their tips are aligned with one another and accordingly a line drawn through the tips is perpendicular to the axis of the filaments. In the other longitudinal plane (XZ), however, neighboring filaments are markedly staggered with respect to each other; i.e., the tip of one filament is displaced with respect to the tip of the next filament. As a result, a line drawn through these staggered tips forms only a very small angle with respect to the filament axis. It must be emphasized that the striations in the XZ plane when shown diagrammatically usually seem to be at a greater angle than actually occurs (Figs. 2 and 4). In reality, the striations in this

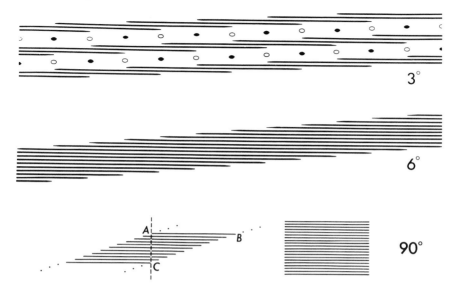

Fig. 3. Diagram showing the relationship of thick myofilaments to angle of striation in the XZ plane. *Top.* Earthworm muscle. Thick filaments are approximately
3 μ long (cf. Fig. 13) and overlap along ~ 60% of their length. The striation angle
is ~ 3°. *Middle. Ascaris* muscle. Thick myofilaments are ~ 6 μ long and overlap
along ~ 90% of their length. The striation angle is ~ 6°. *Bottom, right.* Vertebrate
cross-striated muscle. Thick filaments are 1.5 μ long and about half as thick as either
earthworm or *Ascaris* myofilaments. They overlap along 100% of their length; i.e.,
they are coextensive, and the striation angle is therefore 90°. *Bottom, left.* Relationships used in calculating striation angle in obliquely striated muscle. The tangent
of the angle ABC = AC/AB. AB is the length of one thick filament. AC, for small
striation angles, is approximately equal to the width of the A zone, which can be
measured easily and accurately in transverse sections. In the case of *Ascaris* muscle,
thick myofilaments are ~ 6 μ long; the A zone is ~ 0.6 μ wide; and the calculated
angle is ~ 6°.

plane are nearly longitudinal (Fig. 3), but technical considerations make
it difficult to illustrate this accurately in three-dimensional diagrams.

Thus, the stagger, which is the characteristic feature of obliquely
striated muscle, occurs to a marked degree in one longitudinal plane
but is absent in the perpendicular longitudinal plane, and therefore,
the apparent angle of striation seen in histological sections may be anything from 0° to 90°, depending on the angle of the section with respect
to the three reference planes (Fig. 4). These variations with sectioning
plane may account for some reports of surprisingly large striation angles
in muscles of this kind.

It follows from this architectural arrangement that the bands seen
in the XY plane must be radial in orientation, i.e., in the case of highly

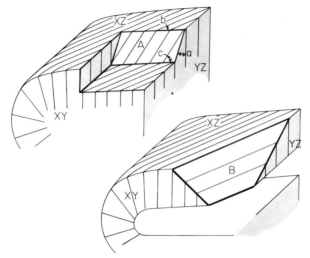

Fig. 4. Diagram showing that the apparent orientation of the striations is dependent upon sectioning angle. In the upper figure, the angle c in plane A can be calculated from the relationship: $\tan c = \tan b / \sin a$. In the lower figure, plane B is cut at an angle that produces longitudinal striations. From Rosenbluth, 1965a.

flattened fibers, perpendicular to the flat side of the ribbon. If filament stagger occurred in the YZ plane as well as the XZ plane, this would not be so. Instead, the bands seen in the XY plane would be oblique to the surface of the fiber, the degree of obliquity depending on the amount of stagger in the YZ plane. In some obliquely striated muscles, a degree of tilting of the bands in this plane does in fact develop on extreme shortening of the fibers.

Ascaris muscle fibers exhibit enough regularity to have been analyzed as just described; however, certain other obliquely striated muscles, notably those of annelids, exhibit an even more regular organization from which additional inferences can be drawn. A typical example is the body musculature of the earthworm (Figs. 9–14). As in the case of *Ascaris* muscle the fibers are again flattened and ribbon shaped. However, the sarcoplasmic core is absent and the contractile sheets of the two sides are directly apposed to one another. Occasionally, a vestigial separation can been seen in the XY plane, but in most cases, the striations extend without apparent interruption directly across the fiber from one side to the other. The bands seen in the XY plane are perpendicular to the flat sides of the fiber, indicating that here, too, the thick filaments are staggered in the XZ plane but not in the YZ plane.

Sections in the XZ plane show thick filament sheaves that are much narrower than those in *Ascaris* muscle. Because there are so few thick filaments across the width of each sheaf and because the individual filaments are only about half as long as those of *Ascaris* muscle, it is sometimes possible to trace individual filaments from end to end and to confirm the inference drawn from transverse sections that the filaments are aligned in such a way as to form staircase patterns in which filaments are displaced or staggered with respect to one another. Thin filaments extend from among the thick filaments within the sheaf into the pale bands that flank the sheaves and can be traced into dense bodies, again confirming inferences drawn from sections in the XY plane.

III. Myofilament Lattice

As in all known muscles, the thin, presumably actin-containing filaments of obliquely striated muscles are always of the same diameter ($\sim$70 Å). One distinctive feature of the thin filaments of some of these muscles (e.g., in *Ascaris*) is the occurrence of an axial period of $\sim$380 Å within the I bands (Fig. 5) (see Huxley, 1967). Thin filament length is difficult to measure directly in longitudinal sections, but the ratio of thin to thick filament length can be inferred from sections in the XY plane. As Fig. 2 indicates, the lengths of A, I, and H bands in the XZ plane are directly proportional to the widths of the A, I, and H zones in the XY plane. Thus, in an extended earthworm muscle (Fig. 9), it can be seen that the width of the region between one H zone and the next in this plane is greater than the width of the A zone by about a third, and therefore the ratio of thin to thick filament lengths in this muscle corresponds approximately to that which obtains in cross-striated muscles (Huxley, 1957), even though the absolute lengths of the filaments are approximately twofold greater in the earthworm. In *Ascaris* muscle, the absolute lengths of the respective filaments are greater still, and in addition, the thin filaments appear to be disproportionately long, as inferred from the unusually large distance from one H zone to the next in the XY plane of extended fibers compared with the width of the A zone. This disproportion is consistent with the wide zone of double overlap that can be seen in the XY plane in maximally shortened *Ascaris* muscle fibers (Figs. 6 and 7).

The thick filaments of annelid muscles unlike those occurring in nematodes sometimes exhibit great variation in length and diameter and, to some extent, in shape as well. This variation occurs not only in muscles from different animals but also within individual muscles. In *Glycera* body wall (Rosenbluth, 1968) most of the fibers contain

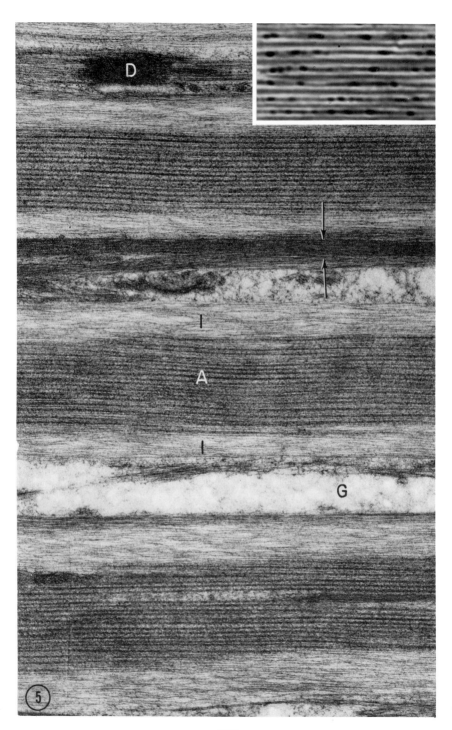

thick filaments ∼300 Å in diameter and approximately 3 μ long (Figs. 15–17). Some fibers, however, (Fig. 18) have much thicker thick filaments (600 Å). Similarly in the earthworm, most thick filaments have a maximum diameter of ∼250–300 Å, but in some fibers they are as thick as ∼700 Å and in their middle regions are flattened rather than cylindrical (Fig. 10). In all cases, the thick filaments taper toward the tips. This point is particularly well seen in obliquely striated muscle in transverse sections, where it is quite clear that the filaments in the middle of the A zone, representing the middle of the sarcomere, are maximal in diameter and those at the edges of the A zone (corresponding to the A–I junction) are thinner.

Such marked variation in thick filament size raises questions about the factors that govern the assembly of the contractile proteins. Among vertebrate striated muscles, myosin filaments are strikingly uniform in size within a given cell, from one cell to another, and indeed from animal to animal. Even extracted myosin can reassemble to form filaments of approximately the same size. Whether the instructions governing filament size are contained within the myosin monomer itself is not clear, however. The fact that in some invertebrate muscles different size thick filaments occur in neighboring cells suggests that the assembly instructions in these cases are more variable or are susceptible to local environmental influences within the individual cells, or perhaps that the protein composition of these filaments varies from cell to cell. Fibers containing larger thick filaments often display broader Z lines and greater stagger and presumably also have longer actin filaments. The mechanisms controlling all of these parameters evidently operate in a coordinated manner within the individual cells.

In longitudinal sections, cross bridges are frequently visible between thick and thin filaments (Hanson and Lowy, 1961) as they are in cross-striated muscles, and cross banding at a period of ∼143 Å can also sometimes be seen especially in the thickest filaments, which presumably contain paramyosin.

Interdigitation of thick and thin filaments and the presence of distinct

---

Fig. 5. Ascaris muscle fiber fixed in extension (XZ plane). Three A bands are shown separated from each other by I bands. The middle region of the I bands contains an assortment of structures, including dense bodies (D), elements of the intracellular skeleton (between arrows), and glycogen (G). In the A band at the bottom, interdigitation of thick and thin filaments is visible. The terminal ends of several thick filaments can also be seen staggered at an interval of ∼0.6 μ and producing a staircase effect. (× 40,000.) *Inset:* Extended *Ascaris* muscle fiber showing striations and dense bodies (×2200). From Rosenbluth 1967.

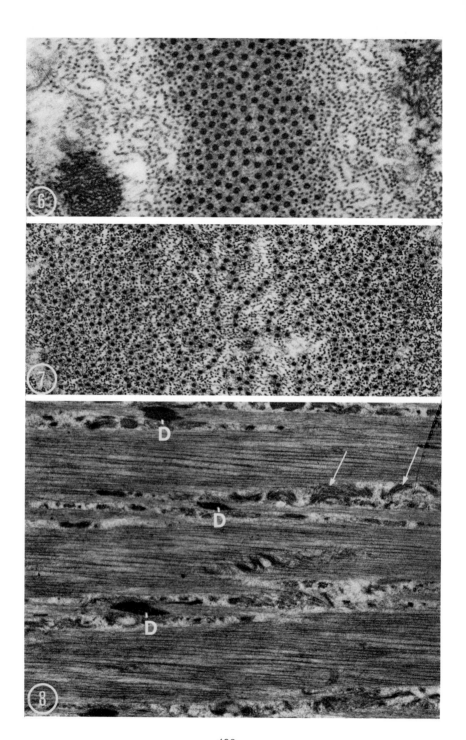

I, A, and H bands can be seen in longitudinal sections and inferred from transverse sections. Because of the oblique arrangement, it is possible to study changes in the spacing and arrangement of myofilaments of different regions of the sarcomere in a single section in the XY plane. As Fig. 6 shows, for example, the average spacing of the thick filaments in the H zone is somewhat closer than in the rest of the A zone. Even more strikingly, when contraction proceeds to the point of double overlap, the thick filaments in the middle of the sarcomere are much more spread apart than those near the A–I junction (Fig. 7). Thus, the well known expansion of the myofilament lattice (Elliott, 1967) that accompanies increased overlap of thick and thin filaments appears to occur nonuniformly in this sarcomere; i.e., only those portions of the thick filaments that overlap thin filaments increase their spacing. As pointed out previously (Rosenbluth, 1967), such nonuniform expansion of the myofilament lattice must entail some bending of the thick myofilaments, and indeed, a sigmoid shape is frequently encountered in longitudinal sections (Ikemoto, 1963). This is especially noticeable when the muscle fibers have shortened to the point of double overlap.

In the A zones (including the H zone) the thick filaments are in hexagonal array. The thin filaments of the I zone occasionally exhibit hexagonal array as well (Fig. 10). In the A zone, one finds orbits of nine to twelve thin filaments around each thick filament; however, it is not possible to assign fixed loci to thin filaments related either to the trigonal points or midpoints between thick filaments. Such order, if it were present, might be very easily disturbed during the preparation of tissues for electron microscopy, however, and X-ray diffraction studies are therefore necessary to decide this point. Considerable order in the thin filament orbits exists in certain nematode muscles (Hirumi *et al.*, 1971).

Fig. 6. Extended *Ascaris* muscle in the XY plane. The A zone in the middle is flanked by I zones. The middle of the A zone contains only thick filaments (H zone) in hexagonal array. On either side of the H zone, thick filaments are surrounded by orbits of nine to twelve thin filaments. (×100,000.) From Rosenbluth, 1967.

Fig. 7. Contracted *Ascaris* muscle in the XY plane. The A zone is very wide and I zones are no longer present. Thin filaments surround thick filaments throughout the entire width of the A zone and in the middle there is a region of double overlap. (×60,000.) From Rosenbluth, 1967.

Fig. 8. Contracted *Ascaris* muscle in the XZ plane. Three sarcomeres are shown. I bands are virtually absent and A bands are wide; thin filaments extending from dense bodies (D) are visibly tilted with respect to the axis of the striations, and the strands of the intracellular skeleton are coiled (arrow). (×15,000.) From Rosenbluth, 1967.

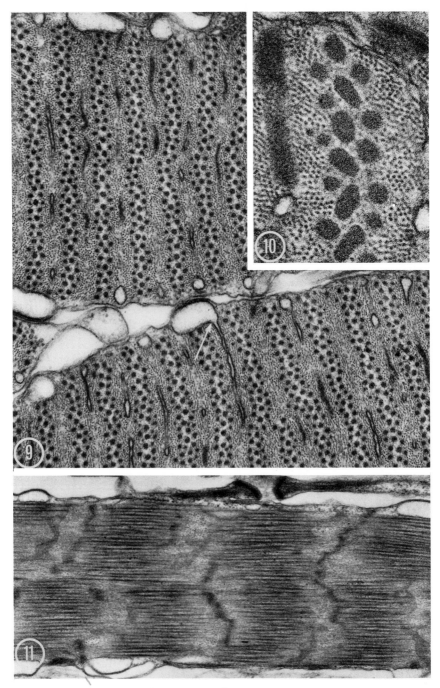

Figs. 9–11. Earthworm muscle.

IV. Z Lines and Dense Bodies

In the annelid muscles, Z lines similar to those of cross striated muscles can be identified in the YZ plane in the middle of the I band (Figs. 11, 15). In fortunate sections, they can be followed transversely across half the width of the fiber in this plane. In the XZ plane, in contrast, where the striation is oblique rather than transverse, the Z lines as such are not present (Figs. 12, 19). Instead, one sees in the middle of the I band, rows of small, discontinuous, spindle-shaped bodies spaced about 0.8 μ apart in the extended muscle and alternating with membrane-limited sarcotubules. Thin filaments can be traced into these bodies but not into the tubules. The simplest interpretation of these findings is that the Z lines in these animals are truly linear and that they extend across the fiber in the YZ plane. Presumably, it is the thin filaments inserting into opposite sides of these Z lines that impart a spindle shape to them when they are seen end on in the XZ plane. The similarity of these Z line cross sections to dense bodies elsewhere justifies referring to them by the same term and suggests strongly that the dense bodies seen in muscles where no Z lines can be found in any plane also correspond to Z lines functionally. Such is the case in vertebrate smooth muscles, invertebrate smooth muscles, and some molluscan and nematode obliquely striated muscles.

In the annelid muscles, Z lines are spaced rather closely and are of small diameter. One Z line cross section (dense body) appears each $\sim$0.8 μ in the XZ plane of extended fibers. This distance is approximately the same as the calculated displacement of adjacent thick filaments in the XZ plane.* Thus, the ratio of Z line spacing to thick filament dis-

---

* Because of their hexagonal arrangement, "adjacent" thick filaments in the XZ "plane" are actually not quite coplanar. They are nevertheless seen together in electron micrographs since the section thickness is several times greater than the interfilament distance.

Fig. 9. In the XY plane, rows of thick and thin filament profiles resemble those of *Glycera* muscle. Thick filaments exhibit hexagonal array. In the lower fiber, a sarcotubule can be seen arising (arrow) from a terminal cisterna at the cell surface. ($\times$58,000.)

Fig. 10. Some fibers have thick filaments of much greater caliber. In this instance, they measure 400 Å $\times$ 700 Å at their widest. Z lines are also greatly thickened in such cells. Hexagonal array can be seen in the thin filaments. ($\times$150,000.)

Fig. 11. In the YZ plane, striations are nearly transverse. At the top, a fibrillar bundle leads to a desmosome. ($\times$25,000.)

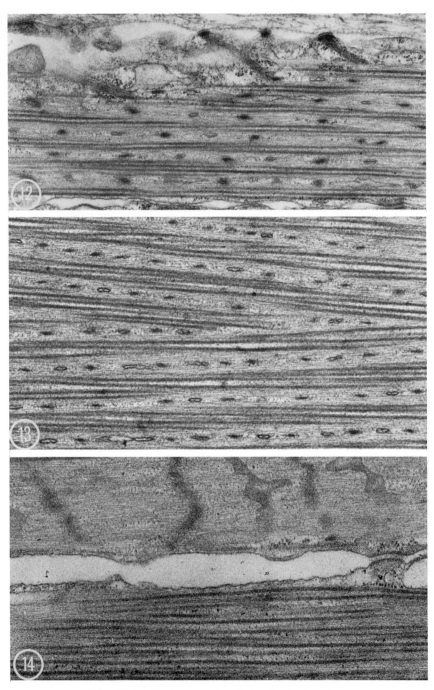

Figs. 12–14. Earthworm fibers in longitudinal section.

See facing page for legends.

placement is one to one, or, to put it differently, for every thick filament seen in the XZ plane, there is a separate Z line cross section. Whether the entire thin filament orbit of each thick filament converges on the same Z line is, however, not known.

The ratio in nematode muscle appears to be greater. In *Ascaris* muscle the displacement of thick filaments in the XZ plane is approximately 0.2–0.6 μ at extreme extension and contraction, respectively, while the spacing of the dense bodies in this plane is roughly ten times as much. Part of this discrepancy derives from the fact that the dense bodies in this animal are discontinuous in the XY plane; i.e., the thin filaments, instead of attaching along a continuous line in the XY plane, converge, as it were, onto segments of the line which have been compressed together. As a result sections in the XZ plane are apt to pass between dense bodies and therefore provide an artificially low estimate of their frequency. However, in another nematode muscle (Hirumi *et al.*, 1971), where the stagger of the thick filaments is comparable to that in *Ascaris*, Z lines are not compressed into dense bodies in the XY plane, and therefore, as in *Glycera*, all of them should intersect the XZ plane. In this case the spacing of these "Z plates" in the XZ plane is still at least several times greater than the average displacement of the thick filaments in this plane. Thus, in contrast to annelid muscle, it seems likely that in nematode muscle thin filaments related to multiple thick filaments converge onto each end of a dense body in the XZ plane. This probably accounts for the relatively large size of nematode dense bodies in comparison with those of annelids, and may also underlie their sometimes extraordinary longitudinal extent. In *T. christiei*, for example (Hirumi *et al.*, 1971), the dense bodies are so elongated in the longitudinal direction that in transverse sections two or three may appear side by side in the middle of each I zone. These Z line counterparts in this instance are neither linear nor spindle shaped but in the form of flattened plates, which by virtue of their shape are presumably able to convey tension generated deep from the cell surface to the connective tissue adherent to the cell membrane.

Fig. 12. Extended fiber in the XZ plane. Dense bodies are widely spaced. Two fibrillar bundles extend to hemidesmosomes at the top. (×28,000.)

Fig. 13. Fiber at rest length. The plane of section is probably tilted slightly with respect to the XZ plane so that it passes from one-half of the fiber (top) to the other half (bottom). As it does so, the direction of thick filament stagger reverses. Dense bodies are much more closely spaced than in Fig. 12. (×34,000.)

Fig. 14. Section running parallel to striations. In the fiber at the top the section remains within the I band cutting Z lines and sarcotubules but no thick filaments. In the lower fiber, the section remains within the A band. (×39,000.)

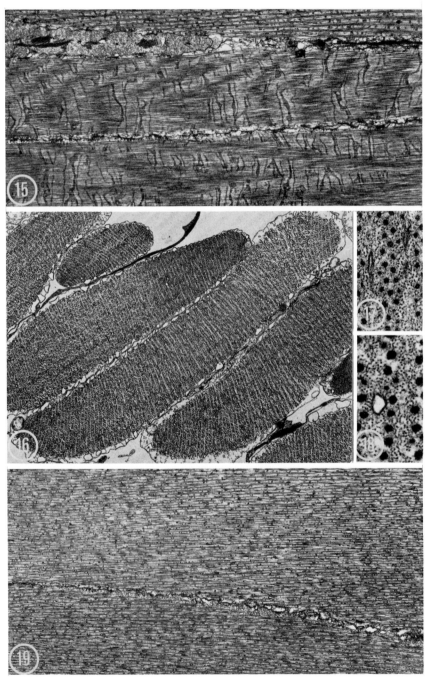

Figs. 15–19. Extended *Glycera* muscle. From Rosenbluth, 1968.

See facing page for legends.

In the muscles so far described it appears that the Z structures, regardless of their configuration, are probably responsible for maintaining the orderly arrangement of the contractile elements by organizing them into groups. In annelid muscles these groups are attached together in the YZ plane, while in nematode muscles, as discussed above, there is evidence for some grouping in the XZ plane as well. In a situation where the contractile material extending across the fiber in the YZ plane was not bound together, great irregularity in the muscle architecture would result. This may perhaps account for the appearance of certain molluscan muscles such as the buccal muscle of *Aplysia*, which possesses dense bodies that are circular rather than ovoid in cross section and which has no Z lines in any plane. This muscle exhibits obliquely striated patterns in only some regions (Fig. 21). In nematodes the cytoskeleton probably also serves to maintain the oblique arrays.

V. Cytoskeleton

In some of the earliest histological descriptions of *Ascaris* muscle (Apathy, 1894), it was noted that a fibrillar network, apparently separate from the contractile apparatus, penetrates throughout the cell including the muscle belly and arm regions and the sarcoplasmic core of the muscle fiber as well. These skeletal fibrils apparently travel in the I bands (Reger, 1964; Rosenbluth, 1965a) along with the dense bodies and also occasionally in the A bands. However, because of the similarity between skeletal fibrils and Z components, it has been difficult to distinguish these two elements in electron micrographs. Presumably, the skeletal fibrils are connected to the dense bodies as well as to the cell surface.

The skeletal fibrils may serve to transmit the tension of the contractile components to the surrounding connective tissue and act in effect as intracellular microtendons. Thus, the contractile elements may be able

---

Fig. 15. Three fibers are shown. The middle one is cut in the YZ plane. Transverse Z lines and sarcotubules are visible. (×5000.)

Fig. 16. Fibers cut in the XY plane. Each fiber is flattened from side to side and is traversed across its width by fine stripes composed of rows of filament profiles. (×5000.)

Fig. 17. Detail of filaments in XY plane. (×58,000.)

Fig. 18. Detail of large caliber thick filaments exhibited by some fibers in this muscle. (×58,000.)

Fig. 19. Two fibers cut in the XZ plane. Striations are almost longitudinal and are so fine that they would not be seen by light microscopy. (×5000.)

to exert tension on the surrounding connective tissue either through this network of skeletal fibrils or through the myofibrils themselves as in cross-striated muscles. In an extended cell if only a small portion of a myofibril were to contract, the slack sarcomeres in series with it would merely stretch slightly and would be ineffective in transmitting the tension generated to the ends of the fiber. If the dense bodies related to the contracting portion were attached to oblique strands of the intracellular skeleton, however, the latter could convey the tension developed to widely separated points on the fiber surface. The force developed by a single sarcomere is just as great as that generated by a chain of sarcomeres, although the amount of shortening is much less (Rosenbluth, 1967). Thus, a mechanism for linking short segments of the contractile apparatus independently to the surrounding connective tissue could permit efficient maintenance of steady isometric tone. In cross striated muscles the only structures comparable to the cytoskeletal fibrils are the Z discs which also extend to the sarcolemma. Since they are perpendicular to the direction of the force generated by the myofilaments, they are, however, in a poor position to transmit this force to the cell surface. Thus, in cross striated muscle the forces produced within the sarcomeres must be transmitted through the myofibrils themselves to the myotendinous junctions near the ends of the fiber. Such an arrangement, which requires synchronous contraction of the whole series, is probably less efficient for the maintenance of isometric tone. In short, the value of an intracellular skeleton in obliquely striated muscles (and perhaps in certain smooth muscles as well) may be in permitting localized or asynchronous contractions, involving only a small part of the contractile apparatus at any one time, to produce tension, without shortening, effectively and economically. The presence of the cytoskeleton would not prevent synchronous, phasic contraction of the entire cell from occurring as well and producing rapid shortening. The same muscle may thus be able to subserve both tonic and phasic functions.

VI. Stagger

The degree to which adjacent thick filaments are displaced with respect to each other in the XZ plane varies from one animal to another and to some extent from one fiber type to another within a given muscle. In nematodes, a typical cross section of an extended fiber shows an A zone having approximately ten rows of thick filaments. It can be inferred from this observation that in the XZ plane, adjacent thick fila-

ments are displaced by ~10% of their length (~0.6 μ). From the individual length of the thick filaments, and the thickness of the A zone
as seen in the XY plane, one can calculate the angle of striation with
respect to the filament axis (Fig. 3). Thus in *Ascaris*, the angle is ~6°.
This is not necessarily equal to the angle of striation with respect to
the muscle fiber axis, however, (see below) since the filaments may
tilt somewhat.

In the case of annelid muscle, the stagger is distinctly greater. In
Glycera, for example, transverse sections through the fibers show A zones
having only about two to six rows of thick filament profiles indicating
a ~20–40% displacement of thick filaments with respect to their neighbors, or ~0.6–1.2 μ (Figs. 16, 17). One class of fibers in *Glycera*, however, has thick filaments that are much thicker (Fig. 18) and presumably longer and in which the displacement of adjacent thick filaments
and dense bodies is roughly twice as much. In longitudinal sections,
the oblique arrangement is therefore not as conspicious, and the fibers
look rather like those of invertebrate smooth muscles. In earthworm
muscle, the stagger, the thick filament length, and the striation angle
are comparable to their counterparts in *Glycera* muscle. At these small
angles, the spacing of dense bodies can be used as a means of assessing
thick filament length provided that the degree of thick filament stagger
is known (from cross sections) and provided each thick filament has
one and only one dense body corresponding to it in the XZ plane,
as appears to be the case in the annelid muscles studied. In contrast
to the remarkably small striation angles of 1° or 2° in earthworm and
Glycera, the angle in molluscan muscles appears to correspond more
closely to the larger angles seen in nematode muscles (Hanson and
Lowy, 1961; Bowden, 1958), except on shortening below minimum body
length, when the angle may become very much greater, presumably
because of bending of the filaments (Hanson and Lowy, 1961). It is
of course essential that comparisons of striation angle in these various
forms be carried out with the fibers extended to the same degree, since
the striation angle changes so much on contraction.

The direction of striation has been shown to be opposite in the two
halves of a molluscan muscle fiber when both halves are seen superimposed from the same side (Hanson and Lowy, 1961). However, even
in the earthworm, where the fibers are flattened and opposite halves
are in direct contact and where there is no obvious transition zone separating the two halves, Hanson (1957) has shown by light microscopy
that the striation angle is still opposite in the two halves. This observation is confirmed in nearly midsagittal sections through such fibers (Fig.
13) in which the direction of overlap reverses as the section passes

from one half to the other, resulting in arrowhead formations. Hope (1969) points out that in transverse sections the bands of opposite halves of *Deontostoma* muscle are aligned most of the time, suggesting that the striations of the two halves of these fibers are in the same direction and are congruent. However, the same observation also applies to earthworm muscle, where there is clear evidence for a reversal of angle. Perhaps the filaments are not always parallel to the muscle fiber; i.e., they may be tilted oppositely on the two sides of the fiber by one or two degrees such that the striations of both sides become parallel to the fiber axis and congruent with each other. Some evidence for tilting of the filaments of the two halves is shown in Fig. 13, where it is obvious that the filaments in the two halves of the muscle are not quite parallel to each other.

VII. Contraction Mechanism

Comparison of several obliquely striated muscles at extended and shortened lengths shows clearly that the thick and thin filaments shift with respect to each other such that their overlap is greater at short lengths and smaller at extended lengths. Cross links have also been observed between the two types of filaments, and double overlap in the middle of the sarcomere occurs at extreme shortening. These observations conform entirely to those made previously on cross-striated muscles (Huxley, 1957) and are consistent with the view that in obliquely striated as in cross-striated muscles, the elementary contractile process consists of an interaction between the thick and thin filaments causing them to move in opposite directions (Hanson and Lowy, 1961). Moreover, the shape of the sarcomere in the YZ plane and the taper of the thick myofilaments are consistent with an antiparallel organization of the contractile proteins equivalent to that of cross-striated muscles. In addition, the thick myofilaments appear to be in hexagonal array and to undergo spacing increases as thin filaments penetrate into this lattice. In this respect also, the behavior of obliquely striated muscles corresponds to that of cross-striated muscle (Elliott, 1967).

The one phenomenon exhibited by obliquely striated muscle that does not occur in cross-striated muscles is "shearing" of the contractile elements, i.e., slippage of like elements with respect to each other. Transverse sections show that in a shortened muscle more rows of thick filament profiles appear in each A zone than in an extended muscle, thus indicating a smaller degree of stagger and a greater overlap of the thick filaments in the former case. Similarly, in longitudinal sections in the XZ plane, it is clear that in the shortened muscle the adjacent dense bodies within each line are closer together, and the striation angle is larger than in the extended muscle. It should be emphasized that a change in striation angle could result merely from an increase in the lateral spacing of the thick myofilaments as contraction proceeds, but in that case, dense bodies would become further apart rather than closer together.

The cause of shearing is unclear, but it probably does not depend on an imbalance in the forces developed within the individual sarcomeres arising out of the stagger of myofilaments. Indeed, the asymmetry of the forces within the oblique sarcomeres would tend to produce shear in the direction opposite to that which in fact occurs; i.e., like filaments would tend to move apart rather than together during contraction. Perhaps the simplest explanation for shearing is that the angle of striation to the muscle fiber axis is very small indeed, and therefore the sarcomeres are in effect in parallel with the axis along which length changes occur. Thus, if the line of dense bodies within an I band extends virtually longitudinally from end to end in such a fiber, then as the fiber shortens the line of dense bodies must either shorten to the same extent or become coiled. Since the relative position of the dense bodies is fixed with respect to the other contractile elements, coiling may be impossible. Shortening of the line necessarily decreases the space between the dense bodies, since their number is constant. For very small angles of striation, the change in the spacing of dense bodies would be almost directly proportional to the degree of length change in the fiber. The same considerations apply to changes in the stagger of the thick filaments in the A bands (Fig. 20).

Resistance to shearing may depend upon the interaction of thick filaments with one another, especially in the H zone where no thin filaments intervene and where an unidentified material of moderate density surrounds the thick filaments (Rosenbluth, 1967). Strong interaction among the thick filaments in this region could account for a "catch state" in view of the fact that the sarcomeres instead of extending transversely across the fibers, as they do in cross-striated muscles, are virtually longitudinal and coextensive with the fibers containing them. Under certain conditions, the thick filaments could become locked together and thereby impede lengthening of the fibers with which they are in parallel.

Sliding and shearing movements are in principle dissociable provided the muscle fibers are capable of being deformed in gross shape. For example, if an obliquely striated muscle fiber were rhomboidal in the XZ plane with the myofilaments parallel to the base and the striations parallel to the oblique sides, and if the striations and filaments had the same orientation throughout the entire thickness of the fiber (see Hope, 1969), then unlimited changes in striation angle could occur without change in the length of the base; i.e., shearing could occur independent of sliding, provided the oblique sides of the figure could change in length freely. Thus far, however, no clear examples of such dissociation have been reported with the possible exception referred to by Knapp and Mill (1971) of a temporal separation between sliding and shearing movements observed in a glycerinated muscle.

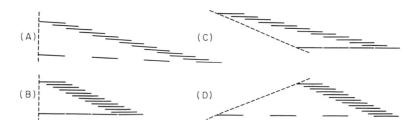

Fig. 20. Diagram illustrating sliding and shearing movements in annelid muscle. (A) Extended muscle in the XZ plane. An oblique sheaf of thick filaments from one sarcomere is coextensive with a series of thick filaments from adjacent sarcomeres. (B) Shortened muscle. The spacing of the thick filaments in series has decreased as a result of sliding movements within the sarcomeres. A proportionate reduction in the length of the oblique sheaf has occurred as a result of shearing movements between the thick filaments. (C) Sliding without shearing. The series of sarcomeres has shortened. Without concommitant shearing, the sheaf of thick filaments remains disproportionately long. (D) Shearing without sliding. If the series of sarcomeres remained extended, shearing together of the thick filaments would cause the sheaf to become disproportionately short.

VIII. Sarcoplasmic Reticulum

Studies of cross-striated muscle have demonstrated, in addition to the contractile elements, a membranous system concerned with the control of muscular contraction. One component of this system consists of the T tubules (Andersson-Cedergren, 1959), which are in essence inpocketings of the plasma membrane reaching transversely into the depths of the cell. Electrical potential changes across the cell membrane are presumed to follow these tubules, which thus carry signals from the cell surface into its depths. The other major component is the sarcoplasmic reticulum, which consists of an intracellular compartment limited by membranes able to transport calcium ions from the sarcoplasm into the lumina of the cisternae and concentrate it there. Specialized junctions between the membranes of the sarcoplasmic reticulum and T tubules, referred to as triads ,or dyads, depending on the precise geometry of the membranes, occur at very regular intervals in some muscles and are characterized by a marked parallelism between the respective membranes and by the presence of periodic deposits of amorphous material in the space between them (Franzini-Armstrong, 1970; Kelly, 1969).

The primary elements of this control system have now all been identified in obliquely striated muscles. Their configuration is variable, however, depending presumably on the physiological properties of different muscle fibers. In the case of *Ascaris* muscle, T tubules extend across the width of the contractile material almost to the sarcoplasmic core of the fiber. The continuity of the tubule membrane with the plasma membrane is obvious in this animal and the penetration of connective tissue components into the tubule is clear, as it is in the case of certain fish muscles (Franzini-Armstrong and Porter, 1964). Along the length of the tubule, highly flattened sacs of sarcoplasmic reticulum are apposed to the cytoplasmic surface of the tubule membrane, separated from it by a uniform distance containing regularly spaced dense plaques which have an affinity for phosphotungstic acid (Rosenbluth, 1969b). A serrated dense material is also located within the lumina of the sacs. These terminal cisternae represent nearly all of the sarcoplasmic reticulum present in these cells in contrast to the muscle cells of other animals where the terminal cisternae lead into extensively ramifying channels of sarcoplasmic reticulum (Figs. 22 and 23).

A different arrangement occurs in annelid muscle (Fig. 9). T tubules are entirely absent, and dyads occur at the cell surface where they are formed by junctions between terminal cisternae of the sarcoplasmic reticulum and the plasma membrane itself. A periodic density between the respective membranes is also visible here. These dyads bear a striking resemblance to the subsurface cisternae of nerve cells (Rosenbluth, 1962). Although T tubules are missing, the sarcoplasmic reticulum in these annelids is prominent, extending from the terminal cisternae at the cell surface toward the middle of the cell. Heumann and Zebe (1967) have shown that calcium oxalate accumulates in these cisternae, thereby establishing their equivalence to the sarcoplasmic reticulum of cross-striated muscle.

Because of the geometry of obliquely striated fibers, the relationship of the sarcoplasmic reticulum and T systems to the contractile apparatus is not quite the same as in cross-striated muscles. In the latter, the

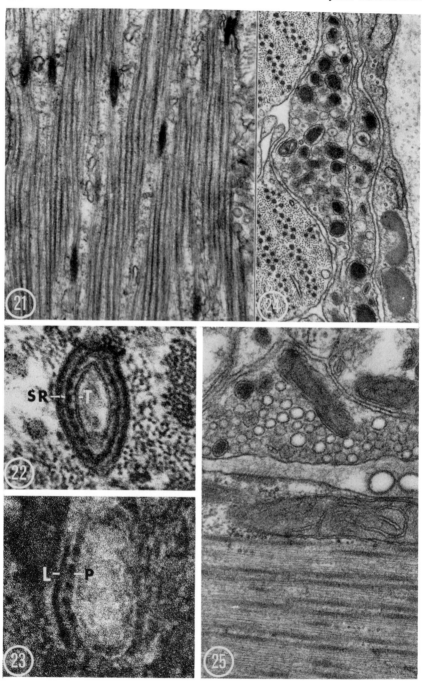

See facing page for legends.

myofilaments are organized into fibrils which are surrounded by sleeves of membranous elements. Only rarely (Reger, 1967) are the two interspersed. In obliquely striated muscles, in contrast, the contractile material of each cell forms a continuous system in which the sarcoplasmic reticulum and T components are interwoven.

In the case of annelid muscle, the sarcotubules, which extend transversely across the fibers from the terminal cisternae at the surface, correspond functionally to the lateral sacs of sarcoplasmic reticulum which run longitudinally in cross-striated muscles. Because of the marked obliquity of the myofilaments in the XZ plane, these sarcotubules, which are strictly speaking located in the middle of the I bands, approximate all other regions of the sarcomere as well. As Fig. 12 shows, the distance of any part of a myofilament from the nearest sarcotubule depends upon the spacing of the tubules from each other along the length of the bands and upon their spacing across each band in the XZ plane. Both of these distances happen to be a function of the degree of myofilament stagger. Thus, increasing the stagger of the myofilaments during extension of the fiber leads to an increasing separation of dense bodies from each other and of the sarcotubules with which they alternate along the length of the bands. Conversely, as the stagger of the myofilaments increases, the number of filaments across the width of the band of myofilaments is reduced, and the lateral spacing of the myofilaments decreases. The separation of the sarcotubules facing each other across each band is correspondingly reduced. In earthworm muscle, the spacing of sarcotubules perpendicularly across the width of the bands is $\sim 0.3\,\mu$ in contracted fibers and $\sim 0.2\,\mu$ in extended fibers, and the spacing of

---

Fig. 21. Aplysia buccal muscle. Oblique striation is visible but is not as regular as in *Ascaris* and annelid muscles. Thin filaments can be seen arising from spindle shaped dense bodies. (×29,000.)

Fig. 22. Dyad in *Ascaris* muscle. A cross sectioned T tubule in the center is enwrapped by a flattened terminal cisterna. Toothlike projections extend from the cisternal membrane (SR) to the T tubule membrane (T). (× 140,000.) From Rosenbluth, 1969b.

Fig. 23. Dyad from *Ascaris* muscle fixed with glutaraldehyde and phosphotungstic acid. The toothlike projections (P) are stained by the PTA and so is the material (L) within the lumen of the flattened terminal cisterna. The membranes themselves are unstained by this method. (×190,000.) From Rosenbluth, 1969b.

Fig. 24. Earthworm myoneural junction. A nerve fiber containing predominantly large, dense-cored vesicles comes into close contact with several muscle fibers. Membrane specializations are minimal at this type of myoneural junction. (×49,000.)

Fig. 25. Earthworm myoneural junction. At this type of junction, the nerve fiber contains predominantly clear vesicles and is separated from the muscle fiber by a wide gap containing amorphous material. Hairlike projections traverse the muscle cell membrane at the site of the junction. (×68,000.)

the tubules along the length of the bands is $\sim 1.6\,\mu$ and $\sim 0.8\,\mu$ in extended and contracted fibers, respectively (allowing for the fact that sarcotubules do not alternate regularly with dense bodies but occur only about half as often). On the basis of these figures the furthest distance of any part of a sarcomere from a sarcotubule is $\sim 0.4\,\mu$ in shortened fibers and $\sim 0.8\,\mu$ in extended fibers. If a typical vertebrate twitch muscle is assumed to be composed of fibrils $\sim 1\,\mu$ in diameter and to have a sleeve of sarcoplasmic reticulum around the entire length of each sarcomere, then the maximum distance of sarcoplasmic reticulum from the middle of the sarcomere would be $0.5\,\mu$. In the case of the lobster remotor, which is unusually fast, the fibrils are flattened to $\sim 0.5\,\mu$ in width, and in this case the maximum distance of filaments from the sarcoplasmic reticulum is $\sim 0.25\,\mu$ (Rosenbluth, 1969a). Fibrils of fish swim bladder muscles (Fawcett and Revel, 1961) are even more flattened (0.3–$0.4\,\mu$). Thus, the oblique arrangement appears to have very little disadvantage over the fastest cross-striated muscles with respect to the proximity of sarcotubules to myofilaments. Since the sarcotubules are located within the myofilament lattice, however, there may be a stringent limit on the volume of sarcoplasmic reticulum that can be accommodated in these interstices. Thus, an obliquely striated muscle having as large a relative volume of sarcoplasmic reticulum as the lobster remotor, or even a more typical fast-acting cross-striated muscle, may represent an architectural impossibility and this may constitute one factor limiting the maximum speed of obliquely striated muscles.

IX. Innervation

The unique innervation of nematode body muscles has been studied extensively by histological methods (see DeBell, 1965, for review) and has recently been reexamined by electron microscopy (Wright, 1966; Rosenbluth, 1965b). In the case of *Ascaris*, two nerve cords travel the length of the animal partly surrounded by a trough of supporting cells. Muscle innervation is accomplished not by way of peripheral nerve fibers arising from the nerve cords and ramifying circumferentially, but rather by way of elongated innervation processes that arise from the muscle cells themselves and converge on the nerve cords. Thus, the counterparts to peripheral nerve fibers in this animal are the innervation processes which are extensions of muscle cells. Stimulation of the nerve cords in these animals results in activation of large groups of muscle fibers. It has also been shown (DeBell *et al.*, 1963) that the muscle fibers are electrically interconnected and that the connections probably occur in the vicinity of the distal ends of the innervation processes. Morphological examination of this bizarre myoneural junction shows that the innervation processes and nerve fibers come into contact with each other separated by a space of ~500 Å, with no other cellular processes intervening. At these regions of contact, large accumulations of vesicles along with giant mitochondria occur intermittently in the nerve fibers facing the muscle processes. Thus, despite the reversal in size of the pre- and postjunctional processes compared with what is found in vertebrate myoneural junctions, the typical junctional organelles are present and are polarized in the same way. A second morphological specialization found at this junction consists of an unusually close apposition between the branches of the innervation processes near their termination. Equivalent junctions which have been described in a variety of locations have been given the name "gap junctions." They occur also in vertebrate smooth and cardiac muscles and may correspond to sites at which these muscles are electrically interconnected (Dewey and Barr, 1964). Syncytial connections between nematode muscle fibers have also been reported (Wright, 1966).

In the earthworm, myoneural junctions occur resembling those of both cross-striated and smooth muscles. The first type is characterized by nerve endings crowded with vesicles, mostly clear, separated by a wide gap from the underlying muscle cell membrane (Mill and Knapp, 1970). The postjunctional membrane exhibits no junctional folds but has projections extending from it which in tangential sections exhibit hexagonal patterns (Rosenbluth, 1972). The second type of ending consists of a nerve fiber containing large dense-cored vesicles in very close apposition to the muscle membrane with only minimal specialization of the latter (Rosenbluth, 1972). Both kinds of endings may occur on the body of the muscle fiber immediately over myofilaments, or on finlike extensions of the edges of the cells that contain sarcoplasm and bear some resemblance to the innervation processes of nematodes (Figs. 24 and 25).

X. Relationship to Other Muscle Types

It may be no coincidence that a number of obliquely striated muscles, notably that of *Ascaris,* were long considered to be smooth. Quite apart from its superficial similarity to smooth muscle with respect to the shape and arrangement of fibers, the movements of the worm are also relatively slow compared with those of vertebrate skeletal muscles, and the waves of contraction that pass down the animal are vaguely reminiscent of peristaltic waves. The similarity extends also to the relationship of the fibers to connective tissue. In contrast to cross-striated muscles, which exert their force at prominent myotendinous junctions at the ends of the fibers, obliquely striated muscle fibers, like those of *Ascaris,* as well as many smooth muscle fibers, have a prominent connective tissue investment all along their sides bound to the contractile elements at frequent intervals. This suggests that the force developed within the fibers is transmitted to connective tissue uniformly throughout the length of the muscle. Partial or asynchronous contraction within such a muscle and indeed even within individual fibers could therefore produce tension, although not shortening, effectively (Rosenbluth, 1967). A muscle of this kind would be well adapted for the maintenance of isometric tone economically, as discussed in Section V.

The relationship of contractile elements to connective tissue is more complex in other obliquely striated muscles, e.g., that of the earthworm. Here the ribbon-shaped fibers are organized into bundles in such a way that the narrow edges of the ribbons, but not the flat sides, are attached to connective tissue elements all along the length of each fiber. The force of contraction is transmitted to the connective tissue along the narrow edges of the fibers partly by way of intracellular fibrillar "tendons." This arrangement, too, could permit asynchronous or partial contraction within the bundle to exert tension effectively on the investing connective tissue sheaths (Figs. 11 and 12).

In either smooth or obliquely striated muscles, oblique sarcomeres effectively in parallel with the muscle fibers may constitute a parallel impedance which will resist sudden stresses but can still adapt gradually. As discussed earlier, a locking mechanism in such sarcomeres could also underlie a catch mechanism. Thus, the impedance attendant upon the shearing of myofilaments in obliquely striated muscles may resist sudden stretches, even though it may also reduce the speed of contraction. The impedance to shear offered by the sarcomeres is probably a function of the interaction of like filaments with each other, which is in turn dependent upon the length along which they overlap each other, as well as upon the nature of any interconnections (e.g., in the M region) among them. Thus, *Ascaris* fibers, in which the thick filaments are longer and the stagger less, might be expected to resist shear more than annelid fibers (Rosenbluth, 1967, 1968).

As pointed out earlier, the significance of the dense bodies in smooth muscles has been clarified primarily through studies of the same structures in obliquely striated muscles. Another feature of certain smooth muscles may also be understood by comparison with obliquely striated muscles. The anterior byssus retractor of *Mytilus* (ABRM), for example, exhibits bundles composed exclusively of either thin filaments or thick filaments running longitudinally within the cells and apparently unrelated to each other. These paradoxical observations are best understood in light of the fact that equivalent bundles have also been seen in both *Ascaris* (Fig. 5) and earthworm muscle (Fig. 14), but it is now clear, especially from electron micrographs of transverse sections, that these bundles represent I bands and A bands that have a very oblique, nearly longitudinal orientation but which nevertheless interdigitate with each other to a degree that depends on the length of the fiber containing them (Figs. 5–8). Further studies of obliquely striated muscles may be expected to shed still more light on smooth muscles.

Obliquely striated muscles are comparable to cross-striated muscles in several respects: both exhibit a double array of myofilaments which are segregated but interdigitate; both exhibit a regular pattern of membranous structures involved in excitation contraction coupling; and, most strikingly, both exhibit sliding of thin filaments toward the midportions of the thick filaments on muscle shortening. Thus, despite differences in cell size and shape, orientation of the sarcomeres, size and shape of myofilaments and Z equivalents, occurrence of "shearing," and presence of intracellular skeletal fibrils, both cross and obliquely striated muscles seem to employ the same basic contraction and control mechanisms and both have been adapted, although perhaps not with equal success, to tonic as well as phasic functions. The phylogenetic distribution of obliquely striated muscles suggests, however, that they are better suited than cross-striated muscles to soft-bodied invertebrates.

REFERENCES

Andersson-Cedergren, E. (1959). *J. Ultrastruct. Res., Suppl.* 1.
Apathy, S. (1894). *Arch. Mikrosk. Anat.* 43, 886.
Bouligand, Y. (1966). *J. Microsc.* 5, 305.
Bowden, J. (1958). *Int. Rev. Cytol.* 7, 295.
DeBell, J. T. (1965). *Quart. Rev. Biol.* 40, 233.
DeBell, J. T., del Castillo, J., and Sanchez, V. (1963). *J. Cell. Comp. Physiol.* 62, 159.
Dewey, M. M., and Barr, L. (1964). *J. Cell Biol.* 23, 553.
Elliott, G. F. (1967). *In* "The Contractile Process," p. 171. Little, Brown, Boston, Massachusetts.
Englemann, T. W. (1881). *Arch. Gesamte. Physiol. Menschen Tiere* 25, 538.
Fawcett, D. W., and Revel, J. P. (1961). *J. Biophys. Biochem. Cytol.* 10, No. 4, Suppl., 89.
Franzini-Armstrong, C. (1970). *J. Cell Biol.* 47, 488.

Franzini-Armstrong, C., and Porter, K. R. (1964). *J. Cell Biol.* **22**, 675.

Hanson, J. (1957). *J. Biophys. Biochem. Cytol.* **3**, 111.

Hanson, J., and Lowy, J. (1961). *Proc. Roy Soc., Ser.* B **154**, 173.

Heumann, G.-G., and Zebe, E. (1967). *Z. Zellforsch. Mikrosk. Anat.* **78**, 131.

Hirumi, H., Raski, D. J., and Jones, N. O. (1971). *J. Ultrastruct. Res.* **34**, 517.

Hope, W. D. (1969). *Proc. Helminthol. Soc. Wash., D.C.* **36**, 10.

Hoyle, G. (1964). *In* "Physiology of Mollusca" (K. M. Wilbur and C. M. Yonge, eds.), pp. 313–351. Academic Press, New York.

Huxley, H. E. (1957). *J. Biophys. Biochem. Cytol.* **3**, 631.

Huxley, H. E. (1967). *In* "The Contractile Process," p. 71. Little, Brown, Boston, Massachusetts.

Ikemoto, N. (1963). *Biol. J. Okayama Univ.* **9**, 81.

Kawaguti, S. (1962). *In* "Proceedings of the Fifth International Congress on Electron Microscopy" (S. S. Breese, ed.), Vol. 2, p. M–11. Academic Press, New York.

Kelly, D. E. (1969). *J. Ultrastruct. Res.* **29**, 37.

Knapp, M. F., and Mill, P. J. (1971). *J. Cell Sci.* **8**, 413.

Marceau, F. (1909). *Arch. Zool. Exp. Gen.* **2**, 295.

Mill, P. J., and Knapp, M. F. (1970a). *J. Cell Sci.* **7**, 233.

Mill, P. J., and Knapp, M. F. (1970b). *J. Cell Sci.* **7**, 263.

Plenk, H. (1924). *Z. Anat. Entwicklungsgesch.* **73**, 358.

Reger, J. (1964). *J. Ultrastruct. Res.* **10**, 48.

Reger, J. (1967). *J. Cell Biol.* **33**, 531.

Rosenbluth, J. (1962). *J. Cell Biol.* **13**, 405.

Rosenbluth, J. (1963). *J .Cell Biol.* **19**, 82A.

Rosenbluth, J. (1965a). *J. Cell Biol.* **25**, 495.

Rosenbluth, J. (1965b). *J. Cell Biol.* **26**, 579.

Rosenbluth, J. (1967). *J. Cell Biol.* **34**, 15.

Rosenbluth, J. (1968). *J. Cell Biol.* **36**, 245.

Rosenbluth, J. (1969a). *J. Cell Biol.* **42**, 534.

Rosenbluth, J. (1969b). *J. Cell Biol.* **42**, 817.

Rosenbluth, J. (1972). *J. Cell Biol.* **54**, 566.

Roskin, G. (1925). *Z. Zellforsch. Mikrosk. Anat.* **2**, 766.

Schwalbe, G. (1869). *Arch. Mikrosk. Anat.* **5**, 205.

Wissocq, J. (1970). *J. Microsc.* **9**, 355.

Wright, K. A. (1966). *Can. J. Zool.* **44**, 329.

CRUSTACEAN MUSCLE

H. L. ATWOOD

I. Introduction

Crustacean muscles have several properties that make them of particular interest to researchers in muscle physiology. In the first place, the individual muscle fibers of crustaceans are comparatively huge; they reach diameters of 4 mm in giant barnacles and in the Alaskan king crab (Selverston, 1967). Obviously, fibers of this size are much better for experimental manipulation than the 20–100 μm fibers of most vertebrate striated muscles. Large crustacean fibers can be injected, voltage-clamped, or chemically analyzed with relative ease. Studies using such techniques have provided much useful information about excitation–contraction coupling and membrane excitability and have turned up unique and interesting features of the excitable membranes of crustacean muscle fibers. Second, the great variety of structure and of contraction speed in fibers of crustacean muscles offers fertile ground for correlated studies of structural and functional aspects of striated muscle. Third, the complex multiterminal innervation of crustacean muscles, which is rather different from the usual innervation of vertebrate skeletal muscles, is worth studying because it may serve as a useful model for situations occurring in central nervous systems (Katz, 1966). Elucidation of the synaptic connections and synaptic mechanisms in crustacean muscles may give some insights to similar patterns and processes within the central nervous system. As an example, one can point to the work on presynaptic inhibition in crustacean muscles (Dudel and Kuffler, 1961c).

Several recent reviews have dealt with various aspects of crustacean muscle. Wiersma (1961) has summarized earlier work on the innervation, contractions, and electrical responses of crustacean muscles. Reviews by Kennedy (1967), Atwood (1967a), and Hoyle (1969) cover more recent studies on innervation, electrical and contractile properties, and structural features of crustacean muscle fibers, while another review (Atwood, 1968) deals specifically with peripheral inhibition in crustacean muscles. Reuben et al. (1967a) have reviewed evidence relating to excitation–contraction coupling in crustacean muscle. Other, more general, articles include coverage of membrane responses (Grundfest, 1966, 1967) and synaptic transmission (Grundfest, 1969).

In the present article, an attempt will be made to synthesize some of the more recent work on structure and physiology of crustacean muscles. Since the article by Franzini-Armstrong (1970) on membrane systems and excitation–contraction coupling and the chapter by Pringle (Chapter 10) on contractile mechanisms of arthropod muscles include some of the relevant material on crustacean muscle, these subjects will be treated rather briefly here. The main emphasis will be placed on excitatory events leading to contraction, on neuromuscular synaptic arrangement and activity, and on certain correlations of structural and physiological features.

The general plan of attack will be to consider briefly first the structure, contractile properties, and electrical properties of the muscle fibers, then ionic mechanisms involved in electrical and contractile activity, and finally synaptic control of the muscle fibers. Throughout the article, frequent comparison of crustacean and vertebrate systems will be made, since most readers will undoubtedly be more familiar with the vertebrate situation and will want to know how the crustacean work "fits in."

Research papers reviewed in this article were mostly published prior to July, 1970.

II. Structure of Crustacean Muscle

An appreciation of the structural features of crustacean muscle is necessary for discussion of the physiology. As will be apparent, crustacean and vertebrate fibers differ structurally in many important respects. Thus, there is a limit to the general applicability of information gained from experiments on crustacean muscle, though many fundamental processes are probably similar in different striated muscles.

A. The Crustacean Muscle Unit

Crustacean muscle fibers are less easily defined than the striated muscle fibers of vertebrates. The latter are cylindrical structures, each enclosed in a well defined sarcolemma tube, and each distinct from its neighbors. By contrast, crustacean fibers often appear to branch and anastomose (Alexandrowicz, 1952) and frequently show evidence of electrical continuity with neighboring fibers (Reuben, 1960; Parnas and Atwood, 1966). When a dye, such as fast green (Thomas and Wilson, 1966), is injected into a fiber, it spreads into the electrically connected neighboring fibers (Jahromi, 1968; Mendelson, 1969). Anatomical study reveals the presence of small cytoplasmic bridges that connect adjacent fibers and apparently provide the pathway for spread of injected current or dye (Jahromi, 1968).

Crustacean muscle fibers do not increase or decrease in number during the postlarval growth of a muscle, but the individual fibers increase greatly in size (Bittner, 1968b). It is therefore likely that the appearance of electrotonically connected fibers in certain muscles is actually due to the branching and partial separation of individual muscle fibers. Probably, these become progressively invaded by connective tissue elements during enlargement and come to form subunits which never split entirely from the parent fiber, but remain attached by narrow bridges. Some subunits may ultimately separate completely from their parents. In muscles of very large lobsters, the number of electrically separate units and the number of cases of electrically connected subunits is greater than in the same muscles from smaller animals (Jahromi and Atwood, 1971b). This would not be expected if fusion of originally separate units occurred, as implied by the term "electrotonically coupled muscle fibers" (Reuben, 1960). In the latter case, the number of electrically independent units would decrease.

Since the crustacean muscle fiber can be a complicated affair, with branches, incomplete partitions, and partly separated subunits, it is perhaps preferable to use the term "unit" (Parnas and Atwood, 1966) in reference to a collection of electrically connected branches or subunits. Many crustacean muscle units are certainly far from fiberlike in their gross configuration. Some of the muscle units in crabs, crayfish, and other animals do conform more closely to the traditional idea of a muscle fiber. Since much of the work to be discussed has been done on the fiberlike muscle units of crayfish and crabs, and since also the term "fiber" is familiar and widely used, I will continue to employ it here in subsequent descriptions.

B. General Features of Fine Structure

The generalized morphology of part of a typical crustacean muscle unit is shown diagrammatically in Fig. 1. These features have been established in many independent investigations (e.g., Peachey and Huxley, 1964; Brandt et al., 1965; Peachey, 1967; Selverston, 1967; Hoyle and McNeill, 1968a; Franzini-Armstrong, 1970). There is disagreement about certain details, in particular about the Z tubules (Peachey, 1967).

Each muscle unit, or fiber, is surrounded by a sarcolemma (plasma membrane of 80 Å and amorphous basement membrane of about 150 Å in thickness). Often a filamentous layer, connective tissue cells, blood vessels, and other structures overlie the sarcolemma and are intimately associated with it.

The contractile material is organized into sarcomeres bounded by Z lines and with arrays of thick and thin myofilaments, giving rise to the A and I bands as in other striated muscles. The sarcomeres and

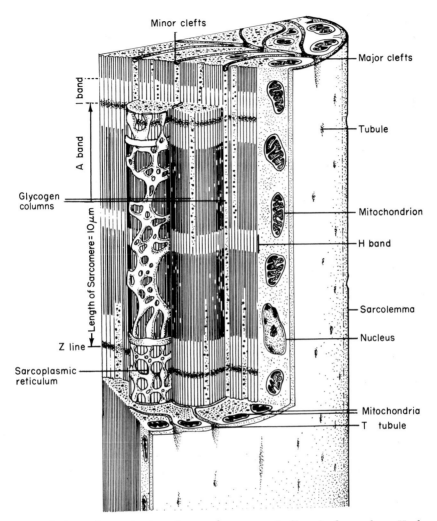

Fig. 1. General organization of a crustacean muscle fiber (redrawn from Hoyle, 1970). Sarcoplasmic reticulum and T tubules are shown only for one myofibril. Major and minor clefts represent parts of the system of sarcolemmal invaginations. The Z tubules are not shown in this diagram, but can be seen in the micrographs of Figs. 4 and 6.

the thick and thin filaments show much more variation in length than in vertebrate striated muscles. This variation occurs from one unit to another and even within a single unit (Franzini-Armstrong, 1970). An H zone is usually present near the center of each sarcomere, though in some cases it is obscured by variations in filament length and position.

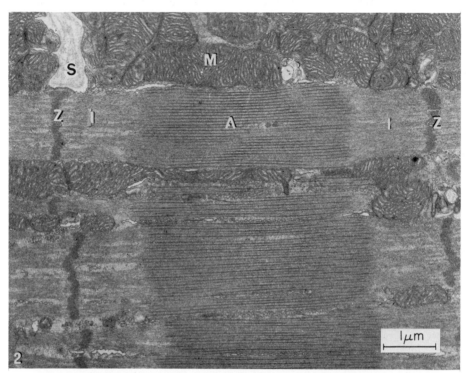

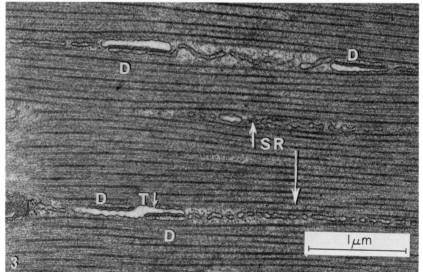

See facing page for legends.

An L line, resulting from an alignment of the regions of the thick filaments, which do not possess cross bridges, can also be seen. The bridge-free region is only 60 nm in crab thick filaments, compared with 120 nm in vertebrate thick filaments. In addition, the thick filaments of crab muscle are somewhat thicker than those of vertebrate muscle, with a more obvious electron-transparent core.

Shortening of the sarcomeres during contraction is associated with movement of thick and thin filaments past each other in accord with the sliding filament model (Huxley and Niedergerke, 1954). In barnacle fibers, extreme shortening (referred to as supercontraction) is accompanied by passage of thick filaments through openings in the Z disk into adjacent sarcomeres (Hoyle *et al.*, 1965).

The contractile material is split longitudinally into myofibrils, usually of irregular shape, surrounded by sarcoplasmic reticulum, which is continuous along the length of the myofibril. The sheet of sarcoplasmic reticulum is sometimes double and is often fenestrated. Glycogen columns also occur between the myofibrils, as do small mitochondria (mostly near the Z line). Larger numbers of mitochondria and the muscle nuclei occur at the periphery (Fig. 1).

One remarkable feature of crustacean muscle fibers is the elaboration of clefts invaded by the sarcolemma. The sarcolemmal invaginations branch extensively within the fiber and penetrate it to great depths, especially in the very large fibers of barnacles and crabs (Hoyle and Smyth, 1963; Selverston, 1967). No portion of the contractile machinery is more than 50 μm from the sarcolemma. Undoubtedly, the cleft system is functionally important in excitation and in the exchange of materials between the muscle unit and its external environment.

The transverse or T tubules, sometimes referred to as excitatory tubules (Hoyle and McNeill, 1968a) or A tubules (Peachey, 1967), arise from the sarcolemmal invaginations and also directly from the surface of the muscle fiber. The tubules consist of plasma membrane without any accompanying basement membrane. They are usually flattened along the longitudinal axis of the sarcomere. Examples of these tubules are shown in the micrographs of Figs. 2–5 and 7.

Fig. 2. Electron micrograph of a long-sarcomere fiber from the carpopodite extensor muscle of the walking leg of a crayfish (*Orconectes*). The section was cut near the surface of the fiber, where a thick layer of mitochondria (M) appears. Other mitochondria lie between the myofibrils. A major cleft or sarcolemmal invagination (S) is seen at the level of the Z line. A and I bands are readily apparent.

Fig. 3. Enlargement of the A band region of Fig. 2, showing dyadic contacts (D) between T tubules (T) and sarcoplasmic reticulum (SR). The T tubules are flattened longitudinally and they run obliquely in this example.

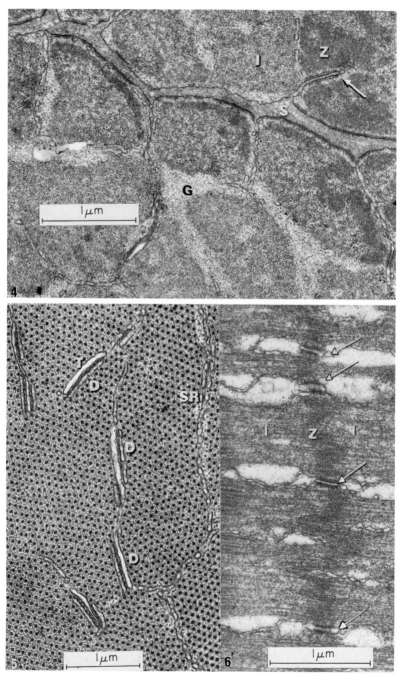

See facing page for legends.

Although the majority of the T tubules are oriented transversely, some run longitudinally between the myofibrils, or obliquely (Brandt *et al.*, 1965; Reger, 1967; Hoyle and McNeill, 1968a). They sometimes traverse the Z line from one sarcomere to the next (Brandt *et al.*, 1965, 1968). In the region of the A band, often near the junction of the A and I bands, the T tubules form dyadic contacts with the sarcoplasmic reticulum. In appearance, these dyads are quite different from the triads of vertebrate muscle, since there is no great elaboration of the sarcoplasmic reticulum into specialized terminal cisternae in crustaceans. Dyadic contacts between the surface membrane or sarcolemmal invaginations and the sarcoplasmic reticulum are not uncommon in crustacean muscle (Brandt *et al.*, 1965) (see Fig. 8).

A second type of tubule, the Z tubule, has been described in certain crustacean muscles (Peachey and Huxley, 1964; Peachey, 1967; Reger, 1967; Hoyle and McNeill, 1968a). These tubules are found in association with the Z lines of many crustacean muscle fibers. They were not described in leg muscles of crayfish (Brandt *et al.*, 1965), but they do in fact occur in these muscles (see Fig. 6). The Z tubules contain elements of the basement membrane, except in their finest ramifications. Probably these tubules are a modified sarcolemmal invagination. When embedded in the Z line, they are surrounded by a characteristic electron-dense material, as are also the sarcolemmal invaginations and T tubules that traverse the Z lines (see Figs. 4, 7, and 8). Brandt *et al.* (1965) believe that only one type of tubule is found in crayfish fibers and that the Z tubule and A tubule of Peachey (1967) are part of the same system, at least in crayfish muscle fibers. However, this is a minority opinion.

Although the Z tubules do not themselves form dyadic contacts with the sarcoplasmic reticulum, they often give rise to T tubules that run longitudinally between the myofibrils to form dyads near the edge of the A band. Reger (1967) has presented a very clear picture of this situation in the muscle of a crab (*Pinnixia*).

Evidence from local stimulation experiments, in which an extracellular microelectrode is pressed against a specific region of the external mem-

Fig. 4. Transverse section of a long-sarcomere fiber from the carpopodite extensor muscle of *Orconectes*. A sarcolemmal invagination (S) gives rise to the T tubules and also to Z tubules (arrow). Z line (Z), I band (I), and a glycogen column (G), are indicated.

Fig. 5. Enlargement of the A band region of Fig. 4, showing dyads (D), sarcoplasmic reticulum (SR), and T tubules (T). Some of the myofibrils are incompletely separated by sarcoplasmic reticulum in this fiber.

Fig. 6. Enlargement of the Z line region in a longitudinal section of the same fiber as shown in Fig. 4, showing Z tubules (arrows). Z line and I band are labeled.

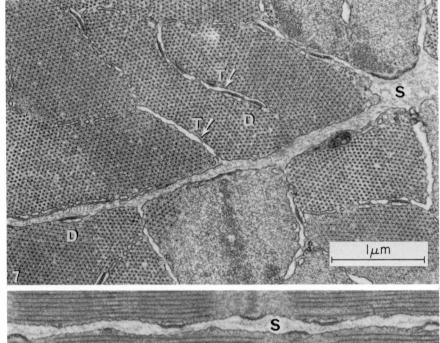

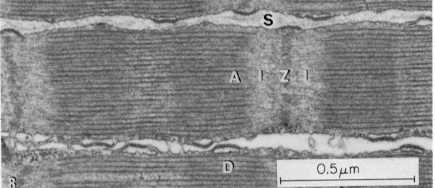

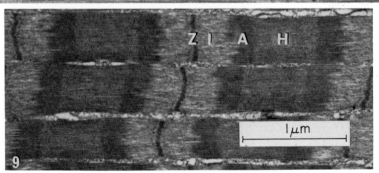

See facing page for legends.

brane while current is applied, suggests that the T tubules, but not the Z tubules, mediate excitation–contraction coupling. Local contractions could be elicited at the edges of the A band where many of the T tubules invaginate, but not at the Z line (Peachey and Huxley, 1964; Huxley and Peachey, 1964). The function of the Z tubules is not clear; they could function in exchange of metabolites (Peachey and Huxley, 1964), to provide mechanical support (Peachey, 1967), or to induce changes in the Z line during excitation or during supercontraction (Hoyle, 1969).

In view of the fact that some of the T tubules originate from Z tubules (Reger, 1967, Fig. 10), one might ask why Huxley and Peachey (1964) failed to observe local contractions at the Z line with external stimulation. One answer could be that the sarcolemmal invaginations, from which most of the Z tubules arise, act to short circuit much of the applied current. Thus, the local stimulation experiments do not conclusively rule out an excitatory function for the Z tubules.

C. Variations in Fine Structure

Considerable variation in fine structure occurs in crustacean muscles. Even within a single muscle, structural differences between fibers are often found (Dorai Raj, 1964; Atwood, 1965; Franzini-Armstrong, 1970; and many other studies). In this section, a brief resume of the variations in structural features encountered to date in crustacean muscles will be given, along with notes on possible functional significance.

1. SARCOMERES

Sarcomere length varies considerably among different crustacean muscle fibers and even within a single fiber (Franzini-Armstrong, 1970).

Fig. 7. Transverse section of a short-sarcomere fiber from the carpopodite extensor muscle in the walking leg of a crayfish (*Orconectes*). As in the long-sarcomere fibers of the same muscle, sarcolemmal invaginations (S) occur, which give rise to T tubules (T), but only rarely to Z tubules. Some of the dyads (D) occur between T tubules and sarcoplasmic reticulum, but others occur between vesicles of the sarcoplasmic reticulum and sarcolemmal invaginations.

Fig. 8. Longitudinal section of the same fiber shown in Fig. 7, showing a sarcolemmal invagination (S) running longitudinally. Note darkening of the walls of this structure in the region of the Z line. A second longitudinal cleft (probably a T tubule) occurs below S and forms dyads (D) in the A band region of the sarcomere.

Fig. 9. Another short-sarcomere fiber, also from the carpopodite extensor muscle of the walking leg of *Orconectes*, showing an unusually large H zone (H). A and I bands and Z line are labeled. The fiber was fixed under slight tension. Long-sarcomere fibers never exhibited this appearance.

In this respect, crustacean (and other arthropod) muscles differ profoundly from the striated muscles of vertebrates, in which the sarcomere
length of almost all fibers is 2–3 μm at rest.

Some crustacean muscles, such as the tonic abdominal muscles of
crayfish and lobsters (Kennedy and Takeda, 1965a,b; Parnas and Atwood, 1966; Jahromi and Atwood, 1967, 1969a), possess muscle units
with uniformly long (6–10 μm) sarcomeres. Other muscles, such as the
phasic abdominal muscles of crayfish and lobsters (Parnas and Atwood,
1966; Jahromi and Atwood, 1967, 1969a), have muscle units with short
(2–4 μm) sarcomeres (see Figs. 2, 3, 8, 9).

Many crustacean leg muscles have mixed fiber populations, as exemplified by sarcomere measurements on the two claw closer muscles of the
American lobster (see Fig. 10). The claws of each lobster are differentiated into a crusher, which is larger and contracts more slowly, and
a cutter, which is smaller and contracts more rapidly (Wiersma, 1955).
As can be seen from the histograms, the fiber populations of the two
homologous muscles differ profoundly. The crusher claw shows a unimodal distribution with a peak at 8 μm, while the cutter claw shows
a bimodal distribution with peaks at 3.4 and 9.4 μm. The short-sarcomere
units, which form the bulk of the cutter claw, do not appear at all
in the crusher claw (Jahromi and Atwood, 1971a).

Variation in sarcomere length is accompanied by variation in the
length of the thick and thin filaments (Swan, 1963; Franzini-Armstrong,
1970). In vertebrate fibers, the thick filaments are always about 1.5
μm long and do not vary within a fiber.

It has now been well established that rate of contraction is inversely
related to sarcomere length in crustacean muscle units. The long-sarcomere fibers give slowly developing contractions, and the short-sarcomere fibers more rapid ones (see Fig. 11). Hoyle (1967) has shown
that twitch duration in crab fibers is proportional to the logarithm of
sarcomere length. Other investigations are reviewed by Atwood (1967a)
and Hoyle (1969).

The relationship between sarcomere length and contraction speed was
postulated on the basis of the sliding filament hypothesis by Huxley
and Niedergerke (1954). Short-sarcomere fibers have more sarcomeres
in series per unit length of muscle than long-sarcomere fibers; hence,
more rapid shortening would be expected in the short-sarcomere fibers.
Huxley and Niedergerke (1954) further postulated that the long-sarcomere fibers should develop more tension per unit of cross-sectional
area, due to greater overlap of thick and thin filaments within the sarcomere. Measurements with several contraction-producing agents in

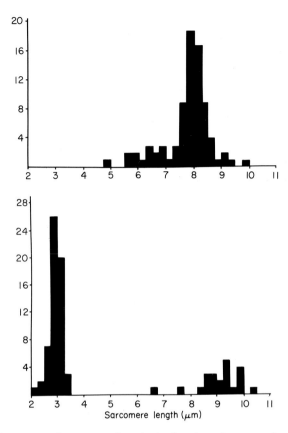

Fig. 10. Histograms of sarcomere lengths in the claw closer muscles of the lobster (*Homarus americanus*), to illustrate variation in muscle fiber properties within homologous but physiologically different muscles. The muscles were fixed in the open position, and a sample of 80 fibers was taken from each. For each fiber, the average of 3 sarcomere readings was taken. The number of muscle fibers with a given sarcomere length was plotted (vertical axis) against the length of the sarcomere. The top graph shows the result for the crusher claw and the bottom graph that for the cutter claw of the same animal (Jahromi and Atwood, 1971a.)

lobster phasic and tonic muscles have substantiated this suggestion for crustacean material (Jahromi and Atwood, 1969a).

2. SARCOPLASMIC RETICULUM

In many crustacean leg and abdominal muscles, the amount of sarcoplasmic reticulum per myofibril does not vary much, even in fibers of

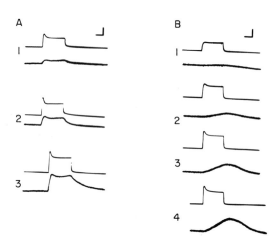

Fig. 11. Fast (A) and slow (B) contractions elicited from two different fibers in the carpopodite extensor muscle of the walking leg of the lobster (*Homarus americanus*). In each series, current was applied intracellularly through a microelectrode with increasing intensity. Upper traces show membrane voltage responses to applied current, and lower traces show tension response of the muscle fiber. Calibrations: voltage, 10 mV; tension, 20 mg; time, 0.2 sec. (Jahromi and Atwood, 1971a.)

markedly different sarcomere length (Jahromi and Atwood, 1967, 1969a; Selverston, 1967; Franzini-Armstrong, 1970). Thus, the pronounced differences in contraction time encountered in these muscles are likely not due to different relative amounts of sarcoplasmic reticulum, as postulated for fast and slow vertebrate muscle fibers (see, for example, Hess, 1965). However, it should be borne in mind that the chemical properties of the sarcoplasmic reticulum may not be the same in all crustacean muscles. Caffeine, which is thought to act on the sarcoplasmic reticulum causing it to release calcium (Ashley, 1967), has quite different effects on lobster fast and slow muscle. Such effects may be due to differences in sarcoplasmic reticulum in the two types of muscle (Jahromi and Atwood, 1969a).

In one extreme case, that of the fast-acting remotor muscle of the lobster antenna, there is a remarkable profusion of sarcoplasmic reticulum (Mendelson, 1969; Rosenbluth, 1969). In this case at least, it is likely that the abundant sarcoplasmic reticulum accounts for the very fast relaxation of the muscle by rapid removal of free calcium from the myofibrils. Variation in sarcoplasmic reticulum probably affects speed of contraction in crustacean muscle, but only in exceptional cases does it appear to set the ultimate limits on contraction speed.

3. T Tubules

In most crustacean muscles so far examined, the T tubules form two to three (sometimes four) rows of dyads in the region of the A band of each sarcomere. Two or more dyads usually occur around the circumference of each myofibril in the dyad-rich regions (see Figs. 5 and 7). The short-sarcomere fibers, other things being equal, possess a larger number of dyads per unit length of myofibril, as illustrated in Fig. 12. Calculations for lobster phasic and tonic abdominal muscles show about 600 dyads per millimeter of myofibril for the latter, 2500 for the former (Jahromi and Atwood, 1969a). Since the activation of calcium release to initiate contraction probably occurs at the dyads, the closer spacing and greater number of dyads in short-sarcomere fibers may contribute to their more rapid contraction rate, because diffusion distances for calcium from its release point to the most distant myofilaments would be less. Some long-sarcomere fibers in the legs of crabs and lobsters appear to have a reduced number of dyads per myofibril in the dyad-rich regions of the sarcomere (Fahrenbach, 1967; Jahromi and Atwood, 1971a). This may contribute to the low speed of contraction in these fibers.

An unusual arrangement of excitatory tubules has been described in the fast-acting muscle of a cyclopoid copepod (Fahrenbach, 1963). The tubules invaginate at the Z line and travel transversely into the fiber. They send longitudinal branches along the sides of the myofibrils, where contacts with cisternae of the sarcoplasmic reticulum are made. In addition, branches of the tubules penetrate the interior of the myofibrils, making further contacts with cisternal elements. This arrangement probably enhances speed of contraction in this muscle by reducing diffusion distances between sites of release of calcium and the myofilaments.

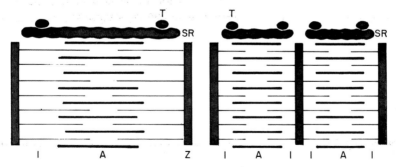

Fig. 12. Relationship between sarcomere length, the spacing of dyads along the myofibril, and the thick and thin filament lengths in crustacean fibers of long and short sarcomere length.

4. MITOCHONDRIA

Most crustacean muscle units have a layer of mitochondria at the periphery (see Fig. 2). In some with long sarcomeres, smaller mitochondria occur throughout the unit. They commonly lie between the myofibrils, near the Z lines. Muscle fibers of a crab eyestalk known to be involved in prolonged, tonic contractions are especially rich in mitochondria; indeed, such units may be pink due to their mitochondrial content (Hoyle, 1968; Hoyle and McNeill, 1968a). Other fibers in the same muscles are white; they contain many fewer mitochondria and fatigue rapidly. One is reminded of red and white muscles of vertebrates, which also show physiological specialization for tonic and phasic activity, respectively.

5. THICK AND THIN FILAMENTS

The fastest of the short-sarcomere crustacean muscle units have a regular arrangement of thick (myosin) and thin (actin–troponin–tropomyosin) filaments similar to that in insect flight muscles (Jahromi and Atwood, 1967; Reger, 1967) (see Fig. 13B). The overall ratio of thin to thick filaments in these muscles is 3:1. In vertebrate muscles (Fig. 13A), the ratio is 2:1.

In fibers of long sarcomere length, the ratio of thin to thick filaments is always greater than 3:1. Sometimes a fairly regular arrangement of twelve (to nine) thin filaments around each thick filament is seen, as in slow abdominal extensor muscles of crayfish and lobster (Jahromi and Atwood, 1967, 1969a) (see Fig. 13C). Here, the ratio of thin to

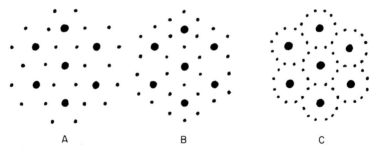

A   B   C

Fig. 13. Arrangements of thick and thin filaments viewed in transverse section in the A band of vertebrate striated muscle (A), some crustacean fast muscles (B), and some crustacean slow muscles (C). The ratios of thin to thick filaments are 2:1 (A), 3:1 (B), and 6:1 (C). In crustaceans, the ratios are often intermediate between B and C.

thick filaments approaches 6:1. In other muscles, such as leg and eyestalk muscles of crabs, the circlets of thin filaments may contain less than twelve, giving thin:thick ratios of 3.6:1 to 5.5:1 (Franzini-Armstrong, 1970). In still other cases, there is not an orderly arrangement of thin filaments around thick ones, and the overall ratio may be 6:1 or greater (Hoyle and McNeill, 1968a). Thick filaments of crayfish and lobster long-sarcomere fibers are slightly thicker (280 Å) than those in short-sarcomere fibers (210 Å) (Jahromi and Atwood, 1969a).

Although the ratio of thin to thick myofilaments tends to increase with sarcomere length, there is no evidence to suggest that this feature is causally related to contraction speed. In coxal muscles of the cockroach, the 3:1 and 6:1 filament ratios appear in different muscles of similar sarcomere length (Jahromi and Atwood, 1969b). These muscles have similar contraction speeds and develop similar amounts of tension per unit area of cross section during a twitch (Usherwood, 1962). It is possible that the total number of reaction sites between thick and thin filaments per unit length of filament does not differ greatly in the two types of muscle (Swan, 1963) and that sarcomere length is a more important determinant of contraction speed than the thin:thick filament ratio.

D. Summary

The above discussion indicates several structural parameters that may be important in determining the speed of contraction, total force, and tension-maintaining ability of crustacean muscle fibers. Of these, sarcomere length, number of dyads per unit length of myofibril, mitochondrial content, and in a few cases, the relative abundance of sarcoplasmic reticulum appear to be the most significant. By contrast, in vertebrate muscles, sarcomere length does not vary much, but the relative amount of sarcoplasmic reticulum and the relative number of triads may be important.

There has been little attempt to determine whether the chemical properties of the myofilaments differ in various crustacean muscles. Such chemical differences may also contribute to contraction speed. Bárány (1967) reported that ATPase activities of myosins extracted from various vertebrate and invertebrate muscles were generally proportional to the speed of shortening of their respective muscles. Constantin et al. (1967) applied calcium ions to frog fast and slow muscle fibers deprived of their sarcolemma and observed that differences in the time course of contraction in the myofibrils were comparable to those in intact fibers.

An excess of available Ca^{2+} could not increase the speed of contraction of the slow fibers to the level seen in the fast fibers. These experiments, among others, suggest that chemical properties of the myofibrils may be as important, or even more important, than structural features in determining the properties of a particular muscle unit. However, Rüegg (1968) argues that the structural features of the myofilaments are of primary importance in determining speed of shortening and economy of tension maintenance. Since there has been much structural but little chemical work on crustacean muscle, this problem remains unresolved and awaits further work.

III. Membrane Electrical Properties

A. *Estimates Based on a Cylindrical Fiber*

The earliest attempts to estimate values for the membrane electrical properties of crustacean muscle fibers used the methods of square pulse analysis previously applied to isolated axons (Hodgkin and Rushton, 1946) and to frog muscle fibers (Fatt and Katz, 1951). Values for membrane resistance and capacitance were calculated in terms of unit surface area of membrane, assuming the crustacean muscle unit to be a cylindrical fiber.

These estimates yielded very large values for membrane capacitance: about 40 $\mu F/cm^2$ in crab fibers (Fatt and Katz, 1953a; Atwood, 1963) and 20 $\mu F/cm^2$ in crayfish fibers (Fatt and Ginsborg, 1958). By comparison, values for frog fibers are 5–8 $\mu F/cm^2$ (Fatt and Katz, 1951); for mammalian fibers, 2.8–5.9 $\mu F/cm^2$ (Boyd and Martin, 1959; Kiyohara and Sato, 1967); and for squid axon, 1 $\mu F/cm^2$ (Curtis and Cole, 1938). Values for specific membrane resistance were comparatively low for most crustacean fibers: 80–1500 Ω cm^2 in crab fibers, compared with over 3000 Ω cm^2 in frog twitch fibers.

The early results for crustacean muscle membrane properties were impossible to reconcile with the known values of surface membrane thickness and dielectric properties. Even before information on fine structure was available, Fatt and Katz (1953a) suggested that the cylindrical model might be invalid due to deep infoldings of the surface membrane. Their evidence for such clefts came from the observation that a fiber's membrane potential would sometimes disappear, then reappear, as the tip of a microelectrode was advanced into the fiber.

B. Estimates Corrected for Total Surface Area

When the existence of sarcolemmal invaginations in crustacean muscle units was established by electron microscopy, further attempts were made to estimate membrane electrical properties, taking into account the total area of the surface plus the sarcolemmal invaginations. Of particular interest were the high values for membrane capacitance.

Peachey (1965) estimated a value of membrane capacitance close to 1 μF/cm² by taking into account the increased surface area contributed by the muscle tubules. From square pulse analysis, Selverston (1967) found values of 50 μF/cm² for crab twitch fibers before correction for increased surface area, and 5.7 μF/cm² after correctio for increased surface area due to sarcolemmal invaginations. He also used a second method of estimating membrane capacitance, based on analysis of the exponentially rising foot of the propagated action potential (see Tasaki and Hagiwara, 1957). Uncorrected and corrected values from this method were 14.7 μF/cm² and 1.5 μF/cm², respectively.

From estimates such as the above, it is evident that the membranes of crustacean muscle fibers do not differ drastically in passive electrical properties from the membrane of other nerve and muscle cells. The apparent high values of membrane capacitance are mainly a consequence of the structural pecularities of crustacean muscle.

C. AC Analyses

Attempts have been made to obtain more exact values for the capacitance and resistance of the transverse tubular system of crustacean muscle units (Falk and Fatt, 1964; R. S. Eisenberg, 1965, 1967). Alternating currents were passed into the fiber through a microelectrode, and changes in membrane impedence and in the phase angle between current and transmembrane voltage were observed over a wide frequency range (1 Hz to 10 kHz). These observations were used to deduce the nature of the equivalent electrical circuit of the muscle fiber.

Falk and Fatt (1964) found that the earlier model for the current path between the inside and the outside of a muscle fiber consisting of resistance and capacitance in parallel (Fig. 14A) could not be used to explain the results of the AC measurements. A separate current path had to be added in parallel with the first to take into account the properties of the tubular system. This pathway consists of resistive and capacitative elements in series (Fig. 14B). The revised "two time constant

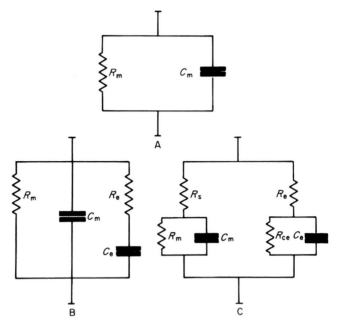

Fig. 14. Proposed electrical models of the muscle fiber membrane. In A, the older, one time constant model is given, with membrane resistance (R_m) in parallel with membrane capacitance (C_m). In B, the two time constant model is shown, with tubular resistance R_e and tubular capacitance C_e added. This model was satisfactory for frog and crayfish fibers. In C, a further elaboration of the model for crab fibers is shown (after R. S. Eisenberg, 1967). Basement membrane resistance R_s is added in series with surface plasma membrane resistance and capacitance R_m and C_m, respectively. R_e and R_{ce} represent tubular core resistance and tubular membrane resistance, respectively, in this model.

model" was successful in accounting for the results of AC measurements in both crayfish and frog muscles. The resistive element in the pathway representing the tubular system could represent the resistivity of the interior of the channels, or it could include the resistance of some structure (such as a dyad) at the ends of the tubules (see Falk, 1968). The capacitative element may represent the capacitance across the walls of the tubules.

For crayfish muscle, the estimated value of R_m (surface membrane specific resistance) is 680 Ω cm²; of C_m (surface membrane capacitance), 3.9 μF/cm²; of R_e (tubular resistive element), 35 Ω cm²; and of C_e (tubular capacitance), 17 μF/cm². All values are referred to unit area of fiber surface. It is noteworthy that the major part of the total capacitance is associated with the tubular system. Of course, there is no way of distinguishing the relative contributions of T tubules and Z tubules electrically.

R. S. Eisenberg (1965, 1967) extended the analysis to crab fibers and proposed a slightly different electrical model to explain the behavior he encountered (see Fig. 14C). Again, the major part of the membrane capacitance is associated with the tubular elements.

D. Comparison of Fibers with Different Structure

In spite of the wealth of observations on diversity of structure and contractile properties of crustacean muscle units, very little work has been done to estimate the electrical properties of different structural types. Falk and Fatt (1964) and R. S. Eisenberg (1965, 1967) give no information on sarcomere length or fine structure of the crayfish and crab fibers they studied. Eisenberg found three qualitatively different types of crab fiber, but it is not possible to identify them with structural types.

A preliminary comparison of crab fibers with long, intermediate, and short sarcomeres has revealed differences in membrane properties among the three types (Atwood, 1963). The long-sarcomere units have significantly lower total membrane capacitance and significantly higher specific membrane resistance than the others. The lower capacitance could be a consequence of the smaller total number of tubules and dyads in the long-sarcomere units (see Section II). The difference in specific membrane resistance could also result from the differing extents of the T system, or it could reflect a difference in ionic permeability analogous to that in frog twitch and slow fibers (Stefani and Steinbach, 1969). Further work is necessary to resolve these questions.

The differences in membrane characteristics mentioned above influence the properties of postsynaptic potentials recorded from the muscle fibers. Those with high membrane resistance have long membrane time constants, and the postsynaptic potentials recorded from such units are correspondingly prolonged.

E. Possible Implications

The electrical properties of the tubular system are thought to be of importance in the control of excitation–contraction coupling. If electrical activity at the surface of the muscle fiber is propagated inward electrotonically along the tubules, as suggested by local stimulation experiments

(Huxley and Taylor, 1958), the voltage changes at the dyads or triads will be attenuated and of slower time course due to tubular resistance and capacitance (Falk and Fatt, 1964; Falk, 1968). In vertebrate fibers, the relatively brief action potential will undergo serious distortion as it reaches the triads (Falk, 1968). However, the large size of this electrical change (125 mV) ensures that the complete cross-sectional area of the muscle fiber is activated (Adrian et al., 1969b).

In crustacean muscle, tubular conduction distances are, if anything, shorter than in larger vertebrate fibers, due to the extensive sarcolemmal invaginations (Selverston, 1967). Furthermore, the electrical changes set up by nerve activity are usually small, slow, summating postsynaptic potentials rather than propagated action potentials. Therefore, the problem of distortion of electrical activity by the properties of the tubular system is not as acute as in vertebrate fibers.

The two time constant model has been invoked to explain some peculiar phenomena in crustacean muscle. In muscles with dual motor innervation, one of the motor axons sometimes produces postsynaptic potentials of briefer time course than those evoked by the other in the same muscle fiber (Fatt and Katz, 1953b). Falk and Fatt (1964) suggested that this situation could arise from a difference in duration of transmitter action in the two axons. If, at synapses of one axon, transmitter action were shorter than $C_m R_e$ (see Fig. 14), not all of the membrane capacitance would experience the necessary charge displacement. The postsynaptic potential would thus have a rapid initial decay that would not involve C_e. If transmitter action were longer than $C_m R_e$ but less than $(C_e + C_m) R_m$ at synapses of the other axon, no rapid initial decay of the postsynaptic potential would occur.

So far, transmitter actions of the required duration have not been found in crustacean muscles. The values of $C_m R_e$ for crayfish fibers average 0.13 msec (Falk and Fatt, 1964) and for crab fibers, 0.32 msec (R. S. Eisenberg, 1967). Values of $(C_e + C_m) R_m$ are 30–100 times higher. Durations of transmitter action measured at crustacean synapses with external microelectrodes range from about 2 msec up to 20 msec (Atwood and Parnas, 1968). The required short durations (less than $C_m R_e$) have not been found. Furthermore, the total range of observed durations is not large enough to account satisfactorily for the difference in shape of the postsynaptic potentials.

Another problem that has not been cleared up yet is the observation by Bittner (1968b) that spontaneous miniature junctional potential amplitudes in crayfish muscle fibers are independent of the effective resistance of the fiber in which they are recorded. Furthermore, the effective resistance of the muscle fiber is not proportional to the two-thirds power of the diameter. Both observations are at variance with those in vertebrate muscle fibers (Katz and Thesleff, 1957). Perhaps the more complex morphology of crustacean units is responsible for this situation. In addition, inhomogeneity of electrical properties and morphology among different units of the same muscle could be partly responsible.

IV. Electrically Excited Responses

The striated twitch muscle fibers of vertebrates normally generate a propagated action potential when a depolarizing stimulus is applied to the membrane. Crustacean muscle fibers may occasionally produce action potentials of this type (Abbott and Parnas, 1965; Selverston, 1967; Ozeki, 1969), but more commonly they produce various graded responses, or delayed rectification (Fatt and Katz, 1953a; Werman and Grundfest, 1961; Atwood, 1963; Atwood *et al.*, 1965). Through experimental treatments, the graded responses can be converted into all-or-nothing action potentials, which have recently been extensively analyzed by voltage clamp techniques in giant barnacle fibers (Hagiwara *et al.*, 1969).

Electrical responses of excitable membranes are generally explained in terms of the ionic theory, based largely on measurements obtained from squid giant axons and subsequently on observations from a great variety of other cells (see Grundfest, 1966, for a review). This viewpoint will be assumed in discussing membrane responses in crustacean muscles. A brief consideration will be given to the ionic nature of extracellular and intracellular environments, which provide the driving forces for the electrical responses. Subsequently, some of the different electrical responses and the membrane permeability changes thought to be responsible for them will be discussed.

A. *Extracellular and Intracellular Environment*

1. TOTAL INORGANIC IONS

Many analyses of ions in blood and muscles of crustaceans have been reported. A few of these of interest for the present discussion are presented in Table I. Earlier results, such as those of Shaw (1955a), did not take into account the binding of ions or of water by muscle proteins. Subsequently, Robertson (1961) analyzed the abdominal flexor muscles

TABLE I

Total Ions in Crustacean Muscle Cells and Blood Plasma

Species	Inorganic ions (mM per kg water)						Reference
	Na$^+$	K$^+$	Ca^{2+}	Mg^{2+}	Cl$^-$	SO$_4^{2-}$	
Shore crab							Shaw (1955a)
(*Carcinus maenas*)							
Muscle cells	54	120	6	16	54	—	
Plasma	552	12	16	28	468	—	
Norwegian lobster,							Robertson (1961)
(*Nephrops*							
norvegicus)							
Muscle cells	24.5	188	3.7	20.3	53	1.0	
Plasma	517	8.6	16.2	10.4	527	18.7	
American lobster,							Dunham and Gainer
(*Homarus*							(1968)
americanus)							
Muscle cells	104	155	—	—	89	—	
Giant barnacle,	21	157	—	—	—	—	Hagiwara *et al.* (1964)
(*Balanus*	56	158	—	—	97	—	McLaughlin and Hinke
nubilus)							(1966)
Muscle cells	21	152	—	—	—	—	Beaugé and Sjodin (1967)
	21	150	—	—	—	—	Brinley (1968)

of *Nephrops* not only for inorganic ions but also for many organic constituents, and he was able to account satisfactorily for the osmotic balance between extracellular and intracellular osmotically active constituents. Some of the organic materials, especially amino acids, are important in osmotic regulation of crab muscle fibers. The fibers can actively transport amino acids to achieve volume regulation (Lang and Gainer, 1969a,b).

Evidence for binding of the inorganic cations by muscle proteins appeared in the work of Robertson (1961). He estimated that about 82% of intracellular sodium, 26% of the potassium, almost 100% of the calcium, and 64% of magnesium is not freely available in the fiber water. There was no evidence for binding of chloride. Evidence for binding of intracellular ions, and also of fiber water, has been forthcoming in recent studies such as those of McLaughlin and Hinke (1966), Dunham and Gainer (1968), and Hays *et al.* (1968). The experiments of McLaughlin and Hinke (1966) on giant barnacle muscle fibers, conducted with intracellular sodium- and potassium-sensitive glass microelectrodes, are of particular interest. Using flame photometry to determine total fiber content of the ions (Table I) and the cation-selective electrodes to measure the activities of sodium and potassium, they estimated that 42–65% of the fiber water was bound, as well as 84–91% of the intracellular sodium and 38% of the intracellular potassium. In a subsequent study, also on giant barnacle fibers (Allen and Hinke, 1970), influx and efflux measurements of radioactive sodium revealed two intracellular compartments for this ion. A rapidly exchanging compartment, identified as myoplasmic free Na^+, was estimated at 51–58% of the total intracellular content of this ion. The slowly exchanging compartment may be the sodium ion bound to muscle proteins.

Although chloride ion is usually assumed to be completely unbound (Robertson, 1961; McLaughlin and Hinke, 1966), there is some evidence that this ion shows "compartmentation" in lobster muscle. Dunham and Gainer (1968) found that about 30 mM per kilogram of cells, out of a total of 70 mM per kilogram of cells, was not exchangeable with extracellular chloride nor with a low-chloride extracellular environment. Possibly, the exchangeable fraction is associated with tubular or sarcolemmal structures. The reason that Robertson (1961) did not observe compartmentation seems to arise from the fact that chloride binding is dependent on the integrity of the cell (Dunham and Gainer, 1968) and that Robertson used a tissue press to extract muscle juice for analysis of unbound ions.

2. Donnan Equilibrium

In work on the euryhaline crab *Carcinus*, Shaw (1955a,b, 1958a,b) postulated that intracellular and extracellular environments were in Donnan equilibrium, since the concentrations of the diffusible potassium and chloride ions were in the required concentrations (see Table I),

$$[K^+_{out}]/[K^+_{in}] = [Cl^-_{in}]/[Cl^-_{out}]$$

However, the data of Robertson (1961) and of Reuben *et al.* (1964) did not show the required distribution, so it seems unlikely that a Donnan equilibrium exists for muscles of *Nephrops* or crayfish.

More recently, Hays *et al.* (1968) have analyzed muscle fibers of another eurhyaline crab, *Callinectes*, under various experimental conditions. After taking into account the bound ion fractions and bound water, they found that the ion distributions agreed well with the Donnan equilibrium condition. Thus, it would seem that various crustacean muscles differ, and that the mechanisms responsible for establishment and maintenance of ion distributions are not the same in all cases. Metabolically driven ion pumps may play a relatively greater role in those species in which a Donnan equilibrium does not exist.

3. Ion Pumps

Work on ion pumps, which have been well studied in frog and other vertebrate muscles, has not been emphasized by investigators of crustacean muscle. Sodium–potassium pumps have been described in crab muscles by Bittar (1966) and by Bittar *et al.* (1967) and in the giant barnacle fibers by Beaugé and Sjodin (1967) and by Brinley (1968). Fibers of these muscles have the ability to extrude sodium against a concentration gradient and to recapture potassium, as in vertebrate muscles. The cardiac glycosides ouabain and strophanthidin inhibit the activity of the pumps. It is not certain whether only one pump is involved or more than one, since there is still some active transport of sodium even after strophanthidin treatment (Brinley, 1968). The possibility exists that some active process (chloride pump?) plays a part in the distribution of chloride in lobster muscle (Motokizawa *et al.*, 1969).

B. Resting Membrane Potential

1. Probable Basis of Resting Potential; Potassium Electrode Behavior

The concentrations of the major inorganic ions of crustacean muscle cells and plasma (Table I) suggest that the inside negative resting membrane potentials of -60 mV to -80 mV commonly recorded by microelectrodes in crustacean muscle fibers could be established by the concentration "batteries" of the potassium and/or chloride ions. The equilibrium potentials for these ion species, estimated from the Nernst equation, are of approximately the right value to account for the observed resting membrane potentials. In *Carcinus*, for example, (Shaw, 1955a) (Table I) the equilibrium potential for potassium ($E_K = -58 \log[K_{out}^+]/[K_{in}^+]$) is -58 mV; the equilibrium potential for chloride ($E_{Cl} = 58 \log[Cl_{out}^-]/[Cl_{in}^-]$) is -55 mV; and Shaw's (1955a) measured resting potentials averaged -58 mV. The above example can be criticized on the basis that no correction was made for bound ion and water fractions and that the value of -58 mV is rather low for *Carcinus* fibers, compared with values obtained by other workers (Fatt and Katz, 1953a; Atwood, 1963). Nevertheless, if the membrane is permeable mainly to these ions at rest, one might expect the resting membrane potential to be close to the (accurately determined) equilibrium potentials of one or both of these ions. A more complete prediction of the membrane potential would be based on the familiar Goldman Equation:

$$E_\mathrm{m} = \frac{RT}{nF} \ln \frac{P_\mathrm{K}[\mathrm{K^+_{out}}] + P_\mathrm{Na}[\mathrm{Na^+_{out}}] + P_\mathrm{Cl}[\mathrm{Cl^-_{in}}]}{P_\mathrm{K}[\mathrm{K^+_{in}}] + P_\mathrm{Na}[\mathrm{Na^+_{in}}] + P_\mathrm{Cl}[\mathrm{Cl^-_{out}}]}$$

where P_Na, P_K, and P_Cl are the relative membrane permeabilities for sodium, potassium, and chloride ions, respectively; T is the absolute temperature, R the gas constant, F the Faraday constant, and n the valency of the ions. The membrane potential E_m is dependent on the concentrations and membrane permeabilities of all ions in the system. Sodium, potassium, and chloride are the most important in relation to the resting membrane potential. Usherwood (1969) has presented a thorough discussion of this problem with particular reference to insect muscles, and his discussion is largely applicable to other arthropod muscles as well. The reader is referred to his article for more extensive treatment.

It has been shown from osmotic experiments that crayfish and lobster muscle fibers are in effect impermeable to sodium ion under normal conditions (Reuben *et al.*, 1967b; Gainer and Grundfest, 1968), so the sodium term in the above equation can be neglected for these particular fibers. However, when calcium in the external solution is reduced to levels below 1–2 mmoles per liter, crayfish fibers become highly permeable to sodium (Reuben *et al.* 1967b). External calcium is essential for maintenance of the normal properties of the membrane, certainly in relation to sodium permeability.

Most crustacean muscle fibers appear to behave like reasonably good potassium electrodes. When the extracellular potassium ion concentration is changed, the membrane potential changes in accordance with the potassium equilibrium potential (Fatt and Katz, 1953a; Werman and Grundfest, 1961; Atwood and Dorai Raj, 1964; Grundfest, 1966; Hays *et al.*, 1968; and others). Perhaps the best example of such behavior occurs in lobster muscle fibers (Werman and Grundfest, 1961). In some crustacean muscles, the behavior is not perfect; for example, certain crab tonic fibers are rather insensitive to potassium ion at external concentrations lower than 30 mM/liter (Atwood and Dorai Raj, 1964), and muscle fibers of certain species of the crayfish genus *Orconectes* become sensitive only at concentrations greater than twice normal external potassium (Grundfest, 1966). For the most part, however, it would seem that crustacean muscle fibers have a relatively high resting potassium permeability, as do most vertebrate muscle fibers.

The large size and durability of giant barnacle muscle fibers have permitted Hagiwara *et al.* (1964) to study resting potential by injecting various solutions inside the fiber as well as changing external solutions. Injection experiments are not feasible with vertebrate muscle fibers. In barnacle fibers, decrease in the internal potassium ion level is accompanied by a decrease in membrane potential and an increase in $[K_{in}^+]$ by an increase in membrane potential, although the changes are less than expected for a potassium electrode. When internal potassium is increased beyond 250 mM, the membrane potential may even start to show a decrease. This anomalous result is difficult to explain. It has been observed also in internally perfused squid giant axons (Baker et al., 1962).

2. CHLORIDE CONDUCTANCE

Several studies have shown that chloride conductance is high in certain crustacean muscle fibers. In crayfish fibers it is approximately equal to potassium conductance (Zachar et al., 1964). In giant fibers of the crab Maia, chloride efflux was measured after injection of the isotope ^{36}Cl into the fiber (Richards, 1969). Measurement of efflux rather than influx avoids the difficulties of interpretation caused by the large extracellular space of the sarcolemmal invaginations. The results of this study showed that a major part of the membrane conductance was probably attributable to chloride movement. By contrast, lobster muscle fibers are reported to have a low chloride conductance (Grundfest, 1962).

Observations such as these indicate that the situation in many crustacean fibers is somewhat similar to that in frog twitch fibers (Hodgkin and Horowicz, 1959). The membrane is permeable to both potassium and chloride, and the membrane potential should therefore be influenced by the movement and distribution of both ions. One would predict changes in membrane potential following alterations in external chloride concentration. In fact, transient depolarizing changes in membrane potential are produced in fibers of the crab Callinectes when most of the external chloride is replaced by an impermeant anion such as methyl sulfate (Hays et al., 1968). Equilibrium of ion movements and stabilization of membrane potential occur when external and internal products of the diffusible ions are equal. The relatively long duration of chloride transients in crustacean fibers is attributable to the large size of the fibers and consequent slower equilibration of internal and external chloride concentrations.

When the external solution is changed in such a way that the product of external K^+ and Cl^- remains constant, the membrane potential of the barnacle fiber changes with a slope of 58 mV per tenfold change in potassium ion concentration, as expected from the Nernst equation

(Hagiwara *et al.*, 1964). Although in other crustacean muscle fibers the agreement with the Donnan equilibrium condition is questioned, there is no doubt that in almost all cases both K^+ and Cl^- influence membrane potential under certain conditions.

3. Influence of pH on Membrane Conductance

Hagiwara *et al.* (1968a) have made a study of the effects of external and internal pH on potassium and chloride conductances in the giant barnacle muscle fiber membrane following earlier studies of pH effects on frog muscle fibers (Hutter and Warner, 1967) and crayfish fibers (Reuben *et al.*, 1962; De Mello and Hutter, 1966). Chloride conductance increases with decreasing external or internal pH. Potassium conductance normally accounts for about five-sixths of the total membrane conductance in the barnacle fiber at pH 7.7, but as the pH is lowered, the chloride conductance becomes six to nine times greater than potassium conductance. The results are different from those in frog muscle, in which chloride permeability decreases with decreasing external pH (Hutter and Warner, 1967). An interpretation of the results in barnacle fibers is that the membrane is amphoteric, with fixed positive and negative charge groups controlling membrane permeability. At high pH, negatively charged groups predominate in the membrane, facilitating passage of cations and impeding anions. At low pH, positively charged groups predominate and anion permeability increases.

4. Structural Localization of Ion Permeabilities

Attempts have been made to localize structures in the muscle fiber where passage of specific ions occurs. This approach has been extremely fruitful for frog muscle fibers. Glycerol-soaked frog fibers returned to Ringer's solution lose the integrity of the transverse tubular system (Howell and Jenden, 1967; Howell, 1969; B. Eisenberg and Eisenberg, 1968). The entire chloride conductance of the membrane is retained, however, and is thus thought to be localized in the surface membrane of the fiber and not in the transverse tubules (R. S. Eisenberg and Gage, 1969). The transverse tubules account for over one half of the potassium conductance of the intact frog fiber.

Although the glycerol treatment has not yet been successfully applied to crustacean fibers, a number of other attempts have been made to localize chloride conductance (Girardier *et al.*, 1963; Brandt *et al.*, 1968; Reuben *et al.*, 1967a). Their experiments have been done on isolated crayfish fibers. Under conditions in which chloride leaves the fiber, such as with application of inward current through a potassium chloride-filled microelectrode, or return of the fiber to van Harreveld's solution after equilibration in a solution of raised potassium chloride, vesicles appear in the fiber. The vesicles result from swelling of portions of the transverse tubular system near the dyads. Selverston (1967) reports a similar finding in crab muscle.

The hypothesis accounting for these observations is that the terminal portions of the transverse tubular system are selectively permeable to chloride ions. When chloride leaves the fiber, it enters the terminal portions of the T system, while cations enter these same regions from

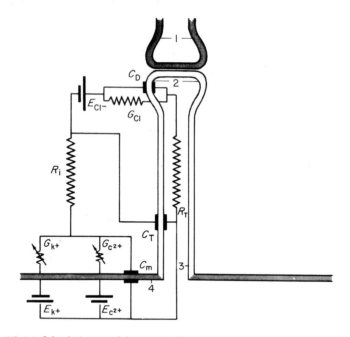

Fig. 15. Model of the crayfish muscle fiber showing regions of selective ion permeability (after Reuben *et al.*, 1967a.) Region 1 is the sarcoplasmic reticulum; region 2, the terminal portions of the T tubules, which are selectively permeable to Cl⁻; region 3, the relatively impermeable walls of the T tubules; and region 4, the surface membrane, which is permeable to K⁺ and to divalent cations (C²⁺). R_t and R_i are resistances of T tubule core and sarcoplasm, respectively. E_{Cl^-}, E_{K^+}, and $E_{C^{2+}}$ are the equilibrium potentials of the above ions, and G_{Cl^-}, G_{K^+}, and $G_{C^{2+}}$ their conductances. Capacitative properties of dyadic membrane, T tubule wall, and surface membrane are represented by C_D, C_T, and C_m, respectively. At rest, $E_{K^+} = E_{Cl^-}$, and no current flows between dyadic regions and the surface; but if the surface membrane becomes depolarized, current flow will occur. At the dyads, it will consist of an inward movement of chloride through G_{Cl^-} (the channeled current hypothesis).

the external medium via the lumens of the T tubules. Since only anions can cross the terminal parts of the T tubules, there will be a net accumulation of salts in the T terminals. Water will then accumulate through normal osmosis, and swelling will result.

The model of the crayfish fiber, based on these observations, is quite different from that proposed for vertebrate fibers (see Fig. 15). According to the crayfish fiber model, the chloride permeability is mainly in the T tubules, not the surface membrane; the potassium permeability is a property of the surface membrane (and of sarcolemmal invaginations?) and not of the T tubules. The major part of the T system is thought to be rather impermeable to ions; that is why swelling occurs only in terminal sacs.

It will be noted that the model provides a current loop through the T system, sarcoplasm, and sarcolemma. Part of the current flowing across the sarcolemma during changes in membrane potential can return to the exterior through the T system. This feature has been made the basis of a "channeled current" hypothesis of excitation–contraction coupling (see Section V and Reuben *et al.*, 1967a). The ultimate mechanisms underlying localization of specific ion permeability in certain structures involve chemical differences in the membranes of these structures. There is little concrete knowledge of such differences at the moment.

C. Types of Electrically Excited Response in Normal Fibers

Crustacean muscle fibers can be made to produce a wide variety of electrically excited responses. Not only do various crustacean fibers normally show a wide variation in their electrical behavior, but they produce many interesting responses when treated with various ions and drugs. A brief resume of some of the "normal" responses with notes on their probable ionic basis is presented here.

1. Responses to Depolarizing Current

A. Rectification. Many fibers, for example the slow or tonic fibers of crabs (Atwood, 1963; Atwood and Dorai Raj, 1964; Atwood *et al.*, 1965) and of lobsters (Jahromi and Atwood, 1969a) show rectification when the membrane is depolarized. The rectification shows up in voltage–current plots, in which it is apparent that the voltage response of the membrane to injected current shows proportionately less increase at stronger depolarizing currents.

Delayed rectification is also often observed; i.e., the voltage response to a maintained current decreases from its initial value to a lower maintained value shortly after application of the current.

As in frog slow muscle fibers (Burke and Ginsborg, 1956), the rectifications in crustacean fibers are probably due mainly to an increase in membrane potassium conductance (potassium activation). Treatments known to block potassium activation, such as exposure to tetraethylammonium or barium ions (Werman and Grundfest, 1961), abolish the rectification.

B. Graded Depolarizing Responses. The most common response to outward current in crustacean fibers is a depolarizing response, the amplitude of which is dependent on the strength of the applied current (hence the term graded response). Usually the response is seen at membrane potentials 10–20 mV more positive than the resting potential. In some fibers, the response is oscillatory during a prolonged depolarizing current. The graded response may be only a few millivolts in height, or it may be as large as 40–50 mV. Most fibers of intermediate to short sarcomere length, and also some of long sarcomere length, show these responses (Atwood, 1967a).

At one time it was postulated that graded responses were due to sodium activation which was highly damped by membrane potassium conductance, as in nonpropagated threshold responses of axons (Werman and Grundfest, 1961). More recently, the technique of intracellular injection of giant muscle fibers has been used to demonstrate that the graded and oscillatory responses are due to calcium rather than sodium activation (Hagiwara and Naka, 1964; Hagiwara and Nakajima, 1966a,b). By varying the amount of free calcium within the muscle fiber, it is possible to get either graded responses or spikes. Graded and oscillatory responses occur when internal free calcium is of the order of 10^{-7} M; spikes occur at 0.7×10^{-7} M or less. Oscillatory responses disappear at internal Ca^{2+} concentration greater than 8.3×10^{-7} M. The amplitude of the graded responses increases with increasing external Ca^{2+}, whereas changes in Na^+ have little effect. Clearly, inward flow of positive current carried by Ca^{2+} seems to be the basic mechanism for the graded response.

C. All-or-Nothing Action Potentials. Some crustacean muscle fibers, more commonly the short sarcomere, rapidly contracting ones, produce propagated, all-or-nothing action potentials under normal physiological conditions (Fatt and Katz, 1953a; Abbott and Parnas, 1965; Parnas and Atwood, 1966; Selverston, 1967; Ozeki, 1969). The spikes are longer in duration, and have a slower conduction rate, than those in vertebrate muscle fibers due to the high membrane capacitance of crustacean fibers (Section III). The normally occurring spikes of crustacean muscle fibers are dependent on an adequate external calcium concentration and are not very sensitive to external Na^+ (Abbott and Parnas, 1965; Ozeki, 1969). Furthermore, these spikes are not blocked by tetrodotoxin (Ozeki, 1969), which normally abolishes the sodium spike of other cells. There is little doubt that these spikes are caused by inflow of positive current carried by Ca^{2+} and that they are essentially similar to the experimentally induced calcium spikes, which will be described later (Section V,D).

2. RESPONSES TO HYPERPOLARIZING CURRENT

A. RECTIFICATION. Grundfest and his co-workers have studied the responses of crayfish and lobster muscle fibers to hyperpolarizing currents very extensively (Grundfest, 1966). Fibers of crayfish walking leg muscles normally show rectification and delayed rectification to hyperpolarizing current; the rectification is more pronounced with stronger currents. Peak and plateau phases occur just as with rectification to depolarizing current. There is little doubt that the source of the rectification is chloride activation. The rectification is abolished by soaking the fiber for a short time in a solution in which chloride has been replaced by an impermanent anion such as propionate.

B. HYPERPOLARIZING RESPONSES. Lobster muscle fibers and chloride-deficient crayfish muscle fibers give rise to a prolonged, large hyperpolarizing response when membrane potential is hyperpolarized by 50–100 mV (Reuben *et al.*, 1961; Grundfest, 1966). This effect is brought on by a regenerative increase in membrane resistance due to a shut-down of potassium conductance (potassium inactivation). The increase in membrane resistance produces a greater hyperpolarization for the same applied current, which in turn leads to higher membrane resistance, in regenerative fashion. The response is terminated by the onset of chloride activation. Hyperpolarizing responses do not occur when barium ion, which inactivates potassium conductance and forecloses further inactivation, has been applied. The responses to hyperpolarizing current are of interest in the study of membrane behavior, but they are probably of little importance to normal physiological functioning of the muscle fibers.

D. Analysis of the Experimentally Induced Calcium Spike in Giant Muscle Fibers

It has long been known that·exposure of crustacean muscle fibers to substituted ammonium ions such as choline or tetraethylammonium chloride (TEA) to certain divalent cations (Ca^{2+}, Sr^{2+}, Ba^{2+}) or to procaine enables these cells to generate large, all-or-none, prolonged action potentials similar in shape to those of the vertebrate heart (Fatt and Katz, 1953a; Fatt and Ginsborg, 1958; Werman and Grundfest, 1961; Takeda, 1967). Other arthropod muscles share this characteristic (Werman *et al.*, 1961). The action potentials persist in the absence of sodium; in fact, choline was originally tried on crustacean muscle fibers because it was known to be an impermeant substitute for sodium in the study of the action potential of the squid axon. It was thought that the crustacean action potential would disappear, like the squid action potential, in a solution containing choline in place of sodium. Instead, the action potential became enlarged and prolonged, and it was obvious that a different ionic mechanism had been found (Fatt and Katz, 1953a).

Calcium ion or one of the closely related divalent cations (Sr^{2+}, Ba^{2+}) must be present externally for these action potentials to occur. Thus, it was apparent that some, if not all, of the inflowing positive current required for the action potential must be carried by divalent cations. It has been suggested that Na$^+$ might contribute some of this current in lobster fibers (Werman and Grundfest, 1961), but recent work has not supported this suggestion for other crustacean muscles. Werman and Grundfest (1961) showed that the prolonged action potential results in part from the loss of the normal potassium conductance and potassium activation of the membrane. Barium ion, TEA, etc., apparently block the potassium channels of the membrane. The spike slowly subsides as potassium conductance sluggishly increases.

With the advent of robust, easily manipulated giant muscle fibers of the large barnacles *Balanus nubilus* and *Balanus aquilus* (Hoyle and Smyth, 1963), techniques for manipulating the internal as well as external ionic environment have been applied. In addition, voltage clamp techniques have provided an analysis of ionic currents during the action potential. Most of these impressive experiments have been carried out by Hagiwara and his co-workers since 1964.

When calcium-binding agents such as citrate, potassium sulfate, or ethylenediamine tetraacetate (EDTA) are injected into the giant barnacle fiber, the normal graded responses become converted into all-or-nothing spikes (Hagiwara and Naka, 1964). Intracellular calcium must be reduced to below 8×10^{-8} M for the spikes to occur at normal external Ca²⁺ levels (Hagiwara and Nakajima, 1966b). The spikes overshoot the zero membrane potential level, making the membrane inside-positive, and the magnitude of the overshoot increases with raised external Ca²⁺ but also with decreased internal K⁺ (Hagiwara *et al.*, 1964) (see Fig. 16). In fact, the magnitude of the inside-positive membrane potential is very closely proportional to the logarithm of the ratio $[\text{Ca}^{2+}_{out}]/[\text{K}^{+}_{in}]$, showing a change of 29 mV for a tenfold change in this ratio, as expected for a biionic potential between a calcium salt on the outside of the membrane and a potassium salt on the inside. Such a possibility was suggested earlier by Fatt and Ginsborg (1958). Voltage clamp studies (see below) have substantiated this mechanism.

Calcium influx during the action potential was studied with radioactive tracer (⁴⁵Ca). In the resting fiber, the influx was 14 pmole/sec per square centimeter of membrane. The influx increased greatly during the spike, and was estimated at 35–85 pmole per spike. This amount corresponds to 7–17 microcoulombs, which is five to ten times the amount of charge required to change the voltage on the membrane capacity (of 13–17 μF/cm²) by 80 mV (Hagiwara and Naka, 1964). Thus, there is ample current carried across the membrane by Ca²⁺ to account for the spike.

The calcium spike is insensitive to tetrodotoxin, which blocks the sodium spike in most cells. However, the calcium spike can be suppressed by manganese, cobalt, or certain other cations (Hagiwara and Nakajima, 1966a; Hagiwara and Takahashi, 1967). The ions which inhibit the calcium spike apparently compete with calcium for sites on the membrane. Different cations are bound at the same membrane sites, but with different dissociation constants. Cobalt and manganese are bound more firmly

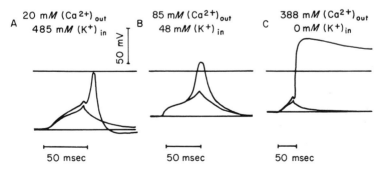

Fig. 16. Effects of variation of internal K⁺ and external Ca²⁺ on resting and spike potentials in a giant barnacle fiber. Note changes in amplitude and duration with increase in the ratio $[\text{Ca}^{2+}_{out}]/[\text{K}^{+}_{in}]$. After Hagiwara *et al.* (1964).

than calcium, which in turn exceeds magnesium or strontium in its binding properties.

The barnacle fiber membrane shows a shift in the membrane potential at which the spike is initiated (the spike threshold) toward more positive values with increasing calcium (or total divalent cation) concentration (Hagiwara and Naka, 1964; Hagiwara and Takahashi, 1967). This phenomenon is similar to the stabilizing action of calcium in squid axons (Frankenhaeuser and Hodgkin, 1957) and is interpreted to mean that calcium is adsorbed at the surface of the membrane by fixed negative charges associated with phospholipid or protein molecules. Depolarization may release calcium from these sites.

Recently, more detailed analysis of the ionic currents of the calcium spike have been carried out by means of voltage clamp techniques (Hagiwara et al., 1969). As in the squid axon, there is an early transient current and a late outward current. When the peak of the early current is plotted against the clamped membrane potential over a range of values for the latter, a result as shown in Fig. 17 curve A is obtained. The late outward current is shown by curve D. The early transient current appears at about −40 mV and is maximal at −10 mV. At more positive membrane potentials, the amplitude of the early inward current declines, and it reverses from inward current to outward current at about +40 mV.

Subdivision of the early transient current into inward and outward components can be accomplished in several ways. When the membrane is given a conditioning depolarization before the spike is elicited, the early transient inward current is suppressed, something in the manner of the sodium current of squid axon (Fig. 17, curves B and C). However, an early outward current persists and is not subject to inactivation. This suggests that the early transient is made up of two components, one subject to inactivation, and the other immune to it. The inward component of the early transient is suppressed by cobalt ion and is augmented by an increase in external Ca^{2+}; it, then, is the calcium current. The outward component is suppressed by procaine, and is thought to be carried by K^+. The late outward current (Fig. 17, curve D) is also carried by K^+.

The conclusion from this work is that the barnacle fiber membrane undergoes an early conductance change for Ca^{2+}, which is subject to inactivation. The outward component of the early transient current can be carried by K^+ through the membrane; the magnitude of this current is in agreement with that expected from the resting membrane conductance; hence, it is unlikely that the Ca^{2+} channel allows K^+ to pass through it. Potassium ion conductance increases following the early

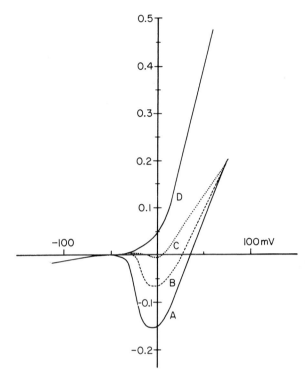

Fig. 17. Voltage clamp experiment on a giant barnacle muscle fiber (from Hagiwara *et al.*, 1969). The ordinate is the value of clamped membrane potential. The abscissa is the peak of the early transient current (A, B, C) and the steady state current 125 msec after the start of the spike-producing test voltage pulse (D). Curve A is the result for the normal fiber, held before the test at a membrane potential of −70 mV. Curves B and C were obtained when the test was preceded by a conditioning voltage pulse of 780 msec. In B, the membrane was conditioned at a membrane potential of −30 mV, and in C, at a membrane potential of −16.5 mV. Conditioning of the membrane at values more positive than −45 mV leads to partial or almost complete abolition of the inward (Ca^{2+}) component of the early transient current.

transient current, and the resulting outflow of K$^+$ (the late outward current) would normally repolarize the fiber.

Since the calcium spike only appears in the barnacle fiber when intracellular calcium is reduced to a low level, it is of interest to enquire whether those crustacean muscle fibers which normally produce a spike have lower free Ca^{2+} than those which do not normally spike. The fast flexor muscle of *Nephrops* (Table I) has only half the total calcium of the slower *Carcinus* leg muscle, while this amount, in turn, is lower than the total calcium (11 mM) in tonic fibers of *Cancer* (Atwood and Dorai Raj, 1964). Rapid contraction often is associated with the ability to spike (Atwood, 1967a). Values for total intracellular calcium tell us nothing about the values for free calcium, but it is at least possible that free calcium is higher in the slow fibers—especially those that maintain steady tension (Hoyle, 1968)—than in the fast ones. This may have something to do with spiking ability.

The analysis of the calcium spike in giant muscle fibers of crustaceans is of interest in relation to other excitable tissues. The action potential of the vertebrate heart, for instance, is partly dependent on calcium ions (Orkand and Niedergerke, 1964; Niedergerke and Orkand, 1966; Hagiwara and Nakajima, 1966a; Beeler, and Reuter, 1970). It is impossible to analyze the action potential in small heart cells with the same thoroughness which has been employed in the case of the giant barnacle fiber. The work on the latter may tell us something about the mechanisms of calcium-dependent spikes in other cells.

Calcium spikes may be present in still other places where they cannot be readily analyzed. The suggestion has been made (Costantin, 1970) that the sarcoplasmic reticulum may be capable of regenerative potential changes during the action potential. Since the calcium gradient across the sarcoplasmic reticulum membrane is probably large at rest, the postulated regenerative changes may be mediated by calcium. If so, the work on the crustacean calcium spike would be of considerable general interest.

V. Ionic Movements and Contraction

Recent work on crustacean muscle has added experiments and information of considerable general interest for physiologists studying excitation–contraction coupling. Once again, the large size of the crustacean muscle fibers has permitted injection experiments and voltage clamp studies of a type which would be difficult or impossible to perform with vertebrate muscle fibers. The studies on crustacean material have emphasized the role of ionic movements in controlling contraction, but have not been directed to any extent toward the fundamental biochemical mechanisms in the contractile filaments.

A. Membrane Depolarization and Contraction

Studies on the relation between membrane depolarization and contraction have been carried out with three basic techniques: (1) direct depolarizations of an impaled fiber by current injected through a glass microelectrode (Orkand, 1962; Atwood *et al.*, 1965; Reuben *et al.*, 1967a); (2) voltage clamp (Hagiwara *et al.*, 1968b; Dudel and Rüdel, 1969); (3) application of drugs and ions, in particular raised potassium, to induce contracture (Zachar and Zacharova, 1966; Reuben *et al.*, 1967a). Most of the reports based on these techniques note a threshold membrane potential at which contraction is first detectable. In crayfish leg muscle fibers, direct electrical stimulation and application of high K^+ have revealed thresholds of -60 mV to -50 mV (Orkand, 1962; Zachar and Zacharova, 1966). Voltage clamping has shown a similar threshold (Dudel and Rüdel, 1969). However, Reuben *et al.* (1967a), using sensitive tension-recording techniques, have shown that the "threshold," if one exists, is much closer to the resting potential; depolarizations of only a few mV are enough to give a slight contraction. The contraction remains small as membrane potential is lowered until the "threshold" described by other authors is reached; then it increases substantially. These observations are supported by Bittner (1968a).

In all such cases of apparent disagreement, it is worthwhile to remember that crustacean fibers often vary enormously both within a muscle and from species to species. Thresholds of different fibers in a crab muscle were quite variable (Atwood *et al.*, 1965). A few supersensitive fibers were found in *Cancer*, in which the threshold practically coincided with the resting potential, but most fibers appeared to require a substantial depolarization of 10–20 mV to develop any tension. The same is true in crayfish abdominal muscles (Parnas and Atwood, 1966; Atwood *et al.*, 1967). In the muscles controlling eyestalk movements in the crab *Podopthalmus*, Hoyle (1968) has even found fibers which are normally in weak contracture at the resting potential. Thus, the "threshold" is extremely variable in different crustacean fibers, and general statements about this parameter based on observations of only one type of fiber (Reuben *et al.*, 1967a) are of limited value.

The relation of the threshold for contraction to the threshold for graded responses or spike activity is also quite variable (Atwood *et al.*, 1965). Some fibers do not contract unless a large graded response or spike occurs; others contract at membrane potentials more negative than those required for electrically excited responses. The apparent threshold for a fiber, and the amount of tension it develops, depend upon both the duration and the strength of an applied stimulus. This is well shown in the data of Reuben *et al.* (1967a) for crayfish fibers, and in the measurements of Edwards *et al.* (1964), of Hagiwara *et al.* (1968b), and of Ashley and Ridgway (1968) on directly stimulated and voltage-clamped barnacle muscle fibers (see Fig. 18).

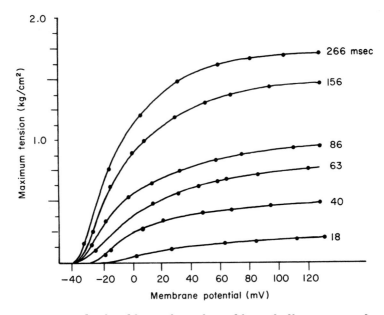

Fig. 18. Tension developed by a voltage-clamped barnacle fiber at various durations of stimulating pulse, and for various membrane potentials (from Hagiwara *et al.*, 1968a). Note shift in apparent contraction "threshold" to more negative membrane potentials with longer durations of the stimulating pulse. Note, however, that the "threshold" does not become more negative than about −40 mV.

The relation between membrane potential and contraction is always an S-shaped curve with a mechanical saturation level (in crayfish) at about −20 mV membrane potential (Zachar and Zacharova, 1966). The result is qualitatively similar to that found in vertebrate fibers (Hodgkin and Horowicz, 1960).

With maintained depolarization, some crustacean muscle fibers relax spontaneously after a few seconds or a few minutes (Atwood and Dorai Raj, 1964; Zachar and Zacharova, 1966), while others (tonic fibers) can maintain tension for at least ½ hr, or even indefinitely (Atwood and Dorai Raj, 1964; Jahromi and Atwood, 1969a; Hoyle, 1968). After spontaneous relaxation in high K⁺, a long rest period must be allowed before the fiber can again develop the same amount of tension in response to the same stimulus. Apparently, repriming of a control system (see Hodgkin and Horowicz, 1960) must occur. There is some doubt as to whether a K⁺-induced contracture can be considered equivalent to depolarization by other means. In the phasic fibers of the lobster abdomen, it is almost impossible to get any potassium contracture, even though a maintained tension appears with a prolonged spike (Jahromi and Atwood, 1969a). Tonic fibers of the same animal develop a long-lasting potassium contracture.

B. Movement of Calcium

As in other striated muscles, calcium is required for contraction in all of the crustacean muscles that have been examined critically. Calcium must reach the region of overlap of thick and thin filaments to cause contraction (Gillis, 1969). Injection of calcium directly into a fiber causes it to contract (Caldwell and Walster, 1963; Ashley, 1967). The free calcium ion level required for contraction is 10^{-7}–10^{-6} M. The calcium-binding agent, ethyleneglycol bis(β-aminoethylether)-N,N'-tetraacetic acid, EGTA, when injected into giant fibers of crab or barnacle, suppresses the caffeine-elicited contraction when present in an amount equal to the free calcium ion level. In barnacle fibers, the amount of EGTA required to suppress a potassium contracture is considerably greater than the amount effective against caffeine. This suggests that caffeine probably acts to release calcium internally from the sarcoplasmic reticulum, but that potassium may in addition cause entry of external calcium into the fiber (Ashley, 1967). Gainer (1968) has also suggested that potassium and caffeine act differently on lobster muscle, the former causing release of calcium from superficial sites, or entry of external calcium, and the latter causing release from deep sites (probably the sarcoplasmic reticulum). There is no doubt that potassium and caffeine potentiate each others' effects (Jahromi and Atwood, 1969a). Thus, it is quite possible that two movements of calcium could be involved in contraction of crustacean muscles—an intracellular movement from the sarcoplasmic reticulum and an inward movement from the exterior or from superficial release sites.

Additional evidence for the inward movement is provided by tracer studies, which show large influxes of calcium during a potassium contracture (Gainer, 1968). Also, Hagiwara et al. (1968b) found enhancement of tension with raised external Ca^{2+} in voltage-clamped barnacle fibers, which they interpret as indicating an inward movement of Ca^{2+}. Cobalt ions, which compete with Ca^{2+} for sites at the membrane, may suppress the inward movement and interfere with tension development.

Evidence for the second movement of calcium, intracellular release from the sarcoplasmic reticulum, has been provided by a number of studies. First of all, the local stimulation experiments (Huxley and Peachey, 1964; Peachey and Huxley, 1964) indicate a participation of the T tubules in excitation–contraction coupling. Second, injected or externally applied caffeine causes contraction as described above, and it is known that crustacean sarcoplasmic reticulum accumulates Ca^{2+} and releases it when treated with caffeine, just as does sarcoplasmic reticulum of vertebrate fibers (Van der Kloot, 1965, 1966).

A third line of evidence has been provided by the experiments of Ridgway and Ashley (1967; see also Ashley and Ridgway, 1968; Hoyle, 1969, 1970). They injected giant barnacle fibers with the bioluminescent protein aequorin, extraction from jellyfish, which emits light when the free calcium ion level is 10^{-8}–10^{-6} M (Hastings et al., 1969). The reaction is highly specific for calcium and very sensitive as an intracellular calcium indicator, since photomultiplier tubes can be used to measure light emission of an injected giant barnacle fiber. It is superior to murexide, which was used by Jöbsis and O'Connor (1966) as an intracellular calcium indicator in toad muscle fibers.

Examples of simultaneous records of light emission, membrane depolarization, and tension development in a single barnacle muscle fiber are given in Fig. 19. A weak electrical stimulus produces a small membrane depolarization and a barely detectable increase in light emission, but no tension (once again, there seems to be "threshold" for mechanical activation). Stronger stimuli elicit a larger light emission and considerable tension. Both light output and tension are graded according to stimulus strength and duration (see Fig. 19, curves A and B). The latent period between onset of stimulation and onset of light output and tension decreases with increasing stimulus intensity.

The peak of the calcium transient coincides with the maximum rate of rise of tension, and the integrated area of the calcium transient (light emission $\times$ time) is proportional to the total amount of force developed (Ridgway et al., 1968). The calcium transient decays exponentially and is terminated by the time relaxation occurs, indicating the lack of availability of free calcium for the aequorin reaction at that time. Probably, calcium is bound to other intracellular components such as troponin and is finally reaccumulated by the sarcoplasmic reticulum.

There is little doubt that the free calcium available in the sarcoplasm for the aequorin reaction comes from the sarcoplasmic reticulum, rather than from the exterior. In the barnacle fiber, a calcium spike does in fact contribute an inflow of calcium sufficient for contraction (see Hagiwara et al., 1968b; Hoyle, 1970), but barnacle fibers do not produce these spikes under normal conditions. The rapid onset of the calcium transient (within 1 msec of depolarization past threshold) is also inconsistent with diffusion of calcium 20–50 μm into the fiber from the surface or from the sarcolemmal invaginations.

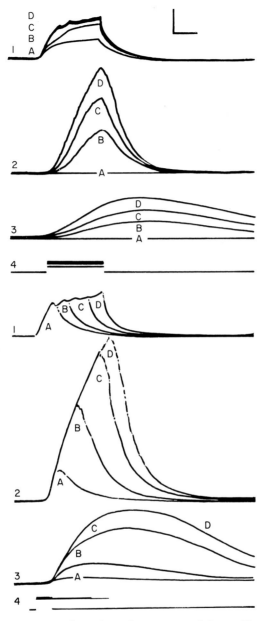

Fig. 19. Simultaneous recording of membrane potential (trace 1), calcium transient (trace 2), tension (trace 3), and stimulating current (trace 4) in a giant barnacle fiber injected with aequorin (from Ashley and Ridgway, 1968). In the upper set of records, the effect of increasing the strength of the stimulating current (from A to D) is shown. In the lower set of records, current strength was held constant, but duration increased. Calibration: (*vertical*) trace 1, 20 mV; trace 2, 3.8×10^{-9} lumens (top) and 1.9×10^{-9} lumens (bottom); trace 3, 5 gm; (*horizontal*) 100 msec.

Depolarization of the fiber with potassium gives light output an order of magnitude greater than that seen during a twitch contraction (Ridgway *et al.*, 1968). In this case, it is quite possible that the contribution of external calcium is appreciable, as suggested earlier.

The work with single giant crustacean fibers has provided new ways of examining calcium movement during contraction and has shown that the active state of muscle is dependent on the duration and amplitude of depolarization, as indicated by availability of free intracellular calcium ion.

C. The Channeled Current Hypothesis

Work on crayfish walking leg muscle fibers (Girardier *et al.*, 1963; Reuben *et al.*, 1967a) has provided the basis for a channeled current hypothesis of excitation–contraction coupling. The selective permeability of the terminals of the T tubules for chloride ions described earlier, combined with relative impermeability of the rest of the T system (see Fig. 15), could provide a "channel" for current carried by chloride ions. At rest, the potassium equilibrium potential (which appears across the surface membrane) and the chloride equilibrium potential (which appears across the T system terminals) are approximately equal. When the surface membrane is depolarized (for example, during synaptic activity or addition of potassium chloride), the potential across the T terminals will be more negative than that at the surface. Consequently, current will flow between these two areas. Inside the fiber, the current could be carried by several ions, but across the T terminals, it can only be carried by Cl⁻. Chloride ions will move into the fiber at the dyadic contact points, creating local regions of high current density which may be effective in causing the sarcoplasmic reticulum to release Ca^{2+}. The current path outside the fiber will be completed via the T system.

The hypothesis predicts that contraction should be a continuous function of membrane potential and may occur without depolarization under certain conditions, provided chloride enters the cell. The concepts of a critical membrane potential, or threshold, at which contraction is initiated and of an electrotonic spread of depolarization down the T system to initiate contraction are rejected by this hypothesis, although these concepts form an integral part of the thinking of many other workers, particularly those working with vertebrate muscle (e.g., Hodgkin and Horowicz, 1960; Orkand, 1962; Falk and Fatt, 1964; Taylor *et al.*, 1969; etc.). Instead of electrotonic depolarization of the dyadic regions, membrane current is postulated as the crucial link in excitation–contraction coupling. The link between chloride movement and release of calcium from the sarcoplasmic reticulum remains disappointingly speculative.

Evidence in support of this hypothesis (Reuben *et al.*, 1967a) derives from experiments in which contraction is elicited in various ways (depolarization with excess potassium, intracellular stimulation, etc.), while the extracellular environment is manipulated through alteration of ion content or application of drugs. Among other findings, it has been shown that the mechanical "threshold" for potassium chloride contracture is very much closer to the resting potential in the crayfish fibers used by Reuben *et al.* (1967a) than was reported for other crayfish fibers by Orkand (1962), Zachar and Zacharova (1966), and others. The discrepancy is ascribed by Reuben *et al.* (1967a) to a difference in sensitivity of tension-detecting techniques; the tension near the resting membrane potential is very slight and does not become readily detectable until a membrane potential of about -60 mV (the "threshold" of other workers) is attained. At this point, the membrane conductance also shows an appreciable increase. The correlation between the increased membrane conductance and the readily detectable tension suggests the possibility of an increased flow of chloride into the fiber and hence, more efficient excitation–contraction coupling. It is worth noting here that in frog fibers, there is evidence that the increased membrane conductance associated with delayed rectification is probably not causally linked to the onset of contraction (Adrian *et al.*, 1969a).

The experiments of Reuben *et al.* (1967a) suggest that conditions which increase chloride flux, such as previous treatment with potassium chloride followed by a return to the normal crayfish solution also increase the effectiveness of excitation–contraction coupling. Less potassium chloride is required to elicit a subsequent contracture following this treatment. Conversely, conditions that decrease chloride movement also frequently decrease contraction. Substitution of the impermeant anion propionate for the chloride in the crayfish solution inhibits contraction, either in response to direct electrical stimulation or with raised potassium. In chloride-free crayfish solution, larger depolarizations are necessary to elicit minimal tension, and the amount of tension for a given depolarization is much more dependent on the level of external Ca^{2+} than in chloride-containing solution. The results could mean that extracellular calcium enters the fiber more readily when chloride is the predominant external anion.

After treatment with procaine, and a period of soaking in chloride-free (propionate) solution, a directly stimulated crayfish fiber will produce a prolonged spike accompanied by relatively little tension (Fig. 20). Addition of chloride externally shortens the spike, but at the same time,

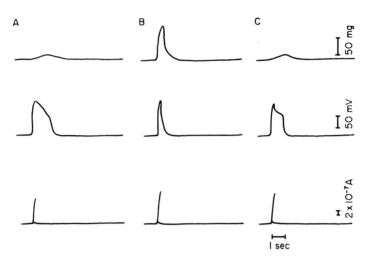

Fig. 20. Tension recorded from a single crayfish muscle fiber (*top traces*), during procaine spike (*middle traces*) elicited by a stimulating current pulse of short duration (*lower traces*). The fiber was equilibrated in a propionate saline with 10^{-3} gm/ml procaine previous to stimulation. In A, a prolonged spike, but little tension, was obtained. In B, chloride-containing saline was introduced. The procaine spike shortened, yet tension increased greatly. Note also that the speed of contraction increased. In C, propionate saline was restored, and the tension declined. (From Reuben *et al.*, 1967a.)

the tension increases severalfold. This experiment, like those quoted above, suggests a role for chloride in excitation–contraction coupling.

The channeled current hypothesis has not gone undisputed. Other workers have put forward observations which they feel do not support it. For example, Gainer (1968) reports that maximal potassium contractures can be obtained in lobster muscle fibers when no chloride is present in the external medium. In this case, he argues, chloride entry could not be contributing to excitation–contraction coupling, since no chloride entry could occur. However, the extensive experiments of Reuben *et al.* (1967a) demonstrate a marked dependence of tension during potassium contracture on both external Cl^- and external Ca^{2+}.

Dudel and Rüdel (1969) studied contraction of voltage-clamped crayfish muscle fibers in chloride-free solutions and observed that the most significant alteration of contractile activity in propionate solutions was a marked slowing of relaxation at the end of a depolarizing pulse. Their interpretation of this effect is based upon the assumption that relaxation is controlled by repolarization of the T system, which may be chloride-dependent. If chloride conductance is localized at the T terminals, as shown in Fig. 15, and if it is reduced in chloride-free solutions, the repolarization of the T system at the end of a stimulating pulse would likely be slowed down, as is the falling phase of the action potential of a vertebrate muscle fiber in chloride-free solutions (Hutter and Noble, 1960; Falk and Landa, 1960). A slower repolarization of the T system could be responsible for the observed slowing of relaxation, if in fact the potential difference across the T terminals is the crucial factor controlling tension.

In another experiment, Dudel and Rüdel substituted methyl sulfate for chloride in the bathing solution. They found that membrane conductance in methyl sulfate increased for membrane potentials at which contraction was initiated and that relaxation became more rapid. They claim that methyl sulfate, unlike propionate, can penetrate the membrane, possibly at the sites normally permeable to chloride (the T terminals). Since methyl sulfate is chemically rather different from chloride, and since contraction remains unimpaired, it can be argued that chloride is not specific for excitation–contraction coupling. The normal function of chloride is repolarization of the T system, according to this view.

Grundfest (1970) questions the assertion that the crayfish muscle membrane is readily permeable to methyl sulfate. He finds similar absence of slowing with the impermeant anions methane sulfonate and isethionate. The slowing of contraction in propionate is obtained not only with direct electrical stimulation and with potassium contracture, but also when calcium is injected intracellularly. The latter mode of application bypasses the membrane control mechanisms and suggests a direct effect of propionate on the contractile proteins. This effect is highly dependent upon pH.

Obviously, there are disagreements not only over the interpretation of the many observations pertinent to the channeled current hypothesis, but even over some of the observations themselves. Much of the evidence supplied by Reuben *et al.* (1967a) can be interpreted in more than one way and thus does not provide unequivocal support for this hypothesis. The anion permeability of the T terminals is accepted by Reuben *et al.* (1967a) and by Dudel and Rüdel (1969). The former authors feel that the contraction is controlled by the conductance of the T terminals and the current passed by them, while the latter authors take the view that the potential difference across the terminals is more important.

Since propionate appears to differ in its effects on contraction from other supposedly impermeant anions, the question arises, is propionate a suitable substitute for chloride in experiments made with chloride-free solutions? If indeed propionate has a direct action on the contractile proteins, as suggested by the calcium injection experiment (Grundfest, 1970), the experiments purporting to show chloride-dependence of contraction by comparison of events in propionate and chloride solutions must be questioned. At the moment, the point that chloride is not a specific activator of excitation–contraction coupling appears to be well taken.

The channeled current hypothesis represents an imaginative attempt to provide a new viewpoint on excitation–contraction coupling in crustacean muscle, but to date the experimental evidence has not been sufficient to displace the more widely held electrotonic spread hypothesis (Huxley and Taylor, 1958; Falk and Fatt, 1964; Falk, 1968). Even so, the channeled current hypothesis has emphasized important differences between crayfish and vertebrate muscle fibers, and there are recent indications that some elements of this hypothesis are being incorporated into the recent versions of the electrotonic model. For example, Costantin (1970) has postulated changes in tubular membrane conductance (possibly for Na^+) that may generate a current responsible for the spread of depolarization along the T system. This process is absent in tetrodotoxin-treated fibers, in which the safety factor for propagation of excitation to the interior of the fiber is low. Thus, changes in tubular membrane conductance and in current flow across the tubular membrane may be important in normal excitation–contraction coupling, although possibly in a way somewhat different from that suggested by Reuben *et al.* (1967a) for crayfish fibers.

VI. Efferent Control of Muscular Contraction

Probably the divergence between crustacean and vertebrate skeletal muscles is most apparent in a comparison of the efferent innervation. In the vertebrates, many parallel motor axons supply a given muscle; hence, any one motor axon makes only a small contribution to the overall performance of the muscle. In crustaceans, economy of innervation is the dominant theme. The number of axons devoted to efferent control is usually small; in fact, a single axon may control an entire muscle in some cases. Since each axon is responsible for a relatively large fraction of the activity in the muscle it innervates, it is not surprising that a compensatory complexity is found in the neuromuscular synaptic organization.

Crustacean muscle fibers differ from those of vertebrates in their possession of both excitatory and inhibitory neuromuscular synapses. Apparently, the occurrence of peripheral inhibitory axons confers additional flexibility of response to crustacean neuromuscular systems. In the vertebrates, peripheral inhibition does not occur in skeletal muscles. In considering neuromuscular synaptic control of crustacean muscles, I will discuss first the two types of transmission and then synaptic organization and synaptic performance.

A. Excitatory Synaptic Transmission

Electrical manifestations of excitatory neuromuscular synaptic transmission in crustaceans include spontaneous miniature potentials (Dudel and Orkand, 1960) and nerve-evoked junctional potentials, or excitatory postsynaptic potentials (EPSPs), as some authors prefer to call them. These may, if sufficiently large or with sufficient summation, trigger secondary electrically excited responses in the muscle fiber membrane (graded responses, spikes, or rectification, as discussed above in Section IV). The sizes and the facilitation rates of different EPSPs vary enormously in different muscle fibers, even when branches of the same motor axon are involved. This variation with be discussed in more detail below in Section VI,D.

The chemical basis for excitatory neuromuscular synaptic transmission in crustaceans has not been as well resolved as has the chemistry of inhibitory synaptic transmission (see Section B). At one time it was suggested that separate fast and slow motor axons which supply certain crustacean muscles might each release a different transmitter. The fast and slow transmitter substances were thought to evoke fast and slow contractions, respectively in fibers throughout the muscle by specific chemical effects on the contractile machinery or on excitation–contraction coupling (Hoyle and Wiersma, 1958a,c). Although the possibility of different excitatory transmitters cannot yet be conclusively ruled out, the effects that suggested it have since been explained in terms of the differences in muscle fiber properties and innervation that commonly occur in crustacean leg muscles (Atwood, 1967a). The slow motor axons evoke large electrical responses in slowly-contracting muscle fibers; the fast motor axons produce their greatest effect in rapidly contracting muscle fibers. No fresh evidence in support of the suggestion of separate transmitter substances has recently appeared (Atwood and Parnas, 1968).

There is, however, a possibility that more than one excitatory chemical may be present in crustacean nerve–muscle preparations. Various amino acids, of which the most effective is l-glutamate, produce excitatory effects in crustacean muscles at relatively low concentrations (Grundfest *et al.*, 1959; Robbins, 1959; Van Harreveld and Mendelson, 1959) and appear in extracts of crustacean nerve (Kravitz *et al.*, 1963). So, too, does a substance termed factor S, which appears to be an amine and which is extracted from crustacean muscle and nerve (Van der Kloot, 1960; Cook, 1967). The latter substance was reported to be released into the perfusion fluid upon stimulation of crustacean motor nerves. Chemically, it appeared to be a derivative of nicotinic acid (Van der Kloot, 1960). However, Armson and Horridge (1964) found that labeled nicotinic acid was taken up equally well by motor and sensory nerve and that the labeled material extracted from the nerve had no effect on crustacean muscle. Thus nicotinic acid is unlikely to be a precursor of the crustacean excitatory transmitter. The results of Cook (1967) provide confirmation for the existence of Van der Kloot's factor S in arthropods and suggest that the substance is a biogenic amine. So far, the physiological experiments with this substance have been crude, and there is no proof that it acts at the excitatory neuromuscular synapse, as would be required if it is a transmitter.

By contrast, the physiological experiments with l-glutamic acid have been carried to a high level of refinement, principally by Takeuchi and Takeuchi (1963, 1964, 1965) and by Ozeki and Grundfest (1967) and Ozeki et al. (1966). Their experiments were performed using the technique of iontophoretic application of glutamate from a microelectrode. The muscle membrane is unresponsive to l-glutamate except at the neuromuscular synapses. The synapses can be located accurately through the use of an extracellular microelectrode, which when positioned correctly, records the localized current flowing through the postsynaptic membrane during synaptic activation. When a glutamate-filled microelectrode is positioned at such a spot, a rapid depolarization of the muscle fiber, correlated with an extracellularly recorded flow of current into the postsynaptic membrane, occurs when glutamate is released from the microelectrode. If the electrode is moved from the site, or pushed into the muscle fiber, the depolarization disappears or is drastically attenuated.

Further observations on the glutamate potential have shown that it exhibits rapid desensitization with steady application of glutamate (Takeuchi and Takeuchi, 1965). The effect is analogous to that seen with acetylcholine application to neuromuscular junctions of vertebrate twitch fibers. EPSPs also show desensitization with glutamate application. A very low maintained dose of glutamate acts to potentiate the EPSPs, however.

The ionic mechanism of glutamate action (and, by inference, of the EPSP) has been studied by Ozeki et al. (1966) and by Ozeki and Grundfest (1967). The glutamate-evoked potential requires sodium in the extracellular medium; lithium is not effective as a substitute, even though it can replace sodium in spike generation of certain axons and muscle fibers (Schou, 1957). Furthermore, movement of potassium through the glutamate-activated synaptic membrane does not occur in the absence of sodium; movements of the two ions are coupled. Tetrodotoxin, which blocks nerve conduction in crustaceans and sodium activation in many nerve and muscle cells, has no effect on glutamate-induced potentials, on EPSPs set up by local depolarization of a nerve terminal, or on spontaneous miniature EPSPs (Ozeki *et al.*, 1966). Sodium activation in the synaptic membrane is thus completely different from that in electrically excitable membranes.

From the above discussion, it can be concluded that there is a strong case for glutamate as an excitatory transmitter in crustacean muscle.

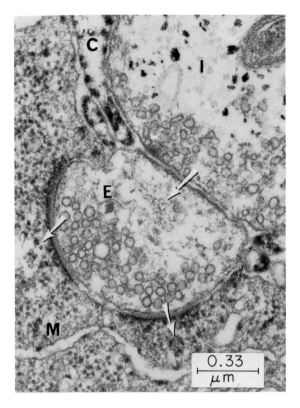

Fig. 21. Electron micrograph of synapses in the opener muscle of the crayfish *Procambarus.* Two nerve terminals (E and I) are in synaptic contact, and E also forms neuromuscular synapses. Direction of transmission is indicated by arrows. Note that the synaptic vesicles in I are less regular in shape than those in E. A Schwann cell finger is indicated by C (Atwood and Morin, 1971).

The accumulating evidence for insect muscles also favors glutamate as an excitatory transmitter (Usherwood, 1967, 1969). However, since the evidence is not completely conclusive, one should perhaps not reject the possibility that other agents may also be released from the nerves, either as transmitters or as trophic agents. Recently, electron micrographs of crustacean neuromuscular synapses have revealed not only the clear synaptic vesicles (Jahromi and Atwood, 1967) (see Fig. 21), but also smaller numbers of dense-cored vesicles reminiscent of the amine storage granules of other systems. Thus, there is morphological evidence suggesting the presence of two agents in the nerve terminals, although as yet it is not possible to say what chemical(s) might be associated with the two morphological entities.

B. Inhibitory Synaptic Transmission

Electrical activity associated with inhibitory innervation in crustacean muscle fibers includes spontaneous miniature potentials (Reuben and Grundfest, 1960) and inhibitory junctional potentials or postsynaptic potentials (IPSPs). The potentials may be hyperpolarizing, depolarizing, or absent, depending on the electrochemical conditions within the muscle fiber and in the external solution (Boistel and Fatt, 1958; Hoyle and Wiersma, 1958b).

In addition to the inhibitory potentials recorded from the muscle fibers, a second manifestation of inhibition is found in some muscles—the reduction of the EPSP by inhibitory stimulation timed to precede it by 1–4 msec (Marmont and Wiersma, 1938; Fatt and Katz, 1953c). Dudel and Kuffler (1961c) have shown that this effect is due to presynaptic inhibition of the excitatory nerve terminals by inhibitory transmitter action. This action reduces output of the excitatory transmitter. Quantum content of the EPSP is reduced, while quantum size remains the same (Dudel and Kuffler, 1961c; Takeuchi and Takeuchi, 1966a).

Recently, morphological evidence for synapses between adjacent nerve terminals (presumed excitatory and inhibitory) has been found in crustacean muscle (Atwood and Jones, 1967) (see Fig. 21). These synapses are probably responsible for the presynaptic inhibitory effect (Atwood and Morin, 1971).

The chemical basis of inhibitory synaptic transmission has been thoroughly studied in crustacean muscles and is well understood (see review by Atwood, 1968). There is now little doubt that γ-aminobutyric acid (GABA) is the transmitter. This substance duplicates all of the physiological effects of inhibitory nerve stimulation. It acts at the inhibitory synaptic regions (identified by external recording) when released ionto-phonetically (Takeuchi and Takeuchi, 1965, 1966a,b). It produces both pre- and postsynaptic inhibition. It is also released into the surrounding extracellular fluid when inhibitory nerves are stimulated (Otsuka *et al.*, 1966). Furthermore it is present at high concentration in the inhibitory neurons, but not in excitatory or sensory neurons (Kravitz *et al.*, 1963; Otsuka *et al.*, 1967). All of the requirements for acceptance of GABA as a transmitter substance seem to have been satisfied. Isolated inhibitory neurons have a high concentration of the enzyme glutamate decar-boxylase, which converts glutamate into GABA (Kravitz *et al.*, 1965). Excitatory neurons have a lower concentration of this enzyme. Since both types of neurons contain glutamate, it is likely that the crucial difference between the two types is the higher level of the decarboxylase enzyme in the inhibitory neurons.

Properties of the postsynaptic receptors at the inhibitory neuromuscu-lar synapses have been well studied. The ionic mechanism of inhibition involves an increase in chloride conductance of the activated postsynap-tic membrane (Boistel and Fatt, 1958). Replacement of the chloride of the external solution by an impermeant anion (e.g., propionate) re-duces or eliminates both pre- and postsynaptic inhibition after a period of equilibration (Takeuchi and Takeuchi, 1966b). In both normal and low-chloride solutions, the IPSPs and GABA-induced membrane poten-tial changes reverse polarity from hyperpolarizing to depolarizing ap-proximately at or near the chloride equilibrium potential

$$(E_{Cl} = 58 \text{ mV} \times \log[Cl_{in}^-]/[Cl_{out}^-])$$

(Boistel and Fatt, 1958). The agreement is closer if account is taken of the mobile and immobile fractions of intracellular chloride mentioned previously (Motokizawa *et al.*, 1969).

The result of inhibitory action on the muscle fiber membrane is to drive the membrane potential toward the chloride equilibrium potential, which normally is close to the resting membrane potential. Also, in-creased membrane conductance leads to reduction of EPSPs. In the excitatory nerve terminal, the increase in chloride conductance short circuits the nerve terminal potential set up by the action potential, thereby reducing the terminal potential change and hence the effective-ness of transmitter release (Dudel, 1963, 1965c).

Some differences have been found between the GABA receptors of the muscle fibers and those of the nerve terminals. In the crayfish, the latter are activated by β-guanidinopropionic acid (β-GP) but the former are not (Dudel, 1965b). However, β-GP and GABA seem to compete for the muscle membrane receptors, since the IPSP is reduced under the influence of β-GP. In the crayfish, the GABA receptors do not become desensitized with prolonged application of GABA (Takeuchi and Takeuchi, 1965). Thus, the GABA and glutamate receptors are distinct, although with iontophoretic techniques they appear to occur in the same locations (due presumably to the closeness of excitatory and inhibitory synapses) (Atwood and Morin, 1971). In crabs, the GABA receptors on the muscle membrane become partly desensitized in GABA solutions, but the receptors on the excitatory nerve terminals do not exhibit desensitization (Grundfest, 1969).

Further observations by Takeuchi and Takeuchi (1967) show that the GABA receptors of crustacean muscle fibers are affected by pH of the external solution. The chloride conductance change increases at low pH. The synaptic GABA receptors, or the membrane channels they control, may have a fixed positive charge sensitive to pH. It is also likely that each receptor combines with two GABA molecules to produce the conductance change. Picrotoxin may combine with one site on each GABA receptor to cause noncompetitive depression of the chloride conductance increase (Takeuchi and Takeuchi, 1969).

Inactivation of GABA following release and reaction with the muscle synaptic membrane probably does not involve a breakdown enzyme analogous to the acetylcholinesterase of the vertebrate neuromuscular synapse. Instead, the released GABA is very likely taken up by either the nerve terminals or postsynaptic structures (Iversen and Kravitz, 1968; Morin and Atwood, 1969). Nerve–muscle preparations can rapidly accumulate GABA against a concentration gradient. This ability is related to the extent of inhibitory innervation, being less pronounced in muscles lacking inhibitory synapses.

C. Synaptic Organization

1. IDENTIFICATION OF EXCITATORY AND INHIBITORY SYNAPSES

Electron microscopic studies have recently revealed the presence of two classes of nerve terminals in crayfish muscles—those in which the synaptic vesicles are predominately round, about 500 Å across, and those in which the vesicles are less regular, often flattened, and slightly smaller (Uchizono, 1967; Atwood and Jones, 1967; Atwood and Morin, 1971). The difference between vesicle populations is sometimes only apparent after statistical comparisons have been made (see Fig. 21). Both types of nerve terminal make neuromuscular synaptic connections.

Confirmation of the identity of the nerve terminals comes from the observation that axoaxonal synapses occur between the two types. They are morphologically polarized, with aggregations of synaptic vesicles on the side of the synapse formed by the nerve terminal containing the less regular, slightly smaller vesicles (see Fig. 21). Since presynaptic inhibition requires transmission from inhibitory to excitatory nerve terminals, this configuration is evidence that the terminals with the less regular synaptic vesicles are inhibitory.

Further evidence has been provided by experiments in which the excitatory and inhibitory axons are stimulated in the presence of dinitrophenol (Atwood and Morin, 1971). In this experiment, transmission gradually declines in the stimulated axon, but remains intact for a long period of time in the unstimulated axon. Stimulation of the excitatory axon alone depletes some of the terminals containing the round vesicles, leaving those with the less regular vesicles full. Stimulation of the inhibitory axon alone depletes the terminals with the less regular vesicles but not the others. Dinitrophenol apparently interferes with regeneration of vesicles in the stimulated axon.

In marine crustaceans, the vesicles in excitatory and inhibitory axons are not readily distinguishable (Cohen and Hess, 1965; Hoyle and McNeill, 1968b; Sherman and Atwood, 1971). Possibly, the higher osmotic concentration of the intracellular medium in the marine forms leads to less differentiation of the vesicles during fixation. The basis for the differentiation observed in crayfish is not known.

2. INNERVATION PATTERNS

Innervation patterns of crustacean leg muscles were worked out by Wiersma and co-workers (Wiersma, 1961). More recently, innervation patterns of abdominal muscles (Kennedy and Takeda, 1965a,b; Parnas and Atwood, 1966; Parnas and Dagan, 1969), antennal muscles (Mendelson, 1969), uropod muscles (Larimer and Kennedy, 1969), stomach muscles (Maynard and Atwood, 1969), and a few others, have been studied. The techniques used in these studies usually include recording from individual muscle fibers with intracellular electrodes during stimulation of various axons in the nerve or nerves supplying the muscle under investigation.

The number of excitatory axons supplying a single muscle varies (so far) from one to five. Inhibitory axons may be entirely absent, as in certain stomach muscles (Maynard and Atwood, 1969). Some muscles, such as the opener and stretcher muscles of crabs, have two inhibitory axons. More commonly, a single inhibitory axon is present in many of the leg and abdominal muscles, although it may not innervate all fibers within the muscle (Atwood *et al.*, 1967). Quite frequently, the innervation of single muscles, or parts of muscles, is shared with other muscles, as in the opener–stretcher system of the leg (see Fig. 22) or in abdominal, uropod, and stomach muscles.

The distribution of synapses to fibers within a muscle increases in complexity as the number of motor axons becomes larger. In muscles such as the crayfish opener, the single motor axon branches to supply each of the muscle fibers with at least fifty synapses (Fig. 22). The

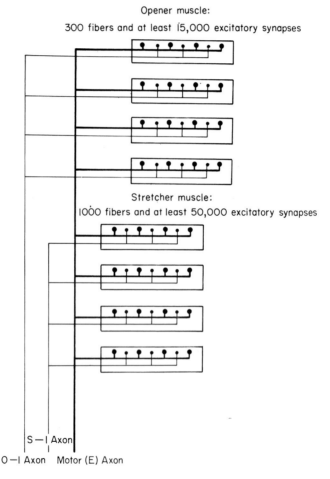

Opener muscle:

300 fibers and at least 15,000 excitatory synapses

Stretcher muscle:

1000 fibers and at least 50,000 excitatory synapses

S—I Axon

O—I Axon   Motor (E) Axon

Fig. 22. Neuromuscular synaptic distribution in opener and stretcher muscles of the crayfish claw. The fiber estimates are those Bittner (1970). The numbers of synapses estimated for the motor axon may be low by a factor of 2 to 4.

estimate of fifty is derived from the ratio of the total number of spontaneous miniature EPSPs recorded intracellularly from a short muscle fiber (representing activity of all synapses) to the number recorded extracellularly at a single synaptic area (Dudel and Kuffler, 1961a). Since an extracellular electrode probably records from at least two to three of the synapses seen with the electron microscope, the figure fifty may be too low (Atwood and Morin, 1971). In any case, the crayfish opener and stretcher muscles receive at least 65,000 synapses from the one motor axon (Bittner, 1970). The excitatory nerve branches are accompanied by inhibitory nerve branches, which probably supply fewer neuromuscular synapses per muscle fiber than the excitor, as judged from electron micrographs. At least ten times fewer axoaxonal synapses (Fig. 21) than neuromuscular synapses are formed by the inhibitor (Atwood and Morin, 1971).

In doubly motor-innervated leg muscles, the two motor axons, sometimes termed "fast" and "slow" because of the different contractions they evoke, are often distributed in roughly reciprocal fashion to the fibers within the muscle. Those fibers that receive a heavy innervation from the slow axon do not normally receive a heavy fast axon innervation (Atwood, 1963, 1965, 1967a; Atwood and Hoyle, 1965). The inhibitory axon in doubly motor-innervated crab muscles supplies more innervation to the fibers receiving a heavy slow input and less to fibers with a mixed or exclusively fast input (Atwood et al., 1967; Atwood, 1967a). Some of the fibers with a fast axon input do not have any inhibitory innervation. This explains the greater efficacy of inhibitory stimulation against the slow contractions.

In multiply motor-innervated muscles, the individual fibers may receive innervation from a single motor axon or from any or all of the axons supplying the muscle (Hoyle and Wiersma, 1958a; Kennedy and Takeda, 1965b; Larimer and Kennedy, 1969). In the tonic flexor muscles of the crayfish abdomen, for example, one of the five motor axons supplies over 90% of the muscle fibers, while two others supply less than 20% of the fibers (Kennedy and Takeda, 1965b). Some of the motor axons supply specific regions of the muscle. The single inhibitory axon gives most innervation to those fibers that receive three or more motor axons.

Interesting variations in extent of pre- and postsynaptic input from inhibitory axons to different muscles have been described. The opener muscle of the crayfish shows pronounced presynaptic inhibition; at low frequencies of stimulation, almost all of the attenuation of the EPSPs is due to the presynaptic effect. The abdominal muscles of crayfish show almost no presynaptic inhibition (Kennedy and Evoy, 1966). In crabs, presynaptic inhibition in the stretcher muscle is much stronger than that in the opener muscle, which is supplied by the same motor axon but a different inhibitor. The impulses in the stretcher inhibitor during normal activity are timed to make use of the available presynaptic mechanism and arrive consistently just before the excitatory impulses when the muscle is inhibited. There is no comparable timing sequence in the axons to the opener muscle, where inhibition is mainly postsynaptic (Spirito, 1969).

The mechanisms responsible for the establishment of specific synaptic connections between nerve and muscle and between one axon and another in crustaceans are completely unknown; not even a convincing start has been made on this problem.

D. Variations in Synaptic Performance

It has recently become evident that the individual nerve terminals of a given crustacean axon may show almost as much physiological variation as the muscle fibers themselves. In addition, there are well marked overall physiological differences between axons. The physiological differences include variation in the amount of transmitter released by a nerve impulse, in the facilitation during repetitive stimulation, and in rate of fatigue with repetitive stimulation. Documentation of these features is provided in numerous studies, including those by Hoyle and Wiersma (1958a), Dudel and Kuffler (1961b), Atwood (1963, 1965, 1967a,b), Atwood and Parnas (1968), Kennedy and Takeda (1965b), Bittner (1968a,b), and Bittner and Harrison (1970).

Both intracellular and extracellular records have shown that one factor in the variation of EPSP size is the different quantal content of the EPSP produced by different nerve terminals (Atwood, 1967b; Bittner, 1968a,b). A second factor, which is also of importance, is the membrane resistance of the muscle fiber, which is quite variable in crustaceans. The larger EPSPs are associated with muscle fibers of high membrane resistance in the singly motor-innervated muscles of crayfish and crabs (Bittner, 1968a; Atwood and Bittner, 1971).

Some synapses examined with extracellular electrodes are noteworthy for the lack of agreement with the Poisson release mechanism of vertebrate synapses (Atwood and Parnas, 1968; Bittner and Harrison, 1970). The reason for this is unknown.

An example of variation in EPSPs produced by different terminals of one axon (supplying a crab stretcher muscle) is given in Fig. 23. From the data in this figure, it is evident that an inverse correlation

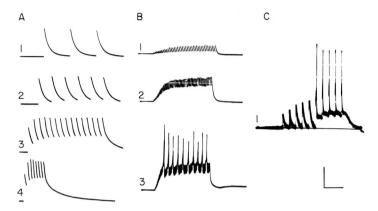

Fig. 23. Electrical responses to stimulation of the motor axon in three muscle fibers (A, B, C) of the stretcher muscle of a crab *Grapsus*. In the first fiber (A) the EPSPs are large and show little facilitation. In the second fiber (B), the EPSPs are smaller but facilitate strongly. Graded membrane responses appear at high frequencies (B3). The third fiber (C) is intermediate and generates large, graded membrane responses at low frequencies of stimulation. The frequencies of stimulation employed were: in A, 1 Hz (1), 2 Hz (2), 5 Hz (3), 10 Hz (4); in B, 20 Hz (1), 40 Hz (2), 60 Hz (3); in C, 10 Hz. Calibration: A, B, 20 mV and 1 sec; C, 20 mV and 0.5 sec. (Atwood and Bittner, 1971.)

exists between the initial size of an EPSP and the extent of facilitation. In singly motor-innervated crab and crayfish leg muscles, the relationship is highly significant statistically when large samples of muscle fibers are examined. This is shown in Fig. 24, in which initial EPSP amplitude is plotted against a facilitation index—taken as the ratio of EPSP size at 10 stimuli per second to EPSP size at 1 stimulus per second (Atwood and Bittner, 1971).

A further point of interest concerns the inhibitory axons to these muscles. The specific inhibitors show differentiation of the nerve terminals parallel to that of the excitatory axons. The correlation between facilitation index for the EPSP and that for the IPSP recorded from the same fiber is highly significant for a large sample of muscle fibers (Fig. 25).

Such observations enable one to construct more complete "wiring diagrams" for the synaptic connections within a muscle. Figure 26 shows an outline of the scheme worked out for a crab stretcher muscle, which receives two inhibitory axons along with one excitatory axon. The specific stretcher inhibitor shows synaptic differentiation parallel to that of the excitor. Both pre- and postsynaptic inhibitory effects show facilitation properties like those of the EPSPs from the same muscle fiber. The common inhibitor, on the other hand, does not show good correlation

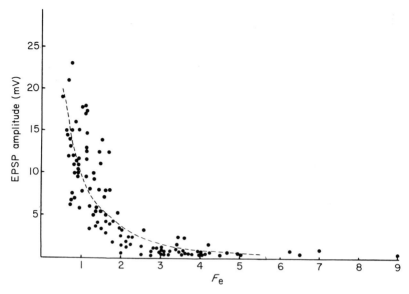

Fig. 24. Relationship between amplitude of the EPSP at 1 Hz, and facilitation index F_e, taken as the ratio of EPSP amplitude at 10 Hz to EPSP amplitude at 1 Hz, for 111 fibers in stretcher muscles of the crab *Grapsus*. The fitted regression line is hyperbolic in form. (Atwood and Bittner, 1971.)

of facilitation properties with the other axons. Its synapses produce facilitating IPSPs in fibers with poorly-facilitating EPSPs. Also, it does not innervate all of the fibers in the muscle, nor does it produce much presynaptic inhibition.

The mechanisms underlying the differences between the nerve terminals are presently only speculative. One guess is that the axon spike invades some terminals more completely than others and causes them to release comparatively more transmitter (Atwood, 1967b; Bittner, 1968a). There is good evidence that the axon spike propagates electrotonically into the synapse-bearing terminals of crayfish motor axons (Dudel, 1963, 1965a,c). Variations in location of synapses relative to the spike-conducting region of the terminal could thus result in a different probability of transmitter release at each synapse. Terminals with many synapses close to the spike-conducting region of the axon would produce a large transmitter output and a good-sized EPSP. The experiments of Bittner (1968a) have ruled out some of the other possibilities, including different sizes of transmitter quanta, differences in rate of transmitter mobilization, and differences in the relationship between terminal depolarization and transmitter release.

The facilitation properties of the nerve terminals may be influenced by several factors, including difference in the change of amplitude of the nerve terminal potential observed during repetitive stimulation (see Dudel, 1965a), and differences in the relationship between mobilized transmitter quanta and available release sites (see Bittner and Harrison, 1970). The data are too few at the moment to permit formulation of a substantial hypothesis. For example, Dudel (1965a) showed that an increase occurs in the amplitude of the externally recorded nerve terminal potential during repetitive stimulation and ascribed facilitation to this phenomenon. However, in mammalian neuromuscular synapses, Gage and Hubbard (1966) have ruled out this phenomenon as a causative agent in post-tetanic potentiation of the end plate potential. There is no real evidence, apart from Dudel's correlation, to show whether

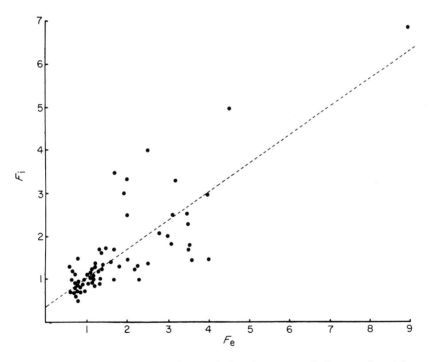

Fig. 25. Plot of facilitation index (F_i) for the IPSPs of the specific inhibitor axon, against the facilitation index (F_e) for the EPSPs of the motor axon recorded in the same muscle fiber. The records were made in 70 fibers of the stretcher muscle of the crab *Grapsus*. Facilitation index was determined as in Fig. 14. The correlation between F_e and F_i is highly significant and shows that excitatory and inhibitory synapses have "matched" physiological properties in a given muscle fiber. (Atwood and Bittner, 1971.)

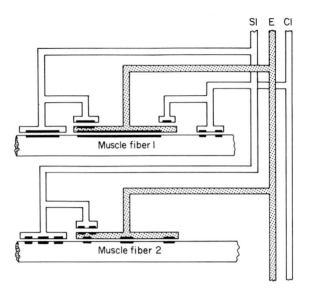

Fig. 26. Synaptic connections in the stretcher muscle of the crab *Grapsus*. The motor axon (E) and the specific inhibitor (SI) innervate all fibers. The common inhibitor (CI) innervates only fibers with low F_e (muscle fiber 1), not those with high F_e (muscle fiber 2). Poorly facilitating synapses are shown with long continuous synaptic membranes, while highly facilitating synapses are shown with short patches of synaptic membrane. Note that axoaxonal synapses have similar facilitation properties to neuromuscular synapses on the same muscle fiber. (Atwood and Bittner, 1971.)

it is of any significance in the crustacean neuromuscular systems. It may well be, since nerve terminals the action potential does not actively invade may behave differently in many respects when compared with the fully invaded vertebrate nerve terminals.

Since secretion of the transmitter substance during neuromuscular synaptic transmission requires free external calcium ions in crustaceans as well as in vertebrates (Otsuka *et al.*, 1966; Bracho and Orkand, 1970), differences in density of membrane calcium receptors at the presynaptic terminal may determine the amount of transmitter released at different terminals. The observed differences in time course of facilitation at different crustacean motor nerve terminals (Bittner, 1968b) could perhaps be explained by small differences in kinetic properties of the calcium receptors. Studies on excitation–secretion coupling in crustacean motor nerve terminals have not advanced far enough to rule out any of these possibilities.

Norman and Maynard (1968) have shown mathematically that growth of certain EPSPs by facilitation must involve more than one factor.

Thus, calcium mobilization, nerve terminal potential changes, and possibly other factors may all be involved in determining the time course of facilitation during repetitive activity.

The overall physiological differences between various crustacean motor axons are also worth noting. Some of the fast axons of crustacean leg muscles, which are apparently used only occasionally by the animal to produce rapid movements, show rather high quantal content of transmitter release at their terminals and very rapid fatigue with repetitive stimulation (Atwood and Johnston, 1968). By contrast, axons to tonic muscles are often continuously active (Kennedy and Takeda, 1965b), and they can be stimulated at rates of 10/sec or more for well over an hour without showing any fatigue. Electron micrographs of phasic and tonic axons have shown a much higher density of synaptic vesicles in the latter (Atwood and Johnston, 1968). Possibly, there is feedback from processes set in motion by maintained activity to regulate the supply of vesicles.

E. Matching of Axons and Muscle Fibers

In many crustacean muscles, such as those in the legs of some crabs, a mixed population of fibers exists, some being of the short sarcomere, fast-contracting type, with high probability of producing large graded membrane responses, while at the other end of the spectrum are long sarcomere, slow-contracting fibers. Some muscles (e.g., the crayfish claw opener muscle) have only the long sarcomere type. In the muscles of the crayfish abdomen and uropod and in the lobster antenna, fast and slow components are typically separated into discrete muscles or well-defined parts of muscles.

The innervation, with variation in axonal terminals, is superimposed upon the muscle fiber substrate, often in a rather complex fashion. As a general rule, the fast muscles or muscle fibers are innervated by phasic axons with no spontaneous activity and large, easily fatigued EPSPs. This situation is well exemplified by the crayfish abdominal muscles. In some singly motor-innervated crab leg muscles, this rule does not hold, for here the faster fibers show small facilitating EPSPs, which trigger contraction only at high frequencies through generation of large graded responses or spikes (Atwood, 1965) (see Fig. 23).

Slow fibers usually receive innervation from tonic, spontaneously active motor axons, but the EPSPs may show almost any of the possible variations in amplitude and facilitation, as shown in Fig. 24 for a singly motor-innervated crab leg muscle. The different fibers of such a muscle are differentially recruited by different frequencies and patterns of stimulation, as directly demonstrated by Bittner (1968b) in the crayfish opener muscle.

Variation in both muscle fibers and nerve terminals obviously allows for fine control of muscular contraction, thereby counterbalancing the effects of the paucity of motor axons. Pattern sensitivity of certain muscles (Wilson and Larimer, 1968; Gillary and Kennedy, 1969) allows further refinement in control of muscular contraction.

VII. Conclusion

The unique size of crustacean muscle fibers has made them valuable tools in the study of some fundamental problems in membrane electrical behavior and excitation–contraction coupling. Although many of the ionic mechanisms and some of the details of excitation–contraction coupling are different from vertebrate skeletal muscle, the findings from crustacean material will undoubtedly continue to provide insight into basic principles of muscle operation and membrane excitability.

The complexities of the innervation of crustacean muscle fibers have been of intrinsic interest in providing a demonstration of the mechanism of a finely adjustable neuromuscular system in which maximal performance is wrung out of each component. This is in contrast to the vertebrate system, in which there is by comparison a redundancy of components. In addition, the crustacean system has provided a lot of information about the chemistry and mechanisms of inhibition and of noncholinergic transmission. This knowledge may be of use in the study of the central nervous system.

One of the more important problems yet to be solved in the crustacean systems is the mechanism of formation of specific axon-to-muscle and axon-to-axon contacts. In addition, the mechanism of nerve terminal differentiation remains to be worked out, as well as mechanisms accounting for differences in synaptic performance.

REFERENCES

Abbott, B. C., and Parnas, I. (1965). *J. Gen. Physiol.* **48**, 919.
Adrian, R. H., Chandler, W. K., and Hodgkin, A. L. (1969a). *J. Physiol.* (*London*) **204**, 207.
Adrian, R. H., Costantin, L. L., and Peachey, L. D., (1969b). *J. Physiol.* (*London*) **204**, 231.
Alexandrowicz, J. S. (1952). *J. Mar. Biol. Ass. U.K.* **31**, 277.

Allen, R. D., and Hinke, J. A. M. (1970). *Can. J. Physiol. Pharmacol.* **48**, 139.
Armson, J. M., and Horridge, G. A. (1964). *J. Neurochem.* **11**, 387.
Ashley, C. C. (1967). *Amer. Zool.* **7**, 647.
Ashley, C. C., and Ridgway, E. B. (1968). *Nature (London)* **219**, 1168.
Atwood, H. L. (1963). *Comp. Biochem. Physiol.* **10**, 17.
Atwood, H. L. (1965). *Comp. Biochem. Physiol.* **16**, 409.
Atwood, H. L. (1967a). *Amer. Zool.* **7**, 527.
Atwood, H. L. (1967b). *Nature (London)* **215**, 57.
Atwood, H. L. (1968). *Experientia* **24**, 753.
Atwood, H. L., and Bittner, G. D. (1971). *J. Neurophysiol.* **34**, 157.
Atwood, H. L., and Dorai Raj, B. S. (1964). *J. Cell. Comp. Physiol.* **64**, 55.
Atwood, H. L., and Hoyle, G. (1965). *J. Physiol. (London)* **181**, 225.
Atwood, H. L., and Johnston, H. S. (1968). *J. Exp. Zool.* **167**, 457.
Atwood, H. L., and Jones, A. (1967). *Experientia* **23**, 1036.
Atwood, H. L., and Morin, W. A. (1971). Unpublished experiments.
Atwood, H. L., and Morin, W. A. (1971). *J. Ultrastruct. Res.* (in press).
Atwood, H. L., and Parnas, I. (1968). *Comp. Biochem. Physiol.* **27**, 381.
Atwood, H. L., Hoyle, G., and Smyth, T., Jr. (1965). *J. Physiol. (London)* **180**, 449.
Atwood, H. L., Parnas, I., and Wiersma, C. A. G. (1967). *Comp. Biochem. Physiol.* **20**, 1963.
Baker, P. F., Hodgkin, A. L., and Shaw, T. I. (1962). *J. Physiol. (London)* **164**, 355.
Bárány, M. (1967). *J. Gen. Physiol.* **50**, 197.
Beaugé, L. A., and Sjodin, R. A. (1967). *Nature (London)* **215**, 1307.
Beeler, G. W., and Reuter, H. (1970). *J. Physiol. (London)* **207**, 191.
Bittar, E. E. (1966). *J. Physiol. (London)* **187**, 81.
Bittar, E. E., Caldwell, P. C., and Lowe, A. G. (1967). *J. Mar. Biol. Ass. U.K.* **47**, 709.
Bittner, G. D. (1968a). *J. Gen. Physiol.* **51**, 731.
Bittner, G. D. (1968b). *J. Exp. Zool.* **167**, 439.
Bittner, G. D. (1970). Personal communication.
Bittner, G. D., and Harrison, J. (1970). *J. Physiol. (London)* **206**, 1.
Boistel, J., and Fatt, P. (1958). *J. Physiol. (London)* **144**, 176.
Boyd, I. A., and Martin, A. W. (1959). *J. Physiol. (London)* **147**, 450.
Bracho, H., and Orkand, R. K. (1970). *J. Physiol. (London)* **206**, 61.
Brandt, P. W., Reuben, J. P., Girardier, L., and Grundfest, H. (1965). *J. Cell Biol.* **25**, 233.
Brandt, P. W., Reuben, J. P., and Grundfest, H. (1968). *J. Cell Biol.* **38**, 115.
Brinley, F. J. (1968). *J. Gen. Physiol.* **51**, 445.
Burke, W., and Ginsborg, B. L. (1956). *J. Physiol. (London)* **132**, 586.
Caldwell, P. C., and Walster, G. (1963). *J. Physiol. (London)* **169**, 353.
Cohen, M. J., and Hess, A. (1965). *Amer. J. Anat.* **121**, 285.
Cook, B. J. (1967). *Biol. Bull.* **133**, 526.
Costantin, L. L. (1970). *J. Gen. Physiol.* **55**, 703.
Costantin, L. L., Podolsky, R. J., and Tice, L. W. (1967). *J. Physiol. (London)* **188**, 261.
Curtis, H. J., and Cole, K. S. (1938). *J. Gen. Physiol.* **21**, 757.
De Mello, W. C., and Hutter, O. F. (1966). *J. Physiol. (London)* **183**, 11P.
Dorai Raj, B. S. (1964). *J. Cell. Comp. Physiol.* **64**, 41.
Dudel, J. (1963). *Pfluegers Arch. Gesamte Physiol. Menschen Tiere* **277**, 537.

Dudel, J. (1965a). *Pfluegers Arch. Gesamte Physiol. Menschen Tiere* **282**, 323.
Dudel, J. (1965b). *Pfluegers Arch. Gesamte Physiol. Menschen Tiere* **283**, 104.
Dudel, J. (1965c). *Pfluegers Arch. Gesamte Physiol. Menschen Tiere* **284**, 66.
Dudel, J., and Kuffler, S. W. (1961a). *J. Physiol. (London)* **155**, 514.
Dudel, J., and Kuffler, S. W. (1961b). *J. Physiol. (London)* **155**, 530.
Dudel, J., and Kuffler, S. W., (1961c). *J. Physiol. (London)* **155**, 543.
Dudel, J., and Orkand, R. K. (1960). *Nature (London)* **186**, 476.
Dudel, J., and Rüdel, R. (1969). *Pfluegers Arch.* **308**, 291.
Dunham, P. B., and Gainer, H. (1968). *Biochim. Biophys. Acta* **150**, 488.
Edwards, C., Chichibu, S., and Hagiwara, S. (1964). *J. Gen. Physiol.* **48**, 225.
Eisenberg, B., and Eisenberg, R. S. (1968). *J. Cell Biol.* **39**, 451.
Eisenberg, R. S. (1965). Ph.D. Thesis, University of London.
Eisenberg, R. S. (1967). *J. Gen. Physiol.* **50**, 1785.
Eisenberg, R. S., and Gage, P. W. (1969). *J. Gen. Physiol.* **53**, 279.
Fahrenbach, W. H. (1963). *J. Cell Biol.* **17**, 629.
Fahrenbach, W. H. (1967). *J. Cell Biol.* **135**, 69.
Falk, G. (1968). *Biophys. J.* **8**, 608.
Falk, G., and Fatt, P. (1964). *Proc. Roy. Soc., Ser. B* **160**, 69.
Falk, G., and Landa, J. F. (1960). *Amer. J. Physiol.* **198**, 289.
Fatt, P., and Ginsborg, B. L. (1958). *J. Physiol. (London)* **142**, 516.
Fatt, P., and Katz, B. (1951). *J. Physiol. (London)* **115**, 320.
Fatt, P., and Katz, B. (1953a). *J. Physiol. (London)* **120**, 171.
Fatt, P., and Katz, B. (1953b). *J. Exp. Biol.* **29**, 433.
Fatt, P., and Katz, B. (1953c). *J. Physiol. (London)* **121**, 374.
Frankenhaeuser, B., and Hodgkin, A. L. (1957). *J. Physiol. (London)* **137**, 217.
Franzini-Armstrong, C. (1970). *J. Cell Sci.* **6**, 559.
Gage, P. W., and Hubbard, J. I. (1966). *J. Physiol. (London)* **184**, 353.
Gainer, H. (1968). *J. Gen. Physiol.* **52**, 88.
Gainer, H., and Grundfest, H. (1968). *J. Gen. Physiol.* **51**, 399.
Gillary, H. L., and Kennedy, D. (1969). *J. Neurophysiol.* **32**, 207.
Gillis, J. M. (1969). *J. Physiol. (London)* **200**, 849.
Girardier, L., Reuben, J. P., Brandt, P. W., and Grundfest, H. (1963). *J. Gen. Physiol.* **47**, 189.
Grundfest, H. (1962). In "Properties of Membranes and Diseases of the Nervous System" (M. D. Yahr, ed.), pp. 71–99. Springer Publ., New York.
Grundfest, H. (1966). *Advan. Comp. Physiol. Biochem.* **2**, 1.
Grundfest, H. (1967). *Fed. Proc., Fed. Amer. Soc. Exp. Biol.* **26**, 1613.
Grundfest, H. (1969). In "The Structure and Function of Nervous Tissue" (G. H. Bourne, ed.), Vol. 2, p. 463. Academic Press, New York.
Grundfest, H. (1970). Personal communication.
Grundfest, H., Reuben, J. P., and Rickles, W. H. (1959). *J. Gen. Physiol.* **42**, 1301.
Hagiwara, S., and Naka, K. I. (1964). *J. Gen. Physiol.* **48**, 141.
Hagiwara, S., and Nakajima, S. (1966a). *J. Gen. Physiol.* **49**, 793.
Hagiwara, S., and Nakajima, S. (1966b). *J. Gen. Physiol.* **49**, 807.
Hagiwara, S., and Takahashi, K. (1967). *J. Gen. Physiol.* **50**, 583.
Hagiwara, S., Chichibu, S., and Naka, K. I. (1964). *J. Gen. Physiol.* **48**, 163.
Hagiwara, S., Gruener, R., Hayashi, H., Sakata, H., and Grinnell, A. D. (1968a). *J. Gen. Physiol.* **52**, 773.
Hagiwara, S., Takahashi, K., and Junge, D. (1968b). *J. Gen. Physiol.* **51**, 157.

Hagiwara, S., Hayashi, H., and Takahashi, K. (1969). *J. Physiol.* (*London*) **205**, 115.

Hastings, J. W., Mitchell, G., Mattingly, P. H., Blinks, J. R., and Van Leeuwen, M. (1969). *Nature* (*London*) **222**, 1047.

Hays, E. A., Lang, M. A., and Gainer, H. (1968). *Comp. Biochem. Physiol.* **26**, 761.

Hess, A. (1965). *J. Cell Biol.* **26**, 467.

Hodgkin, A. L., and Horowicz, P. (1959). *J. Physiol.* (*London*) **148**, 127.

Hodgkin, A. L., and Horowicz, P. (1960). *J. Physiol.* (*London*) **153**, 386.

Hodgkin, A. L., and Rushton, W. A. H. (1946). *Proc. Roy. Soc., Ser. B* **133**, 444.

Howell, J. N. (1969). *J. Physiol.* (*London*) **201**, 515.

Howell, J. N., and Jenden, D. J. (1967). *Fed. Proc., Fed. Amer. Soc. Exp. Biol.* **26**, 553.

Hoyle, G. (1967). In "Invertebrate Nervous Systems" (C. A. G. Wiersma, ed.), pp. 151–167. Univ. of Chicago Press, Chicago.

Hoyle, G. (1968). *J. Exp. Zool.* **167**, 471.

Hoyle, G. (1969). *Annu. Rev. Physiol.* **31**, 43.

Hoyle, G. (1970). *Sci. Amer.* **222**, 84.

Hoyle, G., and McNeill, P. A. (1968a). *J. Exp. Zool.* **167**, 487.

Hoyle, G., and McNeill, P. A. (1968b). *J. Exp. Zool.* **167**, 523.

Hoyle, G., and Smyth, T. (1963). *Comp. Biochem. Physiol.* **10**, 291.

Hoyle, G., and Wiersma, C. A. G. (1958a). *J. Physiol.* (*London*) **143**, 402.

Hoyle, G., and Wiersma, C. A. G. (1958b). *J. Physiol.* (*London*) **143**, 426.

Hoyle, G., and Wiersma, C. A. G. (1958c) *J. Physiol.* (*London*) **143**, 441.

Hoyle, G., McAlear, J. H., and Selverston, A. (1965). *J. Cell Biol.* **26**, 621.

Hutter, O. F., and Noble, D. (1960). *J. Physiol.* (*London*) **151**, 89.

Hutter, O. F., and Warner, A. E. (1967). *J. Physiol.* (*London*) **189**, 403.

Huxley, A. F., and Niedergerke, R. (1954). *Nature* (*London*) **173**, 971.

Huxley, A. F., and Peachey, L. D. (1964). *J. Cell Biol.* **23**, 107A.

Huxley, A. F., and Taylor, R. E. (1958). *J. Physiol.* (*London*) **144**, 426.

Iversen, L. L., and Kravitz, E. A. (1968). *J. Neurochem.* **15**, 609.

Jahromi, S. S. (1968). Ph.D. Thesis, University of Toronto.

Jahromi, S. S., and Atwood, H. L. (1967). *Can. J. Zool.* **45**, 601.

Jahromi, S. S., and Atwood, H. L. (1969a). *J. Exp. Zool.* **171**, 25.

Jahromi, S. S., and Atwood, H. L. (1969b). *J. Insect Physiol.* **15**, 2255.

Jahromi, S. S., and Atwood, H. L. (1971a). *J. Exp. Zool.* **176**, 475.

Jahromi, S. S., and Atwood, H. L. (1971b). *Can. J. Zool.* **49**, 1029.

Jöbsis, F. F., and O'Connor, M. J. (1966). *Biochem. Biophys. Res. Commun.* **25**, 246.

Katz, B. (1966). "Nerve, Muscle, and Synapse." McGraw-Hill, New York.

Katz, B., and Thesleff, S. (1957). *J. Physiol.* (*London*) **137**, 267.

Kennedy, D. (1967). In "Invertebrate Nervous Systems" (C. A. G. Wiersma, ed.), pp. 197–212. Univ. of Chicago Press, Chicago.

Kennedy, D., and Evoy, W. H. (1966). *J. Gen. Physiol.* **49**, 457.

Kennedy, D., and Takeda, K. (1965a). *J. Exp. Biol.* **43**, 211.

Kennedy, D., and Takeda, K. (1965b). *J. Exp. Biol.* **43**, 229.

Kiyohara, T., and Sato, M. (1967). *Jap. J. Physiol.* **17**, 720.

Kravitz, E. A., Kuffler, S. W., Potter, D. D., and Van Gelder, N. M. (1963). *J. Neurophysiol.* **26**, 729.

Kravitz, E. A., Molinoff, P. B., and Hall, Z. W. (1965). *Proc. Nat. Acad. Sci. U.S.* **54**, 778.

Lang, M. A., and Gainer, H. (1969a). *J. Gen. Physiol.* **53**, 323.

Lang, M. A., and Gainer, H. (1969b). *Comp. Biochem. Physiol.* **30**, 445.

Larimer, J. L., and Kennedy, D. (1969). *J. Exp. Biol.* **51**, 119.

McLaughlin, S. G. A., and Hinke, J. A. M. (1966). *Can. J. Physiol. Pharmacol.* **44**, 837.

Marmont, G., and Wiersma, C. A. G. (1938). *J. Physiol.* (*London*) **121**, 318.

Maynard, D. M., and Atwood, H. L. (1969). *Amer. Zool.* **9**, 1107.

Mendelson, M. (1969). *J. Cell Biol.* **42**, 548.

Morin, W. A., and Atwood, H. L. (1969). *Comp. Biochem. Physiol.* **30**, 577.

Motokizawa, F., Reuben, J. P., and Grundfest, H. (1969). *J. Gen. Physiol.* **54**, 437.

Niedergerke, R., and Orkand, R. K. (1966). *J. Physiol.* (*London*) **184**, 291.

Norman, R., and Maynard, D. M. (1968). Personal communication.

Orkand, R. K. (1962). *J. Physiol.* (*London*) **161**, 143.

Orkand, R. K., and Niedergerke, R. (1964). *Science* **146**, 1176.

Otsuka, M., Iversen, L. L., Hall, Z. W., and Kravitz, E. A. (1966). *Proc. Nat. Acad. Sci. U.S.* **56**, 1110.

Otsuka, M., Kravitz, E. A., and Potter, D. D. (1967). *J. Neurophysiol.* **30**, 725.

Ozeki, M. (1969). *Science* **163**, 82.

Ozeki, M., and Grundfest, H. (1967). *Science* **155**, 478.

Ozeki, M., Freeman, A. R., and Grundfest, H. (1966). *J. Gen. Physiol.* **49**, 1319.

Parnas, I., and Atwood, H. L. (1966). *Comp. Biochem. Physiol.* **18**, 701.

Parnas, I., and Dagan, D. (1969). *Comp. Biochem. Physiol.* **28**, 359.

Peachey, L. D. (1965). *Fed. Proc., Fed. Amer. Soc. Exp. Biol.* **24**, 1124.

Peachey, L. D. (1967). *Amer. Zool.* **7**, 505.

Peachey, L. D., and Huxley, A. F. (1964). *J. Cell. Biol.* **23**, 70A.

Reger, J. F. (1967). *J. Ultrastruct. Res.* **20**, 72.

Reuben, J. P. (1960). *Biol. Bull.* **119**, 334.

Reuben, J. P., and Grundfest, H. (1960). *Biol. Bull.* **119**, 335.

Reuben, J. P., Werman, R., and Grundfest, H. (1961). *J. Gen. Physiol.* **45**, 243.

Reuben, J. P., Girardier, L., and Grundfest, H. (1962). *Biol. Bull.* **123**, 509.

Reuben, J. P., Girardier, L., and Grundfest, H. (1964). *J. Gen. Physiol.* **47**, 1141.

Reuben, J. P., Brandt, P. W., Garcia, H., and Grundfest, H. (1967a). *American Zoologist* **7**, 623.

Reuben, J. P., Brandt, P. W., Girardier, L., and Grundfest, G. (1967b). *Science* **155**, 1263.

Richards, C. D. (1969). *J. Physiol.* (*London*) **202**, 211.

Ridgway, E. B., and Ashley, C. C. (1967). *Biochem. Biophys. Res. Commun.* **29**, 229.

Ridgway, E. B., Ashley, C. C., and Hoyle, G. (1968). *Fed. Proc., Fed. Amer. Soc. Exp. Biol.* **27**, 375.

Robbins, J. (1959). *J. Physiol.* (*London*) **148**, 39.

Robertson, J. D. (1961). *J. Exp. Biol.* **38**, 707.

Rosenbluth, J. (1969). *J. Cell Biol.* **42**, 534.

Rüegg, J. C. (1968). *Symp. Soc. Exp. Biol.* **30**, 45.

Schou, M. (1957). *Pharmacol. Rev.* **9**, 17.

Selverston, A. (1967). *Amer. Zool.* **7**, 515.

Shaw, J. (1955a). *J. Exp. Biol.* **32**, 383.

Shaw, J. (1955b). *J. Exp. Biol.* **32**, 664.
Shaw, J. (1958a). *J. Exp. Biol.* **35**, 902.
Shaw, J. (1958b). *J. Exp. Biol.* **35**, 920.
Sherman, R. G., and Atwood, H. L. (1971). *J. Exp. Zool.* **176**, 461.
Spirito, C. (1969). *Amer. Zool.* **9**, 1106.
Stefani, E., and Steinbach, A. B. (1969). *J. Physiol.* (*London*) **203**, 383.
Swan, R. C. (1963). *J. Cell Biol.* **19**, 68A.
Takeda, K. (1967). *J. Gen. Physiol.* **50**, 1049.
Takeuchi, A., and Takeuchi, N. (1963). *Nature* (*London*) **198**, 490.
Takeuchi, A., and Takeuchi, N. (1964). *J. Physiol.* (*London*) **170**, 296.
Takeuchi, A., and Takeuchi, N. (1965). *J. Physiol.* (*London*) **177**, 225.
Takeuchi, A., and Takeuchi, N. (1966a). *J. Physiol.* (*London*) **183**, 418.
Takeuchi, A., and Takeuchi, N. (1966b). *J. Physiol.* (*London*) **183**, 433.
Takeuchi, A., and Takeuchi, N. (1967). *J. Physiol.* (*London*) **191**, 575.
Takeuchi, A., and Takeuchi, N. (1969). *J. Physiol.* (*London*) **205**, 377.
Tasaki, I., and Hagiwara, S. (1957). *Amer. J. Physiol.* **188**, 423.
Taylor, S. R., Preiser, H., and Sandow, A. (1969). *J. Gen. Physiol.* **54**, 352.
Thomas, R. C., and Wilson, U. J. (1966). *Science* **151**, 1538.
Uchizono, K. (1967). *Nature* (*London*) **214**, 833.
Usherwood, P. N. R. (1962). *J. Insect. Physiol.* **8**, 31.
Usherwood, P. N. R. (1967). *Amer. Zool.* **7**, 553.
Usherwood, P. N. R. (1969). *Advan. Insect. Physiol.* **6**, 250.
Van der Kloot, W. G. (1960). *J. Neurochem.* **5**, 245.
Van der Kloot, W. G. (1965). *Comp. Biochem. Physiol.* **15**, 547.
Van der Kloot, W. G. (1966). *Comp. Biochem. Physiol.* **17**, 75.
Van Harreveld, A., and Mendelson, M. (1959). *J. Cell. Comp. Physiol.* **54**, 85.
Werman, R., and Grundfest, H. (1961). *J. Gen. Physiol.* **44**, 997.
Werman, R., McCann, F. V., and Grundfest, H. (1961). *J. Gen. Physiol.* **44**, 979.
Wiersma, C. A. G. (1955). *Arch. Neer. Zool.* **11**, 1.
Wiersma, C. A. G. (1961). *In* "The Physiology of Crustacea" (T. H. Waterman, ed.), Vol. 2, p. 191. Academic Press, New York.
Wilson, D. M., and Larimer, J. L. (1968). *Proc. Nat. Acad. Sci. U.S.* **61**, 909.
Zachar, J., and Zacharova, D. (1966). *J. Physiol.* (*London*) **186**, 596.
Zachar, J., Zacharova, D., and Hencek, M. (1964). *Physiol. Bohemoslov.* **13**, 129.

10

ARTHROPOD MUSCLE

J. W. S. PRINGLE

I. Introduction

The phylum Arthropoda, embracing the insects, crustacea, spiders, scorpions, king crabs, and many less known groups, contains most of the animals other than vertebrates that have the ability to move rapidly in the water, on land, or in the air. Since this stock diverged early in evolutionary history from the line that gave rise to the vertebrates, the well developed muscles of arthropods present physiologists with the opportunity to study an alternative type of contractile tissue. The many common features of structure, physiology, and composition between arthropod and vertebrate muscle make it easy to forget the independent evolution of the two phyla. Zoologists will wish to consider whether the similarities indicate a common basis for contractility in the animal kingdom or are the result of convergence in response to similar demands of the way of life; physiologists, biochemists, and biophysicists may be content to accept the additional wealth of experimental material thus provided and to utilize some of the special features of arthropod muscle as an aid in the elucidation of fundamental mechanisms.

Although much less work has been done on arthropod than on vertebrate muscle, the volume of information is sufficient to make it impossible to cover all the ground in a single chapter. We shall be concerned primarily with the structure, biochemical composition, and properties of the myofibrils; excitation and the intramuscular coupling process will be discussed only to the extent that it is necessary in order to give an account of the properties of the contractile material. For personal reasons and because more work has been done on them, the emphasis will be on insect muscles. Some of the other types of arthropod are either very small or are less readily available, but workers in tropical countries have a great opportunity here to make contributions to comparative physiology.

II. Structure

A. Anatomy and Histology

The somatic muscle fibers of arthropods are normally attached at both ends to an exoskeleton which is continuous over the surface of the body. Folding of this cuticle and the presence of soft, elastic regions produce the segments of the body and the joints of limbs; the muscles span these flexible regions. Invaginations of the cuticle may increase the area of attachment and can generate long internal apodemes, which are the analogs of vertebrate tendons. A pinnate arrangement of fibers inserting on a central apodeme is a common pattern in limb muscles. Connective tissue is sparse in arthropods except in the Chelicerata, which possess a cartilaginous endoskeleton serving for the attachment of many somatic muscles in the trunk.

The visceral and cardiac muscles consist of irregularly arranged fibers round the lumen and muscles spanning the body cavity in thin connective tissue sheets, which act as dilators. The most important muscles of the latter type are the segmental alary muscles dilating the pericardial sinus or the contractile dorsal blood vessel; details of the arrangement in individual insects are given by Wigglesworth (1965).

Textbooks usually state that all the muscles of arthropods, both somatic and visceral, are striated. In Onychophora (*Peripatus*), the body wall musculature consists of longitudinal and circular fibers 40–50 μm in diameter with well defined myofibrils which show no signs of striation in the light microscope. There may be other exceptions in the visceral musculature, but in general, the statement is true.

B. Ultrastructure

1. LIMB AND GENERAL TRUNK MUSCLES

The characteristic arrangement of intracellular organelles in arthropod somatic muscles is shown in Figs. 1 and 2. Fiber diameter is 40–200 μm.

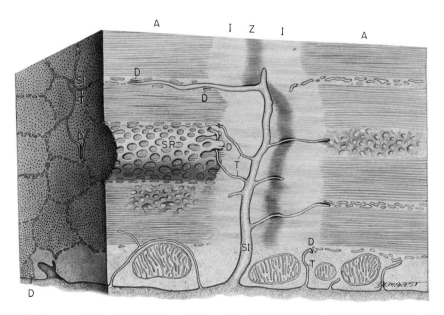

Fig. 1. Composite representation of the fine structure of a crustacean muscle fiber; SI = sarcolemmal invagination, T = finer tubules, SR = sarcoplasmic reticulum, D = dyads. From Brandt *et al.* (1965).

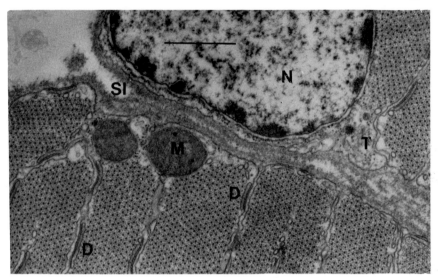

Fig. 2. Transverse section of tergocoxal muscle of *Ilyocoris* (Hemiptera). Scale 1 μm; M = mitochondria, N = nucleus, other labeling as in Fig. 1. Original by Dr. M. J. Cullen.

The many nuclei are located either immediately below the plasma membrane (Crustacea, close-packed fibers of insects) or in a central column of sarcoplasm ("tubular" fibers of insects and arachnids). Deep invaginations of the plasma membrance—Brandt *et al.* (1965), estimate 50–200 per sarcomere in a crayfish leg muscle—penetrate nearly to the center of the fiber, and from these major clefts minor branches ramify longitudinally and again transversely between the myofibrils. The final transverse elements of this system nearly always lie halfway between the Z line and the middle of each sarcomere (Smith, 1966a), where they form dyadic or triadic contacts with the sarcoplasmic reticulum. Mitochondria may be large and numerous, with well aligned cristae. The thick myosin filaments forming the A band interdigitate with thin actin filaments as in vertebrate muscle and the I band is bisected by a Z line.

Many variations on this general pattern have been described in Crustacea (Peterson and Pepe, 1961; Bouligand, 1962; Brandt *et al.*, 1965; Hoyle *et al.*, 1965; Reger, 1967; Hoyle and McNeill, 1968; Rosenbluth, 1969), insects (Smith, 1966b; Hagopian, 1966; Hagopian and Spiro, 1967, 1968; Osborne, 1967; Reger and Cooper, 1967; Saita and Camatini, 1967; Caveney, 1969; Cochrane *et al.*, 1969; Jahromi and Atwood, 1969a,b; Pasquali-Ronchetti, 1969), *Xiphosura* (de Villafranca and Philpott, 1961), arachnids (Auber, 1963a; Zebe and Rathmeyer, 1968), and myriapods (Camatini and Saita, 1967). It is convenient to discuss the diversity in relation to particular topics.

A. FILAMENT RATIO AND LATTICE. The ratio of actin filaments to myosin filaments in arthropod muscles is always greater than the 2:1 found in adult vertebrate muscles. The largest ratio described is about 7:1 in certain white fibers of an eyestalk muscle in a crab (Hoyle and McNeill, 1968). In these fibers and often when the ratio is large, the myosin filaments are not arranged in a regular hexagonal lattice, but are mixed irregularly with the actin filaments, groups of up to 30 of which may occur in a nearly square lattice at some distance from myosin filaments. In other fibers, however, a ratio of 6:1 is found with the myosin filaments in a regular hexagonal lattice, each surrounded by a ring of 12 actin filaments—cockroach femoral muscles (Hagopian, 1966; Hagopian and Spiro, 1967), Apterygota (Caveney, 1969), Myriapoda (Camatini and Saita, 1967), lobster superficial abdominal extensors (Jahromi and Atwood, 1969a). Nonintegral ratios less than 6:1 have been described in many muscles (Auber, 1967a; Walcott and Burrows, 1969; Franzini-Armstrong, 1970a; Pasquali-Ronchetti, 1970) down to a 3:1 ratio with a regular lattice of 6 actin filaments surrounding each myosin filament (*Cyclops*, Bouligand, 1962; lobster antenna, Rosenbluth, 1969; insect flight muscle). Hoyle (1969) and Franzini-Armstrong (1970a) give further references.

In muscles where there is a regular lattice, the location of actin filaments relative to myosin filaments in transverse sections is as shown in Fig. 3. When there is a ring of six actin filaments, these are located midway between two myosin filaments giving a 3:1 filament ratio and not in the trigonal position as in vertebrate muscle (filament ratio 2:1). A lattice similar to that of Fig. 3b is found in some other fast invertebrate muscles such as those of *Sagitta* (Halvarson and Afzelius, 1969). When there is a ring of twelve actin filaments, the arrangement is usually as shown in Fig. 3d (filament ratio 6:1) but may be as Fig. 3c or less regular, with a smaller filament ratio. No significance has been established for the lattice detail when the filament ratio is greater than 3:1. Since all the information to date from these muscles has been obtained from electron microscopy of fixed material, some distortion of the pattern may be occurring during preparation; the nature and regularity of the lattice pattern in living fibers could only be determined by X-ray diffraction.

The interdigitating arrays of filaments are usually divided into discrete myofibrils by fenestrated sheets of sarcoplasmic reticulum; radially arranged blocks of filaments separated by invaginations of the plasma

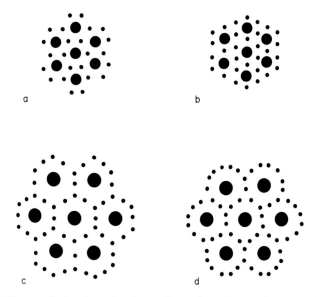

Fig. 3. Filament lattice in striated muscles with various actin-to-myosin filament ratios; (a) vertebrate skeletal muscle, 2:1; (b) insect flight muscle, 3:1; (c) arthropod leg and trunk muscles, 5:1; (d) arthropod leg and trunk muscles, 6:1. From Toselli and Pepe (1968a).

membrane is a common pattern (Fig. 2). Occasionally, the entire cross section of the fibers is occupied by a continuous filament array with isolated tubules (Fahrenbach, 1964). The myofibril is not a real unit in the substructure of many of these muscles.

B. LONGITUDINAL ARRANGEMENT. A great diversity of sarcomere lengths is found in arthropod muscles, and many crustacean limb muscles contain fibers of different sarcomere length intermingled apparently at random (Atwood *et al.*, 1965). The range is 3–14 μ. In *Portunus* leg muscles, there is variability of length in different sarcomeres of the same fiber (Franzini-Armstrong, 1970a). Sometimes, as in the flexors and extensors of the lobster and crayfish abdomen (Kennedy and Takeda, 1965; Parnas and Atwood, 1966) and in some limb muscles of certain species (Hoyle, 1969), long- and short-sarcomere fibers are grouped into separate bundles. In all cases where it has been studied, sarcomere length is greater in slow tonic muscles.

The length of the myosin and actin filaments varies in correspondence with the length of the sarcomere but is usually fairly constant for a given fiber type. In many slow muscles, however, the myosin filaments are variable in length giving an irregular A band boundary (Walcott and Burrows, 1969), and in *Portunus* leg muscles, the mean length may even be significantly different in the two halves of a single sarcomere (Franzini-Armstrong, 1970a). In this genus, the pseudo-H zones devoid of cross bridges are of constant length and are in good lateral register across the fiber, but in other cases, in fibers with a high filament ratio, the H zone may be indistinct (Hoyle and McNeil, 1968). In contrast, sarcomere length, filament length, and lateral register are very precise in fibers with a 3:1 filament ratio.

2. Insect Flight Muscle

The fibrillar flight muscles of Diptera, Hymenoptera, Psocoptera, Thysanoptera, Coleoptera, and Hemiptera are very different in ultra-structure from other muscles of the trunk, but the phasic flight muscles of the cockroach, locust, Odonata, and Lepidoptera, some of which are bifunctional in that they also move the legs (Wilson, 1962), show grada-tions in their modification from the usual pattern. Thus, in the tergocoxal muscle of a cockroach (Hagopian and Spiro, 1968) and in the flight muscles of *Vanessa* (Auber, 1967a), which has a low wing-beat fre-quency, the filament ratio is approximately 4:1, with a ring of 7–9 actin filaments round each myosin filament in the regular hexagonal array. In the tergotrochanteral muscle of *Calliphora*, which is a nonfibrillar phasic muscle serving both for jumping and for the start of flight (Nach-tigall and Wilson, 1967), the ratio is about 5:1 with a ring of 9–11 thin filaments (Auber, 1967b,c). In the phasic flight muscles of all insects with wing-beat frequencies greater than about 10/sec, the filament ratio is reduced to 3:1 and the lattice shows the regular structure of Fig. 3b (Smith, 1961b; Bienz-Isler, 1968a), but the tonic controlling muscles, which are usually tubular in histology, have the same high filament ratio as other slow trunk muscles. Correlated with their nearly isometric mode of operation in the body, phasic flight muscles show a much re-duced I-band. The myofibrils are discrete due to very well developed sarcoplasmic reticulum and numerous mitochondria. The H zone and M line are prominent.

Fibrillar flight muscles were first recognized as a distinct type of striated muscle by von Siebold (1848), and their ultrastructure has been most fully described by Smith (1961a, 1965). The myofibrils are large (1–5 μ in diameter) and form a genuine unit in the structure; single myo-fibrils of considerable length can be prepared by teasing the fresh tissue. Mitochondria may occupy up to 40% of the volume of the cell (Ashhurst, 1967a). The filament ratio is always 3:1 with the lattice of Fig. 3b

TABLE I

PERCENTAGE OF FIBER VOLUME OCCUPIED BY INVAGINATIONS (T SYSTEM)
AND SARCOPLASMIC RETICULUM(SR)

Muscle	T system	SR	Source
Frog sartorius	0.3	5	Peachey (1965)
Dragonfly flight	2	5	Smith (1966b)
Blowfly flight	0.5−1	0.2	Smith and Sacktor (1970)

and a geometrically regular organization (Worthington, 1961; Shafiq, 1963a). Longitudinally, the myosin filaments occupy virtually the whole of the sarcomere so that the I band is difficult to identify at rest length. Both types of filaments have a precise length and are aligned longitudinally to an accuracy of at least 1 nm (Reedy, 1968), giving a clear H zone. There is a well developed M line.

The sarcoplasmic reticulum of fibrillar muscles is reduced and never takes the form of a curtain surrounding the myofibrils; it may consist of little more than expanded vesicles forming dyadic contact with the well developed tubular system (Table I). This is the most obvious ultrastructural feature distinguishing asynchronous from synchronous flight muscles (Smith, 1961a) and has clear functional significance (see Section IV,C,2).

In many insect muscles, but particularly in the flight muscles with their high rate of metabolism, branches of the tracheal system penetrate down the plasma membrane invaginations and ramify inside the fiber, reducing the liquid diffusion pathway for oxygen (Tiegs, 1955; Smith and Sacktor, 1970); exceptionally, this is not found in dragonfly flight muscles (Smith, 1961b). Compression of these fine branches by the rhythmic contraction during flight aids the movement of gases (Weis-Fogh, 1964). Neuromuscular junctions are also found in the deep clefts (Smith and Sacktor, 1970).

3. VISCERAL MUSCLES

The ultrastructure of visceral muscles has been studied in Crustacea by Anderson and Ellis (1967), in insects by Edwards *et al.* (1956), Kawaguti (1961), Smith *et al.* (1966), Schaefer *et al.* (1967), Sandborn *et al.* (1967), Rice (1970), and Odhiambo (1970), and in the scorpion by Auber (1963c). Heart, blood vessel, and alary muscles have been studied in insects by Edwards and Challice (1960) and Sanger and McCann (1968a,b) and in *Xiphosura* by Dumont *et al.* (1968). These muscles all have short, small-diameter fibers (intestine 1–4 μ, heart 10–20 μ), with radial invaginations of the plasma membrane making irregular dyadic contact with diffuse and relatively sparse sarcoplasmic reticulum. Since the mitochondria are also small and disperse, the filament array is continuous across the cell, giving the appearance of a single myofibril. The myosin filament lattice is only approximately hexagonal with orbits of 10–12 actin filaments. The actin filaments have the usual diameter of about 6 nm, but myosin filament diameter varies from 10 to 32 nm; both types of filament are in poor lateral register so that there is no H zone or M line, and the Z line is often broken into discrete dense bodies (see Section II,B,5). Sarcomere length is 7–8 μ.

The intrinsic heart muscles contain a single layer of helically arranged fibers, characterized by the presence of intercalated disks resembling those of the vertebrate heart, with deep folding of the plasma membrane and septate desmosomes giving the appearance of branching fibers under the light microscope. Nerves and neuromuscular junctions are present in the neurogenic cockroach heart (Edwards and Challice, 1960) but not in the moth heart, which has a myogenic mechanism of rhythmicity (Sanger and McCann, 1968a). The extrinsic alary muscles show a further diminution of the sarcoplasmic reticulum, and the mitochondria are concentrated in protrusions of the sarcolemma; their fibers (5–15 μ in diameter) have long (5.5 μ) A bands and form direct intercalary junctions with the intrinsic heart muscles fibers. Such knowledge as we have of the physiology of insect hearts has been reviewed by McCann (1970).

4. ATTACHMENT TO CUTICLE AND OTHER TISSUES

At their ends, the somatic muscle fibers of arthropods normally form a characteristic type of close junction with an epidermal cell which secretes the cuticle; there are no intervening collagen fibrils, as is typical in vertebrates. The arrangement in dipteran flight muscle is shown in Fig. 4 (Auber, 1963b). At the myoepidermal junction, the plasma membranes of muscle fiber and epidermal cell are folded and apposed. The last sarcomere of each myofibril is joined to the membrane by means of a dense felting of filaments embedded in electron-dense material similar to that of the Z line, but of greater width. The intercellular space is transversely striated by lamellar structures perpendicular to the membranes, as in other types of desmosomes, and from the membrane of the epidermal cell, intracellular tonofilaments radiate toward and into the secreted cuticle. Each myofibril thus has a direct mechanical connection at each end with the skeleton, and there is a minimum of material which could give rise to series elasticity. A similar arrangement is found

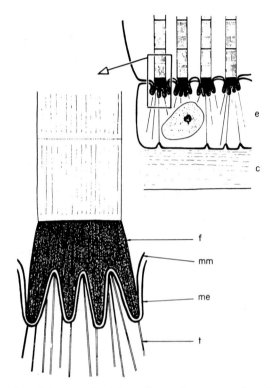

Fig. 4. Myoepidermal junction in arthropod muscles; e = epidermal cell; c = cuticle; f = junctional filaments; mm, me = plasma membrane of muscle and epidermal cells; t = tonofilaments. From Auber (1963b).

throught the phylum in the muscles involved in locomotion—*Cyclops* (Bouligand, 1962), spiders (Smith *et al.*, 1969), Apterygota (Caveney, 1969).

The problem of preserving mechanical continuity during ecdysis is discussed by Lai-Fook (1967) and by Caveney (1969). The attachment of the muscle fiber to epidermal cell is not disturbed; in the epidermal cell, microtubule-like tonofilaments likewise remain and "muscle attachment fibers" in the pore canals become attached to the new epicuticle before their terminal portions are shed with the old cuticle.

In visceral and other internal muscles, such as those of the venon gland of a spider (Smith *et al.*, 1969), the mechanical connection is with connective tissue. In such cases, the terminal actin filaments are again embedded in electron-dense substance at the cell membrane, but collagen fibrils instead of desmosome elements form the link on the outer surface of the membrane. Dense bodies of this type may occur not only at the ends but also along the sides of the muscle fibers.

5. THE Z LINE AND I BAND

The Z line in arthropod muscle usually forms a well defined boundary to the sarcomeres, as in vertebrate striated muscle, but its width (dimension longitudinal to the fibers) varies greatly in different types of fiber. In the fibrillar flight muscles of the water bug *Lethocerus*, the width is 100–140 nm; in the first antennal muscle of a crab, Reger (1967) found that fibers with a filament ratio of 6:1 had a thick, wavy Z line but that it was thin and straight in fibers with a 3:1 ratio. A discontinuous Z line consisting of a transverse row of dense bodies was first described by Hoyle *et al.* (1965) in the contracted giant fibers of the barnacle *Balanus nubilis*, where the arrangement permits the myosin filaments to pass through into the next sarcomere to give a super contraction down to 30% of rest length; they described the opening of these pores as an active process. A similar perforated Z disk permitting supercontraction has also been found in the trunk muscles of the blowfly larva (Fig. 5) (Osborne, 1967), in the slowest type of fiber from a crab antennal muscle (Hoyle and McNeill, 1968), and in most types of visceral muscle, where the dense bodies with bundles of actin filaments on each side may be poorly aligned across the fiber. In their lack of lateral alignment of filaments and dense bodies, some of these muscles resemble the smooth muscles of some invertebrates. Penetration

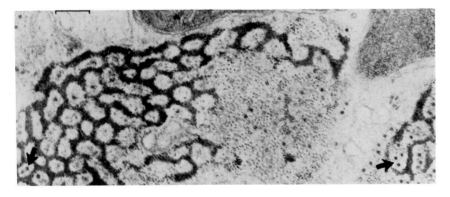

Fig. 5. Transverse section through Z line of ventral abdominal muscle of blowfly larva, showing myosin filaments (arrow) penetrating the perforations. Scale 0.25 μm. From Osborne (1967).

of the Z line by myosin filaments is the probable explanation of apparent A band shortening observed by phase contrast microscopy by von Hehn and Schlote (1964) in the oviduct muscle of *Carausius*.

Except where a dense body or thickened Z line joins actin filaments to the plasma membrane, this material always links actin filaments from adjacent sarcomeres. In many cases, the material is structureless in thin sections stained with heavy metals (Hoyle and McNeill, 1968; Franzini-Armstrong, 1970a; developing Dipteran flight muscles, Auber, 1969); but in the well ordered myofibrils of mature fibrillar flight muscle, Auber and Couteaux (1962, 1963) found that transverse sections of the Z line showed a regular pattern of "tubes" 30 nm in diameter. Ashhurst (1967b) reinvestigated this structure in the thick Z line of water bug flight muscles and showed that the pattern is generated by overlap of the two sets of actin filaments without distortion of the hexagonal lattice (Fig. 6). The pattern is, in fact, formed of small hexagons with filaments from adjacent sarcomeres occupying alternate corners; the space between the hexagons is filled with dense material. The Z line of a moth flight muscle has a similar structure (Bienz-Isler, 1968a). In this lattice, the myosin filaments meet a barrier of dense material when they touch the Z line as the muscle shortens, and it is therefore understandable that they buckle, forming the contraction bands of shortened fibrillar muscle visible in the phase contrast microscope (Hanson, 1956). The structure is quite different from that of the perforated Z disk of larval blowfly muscle (Osborne, 1967) in which larger "holes" occur between the regions where bundles of actin filaments are clumped together and leave room for the myosin filaments to penetrate (Fig. 5). Pringle (1968a) has shown how the different fine structure of the Z line in vertebrate muscle can be generated by a similar type of interdigitating overlap of actin filaments.

The region of the I band near to the Z line is of particular interest. The I band of insect fibrillar muscle is very short, and in dipteran and bee flight muscle, respectively, Auber and Couteaux (1962, 1963) and Garamvölgyi (1963) stated that the tapered myosin filaments appear to be prolonged to the border of the Z line by less dense material, which is nevertheless more dense than the background (Fig. 7). This observation gave rise to the idea that in this tissue there is a mechanical connection between myosin filaments and Z line (Garamvölgyi, 1965a), which Pringle (1967a) and Garamvölgyi and Belágyi ,1968a) named C filaments. In her study of the I band of water bug flight muscle, Ashhurst (1967b) did not find material of higher density at the appropriate points of the lattice and cast doubt on the reality of this element; the appearance could easily be created by superposition of the images

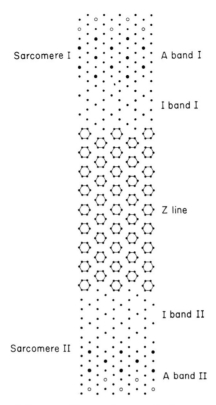

Fig. 6. Interposition of the actin filament lattices of adjacent sarcomeres to produce the hexagonal lattice of the Z line of insect fibrillar flight muscle. From Ashhurst (1967b) (corrected diagram).

of a myosin and an actin filament in a longitudinal section that is not thin enough to cut one layer only of the lattice (Ashhurst, 1971). The existence of a structural connection between the myosin filaments and the Z line is, however, still supported by Garamavölgyi (1969) in spite of the fact that Garamvölgyi and Belágyi (1968b) showed that bee flight muscle can be extended reversibly to 300% without structural damage. A different picture of the mode of termination of the myosin filament in moth flight muscle is given by Bienz-Isler (1968a), who states that the tapered region is connected to the actin filament by long bridges. The problem, which is further discussed in Section IV,B, can probably not be resolved by straightforward electron microscopy.

In leg and trunk muscle fibers, where the I band is of greater length, there is often a band of higher density (N line) a short distance from

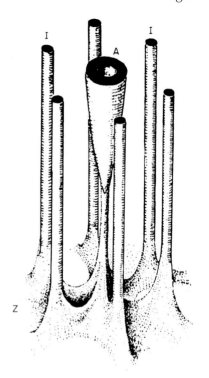

Fig. 7. Connection between myosin filaments (A) and Z line in blowfly flight
muscle, according to Auber and Couteaux
(1963).

the Z line, in which the actin filaments are clumped together (Franzini-
Armstrong, 1970b). Between the N and Z lines, the longitudinal order
of the actin filaments is always greater than elsewhere in the I band,
and this region is birefringent (Aronson, 1961). In crab leg muscle fibers
contraction bands involving bending of the ends of the myosin filaments
appear in the N region, 1.8 μm away from the Z line (Gillis, 1969).

6. The Myosin Filament and M Line

The myosin filaments of vertebrate striated muscle consist of a polymeric arrangement of myosin molecules with staggered lateral association of the light meromyosin portion of the molecule (H. E. Huxley,
1963). A detailed model for the internal construction of the filament
has been proposed by Pepe (1967) in which, in its simplest form, there
are twelve subunits in the central M line region and eighteen subunits
in the main portion of the filaments; the filament tapers at the ends
due to gradual reduction in the number of subunits. A similar but slightly
more complicated model can be constructed with twice the number
of subunits, consistent with Huxley's estimate of the amount of myosin

present (two molecules per cross bridge). This model generates a filament with the approximate diameter (10–15 nm) found in negatively stained preparations (H. E. Huxley, 1963) and accounts for the triangular cross section found in the M line region of vertebrate muscles; it supposes that the M filaments, which unite the myosin filaments at this region of the sarcomere of twitch fibers (Page, 1965), arise from the points at which myosin molecules abut tail-to-tail. In longitudinal sections of vertebrate muscle, there is a cross-bridge-free region 150 nm long in the middle of the filaments (the pseudo-H band), and up to six sets of M bridges are often visible.

Arthropod muscles show a much greater variety of myosin filament length and diameter and more varied M line structure than has been described in vertebrate muscles. No accurate comparative study of diameters has been made, but measurements from published transverse sections give diameters from 12 to 25 nm. The filaments often appear to have a less dense core, especially with osmium fixation. In some crab fibers, it is possible to see a subunit structure within the myosin filament in high-power electron micrographs (Gilev, 1965, 1966). Negatively stained transverse sections show an outer ring of 10–12 subunits 2–2.5 nm in diameter and an inner group of 1–6 with a diameter of 2.5–3 nm (Gilev *et al.*, 1968); the maximum total visible is 18, which corresponds to Pepe's (1967) model. On the other hand, Bacetti (1965) states that in the muscles of a polycheate worm, the housefly, and man, the substructure is a ring of 9 units with 2 in the central core. Negatively stained filament preparations from the crayfish show longitudinal strands 2–2.5 nm wide, with a fine striation at 4 nm formed of dense spots 2 nm in diameter spaced by 2 nm (Peterson, 1963). This author argues that this striation must indicate the presence of a repeat structure within the light meromyosin shaft of the myosin molecules, since there is no diagonal banding such as would be expected from stagger of the ends of the molecules. Arthropod myosin filaments seem to be more promising material than vertebrate filaments for the direct study of the internal molecular architecture.

The packing of molecules in the filaments is presumably responsible for the lattice in which the heavy meromyosin heads project on the outside of the filament. In vertebrate muscle, the cross bridges arise symmetrically on opposite sides of the filament with a longitudinal spacing of 14.5 nm, each successive pair being rotated in azimuth by 60° (H. E. Huxley *et al.*, 1965). In a detailed electron microscope study of water bug flight muscle in rigor, Reedy (1967, 1968) was able to use the precise organization of this tissue to show that cross bridges are double and arise in pairs on opposite sides of the filament with

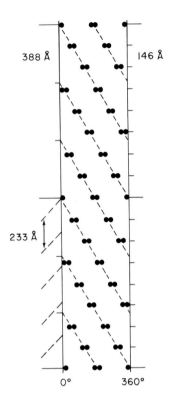

Fig. 8. Radial projection (unrolled cylinder) of the myosin filament of insect fibrillar flight muscle, showing the location of the origins of the double cross bridges. From Reedy (1967).

a longitudinal spacing of 14.5 nm; the azimuthal rotation is 67.5°, giving a nonintegral right-handed helix with a half pitch of 38.8 nm (Fig. 8), which is the same as the pitch of the left-handed actin helix. This model has recently been confirmed and extended in a detailed study by X-ray diffraction (Miller and Tregear, 1971). The nonintegral cross bridge helix is present in relaxed muscle and has 16 axial repeats every 3 turns, each complete turn occupying 77 nm. The double bridges on each side of the filament are at the same axial level but are spread in azimuth as if wrapped round the filament. Each myosin filament is surrounded by a helical array of thin filaments, the pitch being 77 nm, and the thin filaments must contain some nonactin material to account for all the observed X-ray reflections. In rigor, the mismatch between the pitches of the myosin and actin helices produces a pertubation which is sufficiently regular to generate diffraction spots. Recently, Squire (1971, 1972) has developed a general theory of the structure of the myosin filament. He suggests that in insect fibrillar muscle the molecules are assembled in a six-stranded helix of pitch 231 nm with 16 residues in 1 turn, and maintains that this structure explains equally well certain features of the X-ray diffraction picture. It also becomes possible to relate the myosin filament structure in insect fibrillar muscle to that in other muscle types. An apparent difficulty is Reedy's (1968) failure to visualize more than 4 bridges every 14.5 nm by electronmicroscopy of muscles in rigor, but Squire explains this in terms of the limited steric possibilities of interaction with the actin filaments. More work on other types of flight muscle is needed to resolve the disagreements.

The relationship between the amount of myosin present and the number of cross bridges is currently unresolved. H. E. Huxley (1960) suggested that in rabbit skeletal muscle there are about two molecules per cross bridge. Chaplain and Tregear (1966) found a figure nearer to three for the number of molecules of molecular weight 500,000 on the assumption that there are two bridges every 14.5 nm along the A band region of water bug flight muscle. Reedy's (1968) visualization of only four bridges per 14.5 nm is inconsistent with this, but, in a brief note, Reedy *et al.* (1969) give a new estimate of one molecule per bridge. Squire's (1972) model of a six-stranded helix with one molecule per bridge would be consistent with Chaplain and Tregear's (1966) figures.

The pseudo-H zone in the center of the myosin filament (the region devoid of cross bridges) is visible in all arthropod fibers where the filaments are in good lateral register; its length is 60 nm in crustacean leg fibers, which is much less than that in vertebrate muscles (Franzini-Armstrong, 1970a). In this region, the myosin filament always has a different cross section from that of the overlap region, but the detail varies (Auber, 1967b). In Crustacea and in dipteran flight muscles it is circular but solid; in the flight muscles of Hymenoptera it is quadrangular (Fig. 9a), in water bugs elliptical (Fig. 9b), and in *Melolontha, Aeshna,* and Lepidoptera it is flattened or composed of two circular profiles (Fig. 9c,d). If the model proposed by Pepe (1967) is broadly correct, there must be considerable differences in the detail of packing of subunits in the myosin filaments of different arthropods.

Between the myosin filaments in the pseudo-H region, transverse M filaments are found in all muscles with a 3:1 filament ratio, and in water bug flight muscle these link to a short longitudinal M filament occupying the trigonal position in the lattice (Fig. 9b) and therefore so placed as not to interfere with the sliding of actin filaments (Ashhurst, 1967a). Longitudinal sections show six evenly spaced sets of M filaments. In a *Cyclops* muscle, there are four pairs of M bridges each 12 nm apart, but not evenly spaced (Bouligand, 1962). Some of the density of the M line, however, is not formed by M filaments. In *Bombylius*

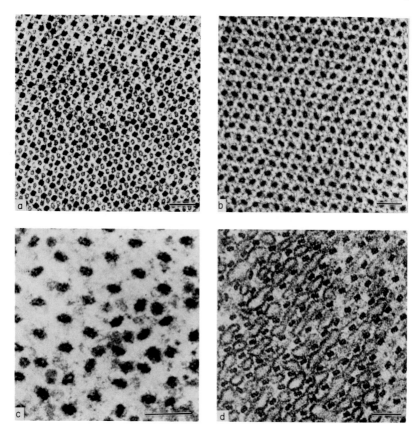

Fig. 9. Transverse sections through the M line region of insect flight muscles. Scale 0.1 μm. (a) *Apis* (Hymenoptera) (from Auber, 1967b); (b) *Lethocerus* (Hemiptera) (from Ashhurst, 1967a); (c) *Melolontha* (Coleoptera) (from Auber, 1967b); (d) *Achalarus* (Lepidoptera). From Reger and Cooper (1967).

flight muscle, glycogen granules occur in this region (Auber, 1967b) and in *Apis* and some Lepidoptera there is an elaborate pattern of membranous vesicles which appears to be an intramyofibrillar extension of the sarcoplasmic reticulum (Fig. 9d) (Auber, 1967a; Reger and Cooper, 1967). It is clearly not possible to generalize about the organization of the arthropod M line except that, as in vertebrates, it is a region of lateral association between the myosin filaments.

7. OTHER FILAMENTOUS STRUCTURES

A number of authors have revived the possibility that there may be longitudinal structures in muscle other than the myosin and actin filaments and have derived much of their evidence from arthropods. McNeill and Hoyle (1967) described ultrathin filaments 2–2.5 nm in diameter visible both in longitudinal and transverse sections through the H zones of various crustacean muscles and in the gap zone between myosin and actin filaments in very stretched fibers. The ultrathin filaments in the H zones are of similar diameter to the longitudinal M filaments visible in some insect fibrillar flight muscle, but are more numerous in the muscles studied by McNeill and Hoyle which have a 6:1 actin to myosin filament ratio; it is claimed, however, that the ultrathin filaments extend throughout the length of the sarcomere.

Toselli and Pepe (1968a) described two distinct diameters of filaments (modes 5.5 nm and 7.5 nm) attaching to the dense bodies at the desmosomes of mature *Rhodnius* abdominal muscles and identified only the larger of the two with actin filaments. Further evidence for the existence of small-diameter filamentous material is given in Section II,C. Although the existence of these elements has been challenged or discounted by some workers, it would be as well to suspend final judgement, since the mechanical properties of many types of fiber are not fully explained by the presence only of myosin and actin filaments (Section IV,B).

8. Correlation of Structure and Function in Nonfibrillar Muscles

The diversity of structure found in arthropod muscles should provide opportunities to correlate fiber organization with physiological properties, and certain muscles have been studied with this particularly in mind. The extensor trochanteris complex of the cockroach, situated in the coxa, consists of muscles 135a,b,c, which also serve as wing levators, and muscles 136 and 137, which are purely leg muscles (Jahromi and Atwood, 1969b; numbering of Carbonell, 1947). Muscle 135 is more resistant to fatigue and has a higher mitochondrial content (Smit *et al.*, 1967). This is the first and most general correlation. The small bundle of fibers in muscle 135b has a mean sarcomere length of 7.3 μm and is classified as slow by Becht and Dresden (1956). The mean sarcomere length of the main bundles 135a,c is 3.7 μm, slightly though significantly different from the value of 4.2 μm for muscle 137. The speed of contraction (time to twitch peak) and maximum tension per unit section area are not greatly different for these two muscles (Usherwood, 1962), and they have a similar density of dyadic contacts between the plasma membrane invaginations and the sarcoplasmic reticulum. On the other hand, muscle 137 has a filament ratio of 6:1 compared with a 3:1 ratio in muscle 135. Data for these muscles and for the deep and superficial abdominal extensors of the lobster are summarized in Table II. Jahromi and Atwood (1969a,b) conclude that in arthropod skeletal muscle, the length of the sarcomere and the extent of overlap of myosin and actin filaments are more important in determining contractile properties than the filament ratio or the absolute number of myosin filaments per unit area of cross section. The relationship between sarcomere length, speed of contraction, and maximum tension is in qualitative agreement with that proposed for vertebrate muscles by A. F. Huxley and Niedergerke (1954).

Some similar data for the four fiber types in a crab eyestalk muscle are shown in Table III. Hoyle and McNeill (1968) drew attention to the broad correlation between speed of contraction and relaxation and both the density of dyads and the length of the sarcomeres, but are unable to sustain any explanation for the high filament ratio of the intermediate fiber group; these are the fibers in which clumps of actin filaments often occur at some distance from myosin filaments. The same type of comparison has been made by Cochrane *et al.* (1969) in certain tonic and phasic muscles of the locust and by Huddart and Oates (1970) for the homologous muscles of the locust and the stick insect whose contraction and relaxation times are very different; the conclusion in each case is that the best correlation is with the density of sarcoplasmic reticulum. Rosenbluth (1969) describes an extreme example in a fast muscle of the lobster second antenna, where nearly three-quarters of the fiber is occupied by reticular elements. Another good example of the same correlation was found by Tyrer (1969) in a parallel study of contraction and relaxation times and the growth of the sarcoplasmic reticulum in developing locust intersegmental muscle.

In the myofibrils themselves, correlations between structure and function are more difficult. The sliding filament theory of muscular contraction makes the following relevant predictions A. F. Huxley (1965):

a. Maximum tension per unit cross-sectional area of myofibril should be greater in fibers with a long length of overlap between the myosin and actin filaments. In general, overlap will be greater with long sarcomeres, but for quantitative comparisons it is important to determine the fraction of fiber area occupied by the filament array, and the widths of the I band and H zone under the conditions with which the tension is measured.

b. The velocity of sarcomere shortening is independent of the length of overlap, but since short-sarcomere fibers will have more sarcomeres per unit length of fiber, they should, overall, have a higher speed of contraction. Time to peak twitch tension is not, however, a valid measure

TABLE II

STRUCTURAL AND PHYSIOLOGICAL MEASUREMENTS ON COCKROACH COXAL AND LOBSTER ABDOMINAL EXTENSOR MUSCLES[a]

Muscle	Sarcomere length (μm)	Overlap length (μm)	Dyads/ sarcomere	Filament ratio	Myosin filaments per μm²	Peak twitch time (msec)	Peak tension (kg·cm⁻²)
Cockroach 137	4.2	3.2	2–4	6:1	600	13 (twitch)	0.85 (twitch)
Cockroach 135c	3.7	2.5	2–4	3:1	650	10 (twitch)	0.70 (twitch)
Lobster deep	4	3	2–4	3:1	450–600	50 (twitch)	0.8–1.4 (TEA)[b]
Lobster superficial	9	6	2	6:1	350–400	400 (TEA)[b]	5–14 (TEA)[b]

[a] From Jahromi and Atwood (1969a,b).
[b] TEA = tetraethylammonium chloride.

TABLE III

STRUCTURAL AND PHYSIOLOGICAL MEASUREMENTS ON FIBER BUNDLES FROM CRAB EYESTALK MUSCLE[a]

Muscle bundle	Sarcomere length (μm)	A band length (μm)	Mitochondria	Sarcoplasmic reticulum	Filament ratio	Twitch speed (cm·sec⁻¹·cm⁻¹)	Half relaxation time (sec)
Pinkest	13.0	7.5	++++	++	4–5:1	0.04	>2.0
Gray	13.0	7.5	+++	++	4:1	0.25	2.0
Intermediate	9.8	6.6	++	+++	7:1	1.0	1.0
Whitest	7.9	5.9	+	++++	3–5:1	4.0	0.2

[a] From Hoyle and McNeill (1968).

of the speed of action of the contractile machinery, owing to possible limitations by the speed of the intramuscular coupling process; in order to eliminate this, the redevelopment of tension after a quick release must be measured during sustained tetanic excitation.

c. The velocity of filament sliding correlates over a wide range with the turnover rate of the myosin ATPase activity (Barany, 1967).

It must be concluded that the comparative studies which have so far been made of nonfibrillar arthropod muscles are not yet sufficiently precise to enable any more than broad conclusions to be drawn about variation in parameters of the contractile machinery itself. The opportunity remains.

C. Ontogeny

The muscles of arthropods, as of vertebrates, arise in the embryo from mesodermal somites, which are the first parts of the body to show segmentation (Manton, 1949). In the unhatched embryo of the marine crustacean *Nebalia*, contractility appears first in the gut muscles, which at that stage show no signs of striation or of fibrils (Manton, 1934); the degree of polymerization of the contractile proteins has not been investigated. Subsequent differentiation has been studied mainly in insects, and particular attention has been paid to the flight muscles, where Tiegs (1955) showed that in Orthoptera, with their direct development, cell cleavage continues after the nymphal fibers are striated and functional, while the fibrillar flight muscles of Diptera differentiate only after massive fusion of myoblasts; both processes operate in Hemiptera. Fusion of myoblasts originating from degenerate larval muscles also occurs in Lepidoptera (Eigenmann, 1965; Bienz-Isler, 1968b). The final location of the muscles is determined by attachment of myoblasts to certain specific areas of the epidermal cell layer, which can be transplanted (Sahota and Beckel, 1967a,b). The full development of arthropod muscles occurs only when they are innervated (Nuesch, 1965; Basler, 1969).

Differentiation has been studied at the ultrastructural level in fibrillar flight muscles during normal ontogeny and in crustacean and insect trunk and limb muscles during the molting cycle. The abdominal intersegmental muscles of *Antheraea pernyi* degenerate within 48 hr of emergence of the adult moth and the progress of lysis has been studied by Lockshin and Williams (1965). They claim that it is initiated by the sudden cessation of the outflow of impulses in the motor nerves under the influence of the hormone ecdysone and can be prevented by electrical stimulation or physostigmine, which excites the ganglia. Mitochondria, myofilaments and Z line are eroded by lysosomes in that order. However, Runion and Pipa (1970) show that the leg muscle degeneration which occurs during metamorphosis in *Galleria* proceeds in spite of continued nervous activity. The muscles of the chela of the land crab *Geocarcinus* are reduced in protein content to 60% during the premolt period and rapidly resynthesized 2 weeks after ecdysis (Skinner, 1966). Degeneration of the flight muscles in later adult life is a common phenomenon in beetles and involves lysosome-mediated lysis of the myofibrils (Bhakthan *et al.*, 1970). In *Rhodnius* the abdominal intersegmental muscles degenerate and redevelop during each molting cycle, and the process has been followed in detail. Ten hours after feeding, ribosomes disperse from the nuclear envelope and form polysomes in the sarcoplasm. Myosin filaments appear within 10–15 hr in the cortex of the cell in association with 5–6 nm filament bundles and dense bodies at the invaginating sarcolemma (Auber-Thomay, 1967); their formation is not affected by disruption of the microtubules by colchicine (Warren and Porter, 1969). Four days after feeding the developing cells contain microtubules, filaments 7.5 nm and 20 nm in diameter (which from their constant 6:1 ratio and lattice are presumably actin and myosin filaments), and "tonofilaments" 5.5 nm in diameter joining the dense bodies of the desmosomes. Both 7.5 nm and 5.5 nm filaments are present at molting after 15 days. The lattice of myosin and actin filaments grows by addition at the periphery and microtubules never become incorporated in the lattice (Toselli and Pepe, 1968b).

The normal process of myofibrillogenesis in dipteran flight muscle has been described by Shafiq (1963b) and most completely by Beinbrech (1968) and by Auber (1969). About 40 hr after pupation the embryonic fibers contain numerous microtubules and bundles of initial filaments 5 nm in diameter grouped in dense bodies at the periphery of the cells. Myosin and actin filaments develop in these bundles at levels where they are less tightly packed, creating the rudiments of sarcomeres and Z lines. As more filamentous material is laid down, the dense bodies divide laterally and longitudinally, and the sarcomere rudiments increase so that by the middle of the third day there is the full number of myofibrillar bundles (Fig. 10a). Between the second and third day, the dense bodies condense longitudinally, producing an overall shortening of the fibers with corresponding elongation of the apodemes (Fig. 10b). There is then a rapid elongation of the fibers due to longitudinal cleavage of dense bodies with the appearance of new sarcomere rudiments (Fig. 10c), but the dense bodies retain their disorganized internal structure and the myosin filaments are not arranged in a regular hexagonal

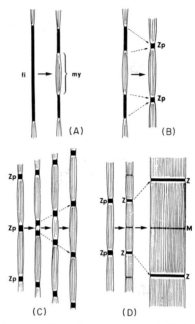

Fig. 10. Stages in the formation of sarcomeres and Z line in the flight muscles of the blowfly; time measured from pupation. (A) 2 days, first appearance of myosin and actin filaments; (B) 2–3½ days, condensation of dense bodies (Zp); (C) 3½–4 days, formation of new sarcomeres; (D) 4–6 days, fine structure in Z–line and thickening of myofibrils. From Auber (1969).

lattice. From the fourth to the seventh day, there is no overall increase in length, but the Z lines and the hexagonal lattice assume their regular structure and the M lines appear. From then until emergence, the myofibrils grow by addition of filaments at their periphery and the sarcomeres increase in length (Fig. 10d); at the edge of the myofibril, the newly added myosin filaments are smaller than their final diameter. Microtubules disappear about the seventh day, but there are still polysomes in the sarcoplasm. The plasma membrane invaginations start to penetrate between the myofibrillar elements during the third day. Differentiation occurs mainly near to the muscle insertions in dorsoventral flight muscle, but is more widespread in the dorsal longitudinal muscle (Beinbrech, 1968).

Auber (1962, 1965) and others state that in arthropod muscles the myosin and actin filaments are preceded by small-diameter "initial" filaments and that the myosin and actin filaments, which appear simultaneously, are laid down either around or between the initial filaments in regions where they are not bound together by dense material. In the mature muscle, the initial filaments persist as the Z filaments and the filaments in the dense bodies associated with desmosomes; they may be related to the ultrathin filaments of McNeill and Hoyle (1967) if these elements have a real existence.

Eigenmann (1965) has shown that actomyosin ATPase activity is demonstrable in developing moth muscle just before the first appearance of myofibrils. We may suppose that myosin and actin, synthesized in the polysomes, undergo their process of polymerization in some way organized by the initial filaments, which could play a role in determining their final length. The myofibrils grow by addition of both types of filament to their outer edge as in vertebrate muscle (Morkin, 1970). The later growth in length of the sarcomeres does not seem to take place in the same way in arthropod and vertebrate muscles. Goldspink (1968) has shown that the sarcomeres of newborn mice extend from 2.3 to the 2.8 μm of the adult by a sliding mechanism without change in filament length and that those in the terminal portions of the young fibers are inextensible. Auber's (1969) careful measurements of sarcomere length in a fly show that the mean sarcomere elongation from the sixth day to emergence is from 2.0 to 3.2 μm. Since there is at no stage any significant I band, some elongation of the filaments must take place with deposition of new protein, and if the myosin and actin filaments both grow at their free ends, some sliding must be occurring during this process. Similarly, Aronson (1961), by direct observation with the polarizing microscope of the dorsal muscle of a developing mite, has also shown a lengthening of the A band which can only be explained by continued deposition of new filamentous material after the full sarcomere organization has been established. The three sarcomeres of these small muscles lengthen from 2.5 μm to 10 μm during a period of 30 hr. The A band increases in length steadily from 1.5 to 4 μm. The I band increases from 1.5 to 6 μm during the middle 7 hr of this period and then shortens to 5.5 μm in the final period before the muscles become contractile.

The formation of new sarcomeres in the blowfly by longitudinal splitting of the dense bodies explains how fibers can elongate without a corresponding increase in the length of the sarcomeres. Such an elongation also occurs in the propodite muscles of a crayfish (Bittner, 1968), but has not been studied by electron microscopy. The proliferation of myofibril rudiments in the fly by lateral division of dense bodies parallels Goldspink's (1970) demonstration of Z line partition in mammalian muscle; the process of lateral growth thus seemed to be similar in all striated muscles. Garamvölgyi (1965b) has described transverse bridges of dense material linking the Z lines of the myofibrils to each other and to the dense bodies at the sarcolemma of the bee muscle; these may be remnants of the ontogenetic process.

D. *Localization of Contractile Proteins*

1. SELECTIVE EXTRACTION

Controlled biochemical degradation of a striated muscle can be achieved in a number of different ways, with progressive elimination of component parts of the intact cell. Disruption of the plasma membrane, allowing leakage of the sarcoplasmic constituents, occurs on soaking in 50% glycerol–water and after treatment with detergents or osmotic shock. Glycerination is the standard method of preserving the contractile apparatus, which retains most of its properties after prolonged storage in 50% glycerol at −20°C. Its effect on the structure of water bug flight muscle has been studied in detail by Ashhurst (1969). The myofibrils are intact after several months. Mitochondria, plasma membrane invaginations and dyads are rapidly disrupted, but some sarcoplasmic reticulum remains and mitochondrial fragments later come together to form mitochondrion-like bodies. These still perform appreciable oxidative phosphorylation, but can be removed without affecting the myofibrils by treatment with the nonionic detergent Tween 80 (polyoxyethylene sorbitan monooleate). About 33% of the total protein is removed by these two treatments, and there is then no detectable oxygen consumption (R. H. Abbott and Chaplain, 1966). Glycerol extraction has also been used on long sarcomere crab fibers to produce a preparation in which it can be shown that iontophoretically applied Ca^{2+} is effective only when it is applied to the overlap region (Gillis, 1969).

Within the myofibrils, selective extraction can be used to determine the location of the different proteins and their contribution to the contractile properties, but all experiments using arthropod muscle have so far been based on the known solubilities of proteins from rabbit muscle. Myosin is extracted from *Limulus* muscles (de Villafranca *et al.*, 1959), *Balanus* muscle (Baskin *et al.*, 1969), and abdominal muscles of the lobster and water bug (Walcott and Ridgway, 1967) by potassium chloride solutions of more than 0.4 *M*, but if the solution is alkaline, the extracts contain an appreciably greater amount of actin than with rabbit muscle. Hanson (1956) found that at pH 6.5 or 7.0, a potassium chloride concentration of not less than 0.7 *M* was required to remove the myosin filaments from blowfly flight muscle and made a careful study of the appearance of the myofibrils by phase contrast and polarization microscopy. The results are not affected by the presence of ATP or pyrophosphate. Extraction starts in the center of the sarcomeres, and at an intermediate stage, some additional density resembling the contraction bands appears near the Z line, provided the actin filaments have not been pulled away from the Z line by excessive stretch. This so-called migration of the A substance is confirmed by Garamvölgyi and Kerner (1966) and Tigyi-Sebes (1966) and is understandable in view of the activation of ATPase activity by very high ionic strength (see Section IV,D). The final state is a fiber consisting of I segments with a persistent M line and a residual low-density filamentous backbone (Garamvölgyi and Kerner, 1966; Walcott and Ridgway, 1967). If these extracted fibers are treated with rabbit myosin, density is restored except in the H zone, and adding ATP then results in contraction bands but no overall sarcomere shortening. The A substance can be removed even at normal ionic strength by lowering the pH to 5.0 (Hanson, 1956).

The actin filaments can be removed from myosin-extracted blowfly myofibrils with a solution containing 0.6 M potassium iodide and 0.006 M sodium thiosulphate at pH 5.5 (Hanson, 1956). In the phase contrast microscope, this reduces the density to a uniform level between the Z lines, but the integrity of the structure is preserved.

The Z line and M line of rabbit muscle can be removed by a solution of very low ionic strength and partially reconstituted by restoring the ionic strength to 0.1 (Stromer *et al.*, 1969). Density can be partially removed from the Z and M lines of the bee flight muscle by treatment with acetone or lipase, but filamentous material remains at Z, linking the actin filaments of adjacent sarcomeres (Garamvölgyi, 1965b).

In summary, there is little doubt that the two main filamentous elements of arthropod muscles are composed of proteins similar to the myosin and actin of vertebrate muscle (Maruyama, 1965), though the myosin has slightly different solubility properties. There seems also to be another set of longitudinal elements which is revealed after extraction of myosin, actin, and Z line of mature muscles and which is present at an early stage in developing muscles. Whether these are the same elements as have been described in the intact Z line and in the H zones after extreme stretch remains to be established. The Z line material of insect flight muscle has different solubility properties from that of rabbit muscle; it may be significant that the greatest structural difference between the two types of muscle is found in this region of the sarcomere.

2. Visualization of ATPase Activity

A number of authors have claimed to demonstrate the localization of ATPase activity in myofibrils by incubating them in a solution containing ATP and a lead salt and noting the position of dense granules in electron micrographs. Using this method on blowfly flight muscle, Tice and Smith (1965) concluded that activity was restricted to the overlap region of the sarcomere; Tice (1969) specifically associated the granules with myosin filaments in experiments designed to distinguish granules produced through enzymic activity from those deposited in inorganic phosphate. On the other hand, Zebe and Falk (1963a,b, 1964) claimed that ATPase activity was also associated with the Z lines or bands narrowly located on each side of them, and Zebe (1966) demonstrated a similar localization in I segments prepared by extraction of myosin. All this work has to be reassessed following Gillis and Page's (1967) demonstration of appreciable movement of lead granules after deposition during the preparatory procedures for electron microscopy, and Franzini-Armstrong's (1970b) study of the localization of inert ferritin granules in skinned frog fibers. The latter author showed that ferritin becomes concentrated in two regions of the I band at the N_1 and N_2 lines where additional material binds together the actin filaments. Since the Z line of arthropod muscle also restricts the longitudinal diffusion of granules, Zebe's evidence cannot be taken as proving localization of ATPase activity in this region.

III. Biochemistry

A. Composition

The biochemistry of the contractile elements of insect flight muscle was fully reviewed by Maruyama (1965), and little further analytical work on arthropod muscles has been carried out since then. Comparative studies are at a very rudimentary stage, and the many opportunities for correlation between protein composition and function remain to be exploited. None of the minor constituents of the myofibril have yet been properly characterized from arthropod muscles.

1. Myosin

All investigators agree that the standard high ionic strength extract of arthropod myofibrils contains appreciable quantities of actin in addition to myosin (Gilmour and Calaby, 1953; de Villafranca et al., 1959; de Villafranca and Naumann, 1964); this must be separated subsequently by dissociation with pyrophosphate or ATP and Mg^{2+}. The amino acid

TABLE IV

Amino Acid Composition of Myosin, Actin, and Tropomyosin from Rabbit
and Adult Blowfly Muscle[a]

	Residues·per 10^5 gm protein					
	Myosin		Actin		Tropomyosin	
Amino Acid	Rabbit	Blowfly	Rabbit	Blowfly	Rabbit	Blowfly
Lysine	85	72	44	58	107	87
Histidine	15	13	18	16	5.5	5.8
Amide NH_3	(86)	(94)	(59)	(67)	(64)	(106)?
Arginine	41	46	45	46	42	67
Aspartic	85	88	81	79	89	93
Threonine	41	30	61	47	26	31
Serine	41	36	53	49	40	37
Glutamic	155	158	95	115	212	229
Proline	22	19	47	44	1.7	2.8
Glycine	39	44	64	66	12	16
Alanine	78	81	68	80	108	102
Valine	42	30	38	41	27	36
Methionine	22	15	36	24	16	10.6
Isoleucine	42	32	48	40	30	16
Leucine	79	88	63	69	95	98
Tyrosine	18	17	37	28	15	5.8
Phenylalanine	27	24	28	27	3.3	8.5
Total groups	832	793	826	829	830	846

[a] From Maruyama (1965).

composition of myosin from whole blowfly is shown in Table IV. The
pure protein is activated by Ca^{2+} and not by Mg^{2+} and is split into
heavy and light meromyosins by tryptic digestion, as with rabbit myosin.
Determination of the molecular weight is complicated by the greater tendency of arthropod myosin to associate into polymeric units; both ultra-
centrifugation (*Limulus*, de Villafranca, 1968) and light scattering
studies (lobster, Woods *et al.*, 1963) give what are probably erroneously
high values. Lobster myosin contains higher percentages of methylated
lysine residues than rabbit myosin but similar amounts of methylhistidine
(Kuehl and Adelstein, 1970).

2. ACTIN

Actin can be prepared from acetone-dried insect material in the same
manner as with rabbit muscle; the amino acid composition is given

in Table IV. Apart from the details in this table, no difference in properties from rabbit actin have been described for actin prepared from any arthropod.

3. Tropomyosin

Crystalline tropomyosin was prepared from whole adult and larval blowflies by Kominz *et al.* (1962) using the Bailey (1948) method and gave significantly different amino acid compositions from rabbit tropomyosin (Table IV). All these amino acid compositions must be viewed with reserve, since the starting material was the whole insect and there may well be differences in the proteins from different muscles. A reinvestigation of the proteins in insect fibrillar flight muscle is in progress in Oxford.

B. ATPase Activity

The ATPase of extracted proteins has been studied in crayfish, crab, and *Limulus* leg muscles and in insect flight muscle preparations of various degrees of purity. Actomyosin from arthropod muscles was originally reported to be activated only by Ca^{2+} and to be inhibited by Mg^{2+} when measured at low ionic strength (0.1 M potassium chloride) (Maruyama, 1965; de Villafranca and Naumann, 1964), but later work (Maruyama, 1966; de Villafranca and Campbell, 1969) showed the usual Mg^{2+} activation. Provided that alkaline conditions have not been used during preparation of the protein, crayfish (Maruyama *et al.*, 1968a) and insect flight muscle (von Brocke, 1966; Maruyama and Allen, 1967) actomyosin MgATPase activity is increased when Ca^{2+} is raised to about 5×10^{-7} M using EGTA* buffers. This sensitivity to micromolar concentrations of Ca^{2+} is due to the presence of a native tropomyosin complex (cf. Hartshorne *et al.*, 1966; Ebashi and Ebashi, 1964), which can be destroyed by short tryptic digestion or extraction at low ionic strength at a pH more alkaline than 7.5 (Meinrenken, 1969). The desensitized actomyosin then shows a high MgATPase activity even in the presence of EGTA, which reduces the Ca^{2+} to 10^{-9} M or less. Enzyme activity is reduced at higher ionic strength (Maruyama and Allen, 1967). In all these respects, the properties of arthropod actomyosin seem to be similar to that from rabbit muscle, but Maruyama *et al.* (1968b) found that there is a significant difference in the relationship between Ca^{2+} and MgATPase activity of the actomyosin complex in fibrillar and non-

* Ethyleneglycol bis(β-aminoethyl ether)-N,N'-tetraacetic acid.

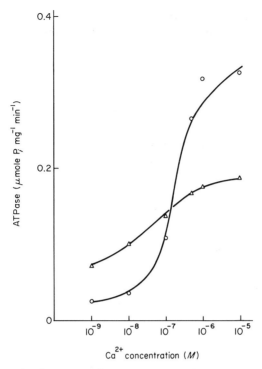

Fig. 11. Relationship between Ca^{2+} concentration and ATPase activity in myofibrils from insect flight and leg muscles. Triangles = flight muscle; circles = leg muscle. From Maruyama *et al.* (1968b).

fibrillar insect muscle (Fig. 11). Whereas the water bug leg muscle proteins show a sharp increase in activity at about 10^{-7} M, the flight muscle proteins show a more gradual increase over a wider range of Ca^{2+} concentrations and a lower maximal activity. After trypsin treatment, both protein complexes are insensitive to Ca^{2+}, but the activity of the flight muscle preparations is not increased to the level of that from leg muscle. This difference may be significant in relation to the functioning of fibrillar muscle (see Section IV,D).

IV. Physiology

In their main physiological properties, the limb and normal trunk muscles of arthropods do not differ qualitatively from those of vertebrates, but the large diameter of the fibers, the length of the sarcomeres, and other special features have been used in some elegant experiments which have helped to elucidate general mechanisms. The exception to this statement is the fibrillar muscle of insects (IFM), discussion of which will therefore occupy much of this section.

A. Calcium Activation

As in other types of striated muscle, the final step in the excitation–contraction coupling process is the release of Ca^{2+} from the sarcoplasmic reticulum into the immediate environment of the myofibrils. The 1 mm diameter fibers of the leg muscles of the crab *Maia squinado* were used by Caldwell and Walster (1963) to prove that only Ca^{2+} and caffeine of various substances tested produced shortening, and by Ashley *et al.* (1965) to inject EGTA in order to determine the Ca^{2+} threshold for contraction in response to potassium depolarization. Recently, Ashley and Ridgway (1970) have used the even larger barnacle fibers and the calcium-sensitive photoprotein aequorin to work out the full time course of depolarization, Ca^{2+} concentration, and isometric tension in a twitch. The 15 μm sarcomeres of glycerinated crab leg muscle enabled Gillis (1969) to show by iontophoretic application that Ca^{2+} must reach the overlap region of the filaments in order to produce contraction.

Experiments with both glycerinated fibers and myofibril suspensions have shown that at pH 7.0 and an ionic strength of 0.1, the Ca^{2+} threshold for tension and increase in ATPase activity is at 10^{-7} M in nonfibrillar arthropod muscles as in vertebrate muscle, but is at about 10^{-8} M in IFM (Meinrenken, 1967; Maruyama *et al.*, 1968b). Two further observations suggest that the calcium-binding protein of IFM has peculiar properties; in contrast to rabbit muscle or insect leg muscle, the protein is removed by 1 min extraction at pH 7.8 in 0.05 M potassium chloride, leaving the preparation active even in the absence of Ca^{2+} (Meinrenken, 1967), and the Ca^{2+}- sensitivity is lowered to a much greater extent by reducing the pH to 6.0 (Schädler, 1967). The Ca^{2+} threshold in IFM is not affected by stretch (Schädler, 1967), and the ATPase activity cannot be saturated by increased Ca^{2+} alone in myofibril suspensions or unstretched glycerinated fibers (Rüegg, 1967; Maruyama *et al.*, 1968b). Both stretch and Ca^{2+} are needed for full activity (see Section IV,D). The amount of calcium bound at a given concentration increases with stretch, suggesting that more binding sites are made available (Chaplain, 1967). These results show that the increase in ATPase activity produced by stretch in IFM (Section IV,D) does not result from a change in the equilibrium constant of the binding protein.

B. Mechanical Properties in Relaxation and Rigor

It is usually assumed that in the normal external environment of the body fluids and in the absence of nervous excitation, an intact muscle is in a fully relaxed state. Hoyle (1968) has shown that certain fibers of a crab eyestalk muscle have a naturally low membrane potential and a maintained tension development in the absence of stimulation; they relax further when the membrane is artifically hyperpolarized. The widespread occurrence of inhibitory innervation in arthropod limb and trunk muscles gives possible functional significance to this observation, since the relaxation time of cockroach leg muscles after stimulation of their excitatory nerves is too long to account for the speed of leg movement during walking and is accelerated by an inhibitory discharge (Iles and Pearson, 1970).

Relaxed crustacean leg muscles have a similar extensibility to those of the frog if stretched very slowly (Zachar and Zacharova, 1966), but relaxed insect flight muscles, particularly fibrillar muscles, are much more stiff (Fig. 12). There are no elements other than the myofibrils that could account for this resting tension (Buchthal and Weis-Fogh, 1956), and a similar inextensibility is shown by glycerinated fibers in

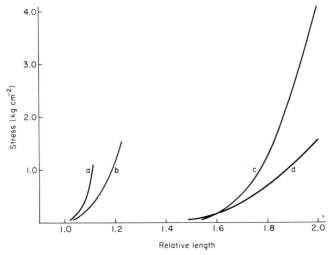

Fig. 12. Tension–length relationship in relaxed muscles, referred to the length at which the fibers are just slack; (a) insect fibrillar flight muscle (beetle) (from Machin and Pringle, 1959); (b) insect nonfibrillar flight muscle (locust) (from Weis-Fogh, 1956); (c) crustacean muscle (crayfish) (from Zachar and Zacharova, 1966); (d) vertebrate skeletal muscle (frog semitendinosus) (from Ramsay and Street, 1940).

relaxing solutions (Pringle, 1967a). Within the myofibril there are three possible locations for the material responsible: (1) from the myosin filaments to the Z line (Section II,B,5), (2) between the actin filaments in the H zone (S filaments), and (3) from Z line to Z line independent of the actin and myosin filaments (Section II,B,7). There is also the possibility of residually active or fixed cross bridges (Bienz-Isler, 1968a). Fine-structural studies appear to eliminate (2), but at the present time no definite choice can be made between the other alternatives, though (1) is favored by most authors. Extraction of myosin by solutions of high ionic strength destroys the inextensibility so that it does not reside in the skeletal background that remains after this procedure.

Relaxed fibers and glycerol-extracted material in relaxing solution are viscoelastic, and tension decays slowly after a step increase in length. Accurate studies of the time course of stress relaxation in IFM have been made by White (1967) and Chaplain *et al.* (1968, 1969); tension decay follows a power law of the form $T = t^{-k} + B$, where B and k are constants for a given amount of stretch, and cannot be explained in terms of a series of lumped linear elements as suggested for other muscles by B. C. Abbott and Lowy (1957). Chaplain *et al.* (1969) suggest a model containing nonlinear elasticities, but their location of the main element in the H zone seems implausible, since Garamvölgyi and Belagyi (1968a) have shown that relaxed bee flight muscle, if stretched very slowly or held extended for a long time, shows a wide I band and can be extended beyond the point of filament overlap. Because of stress relaxation, cyclic length changes at very low frequency produce a marked tension hysteresis.

The requirement for full relaxation of glycerinated fibers at 20°C is the presence of Mg^{2+}, at least 3×10^{-4} M ATP and a Ca^{2+} concentration of less than 10^{-9} M (White, 1970). On the removal of either ATP or Mg^{2+}, the fibers develop tension if isometric and shorten by up to 6% if isotonic; they then pass into rigor and become even less extensible due to the formation of fixed myosin–actin bridges (Section IV,E). The relaxation that occurs on addition of further ATP appears to be an active process, since unstrained myofibrils buckle and the contraction bands present in rigor disappear (Aronson, 1962).

C. Mechanical Properties of Active Muscle

1. NONFIBRILLAR MUSCLES

Studies using intact muscle of the time course of contraction and relaxation following nervous excitation are complicated by the fact that, in arthropods more than in vertebrates, "active state" (visualized as the processes leading to an increase in intracellular Ca²⁺) may be graded and may rise and fall at rates comparable to those of the events in the myofibrils (Hoyle and Abbott, 1967). This phenomenon, even more than the presence of series elasticity, can produce twitch/tetanus ratios in isometric experiments on different muscles that range all the way from 0 to the 0.7 found in frog sartorius muscle at 0°C (Hill, 1951). The most exact studies of contraction and relaxation times have been discussed in Section II,B,8.

Many years ago, Blaschko *et al.* (1931) discovered that if a crab leg muscle developing a small isometric tension in response to a low frequency of nervous excitation was given a single extra stimulus injected into the regular series, the resulting rapid twitch was followed by a maintained tension which terminated only on cessation of the low-frequency stimulation. This catch phenomenon has been reinvestigated by Wilson and Larimer (1968) and Wilson *et al.* (1970), who show that it gives rise to a hysteresis if the frequency of stimulation is varied sinusoidally with a cycle time of several seconds (Fig. 13). The phenomenon is apparent with truly isometric as well as with isotonic recording and is shown by crustacean and insect muscles, particularly but

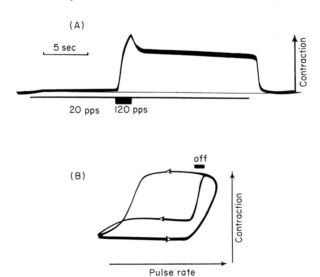

Fig. 13. Catch mechanism in arthropod muscles. (A) Crayfish claw opener muscle stimulated at 20 Hz holds tension after brief stimulation at 120 Hz. (B) Insect leg muscle shows hysteresis loop when stimulated by a pulse train modulated from 10 Hz to 100 Hz at 0.1 Hz; interruption of stimuli produces rapid relaxation. From Wilson and Larimer (1968).

not exclusively by slow muscles. Since a slow cyclic depolarization of barnacle fibers by intracellular electrodes gives the same result, the explanation cannot be neuromuscular, and it may provide further evidence of a long time constant in the intramuscular coupling process or in the binding and release of Ca^{2+} by the myofibrillar proteins. Small length changes imposed during the cycle show that muscle stiffness is always a function of tension, but larger lengthening and restretch resets the tension during the frequency-decreasing part of the cycle; this is difficult to explain except in terms of a direct influence of length steps on the contractile machinery.

The true mechanical properties of the contractile mechanism of intact fibers can best be studied during maximal potassium contracture (Zachar and Zacharova, 1966). In crayfish leg muscle fibers, maximal active tension is developed at a sarcomere length of 10.5 μm, declining linearly at longer lengths and reaching zero at 16.5 μm. Overlap should cease at 14.5 μm, but sarcomere length is not uniform at the ends of stretched fibers. The steeper decline compared with frog fibers (Gordon et al., 1966) is accounted for by the shorter overlap at rest length; the ratio of A band to I band lengths is only 0.38 compared with 0.76 in the frog. This could also explain why the maximum tension (8.2 kg·cm^{-2}) is only twice that of frog fibers although the sarcomere length is more than four times as great. Some authors have claimed that the A band of arthropod muscles changes length during elongation and contraction and have put forward a variety of theories to account for this, but the most critical experiments are entirely consistent with the sliding filament theory.

Nonfibrillar insect flight muscles present a special problem because of the rapid contraction and relaxation needed to achieve an efficient wingbeat cycle. The history of this problem is described by Pringle (1957); it was resolved by Buchthal et al. (1957), who showed that twitch duration is sufficiently short under the nearly isometric conditions and high temperature found in flight; the velocity of shortening is not exceptional for a fast muscle. Tetanic contraction in these highly phasic muscles is an unnatural phenomenon and can only be sustained without internal damage by the use of low temperatures. At 11°C, locust flight muscle develops its maximal isometric tension increment of 1.6 kg·cm^{-2} at $0.96L_b$* but the increment might be as high as 4 kg·cm^{-2} at 35°C (Weis-Fogh, 1956).

Arthropod visceral muscles differ from somatic muscles in that action potentials can be elicited by mechanical stretch, due to a smooth relationship between length and membrane potential (Nagai, 1970). The

* L_b = natural length of muscle in intact insect.

maximum active tension increment of about 1.5 kg·cm^{-2} (bundle cross section) occurs at about $2L_0$*. The cockroach proctodeal muscle on which these measurements were made can also be excited neurally, but a direct excitation mechanism appears to operate alone in the extrinsic alary muscles of the heart, which are not innervated and contract in response to stretch imposed by the intrinsic heart muscle (de Wilde, 1947).

2. FIBRILLAR MUSCLES (IFM)

The oscillatory contraction of IFM was first described by Pringle (1949) from experiments demonstrating the lack of correspondence between electrical and mechanical events in the flight muscles of intact flies; it was confirmed by Roeder (1951) in bees and wasps. The mechanism of the mechanical rhythm was elucidated by Boettiger and his pupils (summary in Boettiger, 1957). Apart from flight muscles, the special property of the myofibrils that gives rise to oscillation is known to be of functional significance only in the timbal sound-producing muscles of certain cicadas (Pringle, 1954; Hagiwara *et al.*, 1954; Hagiwara, 1956). In most of these fibrillar muscles, the frequency of motor nerve impulses necessary to maintain the active state is considerably less than that of the mechanical rhythm (very large beetles are an exception), but the important observation is that there is no temporal correspondence between electrical and mechanical events even when the two frequencies are nearly identical (Boettiger, 1951; Machin and Pringle, 1960, appendix). This feature of the physiology of IFM clearly correlates with the paucity of sarcoplasmic reticulum (Section II,B,2).

When an IFM is studied under isometric conditions, it behaves in much the same manner as a normal striated muscle (Pringle, 1967a); examples are known (bee flight muscle, timbal muscle of *Platypleura*) in which a single impulse produces a rapid twitch and others that are slow in that there is always a smooth tetanus with neuromuscular facilitation (beetle). The frequency of oscillation in the intact insect is unrelated to this categorization and is determined solely by the mechanical resonance of the entire system. The special property giving rise to oscillation is a direct influence of length changes on the contractile machinery which leads to the production of active tension after a delay; Jewell and Rüegg (1966) showed that an identical effect of length changes on active tension persists in glycerol-extracted fibers in an activating solution (Fig. 14), and it is clear that the property resides in the myofibrils. Oscillatory contraction results from this property in the intact insect because of the inertia of the load and can be reproduced in

*L_0 = length at which tension is zero in isolated muscle.

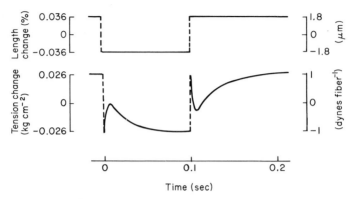

Fig. 14. Tension changes in activated glycerol-extracted IFM following quick release and quick stretch. From Jewell and Rüegg (1966).

an isolated preparation (Fig. 15). The functional aspects of the flight muscle system are discussed by Pringle (1965, 1968b) and the state of knowledge of the contractile mechanism up to that date by Pringle (1967a). This section will focus attention on recent work that has made it necessary to revise parts of the hypothesis put forward in the latter paper.

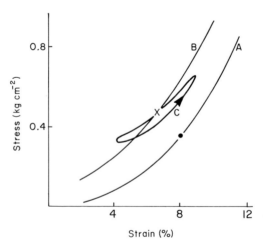

Fig. 15. Stress–strain curves for intact beetle flight muscle (IFM); (A) unstimulated, (B) stimulated under isotonic or damped isometric conditions, (C) oscillation when the load contains an inertia; ● = unstimulated and × = stimulated but non-oscillatory working point in the experiment which gave the oscillatory loop. During oscillation, tension is greater during shortening than during extension. From Pringle (1967a).

A. DELAYED TENSION IN NONFIBRILLAR MUSCLES. Delayed changes in tension after step changes in length have now been found in glycerol-extracted fibers from a variety of muscles in activating solution (MgATP plus Ca^{2+}). Aidley and White (1969) described the effect in a nonfibrillar cicada timbal muscle whose contractions in the intact insect are twitches produced by synchronous motor nerve impulses. Rüegg *et al.* (1970), using glycerol-extracted rabbit psoas, frog semitendinosus, or rabbit heart muscle, found a secondary peak of tension 0.05–0.1 sec after abrupt stretches and also showed that even these vertebrate fibers were capable of doing oscillatory work when subjected to cyclic length changes in activating solution. In an earlier preliminary report Armstrong *et al.* (1966) had reported similar oscillatory responses in living frog muscle fibers, and it seems that direct mechanical activation of the contractile system is not a peculiar property of insect fibrillar muscle, but is merely developed in this tissue to an extent that gives it functional significance.

B. NONLINEAR EFFECTS. For small amplitudes of length change (less than 0.5%), the resulting changes of tension are symmetrical for stretch and release, provided the fibers are initially stretched sufficiently to prevent them going slack (Fig. 14); with sinusoidal forcing of length, the oscillatory tension–length loop is elliptical. At higher amplitudes, the effects of quick stretch and quick release become asymmetrical and oscillatory loops are distorted (Fig. 15). Pringle and Tregear (1969) analyzed the factors that limit the performance of activated fibers at high amplitudes of oscillation. They found that the maximum work per cycle was reached at a smaller amplitude of oscillatory length change as the frequency of oscillation was increased and concluded that performance is ultimately limited by the maximum velocity at which filament sliding can occur. In freshly extracted fibers, the limiting amplitude of 6–8% length change is comparable to that found in intact fibers, and at this limit there was evidence from the work output and the tension increment that most or all of the contractile material is activated by the change of length. It is apparent that the direct mechanical control in insect fibrillar muscle is as effective a way of modulating the activity of the tissue as the control by Ca^{2+} found in normal muscles. Furthermore, provided the Ca^{2+} concentration is high enough to achieve full activation, neither it nor the mean length of the fibers is very critical for optimal performance, since stress relaxation ensures that the mean tension during high amplitude oscillation adjusts itself in a few seconds to the value at which the tension is nearly zero at the shortened phase of each length cycle and is maximal at the lengthened phase. It is thus understandable that IFM has a greatly reduced sarcoplasmic reticulum compared with nonfibrillar flight muscles, for exact and rapid regulation of the Ca^{2+} concentration in the myofibrils is unnecessary.

C. TIME CONSTANT OF DELAYED TENSION. The delay between length change and tension change in active IFM enables the muscle to do oscillatory work when its length is modulated within the correct frequency range. The relationship is conveniently displayed in a quantitative manner as a dynamic modulus or Nyquist plot (Fig. 16a), and information about the nature of the delay can be obtained from the

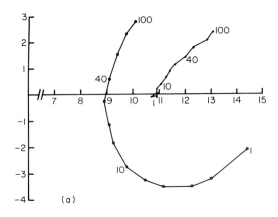

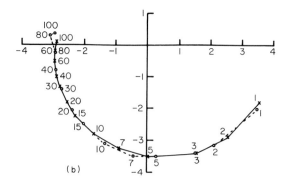

Fig. 16. Nyquist plots of tension in glycerol-extracted IFM. This plot is obtained by forcing small sinusoidal length changes on the fibers and plotting the in-phase and quadrature components of tension as X and Y values for each frequency of oscillation; (a) original plots, figures refer to frequency of oscillation; (b) O = vector difference plot for data from a, ● = theoretical curve for narrow distribution of exponential time constants. From R. H. Abbott (1968).

form of this plot. R. H. Abbott (1968), making the simplifying assumption that the elements responsible for the dynamic properties of relaxed muscle are present and unchanged during activation, has shown that the performance of the residual active element is accurately simulated by a population of exponential delays with a narrow distribution of time constants (Fig. 16b). The mean value of this time constant is the most significant factor defining the frequency limits for oscillation.

The value of the time constant is reduced by rise of temperature (Machin *et al.*, 1962; Steiger and Rüegg, 1969) and by increase in Ca^{2+} concentration over the range from 10^{-8} to 10^{-6} M (R. H. Abbott, 1972), but is unaffected by the mean length of the fibers, increase in which increases the magnitude of the delayed tension. The dependence on temperature explains the need for a warming-up period in many insects, without which the mechanical resonance of the thorax–wing system does not match the optimum for oscillatory power output. The dependence on Ca^{2+} partly explains the dependence of wingbeat frequency on the frequency of action potentials observed by Esch and Bastian (1968) in the honey bee, though the simultaneous increase in power output also contributes to the effect in the intact insect.

D. ISOMETRIC OSCILLATIONS. Schädler *et al.* (1969) have described a new oscillatory phenomenon in glycerinated fibers of IFM. If a bundle of these fibers is subjected to a large, rapid stretch in a strongly activating solution containing adenylate kinase and histidine buffer, the normal delayed rise of tension is followed by damped isometric tension oscillations lasting for several cycles. The frequency of these oscillations is temperature-dependent in the same manner as the time constant of delayed tension. Further experiments are in progress, but it appears that sudden, near maximal stretch activation and the absence of inorganic phosphate are essential conditions for the manifestation of this phenomenon.

D. Enzymic Activity

Rüegg and Tregear (1966) were the first to show that stretching fibers from IFM increases not only their active tension but also their ATPase activity. The effect is long lasting, and stretch in relaxing solution with subsequent Ca^{2+} activation is also effective; in such experiments an increase in ATPase activity of about five times is obtained for 10% extension, and beyond this, ATPase activity declines in the manner expected from reduction of overlap by filament sliding (Fig. 17), as in vertebrate

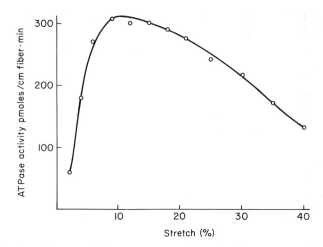

Fig. 17. Glycerol-extracted fibers of IFM. Relationship between stretch in relaxing solution and ATPase activity on subsequent activation by 1.5×10^{-8} *M* Ca^{2+}. From Chaplain (1969).

muscle (Ward *et al.*, 1965; Chaplain, 1969). Rüegg and Stumpf (1969a) showed that the effect is not due to an increase in Ca^{2+} affinity by the contractile proteins, since stretch activation is still present at saturating Ca^{2+} concentrations and the Ca^{2+} concentration required for half-activation is the same in unstrained as in strained fibers. Active tension and ATPase activity are related under all conditions of activation, but because of the time required for chemical estimations, it is not possible to determine whether the stretch activation of ATPase activity also occurs after a delay.

If fibers in activating solution are oscillated cyclically at a frequency at which work is done, there is a further dynamic increase in ATPase activity proportional to the oscillatory work (Rüegg and Stumpf, 1969b). The dynamic increase does not occur if the oscillation frequency is either too low or too high for the performance of oscillatory work. The mechanochemical efficiency of this extra ATPase activity is about 30%. At large amplitudes of oscillation with full Ca^{2+} activation, the enzymic activity in fiber bundles is very high, and diffusion from outside the bundle becomes inadequate to supply and remove metabolites; the properties of the preparation then change. In this condition, which was called the high tension state by Jewell and Rüegg (1966), isometric tension increases regeneratively to a high value, and oscillatory work can only be performed at very low frequencies or not at all. The limiting conditions were explored by Mannherz (1968), who also showed that in contrast to rabbit psoas muscle, where oscillation increases ATPase activity due to better transport only when the external concentration is low, in IFM the increase occurs at all external concentrations high enough to avoid internal starvation.

The precise internal conditions for the high tension state are not clear. It is prevented by having a high (15 mM) concentration of ATP in the solution, by addition of myokinase, or by including in the medium an ATP-regenerating system such as creatine plus creatine kinase; under these conditions an oscillatory power output similar to that in life can be obtained from glycerol-extracted fiber bundles at a temperature of 30°C or higher (Steiger and Rüegg, 1969). But this does not prove that ATP starvation is the immediate cause of the high tension state. ADP (up to 1 mM in the presence of 5 mM MgATP) has an activating effect, and R. H. Abbott and Mannherz (1970) show how accumulation of ADP could be the immediate cause. Because ADP is produced by enzymic activity, the system contains a positive feedback if it is allowed to accumulate. This suggestion would also account for the regenerative nature of the onset of the phenomenon and for the irregularity of results of the addition of ADP to the solutions. The effect does not occur in intact fibers unless oxygen access is prevented, owing to the large concentration of mitochondria in close proximity to the myofibrils (Machin and Pringle, 1959).

Activation by Ca^{2+} in IFM depends on the presence of a minor protein complex which can be extracted from glycerol-extracted fibers by brief treatment with 0.05 M potassium chloride at 7.5–8.0 (Meinrenken, 1969). The fibers then develop the full contractile tension even in the presence of EGTA and are no longer capable of being activated by stretch or of performing oscillatory work. Substitution of ATP by ITP also produces maximal activity uncontrolled by Ca^{2+}, and these solutions can produce supercontraction down to 30% of rest length (Zebe et al., 1968). The control by Ca^{2+} is also removed by brief extraction of the fibers by 0.6 M potassium chloride at pH 7.0 (White, 1967).

E. Biophysics

IFM was used by Reedy et al. (1965) for the first demonstration by a combination of electron microscopy and X-ray diffraction that there is a change in the myosin cross bridges when glycerol-extracted fibers pass from relaxation to rigor. In relaxing solution, the bridge orientation is at right angles to the fiber axis, and their longitudinal spacing of 14.5 nm is that of the myosin filament. In rigor, the bridges are angled at about 45°, and their firm attachment to the actin filaments is shown by the appearance of layer lines with the characteristic actin spacing of 38.8 nm (Section II,B,6). Tregear and Miller (1969) used a correctly located proportional counter to measure the changes in intensity of the 14.5 nm meridional diffraction spot when a fiber bundle in activating solution was oscillated at various frequencies. They found that at low frequencies the intensity was reduced by up to 30% at the same phase of oscillation as the maximum of tension and that at a higher frequency at which no oscillatory work was done there was no reduction. The difficulties of quantitative interpretation of this result are discussed by Miller and Tregear (1971), but a conservative conclusion is that, if all the cross bridges move, then a movement of their ends by 2–3 nm is required; alternatively, if only a fraction of the bridges move, then the smallest fraction is about 16% and the accompanying average displacement would have to be about 10 nm.

In a second dynamic investigation, Miller and Tregear (1970) measured the intensity of the two equatorial reflections in relaxed, activated, and oscillated fibers and found from changes in the intensities that raised Ca^{2+} produces a movement of the myosin bridges outward to the neighborhood of the actin filaments; this movement is not affected by oscillation of activated fibers. Activation by Ca^{2+} and by stretch thus produce different effects on the molecular structure. Since about 15% of the bridges must move outward during Ca^{2+} activation and these are presumably the same bridges that attach and move axially during oscillation, the combined conclusion is that, at any one time, only a relatively small fraction (15%) of the bridges interact with the actin filaments even during maximal enzymic activity and that their axial displacement is at least 10 nm.

F. Models

The original hypothesis about the nature of the molecular events giving rise to oscillatory contraction was that the movement induces a synchronization of the cycles of activity of the myosin cross bridges (Pringle, 1967a). Calculation of the power output and the amount of ATP hydrolyzed suggests that all of the myosin is activated once and possibly more than once in each cycle of oscillation (Pringle, 1967b; Pringle and Tregear, 1969). The dynamic X-ray diffraction studies show that at any instant only a maximum of 15% of the bridges move radially and axially, but in these experiments the power output was limited to about 25% of maximal because of diffusion limitations. If proportionality is maintained, the number in simultaneous movement at maximal power might be 60%. The combined conclusion must be either that only 60% of the bridges are in action and they hydrolyze more than one molecule of ATP per cycle or that all the bridges move but operate in rapid succession during the tension-producing phase of the cycle so that each bridge completes one cycle during each cycle of oscillation, but there is not exact simultaneity. Pringle (1967b) showed that exact simultaneity of all the bridges demands absolute filament inextensibility, which is an improbable condition; the observed departure from simultaneity would be produced by quite a small amount of filament compliance. It thus seems probable that synchronization of all the bridges is achieved in maximal oscillation at the optimum oscillation frequency.

The phenomenon of isometric oscillation (Schädler *et al.*, 1969) suggests that bridges have a natural period of operation, since under these conditions there is no periodic synchronizing mechanical signal; the frequency of these oscillations is about the same as that for maximum oscillatory power and shows the same temperature coefficient. At frequencies below the optimum, some bridges might thus operate twice during an oscillation cycle; this may explain occasional observation of rates of ATP hydrolysis greater than two per bridge per cycle.

Models must also explain the delay between length changes and tension changes in active oscillation. Pringle (1967a) suggested a model of the molecular mechanism that implied that the time constant of the delay was derived from the finite velocity of bridge movement. Thorson and White (1969) showed that this hypothesis is untenable and suggested instead a model, in general similar to that of A. F. Huxley (1957), in which the time constant derives from the probability of detachment of bridges from the actin filament during a cycle of attachment, axial movement, detachment, and recovery. This model was shown to be capable of simulating the dynamic performance of IFM at low amplitudes of oscillation and can also explain the rates of ATP hydrolysis if the probability of bridge attachment is modulated by extension. This is the most complete model that has so far been put forward, and it is capable of extension to account also for the nonlinearities that appear at higher amplitudes of oscillation (Thorson and White, 1972).

Julian's (1969) model, on the other hand, does not satisfactorily simulate the behavior of IFM, in spite of his claim to this effect. In modified form, it is capable of explaining the delayed rise of tension after a quick stretch, but it predicts that tension should then fall again. It cannot explain the fact that tension and ATPase activity are increased if stretch is applied in relaxing solution before Ca^{2+} activation. As described in Section IV,D the effect of stretch cannot be due to a change in the binding constant for Ca^{2+}, and a further feature has to be introduced into the model which is not required for vertebrate muscle.

Formal models of this sort do not make any statements about the chemical nature of the elements responsible for the observed delay, and though they are valuable to a limited extent, the real objective in the study of muscles is to identify the contribution of each protein constituent of the myofibril to the overall performance of the tissue. The minor proteins are probably involved in the mechanical as well as the Ca^{2+} control, and biochemical, biophysical, and physiological techniques will have to be combined in order to advance our understanding. The peculiar structures and properties found in arthropod muscles have, however, already helped to no small extent in the solution of the basic problems of contractility, and it is to be expected that this phylum will continue in the future to supply material for important advances in our knowledge.

REFERENCES

Abbott, B. C., and Lowy, J. (1957). *Proc. Roy. Soc., Ser. B* **146**, 281.

Abbott, R. H. (1968). Ph.D. Thesis, Oxford University.

Abbott, R. H. (1972). *J. Physiol. (London)* (in press).

Abbott, R. H., and Chaplain, R. A. (1966). *J. Cell Sci.* **1**, 311.

Abbott, R. H., and Mannherz, H. G. (1970). *Pfluegers Arch.* **321**, 223.

Aidley, D. J., and White, D. C. S. (1969). *J. Physiol. (London)* **205**, 179.

Anderson, W. A., and Ellis, R. A. (1967). *Z. Zellforsch. Mikrosk, Anat.* **79**, 581.

Armstrong, C. F., Huxley, A. F., and Julian, F. J. (1966). *J. Physiol. (London)* **186**, 26P.

Aronson, J. (1961). *J. Biophys. Biochem. Cytol.* **11**, 147.

Aronson, J. (1962). *J. Cell. Biol.* **13**, 33.

Ashhurst, D. E. (1967a). *J. Cell Sci.* **2**, 435.

Ashhurst, D. E. (1967b). *J. Mol. Biol.* **27**, 385.

Ashhurst, D. E. (1969). *Z. Zellforsch. Mikrosk. Anat.* **93**, 36.

Ashhurst, D. E. (1971). *J. Mol. Biol.* **55**, 283.

Ashley, C. C., and Ridgway, E. B. (1970). *J. Physiol. (London)* **209**, 105.

Ashley, C. C., Caldwell, P. C., Lowe, A. G., Richards, C. D., and Schirmer, H. (1965). *J. Physiol. (London)* **179**, 32P.

Atwood, H. L., Hoyle, G., and Smyth, T., Jr. (1965). *J. Physiol. (London)* **180**, 449.

Auber, J. (1962). *C. R. Acad. Sci.* **254**, 4074.

Auber, J. (1963a). *J. Microsc. (Paris)* **2**, 233.

Auber, J. (1963b). *J. Microsc. (Paris)* **2**, 325.

Auber, J. (1963c). *C. R. Acad. Sci., Ser. D* **256**, 2022.

Auber, J. (1965). *C. R. Acad. Sci., Ser. D* **260**, 668.

Auber, J. (1967a). *C. R. Acad. Sci., Ser. D* **264**, 621.

Auber, J. (1967b). *C. R. Acad. Sci., Ser. D* **264**, 2916.

Auber, J. (1967c). *Amer. Zool.* **7**, 451.

Auber, J. (1969). *J. Microsc. (Paris)* **8**, 197.

Auber, J., and Couteaux, R. (1962). *C. R. Acad. Sci., Ser. D* **254**, 3425.

Auber, J., and Couteaux, R. (1963). *J. Microsc. (Paris)* **2**, 309.

Auber-Thomay, M. (1967). *J. Microsc. (Paris)* **6**, 627.

Bacetti, B. (1965). *J. Ultrastruct. Res.* **13**, 245.

Bailey, K. (1948). *Biochem. J.* **43**, 271.

Bárány, M. (1967). *J. Gen. Physiol.* **50**, 197.

Baskin, R. J., Sanford, W. C., Morse, P. D., and Biggs, M. L. (1969). *Comp. Biochem. Physiol.* **29**, 471.

Basler, W. (1969). *Rev. Suisse Zool.* **76**, 297.

Becht, G., and Dresden, D. (1956). *Nature (London)* **177**, 836.

Beinbrech, G. (1968). *Z. Zellforsch. Mikrosk. Anat.* **90**, 463.

Bhakthan, N. M. G., Borden, J. H., and Nair, K. G. (1970). *J. Cell Sci.* **6**, 807.

Bienz-Isler, G. (1968a). *Acta Anat.* **70**, 416.

Bienz-Isler, G. (1968b). *Acta Anat.* **70**, 524.

Bittner, G. D. (1968). *J. Exp. Zool.* **167**, 439.

Blaschko, A., Catell, McK., and Kahn, J. L. (1931). *J. Physiol. (London)* **73**, 25.

Boettiger, E. G. (1951). *Anat. Rec.* **111**, 443.

Boettiger, E. G. (1957). *In* "Recent Advances in Invertebrate Physiology" (B. T. Scheer, ed.), pp. 117–142. Univ. of Oregon Press, Eugene.

Bouligand, Y. (1962). *J. Microsc. (Paris)* **1**, 377.

Brandt, P. W., Reuben, J. P., Girardier, L., and Grundfest, H. (1965). *J. Cell Biol.* **25**, 233.

Buchthal, F., and Weis-Fogh, T. (1956). *Acta Physiol. Scand.* **35**, 345.

Buchthal, F., Weis-Fogh, T., and Rosenfalck, P. (1957). *Acta Physiol. Scand.* **39**, 246.

Caldwell, P. C., and Walster, G. (1963). *J. Physiol. (London)* **169**, 353.

Camatini, M., and Saita, A. (1967). *Boll. Zool.* **34**, 99.

Carbonell, C. S. (1947). *Smithson. Misc. Collect.* **107**, 1.

Caveney, S. (1969). *J. Cell Sci.* **4**, 541.

Chaplain, R. A. (1967). *Biochim. Biophys. Acta* **131**, 385.

Chaplain, R. A. (1969). *Pflueegers Arch.* **307**, 120.

Chaplain, R. A., and Tregear, R. T. (1966). *J. Mol. Biol.* **21**, 275.

Chaplain, R. A., Frommelt, B., and Pfister, E. (1968). *Kybernetik* **5**, 61.

Chaplain, R. A., Frommelt, B., and Brandt, M. (1969). *Kybernetik* **5**, 177.

Cochrane, D. G., Elder, H. Y., and Usherwood, P. N. R. (1969). *J. Physiol. (London)* **200**, 68P.

de Villafranca, G. W. (1968). *Comp. Biochem. Physiol.* **26**, 443.

de Villafranca, G. W., and Campbell, L. K. (1969). *Comp. Biochem. Physiol.* **29**, 775.

de Villafranca, G. W., and Naumann, D. C. (1964). *Comp. Biochem. Physiol.* **12**, 143.

de Villafranca, G. W., and Philpott, D. E. (1961). *J. Ultrastruct. Res.* **5**, 151.

de Villafranca, G. W., Scheinblum, T. S., and Philpott, D. E. (1959). *Biochim. Biophys. Acta* **34**, 147.

de Wilde, J. (1947). *Arch. Neer. Physiol.* **28**, 530.

Dumont, J. N., Anderson, E., and Chougen, E. (1968). *J. Ultrastruct. Res.* **13**, 38.

Ebashi, S., and Ebashi, F. (1964). *J. Biochem. (Tokyo)* **55**, 604.

Edwards, G. A., and Challice, C. E. (1960). *Ann. Entomol. Soc. Amer.* **53**, 369.

Edwards, G. A., Ruska, H., Santos, P. de S., and Vallejo-Freire, A. (1956). J. Biophys. Biochem. Cytol., Suppl. 2, 143.
Eigenmann, R. (1965). Rev. Suisse Zool. 72, 789.
Esch, H., and Bastian, J. (1968). Z. Vergl. Physiol. 58, 429.
Fahrenbach, W. H. (1964). J. Cell Biol. 22, 477.
Franzini-Armstrong, C. (1970a). J. Cell. Sci. 6, 559.
Franzini-Armstrong, C. (1970b). Tissue & Cell 2, 327.
Garamvölgyi, N. (1963). J. Microsc. (Paris) 2, 107.
Garamvölgyi, N. (1965a). J. Ultrastruct. Res. 13, 409.
Garamvölgyi, N. (1965b). J. Ultrastruct. Res. 13, 435.
Garamvölgyi, N. (1969). J. Ultrastruct. Res. 27, 462.
Garamvölgyi, N., and Belágyi, J. (1968a). Acta Biochem. Biophys. 3, 195.
Garamvölgyi, N., and Belágyi, J. (1968b). Acta Biochim. Biophys. 3, 299.
Garamvölgyi, N., and Kerner, J. (1966). Acta Biochim. Biophys. 1, 81.
Gilev, V. P. (1965). Biochim. Biophys. Acta 112, 340.
Gilev, V. P. (1966). Proc. 6th Int. Congr. Electron Microsc., 1966, p. 689.
Gilev, V. P., Perov, N. A., and Melnikova, E. Y. (1968). Arch. Anat. Histol. Embryol. 54, 41.
Gillis, J. M. (1969). J. Physiol. (London) 200, 849.
Gillis, J. M., and Page, S. G. (1967). J. Cell Sci. 2, 113.
Gilmour, D., and Calaby, J. H. (1953). Enzymologia 16, 23.
Goldspink, G. (1968). J. Cell Sci. 3, 539.
Goldspink, G. (1970). J. Cell Sci. 6, 593.
Gordon, A. M., Huxley, A. F., and Julian, F. J. (1966). J. Physiol. (London) 184, 170.
Hagiwara, S. (1956). Physiol. Comp. Oekol. 4, 142.
Hagiwara, S., Uchiyama, H., and Watanabe, A. (1954). Bull. Tokyo. Med. Dent. Univ. 1, 113.
Hagopian, M. (1966). J. Cell Biol. 28, 545.
Hagopian, M., and Spiro, D. (1967). J. Cell Biol. 32, 535.
Hagopian, M., and Spiro, D. (1968). J. Cell Biol. 36, 433.
Halvarson, M., and Afzelius, B. F. (1969). J. Ultrastruct. Res. 26, 289.
Hanson, J. (1956). J. Biophys. Biochem. Cytol. 2, 691.
Hartshorne, D. J., Perry, S. V., and Davies, V. (1966). Nature (London) 209, 1352.
Hill, A. V. (1951). Proc. Roy. Soc., Ser. B 138, 349.
Hoyle, G. (1968). J. Exp. Zool. 167, 551.
Hoyle, G. (1969). Annu. Rev. Physiol. 31, 43.
Hoyle, G., and Abbott, B. C. (1967). Amer. Zool. 7, 611.
Hoyle, G., and McNeill, P. A. (1968). J. Exp. Zool. 167, 487.
Hoyle, G., McAlear, J. H., and Selverston, A. (1965). J. Cell Biol. 26, 621.
Huddart, H., and Oates, K. (1970). J. Insect. Physiol. 16, 1467.
Huxley, A. F. (1957). Progr. Biophys. Biophys. Chem. 7, 255.
Huxley, A. F. (1965). Intern. Congr. Physiol. Sci. [Abstr.], 23rd, 1961, p. 36.
Huxley, A. F., and Niedergerke, R. (1954). Nature (London) 173, 971.
Huxley, H. E. (1960). In "The Cell" (J. Brachet and A. E. Mirsky, eds.), Vol. 4, p. 419. Academic Press, New York.
Huxley, H. E. (1963). J. Mol. Biol. 7, 281.
Huxley, H. E., Brown, W., and Holmes, K. C. (1965). Nature (London) 206, 1358.

Iles, J. F., and Pearson, K. G. (1970). *J. Physiol. (London)* **210**, 20P.
Jahromi, S. S., and Atwood, H. L. (1969a). *Experimentia* **25**, 1046.
Jahromi, S. S., and Atwood, H. L. (1969b). *J. Insect Physiol.* **15**, 2255.
Jewell, B. R., and Rüegg, J. C. (1966). *Proc. Roy. Soc., Ser. B* **164**, 428.
Julian, P. J. (1969). *Biophys. J.* **9**, 547.
Kawaguti, S. (1961). *Biol. J. Okayama Univ.* **7**, 31.
Kennedy, D., and Takeda, K. (1965). *J. Exp. Biol.* **43**, 221.
Kominz, D. R., Maruyama, K., Levenbook, L., and Lewis, M. (1962). *Biochim. Biophys. Acta* **63**, 106.
Kuehl, W. M., and Adelstein, R. S. (1970). *Biochem. Biophys. Res. Commun.* **39**, 956.
Lai-Fook, J. (1967). *J. Morphol.* **120**, 23.
Lockshin, R. A., and Williams, C. M. (1965). *J. Insect Physiol.* **11**, 123.
McCann, F. V. (1970). *Annu. Rev. Entomol.* **15**, 173.
Machin, K. E., and Pringle, J. W. S. (1959). *Proc. Roy. Soc., Ser. B* **151**, 204.
Machin, K. E., and Pringle, J. W. S. (1960). *Proc. Roy. Soc., Ser. B* **152**, 311.
Machin, K. E., Pringle, J. W. S., and Tamasige, M. (1962). *Proc. Roy. Soc., Ser. B* **155**, 493.
McNeill, P. A., and Hoyle, G. (1967). *Amer. Zool.* **7**, 483.
Mannherz, H. G. (1968). *Pfluegers Arch.* **303**, 230.
Manton, S. M. (1934). *Phil. Trans. Roy. Soc. London Ser. B* **223**, 163.
Manton, S. M. (1949). *Phil. Trans. Roy. Soc. London, Ser. B* **223**, 163.
Maruyama, K. (1965). In "The Physiology of Insecta (M. Rockstein, ed.), Vol. 2, pp. 451–482. Academic Press, New York.
Maruyama, K. (1966). *Comp. Biochem. Physiol.* **18**, 481.
Maruyama, K., and Allen, S. R. (1967). *Comp. Biochem. Physiol.* **21**, 713.
Maruyama, K., Nagashima, S., and Drabikowski, W. (1968a). *Comp. Biochem. Physiol.* **25**, 1107.
Maruyama, K., Pringle, J. W. S., and Tregear, R. T. (1968b). *Proc. Roy. Soc., Ser. B* **169**, 229.
Meinrenken, W. (1967). *Pfluegers Arch. Gesamte Physiol. Menschem Tiere* **294**, 45.
Meinrenken, W. (1969). *Pfluegers Arch.* **311**, 243.
Miller, A., and Tregear, R. T. (1970). *Nature (London)* **226**, 1060.
Miller, A., and Tregear, R. T. (1971). *Proc. Symp. Contractility, Woods Hole, 1971*, Prentice Hall.
Morkin, E. (1970). *Science* **167**, 1499.
Nachtigall, W., and Wilson, D. M. (1967). *J. Exp. Biol.* **47**, 77.
Nagai, T. (1970). *J. Insect Physiol.* **16**, 437.
Nuesch, H. (1965). *Z. Naturforsch. B* **20**, 343.
Odhiambo, T. R. (1970). *Tissue & Cell* **2**, 233.
Osborne, M. P. (1967). *J. Insect Physiol.* **13**, 1471.
Page, S. G. (1965). *J. Cell Biol.* **26**, 477.
Parnas, I., and Atwood, H. L. (1966). *Comp. Biochem. Physiol.* **18**, 701.
Pasquali-Ronchetti, I. (1969). *J. Cell Biol.* **40**, 269.
Pasquali-Ronchetti, I. (1970). *Tissue & Cell* **2**, 339.
Peachey, L. D. (1965). *J. Cell Biol.* **25**, 209.
Pepe, F. A. (1967). *J. Mol. Biol.* **27**, 203.
Peterson, R. P. (1963). *J. Cell Biol.* **18**, 213.
Peterson, R. P., and Pepe, F. A. (1961). *Amer. J. Anat.* **109**, 277.

Pringle, J. W. S. (1949). J. Physiol. (London) **108**, 226.
Pringle, J. W. S. (1954). J. Physiol. (London) **124**, 269.
Pringle, J. W. S. (1957). "Insect Flight." Cambridge Univ. Press, London and New York.
Pringle, J. W. S. (1965). In "The Physiology of Insecta" (M. Rockstein, ed.), Vol. 2, pp. 283–329. Academic Press, New York.
Pringle, J. W. S. (1967a). Progr. Biophys. Mol. Biol. **17**, 1.
Pringle, J. W. S. (1967b). J. Gen. Physiol. **50**, 139.
Pringle, J. W. S. (1968a). Symp. Soc. Exp. Biol. **22**, 67.
Pringle, J. W. S. (1968b). Advan. Insect Physiol. **5**, 163.
Pringle, J. W. S., and Tregear, R. T. (1969). Proc. Roy. Soc., Ser. B **174**, 33.
Ramsay, R. W., and Street, S. F. (1940). J. Cell. Comp. Physiol. **15**, 11.
Reedy, M. K. (1967). Amer. Zool. **7**, 465.
Reedy, M. K. (1968). J. Mol. Biol. **31**, 155.
Reedy, M. K., Holmes, K. C., and Tregear, R. T. (1965). Nature (London) **207**, 1276.
Reedy, M. K., Fischman, D. A., and Bahr, G. F. (1969). Biophys. J. **9**, Soc. Abstr., A95.
Reger, J. F. (1967). J. Ultrastruct. Res. **20**, 72.
Reger, J. F., and Cooper, D. P. (1967). J. Cell Biol. **33**, 531.
Rice, M. J. (1970). J. Insect Physiol. **16**, 1109.
Roeder, K. D. (1951). Biol. Bull. Woods Hole **100**, 95.
Rosenbluth, J. (1969). J. Cell Biol. **42**, 534.
Rüegg, J. C. (1967). Amer. Zool. **7**, 457.
Rüegg, J. C., and Stumpf, H. (1969a). Pfluegers Arch. **305**, 34.
Rüegg, J. C., and Stumpf, H. (1969b). Pfluegers Arch. **305**, 21.
Rüegg, J. C., and Tregear, R. T. (1966). Proc. Roy. Soc., Ser. B **165**, 497.
Rüegg, J. C., Steiger, G. J., and Schädler, M. (1970). Pfluegers Arch. **319**, 139.
Runion, H. I., and Pipa, R. L. (1970). J. Exp. Biol. **53**, 9.
Sahota, T. S., and Beckel, W. E. (1967a). Can. J. Zool. **45**, 407.
Sahota, T. S., and Beckel, W. E. (1967b). Can. J. Zool. **45**, 421.
Saita, A., and Camatini, M. (1967). Biol. Zool. **34**, 170.
Sandborn, E. B., Duclos, S., Messire, P. E., and Roberge, J. J. (1967). J. Ultrastruct. Res. **18**, 695.
Sanger, J. W., and McCann, F. G. (1968a). J. Insect Physiol. **14**, 1105.
Sanger, J. W., and McCann, F. G. (1968b). J. Insect Physiol. **14**, 1539.
Schädler, M. (1967). Pfluegers Arch. Gesamte Physiol. Menschen Tiere **296**, 70.
Schädler, M., Steiger, G., and Rüegg, J. C. (1969). Experientia **25**, 942.
Schaefer, C. W., Vandenberg, J. P., and Rhodin, J. (1967). J. Cell Biol. **34**, 905.
Shafiq, S. A. (1963a). J. Cell Biol. **17**, 351.
Shafiq, S. A. (1963b). J. Cell Biol. **17**, 363.
Skinner, D. M. (1966). J. Exp. Zool. **163**, 115.
Smit, W. A., Becht, G., and Beenakkers, A. M. T. (1967). J. Insect Physiol. **23**, 1857.
Smith, D. S. (1961a). J. Biophys. Biochem. Cytol. **10**, 123.
Smith, D. S. (1961b). J. Biophys. Biochem. Cytol. **11**, 119.
Smith, D. S. (1965). J. Cell Biol. **27**, 379.
Smith, D. S. (1966a). Progr. Biophys. Mol. Biol. **16**, 107.
Smith, D. S. (1966b). J. Cell Biol. **29**, 449.
Smith, D. S., and Sacktor, B. (1970). Tissue & Cell **2**, 355.

Smith, D. S., Gupta, B. L., and Smith, U. (1966). *J. Cell Sci.* 1, 49.
Smith, D. S., Jarlfors, U., and Russell, F. E. (1969). *Tissue & Cell* 1, 673.
Squire, J. M. (1971). *Nature (London)* 233, 457.
Squire, J. M. (1972). (in preparation)
Steiger, F. J., and Rüegg, J. C. (1969). *Pfluegers Arch.* 307, 1.
Stromer, M. H., Hartshorne, D. J., Mueller, H., and Rice, R. V. (1969). *J. Cell Biol.* 40, 167.
Thorson, J., and White, D. C. S. (1969). *Biophys. J.* 9, 360.
Thorson, J., and White, D. C. S. (1972). *J. Gen. Physiol.* 60.
Tice, L. W. (1969). *Tissue & Cell* 1, 97.
Tice, L. W., and Smith, D. S. (1965). *J. Cell Biol.* 25, 121.
Tiegs, O. W. (1955). *Phil. Trans. Roy. Soc. London, Ser. B* 238, 221.
Tigyi-Sebes, A. (1966). *Acta Biochim. Biophys.* 1, 407.
Toselli, P. A., and Pepe, F. A. (1968a). *J. Cell Biol.* 37, 445.
Toselli, P. A., and Pepe, F. A. (1968b). *J. Cell Biol.* 37, 462.
Tregear, R. T., and Miller, A. (1969). *Nature (London)* 222, 1184.
Tyrer, N. M. (1969). *Nature (London)* 224, 815.
Usherwood, P. N. R. (1962). *J. Insect Physiol.* 8, 31.
von Brocke, H. H. (1966). *Pfluegers Arch. Gesamte Physiol. Menschen Tiere* 290, 70.
von Hehn, G., and Schlote, F. W. (1964). *Z. Zellforsch. Mikrosk. Anat.* 63, 459.
von Siebold, C. T. (1848). "Lehrbuch der vergleichenden Anatomie der wirbellosen Thiere." Berlin.
Walcott, B., and Burrows, M. (1969). *J. Insect Physiol.* 15, 1855.
Walcott, B., and Ridgway, E. B. (1967). *Amer. Zool.* 7, 499.
Ward, P. C., Edwards, J. C., and Benson, E. S. (1965). *Proc. Nat. Acad. Sci. U.S.* 53, 1377.
Warren, R. H., and Porter, K. R. (1969). *Amer. J. Anat.* 124, 1.
Weis-Fogh, T. (1956). *J. Exp. Biol.* 33, 668.
Weis-Fogh, T. (1964). *J. Exp. Biol.* 41, 229.
White, D. C. S. (1967). Ph.D. Thesis, Oxford University.
White, D. C. S. (1970). *J. Physiol. (London)* 208, 583.
Wigglesworth, V. B. (1965). "The Principles of Insect Physiology." London, Methuen.
Wilson, D. M. (1962). *J. Exp. Biol.* 39, 669.
Wilson, D. M., and Larimer, J. L. (1968). *Proc. Nat. Acad. Sci. U.S.* 61, 909.
Wilson, D. M., Smith, D. O., and Dempster, P. (1970). *Amer. J. Physiol.* 218, 916.
Woods, E. F., Himmelfarb, S., and Harrington, W. F. (1963). *J. Biol. Chem.* 238, 2374.
Worthington, C. R. (1961). *J. Mol. Biol.* 3, 618.
Zachar, J., and Zacharova, D. (1966). *Experientia* 22, 451.
Zebe, E. (1966). *Experientia* 22, 1.
Zebe, E., and Falk, H. (1963a). *Z. Naturforsch. B* 18, 501.
Zebe, E., and Falk, H. (1963b). *Exp. Cell Res.* 31, 340.
Zebe, E., and Falk, H. (1964). *Histochemie* 4, 161.
Zebe, E., and Rathmayer, W. (1968). *Z. Zellforsch. Mikrosk. Anat.* 92, 377.
Zebe, E., Meinrenken, W., and Rüegg, J. C. (1968). *Z. Zellforsch. Mikrosk. Anat.* 87, 603.

AUTHOR INDEX

Numbers in italics refer to the pages on which the complete references are listed.

A

Abbott, B. C., 443, 452, 453, *484,* 524, 525, *536, 538*
Abbott, J., 242, *296*
Abbott, R. H., 516, 529, 531, 533, *536*
Adams, E. C., 156, 162, 164, 169, *178*
Adams, Jr., E. C., 286, *297*
Adams, R. D., 186, *228*
Adelstein, R. S., *539*
Adrian, R. H., 187, *228,* 442, 465, *484*
Afifi, A. K., 220, *229*
Afzelius, B. F., 495, *538*
Aidley, D. J., 529, *536*
Alam, M., *71*
Albers, G., 291, *295*
Albuquerque, E. X., 290, *293*
Alder, A. B., 184, 185, 212, *228*
Alescio, T., 285, *293*
Alexandrowicz, J. S., 423, *484*
Alexis, S. D., 227, *236*
Alfei, L., 263, 267, *294*
Allbrook, D. B., 4, *21,* 183, 186, 188, *229, 231*
Allen, E. R., 87, 108, 123, 128, *124,* 196, 206, *228*
Allen, R. D., 445, *485*
Allen, S. R., 520, *539*
Allison, J. B., 225, *228*
Aloisi, M., 138, *146,* 221, 222, *233*
Anderson, E., 498, *537*
Anderson, W. A., 498, *436*
Andersson-Cedergren, E., 333, *386,* 412, *419*
Andersson-Cedergren, C., 133, *142*
Andreeva, L. F., 97, *148*

Angevine, D. M., 4, *21,* 108, *145,* 183, 185, *232*
Apathy, S., 407, *419*
Appleton, A. B., 188, *231*
Arese, P., 173, *177*
Armson, J. M., 470, *485*
Armstrong, C. F., 377, *384,* 529, *536*
Aronson, J., 181, *228,* 504, 515, 524, *536*
Aschenbrenner, V., 86, *144*
Ash, J. F., 89, 126, *148*
Ashford, T. P., 86, *146*
Ashhurst, D. E., 497, 502, 503, 507, 508, 516, *536*
Ashley, C. C., 434, 459, 461, 462, 463, 464, *485, 488,* 522, *536*
Asiedu, 220, *234*
Askanas, V., 291, *293*
Asmussen, E., 34, 35, 38, 39, 40, 68, *71*
Astbury, W. T., 345, *384*
Atwood, H. L., 422, 423, 424, 431, 432, 433, 434, 435, 436, 437, 441, 442, 443, 446, 447, 451, 452, 457, 458, 459, 460, 461, 469, 471, 472, 474, 475, 476, 477, 478, 479, 480, 481, 482, *485, 487, 488, 489,* 494, 495, 496, *536, 539*
Auber, J., 123, 128, *142,* 494, 495, 497, 498, 499, 500, 502, 504, 507, 508, 513, 514, 515, *536, 537*
Auber, M. J., 181, 197, *228*
Auber-Thomay, M., 513, *537*
Avery, G., 78, *142,* 240, *293*
Avrameas, F., 137, *142*
Awan, M. Z., 215, *229*

543

522, 523, 524, 526, 527, 528, 529, 531, 533, 534, 535, *539, 540*
Pritchard, J. J., 150, *178*
Provins, K. A., 39, 41, *73*
Przybylski, R., 81, 87, 117, 128, *147*
Przybylski, R. J.., 255, *298*
Puddy, D., 89, *143,* 185, *230*

R

Rabinowitz, M., 109, *147*
Rademacker, M. A., 225, *235*
Ralston, H. J., 26, 62, *73*
Rambourg, A., 137, *147*
Ramsay, R. W., *540*
Ramsey, R. W., 6, *22*
Ramsey, V. W., 221, *230*
Rasch, P. J., 60, *63*
Rash, J. E., 91, 93, 117, 119, *143, 147*
Raski, D. J., 390, 401, 405, *420*
Rathmayer, W., 494, *541*
Rawles, M. E., 76, 77, *148*
Reale, E., 150, 152, 156, 162, 165, 166, 167, 168, *178*
Rebollo, M. A., 174, 175, *178*
Reckard, T., 79, *142,* 204, 215, *229*
Reedy, M. D., 194, *234*
Reedy, M. K., 129, 131, *147,* 359, 363, *386,* 498, 505, 506, 507, 533, *540*
Reger, J., 407, 415, *420*
Reger, J. F., 429, 431, 436, *488,* 494, 501, 508, *540*
Reimold, V. E., 210, *234*
Reiss, D. J., 206, *234*
Reiss, E., 225, *234*
Reiss, I., 325, *387*
Reporter, M. C., *147,* 257, 258, 284, *298*
Resnick, J. S., 220, *234*
Reuben, J. P., 423, 424, 429, 445, 446, 447, 449, 451, 453, 459, 464, 465, 466, 467, 468, 469, 473, *485, 486, 488,* 493, 494, *537*
Reuter, H., 458, *485*
Revel, J. P., 96, 137, *144, 147, 148,* 201, *234,* 416, *419*
Reznik, M., 79, 97, *142, 147,* 273, 275, 278, *294, 298*
Rhodin, J., 498, *540*
Rice, M. J., 498, *540*

Rice, R. V., 129, 131, *148,* 343, 364, *386,* 517, *541*
Rich, A., 87, 111, 123, *144,* 206, *232*
Rich, G. Q., 39, 46, *73*
Rich, M. A., 103, *143*
Richards, C. D., 448, *488,* 522, *536*
Richler, C., 272, *298*
Richter, G. W., 216, *234*
Rickles, W. H., 469, *486*
Ridgway, E. B., 459, 462, 463, 464, *485, 488,* 516, 517, 522, *536, 541*
Rinaldini, L. M. J., 282, 283, 297, *298*
Rinando, M. T., 173, *177, 178*
Robbins, J., 469, *488*
Robbins, N., 79, 139, *147,* 220, 223, *230, 234,* 269, *298*
Roberge, J. J., 498, *540*
Roberts, D. F., 39, *73*
Robertson, D. D., 225, *234*
Robertson, J. D., 443, 444, 445, *488*
Robinson, G., 290, *295*
Rodriguez, M. M., 174, 175, *178*
Roeder, K. D., 527, *540*
Rohmert, W., 34, 35, 53, 54, 60, *73*
Romanul, F. C., *229*
Romanul, F. C. A., 199, 223, *234*
Rona, G., 86, *147*
Ronchetti, I., 354, *384*
Rongey, R. W., 286, *297*
Roodyn, D. B., 109, *147*
Rosa, C. G., 150, *178*
Rosenbluth, J., 390, 391, 393, 394, 396, 397, 399, 401, 406, 407, 408, 411, 413, 415, 416, 417, 418, *420,* 434, *488,* 494, 495, 510, *540*
Rosenbluth, R., 328, *385*
Rosenfalck, P., 526, *537*
Roskin, G., 394, *420*
Rosner, S., 278, *295*
Ross, K. F. A., 257, 258, 277, 291, 292, 293, *298*
Rossi, F., 150, 152, 156, 162, 165, 166, 167, 168, *178*
Rowe, R. W. D., 186, 187, 188, 189, 195, 198, 218, 223, 225, 226, *231, 234*
Roy, H., 81, 93, *143,* 243, *294*
Royce, J., 46, *73*
Rubenstein, D., 225, *229*
Rudnick, D., 77, *147*

SUBJECT INDEX

A

A bands
  description of, 308
  enzymes in, 259
  lengths of, 331–332, 333–335
  long, in arthropods, 382–383
  myosin in, 328
  of obliquely striated muscle, 397, 401
A zones, of obliquely striated muscle, 401
Acetylcholine, effects on
  muscle contraction, 215
  muscle cultures, 260
Acetylcholinesterase, see Cholinesterase
Achalarus muscle, structure of, 508
Acid mucopolysaccharide, in fetal muscle, 174–175
Acid phosphatase, in developing muscle, 167–169
Acid phosphate, in developing cardiac muscle, 165–167
  blood vessels, 165
  myocardium, 166–167
  Purkinje fibers, 165–166
Actin
  activation of myosin ATPase by, 376–377
  in arthropod muscle, 519–520
  biosynthesis of, in embryonic muscle, 87, 111
  in contraction, 337, 339–341
  filaments of, 352–364
    length, 124–125
    molecular arrangement, 352–357
    polarity, 357–364
  location of, in muscle, 325, 328–329
  3-methylhistidine in, 113

–myosin interaction, X-ray studies of, 368–376
  role of, in muscle strength, 24
α-Actinin, M line structure and, 364
Actomyosin, embryonic, calcium sensitivity, 125–126
Adrenaline, effects on muscle cultures, 260
Aeshna muscle, structure of, 507
Age, effects on muscular strength, 38–39
Alaskan king crab, muscle fibers of, 422
Aldolase in muscle tissue cultures, 259
Alkaline phosphatase, in developing cardiac muscle, 162–164
Amino acid metabolism, in muscle cultures, 282–283
Amino groups, in fetal muscle, 173
γ-Aminobutyric acid, in inhibitory nerve stimulation of crustacean muscle, 472–474
5'-AMP deaminase, in developing muscle, 202
Anatomy, of muscles, 1–22
Androgens, effects on muscle fibers, 224
Animals, maximum speeds of, 382
Anisometric contractions, of muscle, 29, 35
Anisotropic bands, see A bands
Anterior tibialis muscle, development of, 188
Antheraea pernyi muscle, ontogeny, 512
Antimycin A, effects on muscle culture respiration, 284
Anus retractor muscles, hormone effects on, 223
Apis muscle, structure of, 508
Aplysia, buccal muscle of, structure, 407
Aponeurotic sheets, of muscles, 3